Linear Programming:
Mathematics, Theory and Algorithms

Applied Optimization

Volume 2

The titles published in this series are listed at the end of this volume.

Linear Programming: Mathematics, Theory and Algorithms

by

Michael J. Panik

University of Hartford

KLUWER ACADEMIC PUBLISHERS

DORDRECHT / BOSTON / LONDON

Library of Congress Cataloging-in-Publication Data

Panik, Michael J.
Linear programming : mathematics, theory and algorithms / by Michael J. Panik.
p. cm. -- (Applied optimization ; vol. 2)
Includes bibliographical references and index.
ISBN 0-7923-3782-4 (HB : alk. paper)
1. Linear programming. I. Title. II. Series.
T57.74.P345 1996
519.7'2--dc20 95-41545

ISBN 0-7923-3782-4

Published by Kluwer Academic Publishers,
P.O. Box 17, 3300 AA Dordrecht, The Netherlands.

Kluwer Academic Publishers incorporates
the publishing programmes of
D. Reidel, Martinus Nijhoff, Dr W. Junk and MTP Press.

Sold and distributed in the U.S.A. and Canada
by Kluwer Academic Publishers,
101 Philip Drive, Norwell, MA 02061, U.S.A.

In all other countries, sold and distributed
by Kluwer Academic Publishers Group,
P.O. Box 322, 3300 AH Dordrecht, The Netherlands.

Printed on acid-free paper

Printed in the Netherlands

In Memory of
Alice E. Bourneuf
and
Ann F. Friedlaender

TABLE OF CONTENTS

ACKNOWLEDGEMENT

Anyone familiar with the area of linear programming cannot help but marvel at its richness, elegance, and diversity. These special qualities of linear programming are by no means accidental. Many extremely talented and dedicated individuals helped cultivate these attributes. I am most grateful to those researchers who pioneered as well as contributed significantly to the development of linear programming theory and methods and whose writings have served as a valuable source of reference and inspiration. Specifically, the classic works of George Dantzig, David Gale, A. J. Goldman, Harold Kuhn, Albert Tucker, and Philip Wolfe, among others, are particularly noteworthy. More recently, my interest in the study of interior point techniques is largely due to the efforts of researchers such as Kurt Anstreicher, Robert Freund, Donald Goldfarb, Clovis Gonzaga, N. Karmarkar, Masakazu Kojima, Nimrod Megiddo, Shinji Mizuno, Michael Todd, Yinyu Ye, and Akiko Yoshise. To each of these authors, as well as to many of their contemporaries, I owe a substantial intellectual debt.

A significant portion of the material presented herein stems from lecture notes developed for courses such as quantitative decision theory and advanced quantitative methods which I taught for a number of years in the A.D. Barney School of Business and Public Administration at the University of Hartford. I am grateful to the University V.B. Coffin Grant Committee and the Barney School Dean's Office for providing financial support.

Finally, a special note of gratitude is extended to Marilyn Baleshiski for her accuracy, dedication, and steadfastness in typing an extremely complicated and lengthy manuscript.

1 INTRODUCTION AND OVERVIEW

My objective in writing this book is to present to students of management science, mathematics, engineering, and economics the salient features of the theory and computational aspects of linear programming. The material is self-contained in that it offers, in considerable detail, the mathematical tools needed to study linear programming theory and to implement a set of algorithms designed to efficiently solve linear programs. The only prerequisites are basic calculus and some exposure to linear algebra and matrix theory.

An important feature of the presentation is that it fully integrates both the primal and dual programs throughout the entire text. In fact, duality theory is introduced early and used extensively as a vehicle for tying together all of the theoretical and computational results.

While many of the computational routines offered herein are simplex based, considerable attention has been given to the particulars of the array of interior-point methods which have come to dominate the algorithmic literature on linear programming since the early 1980's. This includes Karmarkar's work along with an assortment of affine scaling, potential reduction, and path-following techniques.

Linear programming is one of the most versatile and powerful mathematical programming techniques that can be employed for efficiently solving a stylized class of decision problems of the form

$$\begin{aligned} max\ f(\mathbf{X}) &= \mathbf{C}'\mathbf{X} \quad \text{s.t.} \\ \mathbf{AX} \leqq \mathbf{b},\ &\mathbf{X} \geqq \mathbf{0}, \end{aligned} \tag{1.1}$$

where $\mathbf{C}$, $\mathbf{X} \epsilon R^p, \mathbf{b} \epsilon R^m$, $\mathbf{A}$ is a matrix of order $(m \times n)$, and a prime denotes transposition. Here f depicts a linear ***objective function***, $\mathbf{AX} \leqq \mathbf{b}$ represents a simultaneous system of linear ***structural constraints***, and $\mathbf{X} \geqq \mathbf{0}$ will be referred to as the set of ***nonnegativity conditions***. (As will be seen later on, many variations on this structural theme are possible.) If we let $\mathcal{K} = \{\mathbf{X} \mid \mathbf{AX} \leqq \mathbf{b}, \mathbf{X} \geqq \mathbf{0}\}$ represent the region of admissible solutions (or ***feasible region*** for short), then the preceding problem becomes one of finding a point $\mathbf{X}$ in $\mathcal{K}$ for which f is as large as possible.

The range of applications and the wealth of different types of problems that can be solved using linear programming algorithms is staggering, *e.g.*, linear programming has been used with marked success to solve optimization problems in areas such as economics (including banking and finance), business administration and management, agriculture and energy, contract bidding, nutrition (diet) planning, health care, public decision making, facility location, transportation, strategic planning, and so on. The list is virtually endless.

In particular, the elements within $\mathbf{X}$ may represent output levels for a set of p different products with the elements of $\mathbf{C}$ being their associated (constant) unit profit values. The m components of $\mathbf{b}$ can then be thought of as fixed resource availability levels per production run. The columns of $\mathbf{A}$ thus depict a set of p production activity vectors which indicate the amount of each input going into the operation of each production activity at the unit level. What we have just characterized is called an optimum product mix problem. A producing unit seeks to allocate its inputs to the production activities under a given linear technology in order to determine the quantities of the various products to produce to maximize profit.

But this is not the whole story. Associated with the above linear program (which we may call the ***primal problem***) is another such program, obtained directly from the primal, called its ***dual***, *i.e.*, the problem defined as dual to (1.1) is

$$\begin{aligned} min\ g(\mathbf{U}) &= \mathbf{b}'\mathbf{U} \quad \text{s.t} \\ \mathbf{A}'\mathbf{U} &\geqq \mathbf{C},\ \mathbf{U} \geqq \mathbf{0}, \end{aligned} \tag{1.2}$$

where $\mathbf{U} \epsilon \mathbf{R}^m$. While a detailed development of the relationship between these two problems is beyond the scope of this simple introductory discussion, suffice it to say that the interconnections between the primal and dual problems will be emphasized frequently throughout the text; indeed, it is this unifying theme of duality that will permeate most of our theoretical and computational results.

However, we can easily offer, even at this early stage of our exposure to duality, a modest hint as to the importance of the dual program by considering the dual to the production problem specified earlier. To this end we may note that the dual variables or components of $\mathbf{U}$ can be thought of as representing the internal (not market) or imputed unit prices for an organization's inputs. (Such prices can serve to guide the internal allocation of an organization's resources.) Hence g represents the total internal value of all inputs available to the organization. The dual structural constraint system $\mathbf{A}'\mathbf{U} \geqq \mathbf{C}$ then dictates that the total internal value of all inputs going into the operation of a production activity at the unit level must be at least as great as its per unit profit contribution. The dual problem thus has us find the smallest imputed value of the stock of inputs which will fully explain total profit given that each production activity breaks even or is deemed unprofitable and consequently discontinued.

The richness of linear programming lies not only in its diversity of applications but also in the elegance of the mathematical tools used to characterize as well as generate solutions. In this volume the development of the requisite mathematics along with linear programming concepts and methodology will be emphasized.

Chapter 2 reviews the mathematical tools needed to study linear programming. Specifically, it succinctly lays out the essentials of: ***vector*** and ***matrix notation*** and ***operations***; ***vector spaces***, ***subspaces***, and ***projections***; ***point-set theory***; ***simultaneous systems of linear equations*** and ***inequalities***; ***linear transformations***; ***rank*** and ***solution(s) of simultaneous linear systems***; ***linear dependence*** and ***independence***; ***affine functionals, hyperplanes*** and ***half-planes***; ***convex sets***, ***combinations*** and ***hulls***; ***theorems of the altemative*** (including ***Farkas' theorem***); ***convex cones*** along with ***dual*** and ***polar cones***; ***conical hulls***, ***extreme vectors*** and ***extreme solutions of homogeneous linear systems***; ***basic solutions*** to linear equalities; ***extreme points*** and ***directions for convex*** and ***polyhedral convex sets***; the ***representation of convex polyhedra*** using ***extreme points*** and ***extreme directions***; and the notion of a ***simplex***. Proofs of many of the theorems presented in this chapter may be found in Panik (1993).

A formal study of linear programming proper begins with Chapter 3. In this regard, the remainder of this overview outlines the salient features of linear programming to be presented in subsequent chapters.

Chapter 3 presents the ***canonical*** and ***standard forms*** of a linear programming problem. It discusses the concepts of ***feasibility*** and ***optimality*** and illustrates the same with a graphical solution to an example problem. The properties of the ***feasible region*** $\mathcal{K}$ are discussed along with the existence and location of finite optimal solutions. To this end an existence theorem involving a necessary and sufficient condition for a finite optimal solution is presented under a representation of $\mathcal{K}$ in terms of extreme points and extreme directions. Next comes a discussion of the correspondence between ***basic feasible*** and ***extreme point solutions*** to a linear program. This is followed by a presentation of the concepts of ***solutions space*** and ***requirements space.***

In Chapter 4 ***symmetric*** as well as ***unsymmetric dual problems*** are introduced and their properties explored. This is followed by a battery of duality theorems and corollaries. In particular, the ***weak duality theorem*** and the ***sufficient optimality criterion*** are discussed and then followed by the so-called fundamental theorems of linear programming, namely the ***strong duality theorem*** (including a discussion of the ***duality gap***), the ***existence theorem***, and both ***weak*** and ***strong complementary slackness theorems.***

Chapter 5 begins with a procedure for finding basic feasible solutions to the primal linear program. Of particular interest is how the set of basic variables as well as the objective function can be written in terms of the set of non-basic variables. This subsequently leads to the specification of the ***reduced primal problem.*** Next comes the development of the ***primal optimality*** (***dual feasibility***) ***criterion*** and the definition of the vector of dual variables. These two notions then allow us to rationalize the structure of the symmetric dual problem under an application of the ***primal Lagrangian function.*** The dual variables are now interpreted as ***marginal imputed values*** or ***simplex multipliers*** or ***shadow prices*** associated with the right-hand sides of primal structural constraints. ***Optimality evaluators*** or ***relative (reduced) cost coefficients*** are also defined in terms of dual variables. Finally, an iterative solution process called the ***primal simplex method***, with its attendant ***entry*** and ***exit criteria*** for generating improved solutions to a linear program, is outlined and illustrated. This is followed by a description of ***degenerate basic feasible solutions***, ***unbounded solutions***, and ***multiple optimal solutions.***

Looking to Chapter 6, the concept of a ***feasible direction*** along with the set of ***feasible directions*** is introduced and defined within the region of admissible solutions. The set of feasible directions is then related to the notions of ***tangent***, ***dual***, and ***polar support cones.*** These concepts then set the stage for a discussion of the geometry underlying the ***Minkowski-Farkas theorem.*** All of these items are then utilized when we next develop the ***Karush-Kuhn-Tucker necessary conditions*** for linear programs. It is further shown that the ***Lagrange multipliers*** for the same are the dual variables in the dual linear programming problem. An extensive discussion on the existence of dual variables as partial derivatives of the ***primal optimal value function*** follows. Next comes a description of a ***saddlepoint of the primal Lagrangian*** and a set of theorems pertaining to the existence of a saddle point. This is followed by another look at the ***strong duality theorem*** and the ***Kuhn-Tucker equivalence theorem*** which establishes the notion that the existence of an optimal solution to the primal problem is equivalent to the existence of a saddle point of its associated Lagrangian. The chapter ends with another look at the ***duality*** and ***complementary slackness theorems.***

Chapter 7 deals with the computational aspects of linear programming. It reexamines the primal ***simplex method*** in terms of simplex matrix calculations. In particular, we state an optimality criterion and demonstrate how to improve a basic feasible solution through the use of ***pivoting***; *i.e.*, entry and exit criteria are used to "swap" one basic feasible solution for another in an attempt to maximize the primal objective function. We next explore the detection of multiple optimal, unbounded, and degenerate solutions. The chapter ends with a discussion on obtaining the optimal dual solution from the optimal primal simplex matrix.

In Chapter 8 we introduce the notion of ***artificial variables*** and show how they may be used to initiate the simplex method. Here the ***one-phase*** (or ***M-penalty method***) and ***two-phase methods*** are explored in depth along with how an ***inconsistency*** or ***redundancy*** is detected in the constraint system. The last section offers a discussion on how one may obtain the set of optimal primal variables from the optimal simplex matrix when the dual minimization problem is solved using either the one- or two-phase routine.

Chapter 9 presents an assortment of specialized transformations for solving problems involving departures from the general form: $max\, \mathbf{C}'\mathbf{X}$ s.t $\mathbf{AX} \leqq \mathbf{b}$, $\mathbf{X} \geqq \mathbf{0}$. In particular, it considers: ***minimizing an objective*** under conversion to a maximization problem; the cases of unrestricted as well as ***bounded variables***; ***interval linear programming***; and how to handle ***absolute value functionals.***

Chapter 10 considers three important variants of the standard simplex method, namely the ***dual simplex***, ***primal-dual*** and ***complementary pivot routines.*** The theoretical development of these techniques as well as their computational features are presented. ***Dual simplex pivoting*** and a discussion of ***dual degeneracy*** along with the ***artificial constraint method*** (for generating an initial primal-optimal basic solution) is also offered. In addition to developing the standard primal-dual routine (involving the ***extended*** and ***restricted primal problem***, ***restricted basis entry*** the ***restricted dual problem***, and the ***primal-dual optimality theorem***), a ***robust*** or ***hybrid primal-dual method*** is also given. The chapter ends with a presentation of the ***complementary pivot method*** for solving the ***linear complementarity problem***. Included is a discussion of ***variable complements***, ***complementary*** and ***almost complementary basic solutions***, ***complementary cones***, and the ***complementary pivot rule***.

The emphasis in chapter 11 is ***postoptimality analysis***. We first consider ***sensitivity analysis*** — which involves an assessment of the extent to which a problem may be modified without violating the feasibility or optimality of the original solution. Here we consider the impact on feasibility or optimality of a discrete change in an objective function coefficient c_j, a right-hand side value b_i, or in a structural constraint coefficient $a_{ij}, i=1,\ldots,m;\ j=1,\ldots,n$. The thrust of the analysis here is the formulation of lower and upper limits associated with the said discrete changes. Next comes a discussion of ***structural changes*** or the problem of adding or deleting certain variables or structural constraints within the optimal simplex matrix and then reoptimizing.

Chapter 12 offers a second round of postoptimality analysis. In particular, it extends the sensitivity analysis developed in the preceding chapter to what is called ***parametric analysis***. That is, instead of determining, say, the amount by which an objective function coefficient may be altered in some particular fashion before the feasibility or optimality of the current solution is violated, it generates a sequence of basic solutions which in turn become optimal, one after the other, as "all" of the components of C vary continuously in some prescribed direction. In effect, parametric analysis combines sensitivity analysis with simplex pivoting. The chapter ends with the specification of a ***parametric method*** for solving a linear program. In particular, it includes a modification of the usual primal-dual method by working with a special parametric version of the primal problem.

The coverage in Chapter 13 involves the presentation of a relatively new approach to solving linear programming problems, namely ***interior-point techniques*** which have been borrowed from the area of non-linear programming. These methods (and in particular those found to be computationally tractable) received considerable attention after 1984 when N. Karmarkar first published his now famous paper on solving a linear program via a projective rescaling method. The chapter introduces what is called ***Karmarkar's standard form*** of a problem and the concept of a ***roundness restriction.*** It then presents the concept of Karmarkar's ***potential function*** and ***the projective transformation*** of the feasible region used to center the solution. Next comes the specification of the ***transformed or scaled linear program*** and what is called its ***restriction.*** This is followed by the concept of the ***transformed potential function*** and a discussion of the ***complexity*** of the algorithm. A method for transforming a general linear program to Karmarkar standard form is also offered along with a discussion of some important extensions and modifications of Karmarkar's algorithm. All of this is followed by a specification of methods related to Karmarkar's routine, *e.g.*, algorithms that are based on ***primal*** and ***dual affine scaling***, ***projective*** and ***affine potential reduction***, and ***path-following*** or ***barrier methods.***

Chapter 14 extends some of the preceding interior point techniques to the solution of the ***linear complementarity problem*** $LCP(\mathbf{q}, \mathbf{M})$. In particular, ***path-following***, ***potential reduction***, ***predictor-corrector***, and ***large step algorithms*** are considered.

Finally, a set of appendices on ***updating the basis inverse***, ***steepest edge simplex methods***, and deriving ***orthogonal projection matrices*** used to determine feasible step directions for interior point methods serve to round out the text.

2 PRELIMINARY MATHEMATICS

2.1. Vectors in $\mathbf{R}^n$

A ***vector*** is an ordered n-tuple of elements expressed as a column $(n \times 1)$ or row $(1 \times n)$ matrix

$$\mathbf{X} = \begin{bmatrix} x_1 \\ \vdots \\ x_n \end{bmatrix}, \quad \mathbf{X}' = (x_1, \ldots, x_n)$$

respectively. The n elements of a vector are called its ***components*** and a prime indicates transposition. In all of what follows our discussion will take place in an n-dimensional real vector space (denoted $\mathbf{R}^n$).

The ***Cartesian product*** (or ***product set***) $\mathcal{A} \times \mathcal{B}$ of two sets $\mathcal{A}, \mathcal{B}$ in $\mathbf{R}^n$ is the set of ordered pairs (a, b), where $a \in \mathcal{A}$, $b \in \mathcal{B}$. Then the product set $\boldsymbol{R} \times \boldsymbol{R} \times \cdots \times \boldsymbol{R} = \boldsymbol{R}^n = \{\mathbf{X}' = (x_1, x_2, \ldots, x_n) \,|\, x_1, x_2, \ldots, x_n \in \mathbf{R}\}$.

To establish a ***metric*** or concept of distance between two points $\mathbf{X}, \mathbf{Y} \in \mathbf{R}^n$, let us first state that the ***scalar product*** of $\mathbf{X}$, $\mathbf{Y}$ is the scalar

$$\mathbf{X} \cdot \mathbf{Y} = \mathbf{X}'\mathbf{Y} = \sum_{i=1}^{n} x_i y_i. \tag{2.1}$$

Next, a ***norm*** on $\mathbf{R}^n$ is a function which assigns to each $\mathbf{X} \in \mathbf{R}^n$ some number $\|\mathbf{X}\|$ such that:

(a) $\|\mathbf{X}\| \geq 0$ and $\|\mathbf{X}\| = 0$ if and only if $\mathbf{X} = \mathbf{0}$;

(b) $\|\mathbf{X}+\mathbf{Y}\| \leq \|\mathbf{X}\| + \|\mathbf{Y}\|$ (the triangle inequality);

(c) $\|c\mathbf{X}\| = |c|\,\|\mathbf{X}\|$ (homogeneity); and

(d) $|\mathbf{X} \cdot \mathbf{Y}| \leq \|\mathbf{X}\|\,\|\mathbf{Y}\|$ (the Cauchy-Schwarz inequality).

In this regard, the ***distance between points*** $\mathbf{X}, \mathbf{Y} \epsilon \mathbf{R}^n$ induced by the norm "$\|\cdot\|$" on $\mathbf{R}^n$ is

$$\|\mathbf{X}-\mathbf{Y}\| = \left[(\mathbf{X}-\mathbf{Y})'(\mathbf{X}-\mathbf{Y})\right]^{1/2} = \left[\sum_{i=1}^{n} |x_i - y_i|^2\right]^{1/2}. \quad (2.2)$$

An important special case of this definition occurs when we consider the distance between any $\mathbf{X} \epsilon \mathbf{R}^n$ and the origin $\mathbf{0}$, *i.e.*, the ***length (magni-tude)*** of $\mathbf{X}$ or norm of $\mathbf{X}$ is

$$\|\mathbf{X}\| = (\mathbf{X}'\mathbf{X})^{1/2} = \left[\sum_{i=1}^{n} |x_i|^2\right]^{1/2}. \quad (2.3)$$

The ***direction of*** $\mathbf{X}$ ($\neq \mathbf{0}$) is a unit vector $\mathbf{X}/\|\mathbf{X}\|$ which *points in the direction of* $\mathbf{X}$.

Let $\mathbf{X}, \mathbf{Y} \epsilon \mathbf{R}^n$ with $\mathbf{X}, \mathbf{Y} \neq \mathbf{0}$. The **cosine of the angle θ** between $\mathbf{X}, \mathbf{Y}$ is

$$cos\theta = (\mathbf{X}\cdot\mathbf{Y})/\|\mathbf{X}\|\|\mathbf{Y}\|, \quad 0 \leq \theta \leq \pi.$$

The Cauchy-Schwarz inequality implies that $-1 \leq \cos\theta \leq 1$. Also, the angle between $\mathbf{X}, \mathbf{Y}$ is acute ($<\pi/2$), right ($=\pi/2$), or obtuse ($>\pi/2$) according to whether the scalar product $\mathbf{X}\cdot\mathbf{Y}$ is positive, zero, or negative. Conversely, the vectors $\mathbf{X}, \mathbf{Y} \epsilon \mathbf{R}^n$ with $\mathbf{X}, \mathbf{Y} \neq \mathbf{0}$ are said to be ***orthogonal*** (*i.e.*, mutually perpendicular) if $\mathbf{X}\cdot\mathbf{Y} = 0$. Since $\mathbf{X}\cdot\mathbf{Y} = \|\mathbf{X}\|\,\|\mathbf{Y}\|\cos\theta$, $\mathbf{X}\cdot\mathbf{Y} = 0$ if and only if $\theta = \pi/2$ and thus $\cos\theta = 0$.

A vector $\mathbf{X} \epsilon \mathbf{R}^n$ is a ***linear combination*** of the vectors $\mathbf{X}_j \epsilon \mathbf{R}^n$, $j=1, \ldots, m$, if there exist scalars λ_j, $j=1, \ldots, m$, such that

$$\mathbf{X} = \sum_{j=1}^{m} \lambda_j \mathbf{X}_j. \quad (2.4)$$

In this light we can state that a set of vectors $\{\mathbf{X}_j \epsilon \mathbf{R}^n, j=1, \ldots, m\}$ is ***linearly dependent*** if there exist scalars λ_j, $j=1, \ldots, m$, not all zero such that

$$\sum_{j=1}^{m} \lambda_j \mathbf{X}_j = \mathbf{0}, \quad (2.5)$$

i.e., the null vector is a linear combination of the vectors $\mathbf{X}_j$. If the only set of scalars λ_j for which (2.5) holds is $\lambda_j = 0$, $j=1, \ldots, m$, then the vectors $\mathbf{X}_j$ are ***linearly independent***, *i.e.*, the trivial combination $0\mathbf{X}_1 + \cdots + 0\mathbf{X}_m$ is the only linear combination of the $\mathbf{X}_j$ which equals the null vector.

A set of vectors $\{\mathbf{X}_j \in \mathbf{R}^n,\ j=1,\dots,m\}$ is a ***spanning set*** for $\mathbf{R}^n$ if every vector $\mathbf{X} \in \mathbf{R}^n$ can be written as a linear combination of the vectors $\mathbf{X}_j$, *i.e.*, if $\mathbf{X} \in \mathbf{R}^n$, then

$$\mathbf{X} = \sum_{j=1}^{m} \lambda_j \mathbf{X}_j.$$

Here the set $\{\mathbf{X}_j \in \mathbf{R}^n,\ j=1,\dots,m\}$ is said to ***span*** or ***generate*** $\mathbf{R}^n$ since every vector in $\mathbf{R}^n$ is (uniquely) linearly dependent on the spanning set.

A ***basis*** for $\mathbf{R}^n$ is a linearly independent subset of vectors from $\mathbf{R}^n$ which spans $\mathbf{R}^n$. Moreover:

(a) every basis for $\mathbf{R}^n$ contains the same number of basis vectors; and there are precisely n vectors in every basis for $\mathbf{R}^n$; and

(b) any set of n linearly independent vectors from $\mathbf{R}^n$ forms a basis for $\mathbf{R}^n$.

The ***dimension of*** $\mathbf{R}^n$, $dim(\mathbf{R}^n)$, is the maximum number of linearly independent vectors that span the space, *i.e.*, it is the number of vectors in a basis for $\mathbf{R}^n$. Clearly $dim(\mathbf{R}^n) = n$.

A nonempty subset $\mathcal{M}$ of $\mathbf{R}^n$ is termed a ***subspace*** if it is closed under the operations of vector additional and scalar multiplication, *i.e.*, given vectors $\mathbf{X}_1, \mathbf{X}_2 \in \mathcal{M}$, their linear combination $\lambda_1\mathbf{X}_1+\lambda_2\mathbf{X}_2$ is also a member of $\mathcal{M}$ for every pair of scalars $\lambda_1, \lambda_2 \in \mathbf{R}$. Additionally, if $\mathcal{M}$ is a subspace of $\mathbf{R}^n$, then $\mathcal{M}$ must contain *all* linear combinations of its elements. In fact, the set $\ell(\mathcal{M})$ of all finite linear combination of points of $\mathcal{M}$ is called the ***span*** or ***linear hull*** of $\mathcal{M}$ and constitutes a subspace of $\mathbf{R}^n$. Here

$$\ell(\mathcal{M}) = \Big\{\mathbf{X} \mid \mathbf{X} = \sum_{i=1}^{k} \lambda_i \mathbf{X}_i, \mathbf{X}_i \in \mathcal{M},\ \lambda_i \in \mathbf{R} \text{ for all } i, \text{ and } k \text{ is an arbitrary positive integer}\Big\}$$

is the space spanned or generated by $\mathcal{M}$. Moreover, $\ell(\mathcal{M})$ is the smallest subspace containing $\mathcal{M}$. Hence $\mathcal{M}$ is a subspace of $\mathbf{R}^n$ if and only if $\mathcal{M} = \ell(\mathcal{M})$. By convention, $\ell(\phi) = \{\mathbf{0}\}$. A set of vectors $\mathcal{S}$ is a ***spanning set*** of $\mathcal{M}$ in $\mathbf{R}^n$ if $\mathcal{M} = \ell(\mathcal{S})$.

Let $\mathcal{M}$ be a subspace of $\mathbf{R}^n$. A set $\mathcal{S}$ of vectors in $\mathcal{M}$ is a ***basis*** for $\mathcal{M}$ provided:

(1) $\mathcal{S}$ spans $\mathcal{M}$, *i.e.*, $\ell(\mathcal{S}) = \mathcal{M}$; and

(2) $\mathcal{S}$ is linearly independent.

Given that $\mathcal{M}$ is a subspace of $\mathbf{R}^n$, the ***dimension of*** $\mathcal{M}$, $dim(\mathcal{M})$, is $\leq n$ and equals the maximum number of linearly independent vectors that constitute a basis for $\mathcal{M}$. So if $dim(\mathcal{M}) = m < n$, a basis for $\mathcal{M}$ is a subset of m linearly independent vectors selected from a set of n linearly independent vectors in $\mathbf{R}^n$.

If $\mathcal{M}_1$, $\mathcal{M}_2$ are subspaces of $\mathbf{R}^n$, $\mathcal{M}_1 \cap \mathcal{M}_2 = \{\mathbf{0}\}$, and, taken together, $\mathcal{M}_1$ and $\mathcal{M}_2$ **span** $\mathbf{R}^n$ in the sense that each vector in $\mathbf{R}^n$ is uniquely expressible as $\mathbf{Z} = \mathbf{X}+\mathbf{Y}$, where $\mathbf{X} \in \mathcal{M}_1$ and $\mathbf{Y} \in \mathcal{M}_2$, then $\mathbf{R}^n$ is expressible as the ***direct sum of*** $\mathcal{M}_1$, $\mathcal{M}_2$ and written as $\mathbf{R}^n = \mathcal{M}_1 \oplus \mathcal{M}_2 = \{\mathbf{X}+\mathbf{Y} | \mathbf{X} \in \mathcal{M}_1, \ \mathbf{Y} \in \mathcal{M}_2\}$. Moreover, if $\{\mathbf{X}_1, \ldots, \mathbf{X}_r\}$ is a basis for $\mathcal{M}_1$ and $\{\mathbf{Y}_1, \ldots, \mathbf{Y}_s\}$ is a basis for $\mathcal{M}_2$, then the combined set $\{\mathbf{X}_1, \ldots, \mathbf{X}_r, \mathbf{Y}_1, \ldots, \mathbf{Y}_s\}$ is a basis for $\mathbf{R}^n$. In this regard, $dim(\mathbf{R}^n) = n = dim(\mathcal{M}_1 \oplus \mathcal{M}_2) = dim(\mathcal{M}_1) + dim(\mathcal{M}_2) = r+s$. If a subspace $\mathcal{M}_1$ is contained in a subspace $\mathcal{M}_2$, then $dim(\mathcal{M}_1) \leq dim(\mathcal{M}_2)$.

For $\mathcal{M}$ a subspace of $\mathbf{R}^n$, the ***orthogonal complement*** or ***dual subspace of*** $\mathcal{M}$, $\mathcal{M}^{\perp}$, is also a subspace of $\mathbf{R}^n$ and consists of all vectors that are orthogonal to those in $\mathcal{M}$. Hence $\mathcal{M}^{\perp} = \{\mathbf{Y} | \mathbf{Y}'\mathbf{X} = 0, \ \mathbf{X} \in \mathcal{M}\}$. Additionally, $(\mathcal{M}^{\perp})^{\perp} = \mathcal{M}$.

If $\mathbf{X}$ is an arbitrary point in $\mathbf{R}^n$, the ***orthogonal projection*** of $\mathbf{X}$ into $\mathcal{M}$ is the point $\bar{\mathbf{X}} \in \mathcal{M}$ for which $\| \mathbf{X} - \bar{\mathbf{X}} \|^2 = (\mathbf{X} - \bar{\mathbf{X}})'(\mathbf{X} - \bar{\mathbf{X}})$ assumes a minimum. There exists at most one such projection of a point $\mathbf{X}$ into a subspace $\mathcal{M}$. And if $\bar{\mathbf{X}}$ is the projection of $\mathbf{X}$ into $\mathcal{M}$, then $\mathbf{X} - \bar{\mathbf{X}}$ is the projection of $\mathbf{X}$ into $\mathcal{M}^{\perp}$. $\mathcal{M}$ and $\mathcal{M}^{\perp}$ span $\mathbf{R}^n$ (any vector $\mathbf{Z} \in \mathbf{R}^n$ can be uniquely expressed as $\mathbf{Z} = \mathbf{X}+\mathbf{Y}$ for $\mathbf{X} \in \mathcal{M}$, $\mathbf{Y} \in \mathcal{M}^{\perp}$) so that $\mathbf{R}^n = \mathcal{M} \oplus \mathcal{M}^{\perp}$. In this regard, $\mathbf{X}$ and $\mathbf{Y}$ are termed ***orthogonal projections*** of $\mathbf{Z}$ onto the subspaces $\mathcal{M}$, $\mathcal{M}^{\perp}$ respectively. Clearly $dim(M^{\perp}) = n - dim(\mathcal{M})$.

If the set of vectors $\{\mathbf{X}_1, \ldots, \mathbf{X}_m\}$ is a basis for a subspace $\mathcal{M}$ of $\mathbf{R}^n$, then $\mathcal{M}^{\perp} = \{\mathbf{Y} | \mathbf{Y}'\mathbf{X}_i = 0, \ i=1, \ldots, m, \ \mathbf{X}_i \in \mathcal{M}\}$.

2.2. Rank and Linear Transformations

Given an $(m \times n)$ matrix $\mathbf{A}$, the subspace of $\mathbf{R}^m$ spanned by the n column vectors of $\mathbf{A}$ will be called the ***column space*** of $\mathbf{A}$ and denoted $Col(\mathbf{A})$ while the subspace of $\mathbf{R}^n$ spanned by the m row vectors of $\mathbf{A}$ is termed the ***row space*** of $\mathbf{A}$ and written $Row(\mathbf{A})$. These two subspaces are distinct so that, if $m \neq n$, then $Col(\mathbf{A}) \cap Row(\mathbf{A}) = \phi$ and thus they have no vectors in common.

The nonzero row vectors of the echelon form of $\mathbf{A}$ are linearly independent and thus constitute a basis for $Row(\mathbf{A})$. If there are k nonzero vectors in the echelon form of $\mathbf{A}$, then $k = dim(Row(\mathbf{A}))$. Similarly, the column vectors of $\mathbf{A}$ corresponding to the pivot columns of the echelon form of $\mathbf{A}$ are also linearly independent and thus form a basis for $Col(\mathbf{A})$. Here too $k = dim(Col(\mathbf{A}))$. Hence we may unambiguously define the ***rank*** of a matrix $\mathbf{A}$, denoted $\rho(\mathbf{A})$, as $dim(Row(\mathbf{A})) = dim(Col(\mathbf{A})) = k$.

Alternatively, we may specify the rank of a matrix $\mathbf{A}$ as the order of the largest non-singular submatrix of $\mathbf{A}$, *i.e.*, it is the order of the largest nonvanishing determinant in $\mathbf{A}$; moreover, if $\rho(\mathbf{A}) = k$, all submatrices of order $k+1$ are singular. Also, an nth order matrix $\mathbf{B}$ is said to be of ***full rank*** if $\rho(\mathbf{B}) = n$; and this occurs if and only if $\mathbf{B}$ is nonsingular (*i.e.*, $|\mathbf{B}| \neq 0$). Hence, for an nth order matrix $\mathbf{B}$, $\rho(\mathbf{B}) < n$ if and only if $\mathbf{B}$ is singular.

A ***function*** or ***transformation*** T from $\mathbf{R}^n$ to $\mathbf{R}^m$ is a rule which associates with each vector $\mathbf{X} \in \mathbf{R}^n$ a unique vector $\mathbf{Y} \in \mathbf{R}^m$, where $\mathbf{Y} = T(\mathbf{X})$ is called the ***image*** of $\mathbf{X}$ under transformation T. Here $\mathbf{R}^n$ is the ***domain*** of $T, D(T)$, and the collection of those $\mathbf{Y}$'s which are the image of one or more $\mathbf{X}$ under rule T constitutes the ***range*** of $T, R(T)$. In particular, a function $T: \mathbf{R}^n \rightarrow \mathbf{R}^m$ is a ***linear transformation*** if for vectors $\mathbf{X}_1$, $\mathbf{X}_2 \in \mathbf{R}^n$ and scalars $\alpha, \beta \in \mathbf{R}$, $T(\alpha\mathbf{X}_1 + \beta\mathbf{X}_2) = \alpha T(\mathbf{X}_1) + \beta T(\mathbf{X}_2)$. Thus a transformation between $\mathbf{R}^n$ and $\mathbf{R}^m$ is linear if and only if it preserves linear combinations of vectors. Also:

(1) $T(\mathbf{0}) = \mathbf{0}$;

(2) $T(-\mathbf{X}) = -T(\mathbf{X})$;

(3) $\mathbf{X} \in \mathbf{R}^n \mapsto T_0(\mathbf{X}) = \mathbf{0}$ (the ***zero transformation***);

(4) $\mathbf{X} \in \mathbf{R}^n \mapsto T_I(\mathbf{X}) = \mathbf{X}$ (the ***identity transformation***).

The linear transformation T from $\mathbf{R}^n$ to $\mathbf{R}^m$ is termed ***one-to-one*** provided that no two different vectors in $\mathbf{R}^n$ have the same image in $\mathbf{R}^m$, *i.e.*, $T(\mathbf{X}_1) = T(\mathbf{X}_2)$ implies $\mathbf{X}_1 = \mathbf{X}_2$. T is ***onto*** if every vector in $\mathbf{R}^m$ is the image of at least one vector in $\mathbf{R}^n$. That is, for each $\mathbf{Y} \in \mathbf{R}^m$ there exists an $\mathbf{X} \in \mathbf{R}^n$ such that $\mathbf{Y} = T(\mathbf{X})$. Hence T is onto if $R(T) = \mathbf{R}^m$. To underscore the notions of onto and one-to-one linear transformations we state

2.1. THEOREM. For T a linear transformation from $\mathbf{R}^n$ to $\mathbf{R}^m$:

(a) if $m > n$, then T is not onto;

(b) if $m < n$, then T is not one-to-one; and

(c) if $m = n$, then T is one-to-one if and only if it is onto.

By virtue of part (c) of this theorem, if $T: \mathbf{R}^n \rightarrow \mathbf{R}^n$ is both one-to-one and onto, then it is ***invertible***, *i.e.*, there exists a linear transformation $U: \mathbf{R}^n \rightarrow \mathbf{R}^n$ such that $TU = UT = T_I$, the identity transformation. Here U is unique and is denoted as T^{-1}.

The linear transformation T from $\mathbf{R}^n$ to $\mathbf{R}^m$ is said to be ***non-singular*** if $T(\mathbf{X}) = \mathbf{0}$ implies $\mathbf{X} = \mathbf{0}$. Clearly T is one-to-one if and only if T is non-singular (since $T(\mathbf{X}_1 - \mathbf{X}_2) = T(\mathbf{X}_1) - T(\mathbf{X}_2) = \mathbf{0}$ implies $T(\mathbf{X}_1) = T(\mathbf{X}_2)$ and conversely). For $T: \mathbf{R}^n \rightarrow \mathbf{R}^n$, the following are equivalent:

(a) T is non-singular;

(b) T is onto;

(c) T is invertible;

(d) T is one-to-one.

In what follows we shall typically use an $(m \times n)$ matrix $\mathbf{A}$ to define a linear transformation from $\mathbf{R}^n$ to $\mathbf{R}^m$. Specifically,

2.2. THEOREM. Let $T_{\mathbf{A}}: \mathbf{R}^n \rightarrow \mathbf{R}^m$ be a linear transformation. Then there exists a unique $(m \times n)$ matrix $\mathbf{A}$ such that $T_{\mathbf{A}}(\mathbf{X}) = \mathbf{AX}$, $\mathbf{X} \in \mathbf{R}^n$.

Here the aforementioned rule involves (left) multiplication of $\mathbf{X}$ by $\mathbf{A}$ and is given by $\mathbf{X} \in \mathbf{R}^n \mapsto T_{\mathbf{A}}(\mathbf{X}) = \mathbf{AX} = \mathbf{Y} \in \mathbf{R}^m$. Clearly $T_{\mathbf{A}}$ respects linear combinations of vectors. Moreover, it is easily shown that a linear transformation is completely determined under its action on a set of unit column vectors $\{\mathbf{e}_1,\ldots,\mathbf{e}_n\}$ or ***standard basis*** for $\mathbf{R}^n$. That is, if the jth column of $\mathbf{A}$ is the image $T_{\mathbf{A}}(\mathbf{e}_j) = \mathbf{A}\mathbf{e}_j = \mathbf{a}_j$ of the jth unit column vector $\mathbf{e}_j$, then $\mathbf{A} = [T_{\mathbf{A}}(\mathbf{e}_1),\ldots,T_{\mathbf{A}}(\mathbf{e}_n)]$ and thus $T_{\mathbf{A}}(\mathbf{X}) = \Sigma_{j=1}^{n} x_j T_{\mathbf{A}}(\mathbf{e}_j) = [T_{\mathbf{A}}(\mathbf{e}_1),\ldots,T_{\mathbf{A}}(\mathbf{e}_n)]\,\mathbf{X} = \mathbf{AX}$.

We next consider two fundamental spaces associated with a linear transformation. First, given $T_{\mathbf{A}}:\mathbf{R}^n \rightarrow \mathbf{R}^m$, the set of all possible values of $T_{\mathbf{A}}$ or the set of vectors $\mathbf{Y} \epsilon \mathbf{R}^m$ as $\mathbf{X}$ varies over $\mathbf{R}^n$ is called the ***range*** of $T_{\mathbf{A}}$ and denoted $R(\mathbf{A})$ (or $R(T_{\mathbf{A}})) = \{\mathbf{Y} \mid \mathbf{Y} = T_{\mathbf{A}}(\mathbf{X}) = \mathbf{AX},\ \mathbf{X} \epsilon \mathbf{R}^n\}$. Here $R(\mathbf{A})$ is the subspace of $\mathbf{R}^m$ spanned by the columns of $\mathbf{A}$ and thus constitutes the set of linear combinations of the columns of $\mathbf{A}$ or $\mathbf{Y} = \Sigma_{j=1}^{n} x_j \mathbf{a}_j$. Clearly $R(\mathbf{A}) = Col(\mathbf{A}) = \ell\{\mathbf{a}_j\}_{j=1}^{n}$.

If $R(\mathbf{A})$ is finite-dimensional, then the number of vectors in a basis for $R(\mathbf{A})$ is the dimension of $R(\mathbf{A})$ and is called the ***rank*** of $T_{\mathbf{A}}$ and denoted $\rho(T_{\mathbf{A}})$. Here $\rho(T_{\mathbf{A}}) = dim(Col(\mathbf{A})) = dim(R(\mathbf{A})) = \rho(\mathbf{A})$, *i.e.*, the rank of a transformation $T_{\mathbf{A}}$ between $\mathbf{R}^n$ and $\mathbf{R}^m$ equals the rank of its matrix. For $T_{\mathbf{A}}:\mathbf{R}^n \rightarrow \mathbf{R}^n$, $T_{\mathbf{A}}$ is invertible if and only if $\mathbf{A}$ is invertible, *i.e.*, if and only if $\rho(\mathbf{A}) = n$.

If $T_0: \mathbf{R}^n \rightarrow \mathbf{R}^m$ represents the zero transformation, then $R(T_0)$ is the ***zero subspace*** and $\rho(T_0) = 0$. For the identity transformation $T_I: \mathbf{R}^n \rightarrow \mathbf{R}^m$, $R(T_I) = \mathbf{R}^n$ and thus $\rho(T_I) = dim\mathbf{R}^n = n$.

The second fundamental space hinted at above is a subspace of $\mathbf{R}^n$ called the ***null-space*** (or ***kernel***) of $T_{\mathbf{A}}$, $N(T_{\mathbf{A}})$ (or simply $N(\mathbf{A})$), which corresponds to the set of all $\mathbf{X} \epsilon \mathbf{R}^n$ mapped by $T_{\mathbf{A}}$ to the null vector, *i.e.*, $N(\mathbf{A}) = \{\mathbf{X} \mid T_{\mathbf{A}}(\mathbf{X}) = \mathbf{AX} = \mathbf{0},\ \mathbf{X} \epsilon \mathbf{R}^n\}$ is the solution space of the ***homogeneous linear equation system*** $\mathbf{AX} = \mathbf{0}$. Given that $N(\mathbf{A})$ is a subspace, $\alpha\mathbf{X} \epsilon N(\mathbf{A})$ when $\mathbf{X} \epsilon N(\mathbf{A})$, $\alpha \epsilon \mathbf{R}$. If $\mathbf{X} \neq \mathbf{0}$, then clearly $N(\mathbf{A})$ contains infinitely many non-zero vectors. The number of vectors in a basis for $N(\mathbf{A})$ or the dimension of $N(\mathbf{A})$ will be called the ***nullity*** of $\mathbf{A}$ and denoted $n(\mathbf{A})$. A linear transformation with nullity zero is said to be ***of full rank***.

The concepts of rank and nullity of a linear transformation $T_{\mathbf{A}}:\mathbf{R}^n \rightarrow \mathbf{R}^m$ enable us to determine whether or not $T_{\mathbf{A}}$ is onto or one-to-one. That is:

(a) $T_{\mathbf{A}}$ is one-to-one if $n(\mathbf{A}) = 0$;

(b) $T_{\mathbf{A}}$ is onto if $\rho(\mathbf{A}) = m$.

As Figure 2.1 reveals, if $N(T_{\mathbf{A}})$ shrinks to $\{\mathbf{0}\}$, then $T_{\mathbf{A}}$ is one-to-one; and if $R(T_{\mathbf{A}}) = \mathbf{R}^m$, then $T_{\mathbf{A}}$ is onto.

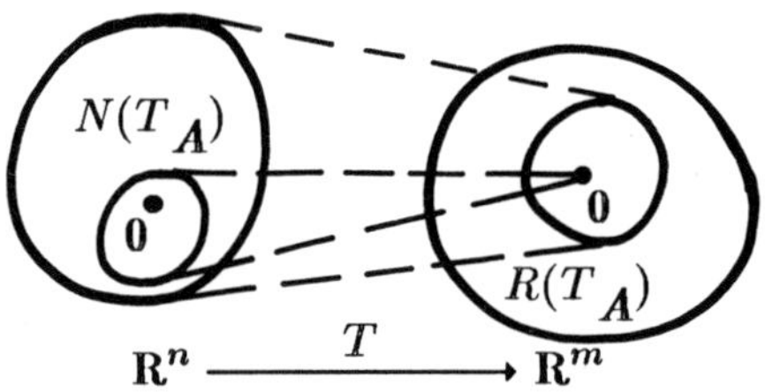

Figure 2.1

We noted earlier that for an $(m \times n)$ matrix $\mathbf{A}$, the subspace spanned by the columns of $\mathbf{A}$ is the range of $\mathbf{A}$, $R(\mathbf{A}) = \{\mathbf{Y} \mid \mathbf{Y} = \mathbf{AX},\ \mathbf{X} \in \mathbf{R}^n\}$. Also, the null-space of $\mathbf{A}$ was defined as $N(\mathbf{A}) = \{\mathbf{X} \mid \mathbf{AX} = \mathbf{0},\ \mathbf{X} \in \mathbf{R}^n\}$. Let $R(\mathbf{A})^{\perp} = \{\mathbf{U} \mid \mathbf{U}'\mathbf{Y} = 0,\ \mathbf{U} \in \mathbf{R}^m,\ \mathbf{Y} \in R(\mathbf{A})\} = \{\mathbf{U} \mid \mathbf{U}'\mathbf{AX} = \mathbf{0},\ \mathbf{U} \in \mathbf{R}^m, \mathbf{X} \in \mathbf{R}^n\}$ and $N(\mathbf{A}') = \{\mathbf{Y} \mid \mathbf{A}'\mathbf{Y} = \mathbf{0},\ \mathbf{Y} \in \mathbf{R}^m\}$. Then it can be shown that $R(\mathbf{A})^{\perp} = N(\mathbf{A}')$. Similarly, $N(\mathbf{A})^{\perp} = R(\mathbf{A}')$.

For the zero transformation $T_0 : \mathbf{R}^n \rightarrow \mathbf{R}^m$, $N(T_0) = \mathbf{R}^n$ and thus $n(T_0) = n$. If $T_I : \mathbf{R}^n \rightarrow \mathbf{R}^m$ depicts the identity transformation, then $N(T_I) = \{\mathbf{0}\}$ and $n(T_I) = 0$.

Next,

2.3. THEOREM. Let $T_{\mathbf{A}}$ be a linear transformation defined by the $(m \times n)$ matrix $\mathbf{A}$. Then $\{\mathbf{a}_1, \ldots, \mathbf{a}_n\}$ is linearly depen-dent if and only if $\mathbf{A}$ has a nontrivial null-space or $N(\mathbf{A}) \neq \{\mathbf{0}\}$.

Clearly this result is consistent with the definition of linear dependence given above: $\{\mathbf{a}_1, \ldots, \mathbf{a}_n\}$ is linearly dependent if and only if $\mathbf{AX} = \mathbf{0}$ for some $\mathbf{X} \neq \mathbf{0}$.

But from the definition of the *linear independence* of the vectors $\{\mathbf{a}_1,\ldots,\mathbf{a}_n\}$, $\mathbf{AX}=\mathbf{0}$ has only the *trivial solution.* Hence the preceding theorem can be rephrased as: $N(\mathbf{A})=\{\mathbf{0}\}$ if and only if $\{\mathbf{a}_1,\ldots,\mathbf{a}_n\}$ is linearly independent. (Note that $T_{\mathbf{A}}$ is non-singular if $N(T_{\mathbf{A}})=\{\mathbf{0}\}$ so that only non-singular linear transformations preserve linear independence.) If $\mathbf{A}$ is $(m\times n)$ and $n>m$, then $\mathbf{A}$ has linearly dependent columns and thus $N(\mathbf{A})\neq\{\mathbf{0}\}$.

The following theorem will enable us to effectively deal with the notions of the existence and uniqueness of solutions to simultaneous linear equation systems. Specifically, we have the

2.4. RANK-NULLITY THEOREM. Let $T_{\mathbf{A}}$ be a linear transfor-mation defined by the $(m\times n)$ matrix $\mathbf{A}$. Then $\rho(\mathbf{A})+n(\mathbf{A})=n$, $\rho(\mathbf{A})\leq m$.

Later on in our discussion of linear programming we shall have occasion to employ two specialized types of transformations, namely projective and affine transformations. Specifically, given an $(n\times n)$ matrix $\mathbf{A}$ with vectors $\mathbf{d}$, $\mathbf{e}\in\mathbf{R}^n$ and a scalar $h\in\mathbf{R}$, a ***projective trans-formation*** of $\mathbf{R}^n$ is a function $\mathrm{T}:\mathbf{R}^n\rightarrow\mathbf{R}^n$ such that

$$\mathbf{X}\in\mathbf{R}^n\mapsto\mathbf{Y}=T(\mathbf{X})=\frac{\mathbf{AX}+\mathbf{d}}{\mathbf{e}'\mathbf{X}+h},\ \mathbf{e}'\mathbf{X}+h\neq 0,$$

where the $(n{+}1\times n{+}1)$ matrix

$$\begin{bmatrix}\mathbf{A} & \vdots & \mathbf{d}\\ \cdots & \vdots & \cdots\\ \mathbf{e}' & \vdots & h\end{bmatrix}$$

is assumed nonsingular.

For $\mathbf{A}$ of order $(n\times n)$ and nonsingular, let us take $\mathbf{d}=\mathbf{0}$, $\mathbf{e}'=\mathbf{1}'(\mathbf{A}-\mathbf{I}_n)$, and $h=1$. Also, let us restrict $\mathbf{X}$ to the simplex $\sigma=\{\mathbf{X}\mid\mathbf{1}'\mathbf{X}=1,\ \mathbf{X}\geqq\mathbf{0}\}$. Then any point $\mathbf{X}\in\sigma$ is taken to a point $\mathbf{Y}\in\sigma$, where

$$\mathbf{Y}=T(\mathbf{X})=\frac{\mathbf{AX}}{\mathbf{1}'(\mathbf{A}-\mathbf{I}_n)\mathbf{X}+1}=\frac{\mathbf{AX}}{\mathbf{1}'\mathbf{AX}}.$$

This restriction of σ to a projective transformation T will be employed in chapter 13 wherein Karmarkar's interior point algorithm for solving a linear program is presented.

Next, for $(n \times n)$ matrices $\mathbf{A}$ and $\mathbf{B}$, the function $T_{\mathbf{B}}: \mathbf{R}^n \rightarrow \mathbf{R}^n$ is termed an ***affine transformation*** if there exists a linear transformation $T_{\mathbf{A}}: \mathbf{R}^n \rightarrow \mathbf{R}^n$ and a vector $\mathbf{C} \in \mathbf{R}^n$ such that

$$\mathbf{X} \in \mathbf{R}^n \mapsto \mathbf{Y} = T_{\mathbf{B}}(\mathbf{X}) = T_{\mathbf{A}}(\mathbf{X}) + \mathbf{C} \quad \text{or}$$
$$\mathbf{Y} = \mathbf{B}\mathbf{X} = \mathbf{A}\mathbf{X} + \mathbf{C}.$$

This expression is essentially a linear transformation followed by a translation, where $\mathbf{C}$ (fixed) specifies the magnitude and direction of the translation.

2.3. The Solution Set of a System of Simultaneous Linear Equations

A ***nonhomogeneous system of m linear equations in n unknowns*** may be written in matrix form as $\mathbf{AX} = \mathbf{C}$, where $\mathbf{A}$ is an $(m \times \mathrm{n})$ matrix of (constant) coefficients a_{ij}, $\mathbf{X}$ is an $(n \times 1)$ matrix of unknown variables x_i, and $\mathbf{C}$ is an $(m \times 1)$ matrix of constants c_i, $i=1, \ldots, m$; $j=1, \ldots, n$. Let us now look to the structure of the solution set for $\mathbf{AX} = \mathbf{C}$. Here we shall be interested in the existence and uniqueness of a solution to this simultaneous linear system. To facilitate this discussion we note that a system of equations is ***consistent*** if it has at least one solution; it is ***inconsistent*** if it does not possess a solution. Moreover, for $\mathbf{X}_0$ a particular solution to $\mathbf{AX} = \mathbf{C}$, the set of all solutions $\mathbf{X}$ to $\mathbf{AX} = \mathbf{C}$ is equivalent to the set of all solutions $\mathbf{X}_0 + \mathbf{h}$, where $\mathbf{h}$ is a member of the set of all solutions of the homogeneous system $\mathbf{Ah} = \mathbf{0}$ associated with $\mathbf{AX} = \mathbf{C}$. In this regard, the *structure of the solution set for* $\mathbf{AX} = \mathbf{C}$, which involves three mutually exclusive and collectively exhaustive cases, appears as:

1. No solution exists ($\mathbf{AX} = \mathbf{C}$ is inconsistent);
2. Exactly one solution exists;
3. An infinity of solutions exists

} ($\mathbf{AX} = \mathbf{C}$ is consistent).

As we shall now see, issues concerning the existence and unique-ness of solutions to $\mathbf{AX} = \mathbf{C}$, $\mathbf{A}$ of order $(m \times n)$, are intimately related to the concepts of the range and null-space of $T_{\mathbf{A}}$. It was mentioned earlier that $R(\mathbf{A})$ consists of linear combinations of the columns of $\mathbf{A}$, *i.e.*, $\mathbf{C} \in R(\mathbf{A}) = \ell\{\mathbf{a}_j\}_{j=1}^{n}$ if

$\mathbf{C} = \Sigma_{j=1}^{n} x_j \mathbf{a}_j$. Thus the question of the existence of solutions to $\mathbf{AX} = \mathbf{C}$ boils down to whether or not $\mathbf{C} \epsilon R(\mathbf{A})$. In this regard we state first that $\mathbf{C} \epsilon R(\mathbf{A})$ if and only if $\mathbf{C} = \mathbf{AX}$, $\mathbf{X} \epsilon \mathbf{R}^n$. So if $\mathbf{C} \epsilon R(\mathbf{A})$, then $\mathbf{AX} = \mathbf{C}$ is consistent and has at least one solution. Moreover, if $R(\mathbf{A}) = \mathbf{R}^m$, then $\mathbf{AX} = \mathbf{C}$ has a solution for all $\mathbf{C} \epsilon \mathbf{R}^m$. The following theorem addresses both the existence and uniqueness of solutions to $\mathbf{AX} = \mathbf{C}$.

2.5. THEOREM. Given an $(m \times n)$ matrix $\mathbf{A}$, $\mathbf{AX} = \mathbf{C}$ has either no solution, exactly one solution, or infinitely many solutions.

An alternative tack for addressing questions concerning the existence and uniqueness of simultaneous linear systems of the form $\mathbf{AX} = \mathbf{C}$ is the direct application of the rank-nullity theorem. In this regard, for $\mathbf{A}$ of order $(m \times n)$:

(a) if $m > n \geq \rho(\mathbf{A})$, it follows that $R(\mathbf{A}) \neq \mathbf{R}^m$. Then for some vector $\mathbf{C} \epsilon \mathbf{R}^m$, $\mathbf{C} \notin R(\mathbf{A})$ and thus $\mathbf{AX} = \mathbf{C}$ has no solution.

(b) if $m < n$, then $n(\mathbf{A}) = n - \rho(\mathbf{A}) = n - m > 0$. In this instance $N(\mathbf{A})$ contains at least one nonzero vector so that the solution to $\mathbf{AX} = \mathbf{C}$ is not unique, *i.e.*, since $\mathbf{Ah} = \mathbf{0}$ has $n - m$ linearly independent columns, arbitrary values may be assigned to $n - m$ of the variables.

(c) if $m = n$, then $\rho(\mathbf{A}) = m$ if and only if $n(\mathbf{A}) = 0$. With $N(\mathbf{A}) = \{\mathbf{0}\}$, $\mathbf{AX} = \mathbf{C}$ has a unique solution for every $\mathbf{C} \epsilon \mathbf{R}^m = R(\mathbf{A})$.

If $[\mathbf{A}, \mathbf{C}]$ depicts the ***augmented matrix*** associated with $\mathbf{AX} = \mathbf{C}$, then we may also address the consistency issue of this equation system by simply comparing the ranks of $\mathbf{A}$, $[\mathbf{A}, \mathbf{C}]$. To this end we state

2.6. THEOREM. Given the system $\mathbf{AX} = \mathbf{C}$, where $\mathbf{A}$ is of order $(m \times \mathrm{n})$, if:

1. $\rho[\mathbf{A}, \mathbf{C}] > \rho(\mathbf{A})$, the system is inconsistent;
2. $\rho[\mathbf{A}, \mathbf{C}] = \rho(\mathbf{A}) =$ number of unknowns n, the system is consistent and possesses a unique solution;
3. $\rho[\mathbf{A}, \mathbf{C}] = \rho(\mathbf{A}) = k <$ number of unknowns n, the system is consistent and possesses an infinity of solutions, where arbitrary values may be assigned to $n-k$ of the variables.

As a particularization of case (2), if $\mathbf{A}$ is $(n \times n)$ and nonsingular, then $\mathbf{X} = \mathbf{A}^{-1}\mathbf{C}$.

If $\mathbf{C} = \mathbf{0}$, then the resulting homogeneous system can never be inconsistent, *i.e.*, $\mathbf{AX} = \mathbf{0}$ always has a solution since $\rho(\mathbf{A}) = \rho[\mathbf{A}, \mathbf{0}]$. Moreover, $\mathbf{X} = \mathbf{0}$ is always a (trivial) solution. What about the existence of nontrivial solutions? The answer is provided by a special case of theorem 2.6, namely

2.7. THEOREM. Given the system $\mathbf{AX} = \mathbf{0}$, where $\mathbf{A}$ is of order $(m \times n)$, if:

1. $\rho(\mathbf{A}) =$ number of unknowns n, the system has a unique (trivial) solution $\mathbf{X} = \mathbf{0}$;
2. $\rho(\mathbf{A}) = k <$ number of unknowns n, the system has an infinity of nontrivial solutions, where arbitrary values may be assigned to $n-k$ of the variables.

If $\mathbf{A}$ is $(m \times n)$ and $m < n$, then $\mathbf{AX} = \mathbf{0}$ always possesses an infinity of nontrivial solutions. In general, for $\mathbf{A}$ of order $(m \times n)$ and $\rho(\mathbf{A}) = k$, the solution set $\mathcal{S}$ of the homogeneous linear system $\mathbf{AX} = \mathbf{0}$, $\mathbf{X} \in \mathbf{R}^n$, is an $(n-k)-$dimensional subspace of $\mathbf{R}^n$, *i.e.*, a basis for $\mathcal{S}$ consists of $n - k$ linearly independent vectors.

We may relate these observations to some of our earlier concepts by noting that for the homogeneous system $\mathbf{AX} = \mathbf{0}$, a vector $\mathbf{X}$ lies in $N(\mathbf{A})$ if and only if it is orthogonal to each vector in $Row(\mathbf{A})$, *i.e.*, $Row(\mathbf{A})$, $N(\mathbf{A})$ are orthogonal complements. Hence $\mathbf{X}$ is a solution to $\mathbf{AX} = \mathbf{0}$ if and only if $\mathbf{X}$ is orthogonal to each row vector of $\mathbf{A}$.

It was mentioned earlier that a consistent system $\mathbf{AX} = \mathbf{C}$ has a unique solution $\mathbf{X}_0$ if and only if $\mathbf{Ah} = \mathbf{0}$ has only the trivial solution $\mathbf{h} = \mathbf{0}$. This result may be strengthened somewhat if $\mathbf{A}$ is taken to be an nth order matrix of full rank n. Then it must always be true that $\mathbf{AX} = \mathbf{C}$ has a unique solution for any $\mathbf{C}$ if and only if the associated homogeneous system $\mathbf{Ah} = \mathbf{0}$ has the unique solution $\mathbf{h} = \mathbf{0}$. To operationalize this result let us consider the following *theorem of the alternative*[1] for equalities

---

[1]A ***theorem of the alternative*** involves two mutually exclusive systems of linear inequalities and/or equalities denoted simply as (I), (II). It asserts that either system (I) has a solution or system (II) has a solution, but never both.

2.8. THEOREM. Given the system $\mathbf{AX} = \mathbf{C}$, where $\mathbf{A}$ is of order $(n \times n)$, either:

(I) $\mathbf{AX} = \mathbf{C}$ has exactly one solution for each $\mathbf{C}$ or

(II) $\mathbf{Ah} = \mathbf{0}$ has a nontrivial solution,

but never both.

So to determine whether or not the nth order system $\mathbf{AX} = \mathbf{C}$ has a unique solution for *every* $\mathbf{C}$, we need only determine whether it has a unique (trivial) solution for *only one* particular $\mathbf{C}$, namely $\mathbf{C} = \mathbf{0}$.

Let us now examine in greater detail the conditions under which the simultaneous linear system $\mathbf{AX} = \mathbf{C}$ has no solutions. Based upon our earlier observations it is easily seen that $\mathbf{AX} = \mathbf{C}$ is inconsistent if $\rho[\mathbf{A}, \mathbf{C}] = 1 + \rho(\mathbf{A})$.

Another approach to this problem involves an application of

2.9. FARKAS' THEOREM OF THE ALTERNATIVE [Farkas (1902)]. For any $(m \times n)$ matrix $\mathbf{A}$ and a vector $\mathbf{C} \in \mathbf{R}^m$, either

(I) $\mathbf{AX} = \mathbf{C}$ has a solution $\mathbf{X} \geqq \mathbf{0}$, $\mathbf{X} \in \mathbf{R}^n$, or

(II) $\mathbf{A'Y} \geqq \mathbf{0}$, $\mathbf{C'Y} < 0$ has a solution $\mathbf{Y} \in \mathbf{R}^m$,

but never both.

(A variety of versions of this theorem as well as its geometric interpretation will be offered in subsequent sections.) In this regard, if the assertion "$\mathbf{AX} = \mathbf{C}$ is consistent" is to serve as Farkas' (I), then what should be the appropriate form of Farkas' (II)? To answer this question and to fully mirror (I), let $\mathbf{X} = \mathbf{U} - \mathbf{V}$ with $\mathbf{U}, \mathbf{V} \geqq \mathbf{0}$. Then

$\mathbf{A}(\mathbf{U}-\mathbf{V}) = \mathbf{C}$ has a solution $\mathbf{U}, \mathbf{V} \geqq \mathbf{0}$ or

(I′) $[\mathbf{A}, -\mathbf{A}]\begin{bmatrix}\mathbf{U}\\ \mathbf{V}\end{bmatrix} = \mathbf{C}$ has a nonnegative solution

has the form of (I). From (II), the alternative to (I′) is

$\begin{bmatrix}\mathbf{A}'\\ -\mathbf{A}'\end{bmatrix}\mathbf{Y} \geqq \mathbf{0}$, $\mathbf{C'Y} < 0$ or

(II′) $\mathbf{A'Y} = \mathbf{0}$, $\mathbf{C'Y} < 0$ has a solution $\mathbf{Y}$.

Actually, the sign of $\mathbf{C'Y}$ is irrelevant in the sense that if $\mathbf{Y}$ is replaced by $-\mathbf{Y}$, then (II′) becomes

(II$''$) $\mathbf{A}'\mathbf{Y} = \mathbf{0},\ \mathbf{C}'\mathbf{Y} \neq 0$ has a solution $\mathbf{Y}$.

In sum, either $\mathbf{AX} = \mathbf{C}$ is consistent or there exists a vector $\mathbf{Y}$ which is orthogonal to the columns of $\mathbf{A}$ but not orthogonal to $\mathbf{C}$. These observations are consolidated in the following

2.10. THEOREM. The simultaneous linear system $\mathbf{AX} = \mathbf{C}$, where $\mathbf{A}$ is of order $(m \times n)$ and $\mathbf{X} \in \mathbf{R}^n$, is inconsistent if and only if the linear system $\mathbf{A}'\mathbf{U} = \mathbf{0},\ \mathbf{C}'\mathbf{U} = \alpha,\ 0 \neq \alpha \in \mathbf{R}$, has a solution $\mathbf{U} \in \mathbf{R}^m$.

Let us relate the results obtained in theorem 2.7 to the concept of linear independence (dependence) presented in section 2.1 above. Specifically, if we redefine the ***rank*** of an $(m \times \mathrm{n})$ matrix $\mathbf{A}$ as the maximum number of linearly independent vectors $\mathbf{X}_j \in \mathbf{R}^m$, j=1, ..., n, which span the columns of $\mathbf{A}$ (*i.e.*, it is at most the number of vectors in a basis for $\mathbf{R}^m$), then a test for the linear independence of a set of vectors of the form $\{\mathbf{X}_j \in \mathbf{R}^m,\ j=1, \ldots, n\}$ may be executed by considering the vector $\mathbf{X}_j$ as the jth column of an $(m \times \mathrm{n})$ matrix $\mathbf{A} = [\mathbf{X}_1, \ldots, \mathbf{X}_n]$. Then:

1. if $m \geq n$ and $\rho(\mathbf{A}) = n$, the set of vectors $\{\mathbf{X}_j \in \mathbf{R}^m, j=1, \ldots, n\}$ is linearly independent (by part 1 of theorem 2.7); but

2. if $m \geq n$ and $\rho(\mathbf{A}) < n$, or if $m < \mathrm{n}$, the set of vectors $\{\mathbf{X}_j \in \mathbf{R}^m,\ j=1, \ldots, n\}$ is linearly dependent (by part 2 of theorem 2.7).

An alternative way of expressing 1 is provided by

2.11. THEOREM. Let $\mathbf{A}' = [\mathbf{a}_1, \ldots, \mathbf{a}_m]$ be of order $(n \times m)$. If the vectors $\mathbf{a}_1, \ldots, \mathbf{a}_m$ are linearly independent in $\mathbf{R}^n$, then there exists a vector $\mathbf{Y} \in \mathbf{R}^n$ such that $\mathbf{AY} = \mathbf{b}$ for any $\mathbf{b} \in \mathbf{R}^m$.

Theorem 2.7 dealt with the existence of solutions to homogeneous linear systems of *equalities*. However, in certain important areas of linear programming (*e.g.*, duality theory), the existence of solutions to homogeneous linear systems of *inequalities* (that possible contain some strict equalities) is of paramount importance. In particular, two systems of homogeneous linear inequalities/equalities are termed ***dual systems*** (Tucker (1956)) if they have the following structure:

$$\left.\begin{array}{c} \mathbf{A}_{11}\mathbf{X}_1 + \mathbf{A}_{12}\mathbf{X}_2 \geqq \mathbf{0} \\ \mathbf{A}_{21}\mathbf{X}_1 + \mathbf{A}_{22}\mathbf{X}_2 = \mathbf{0} \\ \mathbf{X}_1 \geqq \mathbf{0} \\ \mathbf{X}_2 \text{ unrestricted} \end{array}\right\} \longleftarrow\boxed{\text{is dual to}}\longrightarrow \left\{\begin{array}{c} \mathbf{U}_1 \geqq \mathbf{0} \\ \mathbf{U}_2 \text{ unrestricted} \\ -\mathbf{A}'_{11}\mathbf{U}_1 - \mathbf{A}'_{21}\mathbf{U}_2 \geqq \mathbf{0} \\ -\mathbf{A}'_{12}\mathbf{U}_1 - \mathbf{A}'_{22}\mathbf{U}_2 = \mathbf{0} \end{array}\right.$$

The correspondence between these two homogeneous systems possesses the following properties:

(a) each system is dual to the other;

(b) each inequality structural (*fully coefficiented*) constraint is complementary to (is in one-to-one correspondence with) a non-negative variable and conversely;

(c) each equality structural constraint is complementary to a variable unrestricted in sign and conversely;

(d) the coefficient matrices of the linear structural relations of one system are the negative transposes of the coefficient matrices of the linear structural relations of the other system and conversely.

If $\mathbf{A}_{12} = \mathbf{A}_{21} = \mathbf{A}_{22} = \mathbf{0}$ and $\mathbf{A}_{11} = \mathbf{A}$ is ***skew-symmetric*** $(\mathbf{A}' = -\mathbf{A})$, then as a special case of the preceding dual systems we obtain, for $\mathbf{X}_1 = \mathbf{X}$, $\mathbf{U}_1 = \mathbf{U}$, the ***self-dual system***

$$\left.\begin{array}{r} \mathbf{AX} \geqq \mathbf{0} \\ \mathbf{X} \geqq \mathbf{0} \end{array}\right\} \longleftarrow\boxed{\text{is self-dual to}}\longrightarrow \left\{\begin{array}{l} \mathbf{U} \geqq \mathbf{0} \\ -\mathbf{A}'\mathbf{U} \geqq \mathbf{0}\,. \end{array}\right.$$

Relative to this self-dual system we have the so-called skew-symmetric

2.7.a. THEOREM [Tucker (1956)]. The self-dual system

$$\begin{array}{r} \mathbf{AX} \geqq \mathbf{0} \\ \mathbf{X} \geqq \mathbf{0}\,, \end{array}$$

where $\mathbf{A}$ is skew-symmetric, possesses at least one solution $\bar{\mathbf{X}} \geqq \mathbf{0}$ such that

$$\begin{array}{ll} \mathbf{A}\bar{\mathbf{X}} & \geqq \mathbf{0} \\ \mathbf{A}\bar{\mathbf{X}} + \bar{\mathbf{X}} & > \mathbf{0}. \end{array}$$

2.4. Orthogonal Projections and Least Squares Solutions

Consider the simultaneous linear equation system $\mathbf{AX} = \mathbf{C}$, where $\mathbf{A}$ is of order $(m \times n)$ with $m > n = \rho(\mathbf{A})$ (the columns of $\mathbf{A}$ are linearly independent and form a basis for $Col(\mathbf{A})$). A system such as this is termed ***overdetermined*** in that, since $\rho(\mathbf{A}) < m$, there are vectors $\mathbf{C} \epsilon \mathbf{R}^m$ such that $\mathbf{AX} = \mathbf{C}$ is inconsistent. While $\mathbf{AX} = \mathbf{C}$ has no *exact solution*, it does admit an *approximate solution* which we shall call a ***pseudosolution*** or ***least-squares solution***, *i.e.*, a solution $\mathbf{X}$ that comes *as close as possible* to satisfying $\mathbf{AX} = \mathbf{C}$.

Let $\mathcal{P}$ in $\mathbf{R}^m$ be a plane through the origin spanned by the linearly independent columns of $\mathbf{A}$, *i.e.*, $\mathcal{P} = Col(\mathbf{A})$. (For $\mathcal{P}$ in $\mathbf{R}^3$ and $\mathbf{A} = [\mathbf{a}_1, \mathbf{a}_2]$, see Figure 2.2.) Since $\mathbf{AX} = \mathbf{C}$ is inconsistent, $\mathbf{C} \notin \mathcal{P}$. Let $\mathbf{C} = \mathbf{U} + \mathbf{V}$, where $\mathbf{U}$ is the orthogonal projection of $\mathbf{C}$ onto $\mathcal{P}$ and $\mathbf{V}$ is orthogonal to $\mathcal{P}$, *i.e.*, $\mathbf{V} \epsilon \mathcal{P}^{\perp}$. Since $\mathbf{U} \epsilon \mathcal{P}$, there exists a vector $\bar{\mathbf{X}} \epsilon \mathbf{R}^n$ such that $\mathbf{U} = \mathbf{A}\bar{\mathbf{X}}$. The ***orthogonal projection*** $\mathbf{U}$ of the point $\mathbf{C}$ onto $\mathcal{P}$ is the (unique) element of $\mathcal{P}$ for which $\| \mathbf{C} - \mathbf{U} \|$ is as small as possible.

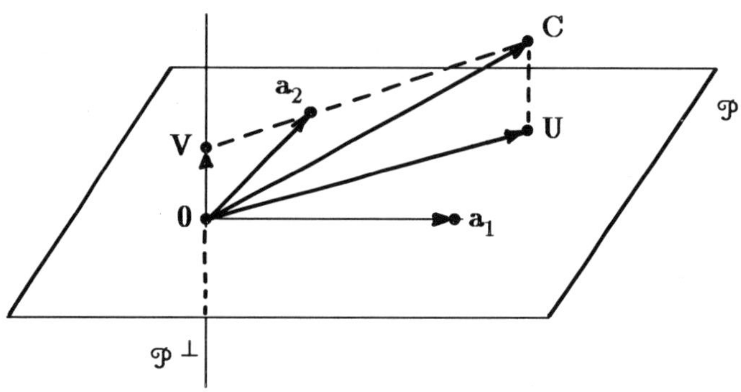

Figure 2.2

Equivalently, we choose to find an $\bar{\mathbf{X}}$ such that

$$g(\bar{\mathbf{X}}) = \min_{\mathbf{X}} \Big\{ \| \mathbf{C} - \mathbf{U} \|^2 = \| \mathbf{C} - \mathbf{AX} \|^2 = \Sigma_{i=1}^m | c_i - \boldsymbol{\alpha}_i \mathbf{X} |^2 = \Sigma_{i=1}^m r_i^2 \Big\}, \tag{2.6}$$

where α_i is the ith row of $\mathbf{A}$ and $r_i = |c_i - \alpha_i \mathbf{X}|$ is the ***residual*** or ***error*** associated with the ith equation in $\mathbf{AX} = \mathbf{C}$. Hence we seek a vector $\bar{\mathbf{X}}$ which renders the *smallest sum of squared residuals* in the sense that $\bar{\mathbf{X}}$ is as close as possible to being a solution of an inconsistent system $\mathbf{AX} = \mathbf{C}$. Thus among all possible $\mathbf{X}$, $\bar{\mathbf{X}}$ comes closest to satisfying $\mathbf{AX} = \mathbf{C}$ in that the residual $\bar{\mathbf{r}} = \mathbf{C} - \mathbf{A}\bar{\mathbf{X}}$ has the smallest norm among all possible residual vectors $\mathbf{r} = \mathbf{C} - \mathbf{AX}$.

To obtain $\mathbf{U}$, the orthogonal projection of $\mathbf{C}$ onto $\mathcal{P}$, let us consider

2.12. THEOREM. The vector $\bar{\mathbf{X}}$ is a ***least squares solution*** to $\mathbf{AX} = \mathbf{C}$ if and only if

$$\mathbf{A}'\mathbf{A}\bar{\mathbf{X}} = \mathbf{A}'\mathbf{C}. \tag{2.7}$$

Here (2.7) is termed the system of ***normal equations*** which coincides with a least squares solution to $\mathbf{AX} = \mathbf{C}$. To derive this system, let us use (2.6) to

$$\min_{\mathbf{X}} g = \min_{\mathbf{X}} \{\mathbf{C}'\mathbf{C} - 2\mathbf{X}'\mathbf{A}'\mathbf{C} + \mathbf{X}'\mathbf{A}'\mathbf{A}\mathbf{X}\}.$$

Setting $\partial g / \partial \mathbf{X} = \mathbf{0}$ then renders (2.7).

Since $\mathbf{A}$ and the $(n \times n)$ matrix $\mathbf{A}'\mathbf{A}$ have the same null-space, $\rho(\mathbf{A}'\mathbf{A}) = n - n(\mathbf{A}'\mathbf{A}) = n - n(\mathbf{A}) = \rho(\mathbf{A}) = n$ (by the rank-nullity theorem). Hence $\mathbf{A}'\mathbf{A}$ has an inverse and thus we obtain, from (2.7),

$$\bar{\mathbf{X}} = (\mathbf{A}'\mathbf{A})^{-1}\mathbf{A}'\mathbf{C}. \tag{2.7.1}$$

In sum, if the $(m \times n)$ matrix $\mathbf{A}$, where $m > n$, has maximal rank n, then $\mathbf{A}'\mathbf{A}$ is invertible and the unique least squares solution of $\mathbf{AX} = \mathbf{C}$ is provided by (2.7.1). Here the matrix $\mathbf{A}^{\dagger} = (\mathbf{A}'\mathbf{A})^{-1}\mathbf{A}'$ is called the ***generalized*** or ***pseudoinverse*** of $\mathbf{A}$ and serves only as a *left-inverse* for $\mathbf{A}$ since $\mathbf{A}^{\dagger}\mathbf{A} = \mathbf{I}_n \neq \mathbf{A}\mathbf{A}^{\dagger}$. (If $m{=}n, \mathbf{A}^{\dagger}{=}\mathbf{A}^{-1}$.) Note that $(\mathbf{A}')^{\dagger} = (\mathbf{A}\mathbf{A}')^{-1}\mathbf{A}$ and that $(\mathbf{A}^{\dagger})' = (\mathbf{A}')^{\dagger}$.

Given $\bar{\mathbf{X}}$ from (2.7.1), the orthogonal projection $\mathbf{U}$ of $\mathbf{C}$ onto $\mathcal{P} = Col(\mathbf{A})$ is

$$\mathbf{U} = \mathbf{A}\bar{\mathbf{X}} = \mathbf{A}(\mathbf{A}'\mathbf{A})^{-1}\mathbf{A}'\mathbf{C}. \tag{2.8}$$

Clearly the orthogonal projection $\mathbf{U}$ of $\mathbf{C}$ onto $\mathcal{P}$ is closer to $\mathbf{C}$ than any other point in $\mathcal{P}$.

Alternatively, given $\mathbf{U} = \mathbf{A}\bar{\mathbf{X}}$, let $\mathbf{Y}$ be an arbitrary vector in $\mathbf{R}^n$. Then

$\mathbf{AY} = \Sigma_{j=1}^{n} y_j \mathbf{a}_j$ is an element of $\mathcal{P}$ and thus orthogonal to $\mathbf{V} = \mathbf{C} - \mathbf{U} = \mathbf{C} - \mathbf{A}\bar{\mathbf{X}} \in \mathcal{P}^{\perp}$. Hence

$$\begin{aligned}(\mathbf{AY}) \cdot \mathbf{V} &= \mathbf{Y}'\mathbf{A}'(\mathbf{C} - \mathbf{A}\bar{\mathbf{X}}) \\ &= \mathbf{Y}'(\mathbf{A}'\mathbf{C} - \mathbf{A}'\mathbf{A}\bar{\mathbf{X}}) = 0.\end{aligned}$$

Since this expression holds for any $\mathbf{Y} \in \mathbf{R}^n$, it must be the case that $\mathbf{A}'\mathbf{C} - \mathbf{A}'\mathbf{A}\bar{\mathbf{X}} = \mathbf{0}$ or (2.7) holds. Then under the preceding rank argument, (2.7) again has a unique solution and thus (2.8) obtains.

2.5. Point-Set Theory: Topological Properties of $\mathbf{R}^n$

Let δ be any positive scalar. A ***δ-neighborhood of a point*** $\mathbf{X}_0 \in \mathbf{R}^n$ (or an ***open ball or sphere of radius δ about*** $\mathbf{X}_0$) is the set $B(\mathbf{X}_0, \delta)$ consisting of all points $\mathbf{X}$ such that $\|\mathbf{X}-\mathbf{X}_0\| < \delta$, *i.e.*, $B(\mathbf{X}_0, \delta) = \{\mathbf{X} \mid \|\mathbf{X}-\mathbf{X}_0\| < \delta, \delta > 0\}$.

A point $\bar{\mathbf{X}}$ is an ***interior point*** of a set $\mathcal{S}$ in $\mathbf{R}^n$ if there exists an open ball of radius δ about $\bar{\mathbf{X}}$ which contains only points of $\mathcal{S}$. The ***interior*** of a set $\mathcal{S}$ in $\mathbf{R}^n$, denoted $\mathcal{S}^o$, is the collection of all its interior points.

A set $\mathcal{S}$ in $\mathbf{R}^n$ is said to be ***open*** if, given any point $\mathbf{X}_0 \in \mathcal{S}$, there exists a positive scalar δ such that $B(\mathbf{X}_0, \delta) \subseteq \mathcal{S}$. Hence $\mathcal{S}$ is open if it contains only interior points. Moreover:

(a) ϕ, $B(\mathbf{X}_0, \delta)$, and $\mathbf{R}^n$ are all open sets; and

(b) $\mathcal{S}$ is an open set if and only if $\mathcal{S} = \mathcal{S}^o$.

Let $\mathcal{S}$ be a set in $\mathbf{R}^n$. Then the ***complementary set of $\mathcal{S}$***, $\mathcal{S}^c$, is the collection of all points of $\mathbf{R}^n$ exclusive (lying outside) of $\mathcal{S}$. A point $\bar{\mathbf{X}} \in \mathcal{S}^c$ is an ***exterior point*** of $\mathcal{S}$ in $\mathbf{R}^n$ if there exists an open ball of radius δ about $\bar{\mathbf{X}}$ which contains only points of $\mathcal{S}^c$. The ***exterior*** of a set $\mathcal{S}$ in $\mathbf{R}^n$, denoted $\mathcal{S}^e$, is the collection of all its exterior points.

A point $\bar{\mathbf{X}}$ is a ***boundary point*** of a set $\mathcal{S}$ in $\mathbf{R}^n$ if every open ball of radius δ about $\bar{\mathbf{X}}$ encompasses points in $\mathcal{S}$ and in $\mathcal{S}^c$, *i.e.*, it is a point which is neither an interior nor an exterior point of $\mathcal{S}$. The ***boundary*** of a set $\mathcal{S}$ in $\mathbf{R}^n$, $\partial(\mathcal{S})$, is the collection of all its boundary points.

A set $\mathcal{S}$ in $\mathbf{R}^n$ is ***bounded*** if there exists a scalar $M \in \mathbf{R}$ such that $\|\mathbf{X}\| < M$ for all $\mathbf{X} \in \mathcal{S}$. Equivalently, $\mathcal{S}$ in $\mathbf{R}^n$ is bounded if it has a finite ***diameter*** $d(\mathcal{S})$, *i.e.*,

$d(\mathcal{S}) = sup\{\|\mathbf{X}-\mathbf{Y}\| \mid \mathbf{X}, \mathbf{Y} \in \mathcal{S}\} < +\infty$. Two additional concepts which utilize the notion of the distance between points in $\mathbf{R}^n$ will at times prove useful. First, for $\mathbf{X}$ a vector in $\mathbf{R}^n$ and $\mathcal{S}$ a subset of $\mathbf{R}^n$, the ***distance between a vector* X *and a set* $\mathcal{S}$** is $d(\mathbf{X}, \mathcal{S}) = inf\{\|\mathbf{X}-\mathbf{Y}\| \mid \mathbf{Y} \in \mathcal{S}\}$. Next, for sets $\mathcal{S}$, $\mathcal{T}$ in $\mathbf{R}^n$, the ***distance between sets* $\mathcal{S}$, $\mathcal{T}$** is $d(\mathcal{S}, \mathcal{T}) = inf\{\|\mathbf{X}-\mathbf{Y}\| \mid \mathbf{X} \in \mathcal{S}, \mathbf{Y} \in \mathcal{T}\}$. If $\mathcal{S} \cap \mathcal{T} \neq \phi$, then $d(\mathcal{S}, \mathcal{T}) = 0$.

A point $\bar{\mathbf{X}}$ is termed a ***point of closure*** or ***adherent point*** of a set $\mathcal{S}$ in $\mathbf{R}^n$ if every open ball of radius δ centered on $\bar{\mathbf{X}}$ contains at least one point of $\mathcal{S}$, *i.e.*, $B(\bar{\mathbf{X}}, \delta) \cap \mathcal{S} \neq \phi$. Hence every open ball of radius δ about $\bar{\mathbf{X}}$ meets $\mathcal{S}$. A subset $\mathcal{S}$ of $\mathbf{R}^n$ is said to be ***closed*** if every point of closure of $\mathcal{S}$ is contained in $\mathcal{S}$, *i.e.*, every point that is arbitrarily close to $\mathcal{S}$ is a member of $\mathcal{S}$. The ***closure*** of a set $\mathcal{S}$ in $\mathbf{R}^n$, denoted $\bar{\mathcal{S}}$, is the set of points of closure of $\mathcal{S}$. Clearly a set $\mathcal{S}$ in $\mathbf{R}^n$ is closed if and only if $\mathcal{S} = \bar{\mathcal{S}}$.

Closely related to the concept of a point of closure of $\mathcal{S}$ is the notion of a ***limit point of a set* $\mathcal{S}$** in $\mathbf{R}^n$ (also termed a ***cluster point*** or ***point of accumulation***). Specifically, $\bar{\mathbf{X}}$ is a limit point of a set $\mathcal{S}$ if each open ball of radius δ about $\bar{\mathbf{X}}$ contains at least one point of $\mathcal{S}$ different from $\bar{\mathbf{X}}$. Hence $B(\bar{\mathbf{X}}, \delta)$ meets $\mathcal{S} - \{\bar{\mathbf{X}}\}$; points of $\mathcal{S}$ different from $\bar{\mathbf{X}}$ *pile up* at $\bar{\mathbf{X}}$. In this regard, if $\bar{\mathbf{X}}$ is a limit point of a set $\mathcal{S}$ in $\mathbf{R}^n$, then for every $\delta > 0$, the set $\mathcal{S} \cap B(\bar{\mathbf{X}}, \delta)$ is an infinite set, *i.e.*, every neighborhood of $\bar{\mathbf{X}}$ contains infinitely many points of $\mathcal{S}$.

On the basis of the preceding discussion we can alternatively conclude that a set $\mathcal{S}$ in $\mathbf{R}^n$ is closed if it contains each of its limit points; and the closure of a set $\mathcal{S}$ in $\mathbf{R}^n$ is $\mathcal{S}$ itself together with its set of limit points. Furthermore:

(a) ϕ, a single vector or point, and $\mathbf{R}^n$ are all closed sets.

(b) Any finite union of closed sets in $\mathbf{R}^n$ is closed; any intersection of closed sets in $\mathbf{R}^n$ is closed.

(c) A subset $\mathcal{S}$ of $\mathbf{R}^n$ is closed if and only if its complementary set $\mathcal{S}^c$ is open.

(d) For $\mathcal{S}$ a set in $\mathbf{R}^n$, the boundary of $\mathcal{S}$, $\partial(\mathcal{S})$, is the closed set $\bar{\mathcal{S}} \cap \bar{\mathcal{S}^c}$. Thus it must be true that $\bar{\mathcal{S}} = \mathcal{S} \cup \partial(\mathcal{S})$.

(e) A subset $\mathcal{S}$ in $\mathbf{R}^n$ is closed if and only if $\mathcal{S}$ contains its boundary.

We previously defined the concepts of a closed set and a bounded set. If a set $\mathcal{S}$ in $\mathbf{R}^n$ is both closed and bounded, then it is termed ***compact.*** We note briefly that:

(a) A closed subset of a compact set $\mathcal{S}$ in $\mathbf{R}^n$ is compact.

(b) The union of a finite number of compact sets in $\mathbf{R}^n$ is compact; the intersection of any number of compact sets in $\mathbf{R}^n$ is compact.

(c) Any finite set $\{\mathbf{X}_1, \ldots, \mathbf{X}_k\}$ in $\mathbf{R}^n$ is compact.

A set $\mathcal{S}$ in $\mathbf{R}^n$ is said to be ***locally compact*** if each of its points has a neighborhood with compact closure, *i.e.*, for each $\mathbf{X} \in \mathcal{S}$ there is an open sphere of radius $\delta > 0$ centered on $\mathbf{X}$, $B(\mathbf{X}, \delta)$, such that $\overline{B(\mathbf{X}, \delta)}$ is compact. It should be evident that any compact space is locally compact but not conversely, *e.g.*, $\mathbf{R}^n$ is locally compact but not compact.

2.6. Hyperplanes and Half-Planes (-Spaces)

A set $\mathcal{L}$ in $\mathbf{R}^n$ is termed a ***linear variety (manifold)*** or ***affine set*** if for vectors $\mathbf{X}_1, \mathbf{X}_2 \in \mathcal{L}$, their ***affine combination***

$$\lambda_1\mathbf{X}_1 + \lambda_2\mathbf{X}_2,\ \lambda_1 + \lambda_2 = 1, \text{ and } \lambda_1, \lambda_2 \in \mathbf{R}$$

is also a member of $\mathcal{L}$. Geometrically, this definition implies that the line passing through any two distinct points of $\mathcal{L}$ lies entirely in $\mathcal{L}$. Alternatively, a set $\mathcal{L}$ in $\mathbf{R}^n$ is a linear variety or affine set if and only if $\mathcal{L} = \mathcal{P} + \mathbf{X}_0$, where $\mathcal{P}$ is a subspace of $\mathbf{R}^n$ and $\mathbf{X}_0$ is an arbitrary point of $\mathbf{R}^n$. Here $\mathcal{P} + \mathbf{X}_0 = \{\mathbf{X}+\mathbf{X}_0 | \mathbf{X} \in \mathcal{P} \subset \mathbf{R}^n, \mathbf{X}_0 \in \mathbf{R}^n\}$ is termed the ***translation of*** $\mathcal{P}$ ***by*** $\mathbf{X}_0$. If $\mathbf{X}_1, \mathbf{X}_2 \in \mathcal{P} + \mathbf{X}_0$, then both $\mathbf{X}_1 - \mathbf{X}_0$ and $\mathbf{X}_2 - \mathbf{X}_0$ are members of $\mathcal{P}$. Conversely, it can be shown that every linear variety is of the form $\mathcal{P} + \mathbf{X}_0$, *i.e.*, if $\mathcal{L}$ is a linear variety, then there exists a subspace $\mathcal{P}$ and a point $\mathbf{X}_0 \in \mathbf{R}^n$ such that $\mathcal{L} = \mathcal{P} + \mathbf{X}_0$.

The ***dimension of an affine set*** $\mathcal{L}$ in $\mathbf{R}^n$ (denoted $dim(\mathcal{L})$) is the dimension of the subspace $\mathcal{P}$ of $\mathbf{R}^n$ obtained by translating $\mathcal{L}$ so that it contains the origin $\{\mathbf{0}\}$. If $\mathcal{L} = \phi$, then $dim(\mathcal{L}) = -1$. Additionally, affine sets of dimension 0, 1, and 2 are points, lines, and planes respectively.

As we shall now demonstrate, for a vector $\mathbf{C}(\neq \mathbf{0}) \epsilon \mathbf{R}^n$ and a scalar $\alpha \epsilon \mathbf{R}$, the ***hyperplane***

$$\mathcal{H} = \{\mathbf{X} | \mathbf{C}'\mathbf{X} = \alpha, \ \mathbf{X} \epsilon \mathbf{R}^n\}$$

is an $(n-1)$-dimensional linear variety or affine subset of $\mathbf{R}^n$. To see this let $\mathbf{X}_1 \epsilon \mathcal{H}$. Translating by $-\mathbf{X}_1$ yields the linear subspace $\mathcal{M} = \mathcal{H} - \mathbf{X}_1$. What about the structure of $\mathcal{M}$? Since $\mathbf{C}'\mathbf{X}_1 = \alpha$, it follows that $\mathbf{C}'\mathbf{X} - \mathbf{C}'\mathbf{X}_1 = \mathbf{C}'(\mathbf{X}-\mathbf{X}_1)$ $= \mathbf{C}'\widehat{\mathbf{X}} = \alpha - \alpha = 0$. Hence $\widehat{\mathbf{X}}(=\mathbf{X}-\mathbf{X}_1) \epsilon \mathcal{M}$ so that $\mathcal{M}$ must be the set of all vectors orthogonal to $\mathbf{C}$. Likewise, $\mathbf{C}$ must be (mutually) orthogonal to the subspace $\mathcal{M}$ so that $\mathbf{C}$ amounts to the one-dimensional subspace $\mathcal{M}^{\perp} = \{\mathbf{C}\}$ and is termed the ***normal*** of $\mathcal{H}$. And since $\mathcal{M}$ and $\{\mathbf{C}\}$ taken together span $\mathbf{R}^n$, it follows that $dim(\mathcal{M}) = dim(\mathcal{H}) = n - dim(\{\mathbf{C}\}) = n-1$.

Any hyperplane $\mathcal{H} = \{\mathbf{X} | \mathbf{C}'\mathbf{X} = \alpha, \ \mathbf{X} \epsilon \mathbf{R}^n\}$ divides $\mathbf{R}^n$ into the two ***closed half-planes (—spaces)***

$$[\mathcal{H}^+] = \{\mathbf{X} | \mathbf{C}'\mathbf{X} \geq \alpha, \ \mathbf{X} \epsilon \mathbf{R}^n\},$$
$$[\mathcal{H}^-] = \{\mathbf{X} | \mathbf{C}'\mathbf{X} \leq \alpha, \ \mathbf{X} \epsilon \mathbf{R}^n\}.$$

If the boundary line $\mathbf{C}'\mathbf{X} = \alpha$ is excluded from the preceding two sets, then we have the two ***open half-planes (—spaces)***

$$(\mathcal{H}^+) = \{\mathbf{X} | \mathbf{C}'\mathbf{X} > \alpha, \ \mathbf{X} \epsilon \mathbf{R}^n\},$$
$$(\mathcal{H}^-) = \{\mathbf{X} | \mathbf{C}'\mathbf{X} < \alpha, \ \mathbf{X} \epsilon \mathbf{R}^n\}.$$

Clearly $[\mathcal{H}^+]$, $[\mathcal{H}^-]$ are closed sets while $(\mathcal{H}^+)$, $(\mathcal{H}^-)$ are open sets.

The following theorem informs us that every affine subset $\mathcal{L}$ of $\mathbf{R}^n$ corresponds to the solution set of a system of m simultaneous linear equations in n unknowns. Specifically, we state

2.13. THEOREM [Rockafellar (1970)]. Given an $(m \times n)$ matrix $\mathbf{B}$ and a vector $\mathbf{b} \epsilon \mathbf{R}^m$, the solution set $\mathcal{L} = \{\mathbf{X} | \mathbf{B}\mathbf{X} = \mathbf{b}, \mathbf{X} \epsilon \mathbf{R}^n\}$ in $\mathbf{R}^n$ is affine.

If in theorem 2.13 the affine set $\mathcal{L}$ is rewritten as $\mathcal{L} = \{\mathbf{X} | \beta_i \mathbf{X} = b_i, i=1, \ldots, m, \mathbf{X} \epsilon \mathbf{R}^n\} = \cap_{i=1}^{m} \mathcal{H}_i$, where β_i is the ith row of $\mathbf{B}$, b_i is the ith component of $\mathbf{b}$, and $\mathcal{H}_i = \{\mathbf{X} | \beta_i \mathbf{X} = b_i, \mathbf{X} \epsilon \mathbf{R}^m\}$, $i=1,\ldots, m$, is a hyperplane, then it is evident that

2.14. COROLLARY. Every affine subset in $\mathbf{R}^n$ is expressible as the intersection of a finite collection of hyperplanes from $\mathbf{R}^n$.

Next, given any set $\mathcal{S}$ in $\mathbf{R}^n$, the ***affine hull of*** $\mathcal{S}$ (or the ***affine subspace spanned by*** $\mathcal{S}$), denoted $aff(\mathcal{S})$, is the set of all affine combinations of points from $\mathcal{S}$ or

$$aff(\mathcal{S}) = \left\{ \mathbf{X} \,\middle|\, \mathbf{X} = \sum_{i=1}^{m} \lambda_i \mathbf{X}_i,\ \mathbf{X}_i \in \mathcal{S},\ \sum_{i=1}^{m} \lambda_i = 1,\ \lambda_i \in R \text{ for all } i \right\}.$$

Here $aff(\mathcal{S})$ is unique and can be thought of as the intersection of all affine sets $\mathcal{A}_i$ in $\mathbf{R}^n$ which contain $\mathcal{S}$, *i.e.*, $aff(\mathcal{S}) = \cap_i \mathcal{A}_i$, where $\mathcal{S} \subset \mathcal{A}_i$ for all i. In this regard, an affine set $\mathcal{T}$ in $\mathbf{R}^n$ is the smallest affine set containing $\mathcal{S}$ in $\mathbf{R}^n$ if and only if $\mathcal{T} = aff(\mathcal{S})$. Then $dim(\mathcal{T}) = dim(aff(\mathcal{S}))$.

Let $\{\mathbf{X}_0, \mathbf{X}_1, \ldots, \mathbf{X}_k\}$ represent a finite set of $k+1$ vectors in $\mathbf{R}^n$. The affine hull of $\{\mathbf{X}_0, \mathbf{X}_1, \ldots, \mathbf{X}_k\}$, denoted $aff\{\mathbf{X}_0, \mathbf{X}_1, \ldots, \mathbf{X}_k\}$, is the set of all affine combinations of its members. The set $\{\mathbf{X}_0, \mathbf{X}_1, \ldots, \mathbf{X}_k\}$ is ***affinely independent*** if $aff\{\mathbf{X}_0, \mathbf{X}_1, \ldots, \mathbf{X}_k\} = aff\{\mathbf{0}, \mathbf{X}_1-\mathbf{X}_0, \ldots, \mathbf{X}_k-\mathbf{X}_0\} + \mathbf{X}_0 = \mathcal{L} + \mathbf{X}_0$ has dimension k. Since $\mathcal{L}$ is the smallest subspace spanned by the ***translated set*** $\{\mathbf{X}_1-\mathbf{X}_0, \ldots, \mathbf{X}_k-\mathbf{X}_0\}$, $dim(aff\{\mathbf{X}_0, \mathbf{X}_1, \ldots, \mathbf{X}_k\}) = dim(\mathcal{L}) = k$ if the translated set of vectors is linearly independent. Hence a set of $k+1$ vectors $\{\mathbf{X}_0, \mathbf{X}_1, \ldots, \mathbf{X}_k\}$ in $\mathbf{R}^n$ is said to be affinely independent if and only if the translated set $\{\mathbf{X}_1-\mathbf{X}_0, \ldots, \mathbf{X}_k-\mathbf{X}_0\}$ is linearly independent.

Additional features of affine sets or linear varieties are:

(a) ϕ, $\mathbf{R}^n$, and the singleton $\{\mathbf{X}\}$ are all affine sets.

(b) The intersection of an arbitrary collection of affine sets is affine.

(c) The dimension of any subset $\mathcal{S}$ of $\mathbf{R}^n$ equals the dimension of its affine hull $aff(\mathcal{S})$.

(d) The set of vectors $\{\mathbf{X}_0, \mathbf{X}_1, \ldots, \mathbf{X}_k\}$ in $\mathbf{R}^n$ is affinely independent if the unique solution of $\Sigma_{i=0}^{k} \lambda_i \mathbf{X}_i = \mathbf{0}$, $\Sigma_{i=0}^{k} \lambda_i = 0$, $\lambda_i \in \mathbf{R}$, is $\lambda_i = 0$ for all i.

(e) Linear independence implies affine independence but not conversely.

(f) The maximum number of affinely independent vectors in $\mathbf{R}^n$ is $n+1$ (consisting of n linearly independent vectors and the null vector).

(g) The affine hull of a finite set of points $\{\mathbf{X}_1, \ldots, \mathbf{X}_k\}$ in $\mathbf{R}^n$ is a subset of its linear hull, i.e., $aff\{\mathbf{X}_1, \ldots, \mathbf{X}_k\} \subseteq \ell\{\mathbf{X}_1, \ldots, \mathbf{X}_k\}$.

(h) Given the simultaneous linear system $\mathbf{AX} = \mathbf{C}$, $\mathbf{X} \epsilon \mathbf{R}^n$, let $\rho(\mathbf{A}) = \rho[\mathbf{A}, \mathbf{C}] = k$. Then the solution set of $\mathbf{AX} = \mathbf{C}$ is an $(n-k)$ - dimensional affine set in $\mathbf{R}^n$.

2.7. Convex Sets

A set $\mathcal{S}$ in $\mathbf{R}^n$ is said to be ***convex*** if for points $\mathbf{X}_1$, $\mathbf{X}_2 \epsilon \mathcal{S}$, their ***convex combination*** (or ***internal average***)

$$\mathbf{X} = \lambda_1\mathbf{X}_1 + \lambda_2\mathbf{X}_2,\ \lambda_1 + \lambda_2 = 1, \text{ and } 0 \leq \lambda_1, \lambda_2 \epsilon \mathbf{R}$$

is also a member of $\mathcal{S}$. Geometrically, this definition implies that the closed line segment joining $\mathbf{X}_1$ and $\mathbf{X}_2$ lies entirely in $\mathcal{S}$. (A convex set differs somewhat from an affine set in that, for the latter, the entire line passing through $\mathbf{X}_1$ and $\mathbf{X}_2$ lies wholly within the set.) A set $\mathcal{S}$ in $\mathbf{R}^n$ containing only a single element is taken to be (trivially) convex.

An important property of a given collection of convex sets is provided by

2.15. THEOREM. Let $\{\mathcal{S}_i\}$ be a family of convex sets in $\mathbf{R}^n$. Then their intersection $\cap_i \mathcal{S}_i$ is a convex set.

Also of particular interest is

2.16. THEOREM. Let $\mathcal{S}$ be a convex set in $\mathbf{R}^n$. Then the closure $\bar{\mathcal{S}}$ of $\mathcal{S}$ is also a convex set.

Some additional properties of convex sets are:

(a) Trivially, ϕ and $\mathbf{R}^n$ are each convex sets.

(b) If $\mathcal{S}$ in $\mathbf{R}^n$ is a convex set, then $\mathcal{S}^o$ (possible empty) is convex.

(c) In general, the union of a collection of convex sets in $\mathbf{R}^n$ is not convex.

(d) A hyperplane along with both open and closed half-spaces in $\mathbf{R}^n$ are all convex sets.

(e) If $\mathcal{S}_1, \ldots, \mathcal{S}_m$ are each convex sets in $\mathbf{R}^n$, then their linear combination $\mathcal{S} = \Sigma_{i=1}^m \lambda_i \mathcal{S}_i$ is also convex for $\lambda_i \epsilon \mathbf{R}$ for all i.

(f) Every affine set is convex, but not conversely.

To set the stage for our next concept let us first consider the following

2.17. THEOREM. Given an $(m \times n)$ matrix $\mathbf{B}$ and a vector $\mathbf{b} \in \mathbf{R}^m$, the solution set $\mathcal{S} = \{\mathbf{X} \,|\, \mathbf{BX} \leqq \mathbf{b}, \mathbf{X} \in \mathbf{R}^n\}$ in $\mathbf{R}^n$ is convex.

In this regard, for β_i, $i=1,\ldots,m$, the ith row of $\mathbf{B}$, let $[\mathcal{H}_i^-] = \{\mathbf{X} \,|\, \beta_i\mathbf{X} \leq b_i, \mathbf{X} \in \mathbf{R}^n\}$, $i=1,\ldots,m$, where b_i is the ith component of $\mathbf{b}$. Clearly $[\mathcal{H}_i^-]$ is a closed half-space. And since each of these closed half-spaces is convex, it follows from theorem 2.15 that $\mathcal{S} = \cap_{i=1}^m [\mathcal{H}_i^-]$ is convex. We can now assert that a set $\mathcal{S}$ in $\mathbf{R}^n$ which can be expressed as the intersection of finitely many closed half-spaces in $\mathbf{R}^n$ is a ***polyhedral convex set.***

A ***convex combination*** of a finite number of points $\mathbf{X}_i$, $i=1,\ldots,m$, in $\mathbf{R}^n$ is the point

$$\mathbf{X} = \sum_{i=1}^m \lambda_i \mathbf{X}_i, \ \sum_{i=1}^m \lambda_i = 1, \quad 0 \leq \lambda_i \in \mathbf{R} \text{ for all } i.$$

The following two theorems relate this construct back to the notion of a convex set presented earlier.

2.18. THEOREM. A set $\mathcal{S}$ in $\mathbf{R}^n$ is convex if and only if every convex combination of finitely many points of $\mathcal{S}$ is an element of $\mathcal{S}$.

The next theorem considers a key property of *the set of all convex combinations of a finite number of points.* Specifically,

2.19. THEOREM. Let $\{\mathbf{X}_1,\ldots,\mathbf{X}_m\}$ be a finite collection of points in $\mathbf{R}^n$. Then the set of all convex combinations of the $\mathbf{X}_i$, $i=1,\ldots,m$,

$$\mathcal{K} = \left\{\mathbf{X} \,|\, \mathbf{X} = \sum_{i=1}^m \lambda_i \mathbf{X}_i, \ \sum_{i=1}^m \lambda_i = 1, \ 0 \leq \lambda_i \in \mathbf{R} \text{ for all } i\right\}$$

is convex.

We note briefly that if $\mathcal{S}_1,\ldots,\mathcal{S}_m$ are each convex sets in $\mathbf{R}^n$, then their convex combination $\mathcal{S} = \Sigma_{i=1}^m \lambda_i \mathcal{S}_i$, $\Sigma_{i=1}^m \lambda_i = 1$, $0 \leq \lambda_i \in \mathbf{R}$ is also convex.

The ***convex hull*** (or ***convex closure***) $co(\mathcal{S})$ of a set $\mathcal{S}$ in $\mathbf{R}^n$ is the collection of all convex combinations of points from $\mathcal{S}$ or

$$co(\mathcal{S}) = \left\{\mathbf{X} \,|\, \mathbf{X} = \sum_{i=1}^m \lambda_i \mathbf{X}_i, \ \mathbf{X}_i \in \mathcal{S}, \ \sum_{i=1}^m \lambda_i = 1, \ 0 \leq \lambda_i \in \mathbf{R} \text{ for all } i, \text{ and } m \text{ is an arbitrary positive integer}\right\}.$$

2.20. THEOREM. A set $\mathcal{T}$ in $\mathbf{R}^n$ is the smallest convex set containing an arbitrary set $\mathcal{S}$ in $\mathbf{R}^n$ if and only if $\mathcal{T} = \text{co}(\mathcal{S})$.

A set $\mathcal{S}$ in $\mathbf{R}^n$ which is the convex hull of finitely many points is said to be *spanned or generated* by those points and is called a ***convex polytope.***

How should the word *smallest* in the preceding theorem be interpreted? The answer is provided by

2.21. THEOREM. The convex hull of a set $\mathcal{S}$ in $\mathbf{R}^n$ is the intersection of all convex sets $\mathcal{A}_i$ in $\mathbf{R}^n$ which contain $\mathcal{S}$, *i.e.*, $co(\mathcal{S}) = \cap_i A_i$, where $\mathcal{S} \subset \mathcal{A}_i$ for all i.

We next have

2.22. THEOREM. Let $\{\mathbf{X}_1, \ldots, \mathbf{X}_m\}$ be a finite collection of points in $\mathbf{R}^n$. Then the convex hull of the $\mathbf{X}_i$, $i = 1, \ldots, m$, is the set of all their convex combinations, *i.e.*, $\mathcal{G} = co\{\mathbf{X}_1, \ldots, \mathbf{X}_m\} = \{\mathbf{X} \mid \mathbf{X} = \Sigma_{i=1}^m \lambda_i \mathbf{X}_i, \Sigma_{i=1}^m \lambda_i = 1, 0 \leq \lambda_i \epsilon \mathbf{R} \text{ for all } i\}$.

We note briefly that:

(a) If a set $\mathcal{S}$ in $\mathbf{R}^n$ is convex, then $co(\mathcal{S}) = \mathcal{S}$.

(b) For $\mathcal{S}$ a subset of $\mathbf{R}^n$, $co(\mathcal{S})$ is a closed set if $\mathcal{S}$ is a finite set.

(c) If a set $\mathcal{S}$ in $\mathbf{R}^n$ is compact, then $co(\mathcal{S})$ is compact.

(d) If $\mathcal{S}$ is an open set in $\mathbf{R}^n$, then $co(\mathcal{S})$ is open. The convex hull of a closed set in $\mathbf{R}^n$ need not be closed.

(e) All affine sets are convex (hence convex sets are more general than affine sets).

(f) For $\mathcal{S}$ a subset of $\mathbf{R}^n$, the ***closed convex hull*** of $\mathcal{S}$ is the closure of $co(\mathcal{S})$ or $\overline{co(\mathcal{S})}$.

(g) In general, the convex hull of a closed set need not be closed. If $\mathcal{S}$ is a bounded subset of $\mathbf{R}^n$, then $\overline{co(\mathcal{S})} = co(\overline{\mathcal{S}})$. So if $\mathcal{S}$ is closed and bounded, then $co(\mathcal{S})$ is closed and bounded.

(h) For $\mathcal{S}(\neq \phi)$ an arbitrary set in $\mathbf{R}^n$, $\mathcal{S} \subseteq co(\mathcal{S}) \subseteq aff(\mathcal{S}) \subseteq \ell(\mathcal{S})$.

(i) For a finite set of points $\{\mathbf{X}_1, \ldots, \mathbf{X}_k\}$ in $\mathbf{R}^n$,

$$co\{\mathbf{X}_1, \ldots, \mathbf{X}_k\} \subseteq aff\{\mathbf{X}_1, \ldots, \mathbf{X}_k\} \subseteq \ell\{\mathbf{X}_1, \ldots, \mathbf{X}_k\}.$$

Earlier in this chapter we encountered the concept of the interior of a set in $\mathbf{R}^n$. Interestingly enough, for convex sets, *interior* is subsumed under the notion of *relative interior.* Specifically, the ***relative interior*** of a convex set $\mathcal{S}$ in $\mathbf{R}^n$, denoted $ri(\mathcal{S})$, is the interior which results whenever $\mathcal{S}$ is considered a subset of its affine hull $aff(\mathcal{S})$. To put this definition in its proper perspective, let us admit another point $\mathbf{Y}$ into our discussion. In this regard, the relative interior of $\mathcal{S}$ consists of those points $\mathbf{X} \in aff(\mathcal{S})$ such that $\mathbf{Y} \in \mathcal{S}$ whenever $\mathbf{Y}$ is near $\mathbf{X}$ and also a member of $aff(\mathcal{S})$ or $ri(\mathcal{S}) = \{\mathbf{X} \in aff(\mathcal{S}) \mid \mathbf{Y} \in B(X, \delta) \cap (aff(\mathcal{S}))$ implies $\mathbf{Y} \in \mathcal{S}\}$.

A convex set $\mathcal{S}$ in $\mathbf{R}^n$ is termed ***relatively open*** if $\mathcal{S} = ri(\mathcal{S})$. Furthermore, the ***relative boundary*** of a convex subset $\mathcal{S}$ of $\mathbf{R}^n$ is $r\partial(\mathcal{S}) = \{\mathbf{X} \mid \mathbf{X} \in \overline{\mathcal{S}}$ and $\mathbf{X} \notin ri(\mathcal{S})\}$.

Additional properties or relationships involving the relative interior of a convex set $\mathcal{S}$ in $\mathbf{R}^n$ are:

(a) A nonempty convex set $\mathcal{S}$ in $\mathbf{R}^n$ always has a nonempty relative interior.

(b) For $\mathcal{S}$ a convex subset of $\mathbf{R}^n$, $ri(\mathcal{S}) \subset \mathcal{S} \subset \overline{\mathcal{S}} \subset aff(\mathcal{S})$.

(c) Any affine set $\mathcal{S}$ in $\mathbf{R}^n$ is relatively open (by definition).

(d) For any convex set $\mathcal{S}$ in $\mathbf{R}^n$ and any scalar $\lambda \in \mathbf{R}$, $ri(\lambda\mathcal{S}) = \lambda ri(\mathcal{S})$.

Finally, a set $\mathcal{S}$ is said to be ***locally convex*** if each point of $\mathcal{S}$ has a neighborhood whose intersection with $\mathcal{S}$ is convex.

2.8. Existence of Separating and Supporting Hyperplanes

We noted in section 2.6 that for a vector $\mathbf{C}(\neq\mathbf{0}) \in \mathbf{R}^n$ and a scalar $\alpha \in \mathbf{R}$, a ***hyperplane***

$$\mathcal{H} = \{\mathbf{X} \mid \mathbf{C}'\mathbf{X} = \alpha,\ \mathbf{X} \in \mathbf{R}^n\}$$

is an $(n-1)$ - dimensional linear variety or affine set and that any hyperplane $\mathcal{H}$ in $\mathbf{R}^n$ generates the two ***closed half-spaces***

$$[\mathcal{H}^+] = \{\mathbf{X} \mid \mathbf{C}'\mathbf{X} \geq \alpha,\ \mathbf{X} \in \mathbf{R}^n\},$$

$$[\mathcal{H}^-] = \{\mathbf{X} \mid \mathbf{C}'\mathbf{X} \leq \alpha,\ \mathbf{X} \in \mathbf{R}^n\}.$$

A point in $\mathbf{R}^n$ lies in $[\mathcal{H}^+]$, $[\mathcal{H}^-]$, or in both of these sets.

If $\mathbf{X}_0$ is taken to be a fixed reference point in $\mathcal{H}$, then $\mathbf{C}'\mathbf{X}_0 = \alpha$ and, for any other $\mathbf{X} \in \mathcal{H}$, $\mathbf{C}'\mathbf{X} = \alpha$ so that $\mathbf{C}'\mathbf{X} - \mathbf{C}'\mathbf{X}_0 = \mathbf{C}'(\mathbf{X}-\mathbf{X}_0) = \alpha - \alpha = 0$. Hence a hyperplane has the alternative representation

$$\mathcal{H} = \{\mathbf{X} \mid \mathbf{C}'(\mathbf{X}-\mathbf{X}_0) = 0,\ \mathbf{X} = \mathbf{X}_0 \text{ fixed},\ \mathbf{X} \in \mathbf{R}^n\}.$$

Clearly $\mathbf{C}$ is orthogonal to $\mathbf{X}-\mathbf{X}_0$ for any $\mathbf{X}\in\mathcal{H}$ and is termed the ***normal*** of the hyperplane. Additionally, for fixed $\mathbf{X}_0\in\mathcal{H}$, we now have an alternative representation of the preceding closed half-spaces:

$$[\mathcal{H}^+] = \{\mathbf{X}\,|\,\mathbf{C}'(\mathbf{X}-\mathbf{X}_0) \geq 0,\ \mathbf{X} = \mathbf{X}_0 \text{ fixed, } \mathbf{X}\in\mathbf{R}^n\},$$

$$[\mathcal{H}^-] = \{\mathbf{X}\,|\,\mathbf{C}'(\mathbf{X}-\mathbf{X}_0) \leq 0,\ \mathbf{X} = \mathbf{X}_0 \text{ fixed, } \mathbf{X}\in\mathbf{R}^n\}$$

In what follows either of these equivalent representations of hyperplanes and half-planes will be used depending upon which version is most appropriate for the discussion at hand.

Let $\mathcal{S}_1, \mathcal{S}_2$ be nonempty sets in $\mathbf{R}^n$. Then the hyperplane $\mathcal{H} = \{\mathbf{X}\,|\,\mathbf{C}'\mathbf{X} = \alpha, \mathbf{X}\in\mathbf{R}^n\}$ ***weakly separates*** $\mathcal{S}_1, \mathcal{S}_2$ if $\mathbf{C}'\mathbf{X} \geq \alpha$ for $\mathbf{X}\in\mathcal{S}_1$ and $\mathbf{C}'\mathbf{X} \leq \alpha$ for $\mathbf{X}\in\mathcal{S}_2$ while $\mathcal{H}$ ***strictly separates*** $\mathcal{S}_1, \mathcal{S}_2$ if $\mathbf{C}'\mathbf{X} > \alpha$ for $\mathbf{X}\in\mathcal{S}_1$ and $\mathbf{C}'\mathbf{X} < \alpha$ for $\mathbf{X}\in\mathcal{S}_2$. Also, $\mathcal{H}$ is said to ***strongly separate*** $\mathcal{S}_1, \mathcal{S}_2$ if there exists $0 < \in\, \in \mathbf{R}$ such that $\mathbf{C}'\mathbf{X} \geq \alpha + \in$ for $\mathbf{X}\in\mathcal{S}_1$ and $\mathbf{C}'\mathbf{X} \leq \alpha$ for $\mathbf{X}\in\mathcal{S}_2$.

For $\mathcal{S}$ a nonempty set in $\mathbf{R}^n$ and $\bar{\mathbf{X}}$ a point in the boundary of $\mathcal{S}$, a hyperplane $\mathcal{H} = \{\mathbf{X}\,|\,\mathbf{C}'(\mathbf{X}-\bar{\mathbf{X}}) = 0,\ \mathbf{X} = \bar{\mathbf{X}} \text{ fixed, } \mathbf{X}\in\mathbf{R}^n\}$ is a ***supporting*** (or ***tangent***) ***hyperplane*** of $\mathcal{S}$ at $\bar{\mathbf{X}}$ if either $\mathcal{S}\subset[\mathcal{H}^+]$ (hence $\mathbf{C}'(\mathbf{X}-\bar{\mathbf{X}}) \geq 0$ for $\mathbf{X}\in\mathcal{S}$) or $\mathcal{S}\subset[\mathcal{H}^-]$ (so that $\mathbf{C}'(\mathbf{X}-\bar{\mathbf{X}}) \leq 0$ for $\mathbf{X}\in\mathcal{S}$). Thus $\mathcal{H}$ is a supporting hyperplane of $\mathcal{S}$ if: (1) $\mathcal{S}$ lies in one of the two closed half-spaces $[\mathcal{H}^+]$, $[\mathcal{H}^-]$ and; (2) $\mathcal{H}$ has a point in common with $\mathcal{S}$. Equivalently, $\mathcal{H}$ is a supporting hyperplane of $\mathcal{S}$ in $\mathbf{R}^n$ at $\bar{\mathbf{X}}\in\partial(\mathcal{S})$ if either $\mathbf{C}'\bar{\mathbf{X}} = \mathit{inf}\{\mathbf{C}'\mathbf{X}\,|\,\mathbf{X}\in\mathcal{S}\}$ or $\mathbf{C}'\bar{\mathbf{X}} = \mathit{sup}\{\mathbf{C}'\mathbf{X}\,|\,\mathbf{X}\in\mathcal{S}\}$. Moreover, $\mathcal{H}$ ***properly supports*** $\mathcal{S}$ in $\mathbf{R}^n$ at $\bar{\mathbf{X}}$ if $\mathcal{S} \not\subset \mathcal{H}$.

If $\mathcal{H}$ is a supporting hyperplane for a convex set $\mathcal{S}$ in $\mathbf{R}^n$ at $\bar{\mathbf{X}}\in\partial(\mathcal{S})$, then the half-spaces

$$[\mathcal{H}_s^+] = \{\mathbf{X}\,|\,\mathbf{C}'\mathbf{X} \geq \alpha = \mathbf{C}'\bar{\mathbf{X}},\ \mathbf{X}\in\mathbf{R}^n\},$$

$$[\mathcal{H}_s^-] = \{\mathbf{X}\,|\,\mathbf{C}'\mathbf{X} \leq \alpha = \mathbf{C}'\bar{\mathbf{X}},\ \mathbf{X}\in\mathbf{R}^n\}$$

are ***supporting closed half-spaces*** for $\mathcal{S}$.

Based upon these definitions, we begin with the

2.23. STRONG SEPARATION THEOREM. Let $\mathcal{S}$ be a non-empty closed convex set in $\mathbf{R}^n$ with $\mathbf{Y}\in\mathbf{R}^n$ a vector not in $\mathcal{S}$. Then there exists a vector $\mathbf{C}(\neq \mathbf{0}) \epsilon \mathbf{R}^n$ and a scalar $\alpha \epsilon \mathbf{R}$ such that $\mathbf{C}'\mathbf{X} \leq \alpha < \mathbf{C}'\mathbf{Y}$ for all $\mathbf{X} \epsilon \mathcal{S}$.

In essence, this theorem states that if $\mathcal{S}$ is a closed convex set in $\mathbf{R}^n$ and if $\mathbf{Y} \epsilon \mathbf{R}^n$ is not a member of $\mathcal{S}$, then there exists a strongly separating hyperplane $\mathcal{H} = \{\mathbf{X} | \mathbf{C}'\mathbf{X} = \alpha, \mathbf{X} \epsilon \mathbf{R}^n\}$ such that $\mathbf{Y}$ is in the open half-space $(\mathcal{H}^+)$ and $\mathcal{S}$ is in the closed half-space $[\mathcal{H}^-]$ determined by $\mathcal{H}$. Note that by implication the hyperplane $\mathcal{H}$ also strictly separates set $\mathcal{S}$ and the vector $\mathbf{Y}$.

The preceding theorem was developed under the assumption that $\mathcal{S}$ is a nonempty closed convex set. However, if $\mathcal{S}$ is not closed and $\mathbf{Y} \notin \bar{\mathcal{S}}$, then the basic result of theorem 2.23 still holds with $\mathbf{C}'\mathbf{Y} > \alpha = sup\{\mathbf{C}'\mathbf{X} | \mathbf{X} \epsilon \mathcal{S}\}$. In this instance theorem 2.23 may be restated in its less restrictive form as

2.23.a. STRONG SEPARATION THEOREM. Let $\mathcal{S}$ be a non-empty convex set in $\mathbf{R}^n$ with $\mathbf{Y} \epsilon \mathbf{R}^n$ an exterior point of $\bar{\mathcal{S}}$. Then there exists a vector $\mathbf{C}(\neq \mathbf{0}) \epsilon \mathbf{R}^n$ and a scalar $\alpha \epsilon \mathbf{R}$ such that $\mathbf{C}'\mathbf{Y} > \alpha = sup\{\mathbf{C}'\mathbf{X} | \mathbf{X} \epsilon \mathcal{S}\}$.

This result reveals that if $\mathcal{S}$ is a convex set in $\mathbf{R}^n$ and $\mathbf{Y} \epsilon \mathbf{R}^n$ is not a member of $\bar{\mathcal{S}}$, then there exists a strongly separating hyperplane $\mathcal{H} = \{\mathbf{X} | \mathbf{C}'\mathbf{X} = \alpha, \mathbf{X} \epsilon \mathbf{R}^n\}$ passing through $\bar{\mathbf{X}} \epsilon \partial(\mathcal{S})$ and having $\mathcal{S}$ on or beneath it with $\mathbf{Y}$ an element of the open half-space $(\mathcal{H}^+)$.

We next have the

2.24. PLANE OF SUPPORT THEOREM. Let $\mathcal{S}$ be a nonempty closed convex set in $\mathbf{R}^n$ and let $\bar{\mathbf{X}}$ be a boundary point of $\mathcal{S}$. Then there exists a vector $\mathbf{C}(\neq \mathbf{0}) \epsilon \mathbf{R}^n$ and a scalar $\alpha \epsilon \mathbf{R}$ such that $\mathbf{C}'\mathbf{X} \leq \alpha = \mathbf{C}'\bar{\mathbf{X}}$ $(= sup\{\mathbf{C}'\mathbf{X} | \mathbf{X} \epsilon \mathcal{S}\})$ for all $\mathbf{X} \epsilon \mathcal{S}$.

If $\mathcal{S}$ is not a closed subset of $\mathbf{R}^n$, then the basic result of theorem 2.24 may be framed as

2.24.a. PLANE OF SUPPORT THEOREM. Let $\mathcal{S}$ be a nonempty convex set with interior points in $\mathbf{R}^n$ and let $\bar{\mathbf{X}}$ be a boundary point of $\mathcal{S}$. Then there exists a vector $\mathbf{C}(\neq \mathbf{0}) \epsilon \mathbf{R}^n$ such that $\mathbf{C}'\mathbf{X} \leq \alpha = \mathbf{C}'\bar{\mathbf{X}}$ $(= sup\{\mathbf{C}'\mathbf{X} | \mathbf{X} \epsilon \bar{\mathcal{S}}\})$ for each $\mathbf{X}$ contained within the closure of $\mathcal{S}$, $\bar{\mathcal{S}}$.

Theorems 2.24, 2.24.a inform us that a convex subset of $\mathbf{R}^n$ (which may or may not be closed) has a supporting hyperplane at each of its boundary points. So given a boundary point of a convex set $\mathcal{S}$ in $\mathbf{R}^n$, there exists a hyperplane that: (1) contains the boundary point, *i.e.*, $\mathcal{H} = \{\mathbf{X} \mid \mathbf{C}'\mathbf{X} = \alpha = \mathbf{C}'\bar{\mathbf{X}},\ \mathbf{X} \in \mathbf{R}^n\}$; and (2) contains $\mathcal{S}$ in one of its supporting closed half-spaces. One important implication of these plane of support theorems is that they allow us to represent a convex set in terms of its associated set of supporting or tangent hyperplanes at boundary points on $\mathcal{S}$. In this regard,

2.25. REPRESENTATION THEOREM. A closed convex set $\mathcal{S}$ in $\mathbf{R}^n$ is the intersection of all supporting closed half-spaces containing it, *i.e.*, $\mathcal{S} = \cap_i[\mathcal{H}_s^+]_i$ (or $\mathcal{S} = \cap_i[\mathcal{H}_s^-]_i$).

This result follows from our earlier discussions on properties of convex sets that the intersection of an arbitrary number of closed and convex sets in $\mathbf{R}^n$ is itself a closed convex set. Hence the intersection of all supporting closed half-spaces containing $\mathcal{S}$ is the smallest closed convex set containing $\mathcal{S}$ so that $\mathcal{S} = co(\mathcal{S}) = \cap_i[\mathcal{H}_s^+]_i$ (by virtue of theorems 2.20, 2.21).

Note that if $\mathcal{S}$ in $\mathbf{R}^n$ is only convex and not also closed, then $\bar{\mathcal{S}} = \cap_i[\mathcal{H}_s^+]_i$ (or $\bar{\mathcal{S}} = \cap_i\, [\mathcal{H}_s^-]_i$).

A somewhat stronger version of the preceding theorem is

2.25.a. REPRESENTATION THEOREM. A closed convex set $\mathcal{S}$ in $\mathbf{R}^n$ is the intersection of all the closed half-spaces tangent to it.

As will now be demonstrated, an important application of the notion of a supporting hyperplane is that it can be used to characterize a closed convex set with nonempty interior as convex. Specifically,

2.26. THEOREM. Let $\mathcal{S}$ be a closed set in $\mathbf{R}^n$ with $\mathcal{S}^o \neq \phi$. If $\mathcal{S}$ has a supporting hyperplane at each of its boundary points, then $\mathcal{S}$ is convex.

If we couple theorem 2.26 with theorem 2.24, then a necessary and sufficient condition for a closed set with nonempty interior to be convex is

2.27. THEOREM. Let $\mathcal{S}$ be a closed set in $\mathbf{R}^n$ with $\mathcal{S}^o \neq \phi$. Then $\mathcal{S}$ is convex if and only if $\mathcal{S}$ has a supporting hyperplane at each of its boundary points.

The preceding separation theorems involved the existence of a hyperplane which separates a convex set $\mathcal{S}$ (possibly closed) in $\mathbf{R}^n$ and a point exterior to $\mathcal{S}$. Let us now turn to the development of a set of theorems which address the separation of two convex sets $\mathcal{S}_1$, $\mathcal{S}_2$ in $\mathbf{R}^n$.

We state first the

2.28. STRONG SEPARATION THEOREM [Minkowski (1911)]. Let $\mathcal{S}_1$, $\mathcal{S}_2$ be two non-empty disjoint convex sets in $\mathbf{R}^n$ with $\mathcal{S}_1$ compact and $\mathcal{S}_2$ closed. Then there exists a vector $\mathbf{C}(\neq \mathbf{0}) \in \mathbf{R}^n$ and a scalar $\alpha \in \mathbf{R}$ such that the following equivalent conditions hold:

(a) $\mathbf{C}'\mathbf{X}_1 < \mathbf{C}'\mathbf{X}_2$ for $\mathbf{X}_1 \in \mathcal{S}_1$, $\mathbf{X}_2 \in \mathcal{S}_2$ or

(b) $inf\{\mathbf{C}'\mathbf{X} \,|\, \mathbf{X} \in \mathcal{S}_2\} > sup\{\mathbf{C}'\mathbf{X} \,|\, \mathbf{X} \in \mathcal{S}_1\}$ or

(c) $\begin{cases} \mathbf{C}'\mathbf{X} < \alpha, \ \mathbf{X} \in \mathcal{S}_1; \\ \mathbf{C}'\mathbf{X} > \alpha, \ \mathbf{X} \in \mathcal{S}_2. \end{cases}$

Next comes the

2.29. WEAK SEPARATION THEOREM [Minkowski (1911)]. Let $\mathcal{S}_1$, $\mathcal{S}_2$ be two non-empty disjoint convex sets in $\mathbf{R}^n$. Then there exists a vector $\mathbf{C}(\neq \mathbf{0}) \in \mathbf{R}^n$ and a scalar $\alpha \in \mathbf{R}$ such that the following equivalent conditions hold:

(a) $\mathbf{C}'\mathbf{X}_1 \leq \mathbf{C}'\mathbf{X}_2$ for $\mathbf{X}_1 \in \mathcal{S}_1$, $\mathbf{X}_2 \in \mathcal{S}_2$ or

(b) $inf\{\mathbf{C}'\mathbf{X} \,|\, \mathbf{X} \in \mathcal{S}_2\} \geq sup\{\mathbf{C}'\mathbf{X} \,|\, \mathbf{X} \in \mathcal{S}_1\}$ or

(c) $\begin{cases} \mathbf{C}'\mathbf{X} \leq \alpha, \ \mathbf{X} \in \mathcal{S}_1; \\ \mathbf{C}'\mathbf{X} \geq \alpha, \ \mathbf{X} \in \mathcal{S}_2. \end{cases}$

Finally, a set of observations concerning the existence of supporting hyperplanes for convex sets are the following:

2.30. THEOREM. Let $\mathcal{S}$ be a nonempty convex set in $\mathbf{R}^n$ and let $\mathcal{T}$ be a nonempty convex subset of $\mathcal{S}$ with $ri(\mathcal{S}) \cap \mathcal{T} = \phi$. Then there exists a proper supporting hyperplane to $\mathcal{S}$ containing $\mathcal{T}$.

2.31. THEOREM. A nonempty convex set $\mathcal{S}$ in $\mathbf{R}^n$ has a normal $\mathbf{C}(\neq \mathbf{0}) \in \mathbf{R}^n$ at each of its boundary points, *i.e.*, there exists a $\mathbf{C}(\neq \mathbf{0})$ such that $\mathbf{C}'\mathbf{X} = 0$ for each $\mathbf{X} \in \partial(\mathcal{S})$.

It is important to note that separability of nonempty sets does not imply disjointness or convexity. But if nonempty sets are convex and disjoint (or at least have no interior points in common), then they are separable.

If we relax the assumption that $\mathcal{S}_1, \mathcal{S}_2$ are convex sets in $\mathbf{R}^n$, then a reformulation of theorem 2.28 appears as

2.28.a. STRONG SEPARATION THEOREM. Let $\mathcal{S}_1, \mathcal{S}_2$ be non-empty compact sets in $\mathbf{R}^n$. Then there exists a vector $\mathbf{C}(\neq \mathbf{0}) \in \mathbf{R}^n$ and a scalar $\alpha \in \mathbf{R}$ such that

$$\mathbf{C}'\mathbf{X} < \alpha, \ \mathbf{X} \in \mathcal{S}_1;$$
$$\mathbf{C}'\mathbf{X} > \alpha, \ \mathbf{X} \in \mathcal{S}_2$$

if and only if $co(\mathcal{S}_1) \cap co(\mathcal{S}_2) = \phi$.

2.9. Convex Cones

A ***cone*** $\mathcal{C}$ in $\mathbf{R}^n$ is a set of points such that if $\mathbf{X} \in \mathcal{C}$, then so is every non-negative scalar multiple of $\mathbf{X}$, *i.e.*, if $\mathbf{X} \in \mathcal{C}$, then $\lambda \mathbf{X} \in \mathcal{C}$ for $0 \leq \lambda \in \mathbf{R}$, $\mathbf{X} \in \mathbf{R}^n$. If we consider the set of points $\mathfrak{X} = \{\mathbf{X}\}$, then the ***cone generated by*** $\mathfrak{X}$ is $\mathcal{C} = \{\mathbf{Y} | \mathbf{Y} = \lambda \mathbf{X}, \ 0 \leq \lambda \in \mathbf{R}, \ \mathbf{X} \in \mathfrak{X}\}$. And if $\mathbf{0} \notin \mathfrak{X}$ and for each $\mathbf{Y}(\neq \mathbf{0}) \in \mathcal{C}$ there are unique $\mathbf{X} \in \mathfrak{X}$ with $\lambda > 0$ such that $\mathbf{Y} = \lambda \mathbf{X}$, then $\mathfrak{X}$ is termed a ***base*** of $\mathcal{C}$. Moreover, the point $\mathbf{0} \in \mathbf{R}^n$ is termed the ***vertex*** of a cone and is an element of every cone since $\mathbf{Y} = 0\mathbf{X} = \mathbf{0} \in \mathcal{C}$. In general, $\mathcal{C}$ is *not a subspace* of $\mathbf{R}^n$ since the definition of $\mathcal{C}$ holds only for $\lambda \geq 0$ and not all λ.

In what follows we shall consider cones which are convex. Generally speaking, a convex cone is one which is a convex set. More specifically, a cone $\mathcal{C}$ in $\mathbf{R}^n$ is termed a ***convex cone*** if and only if it is closed under the operations of addition and multiplication by a non-negative scalar, *i.e.*, $\mathcal{C}$ in $\mathbf{R}^n$ is a convex cone if and only if: (a) for $\mathbf{X} \in \mathcal{C}$, $\lambda \mathbf{X} \in \mathcal{C}$, $0 \leq \lambda \in \mathbf{R}$; and (b) for $\mathbf{X}_1, \mathbf{X}_2 \in \mathcal{C}$, $\mathbf{X}_1 + \mathbf{X}_2 \in \mathcal{C}$.

Equivalently, $\mathcal{C}$ in $\mathbf{R}^n$ is a convex cone if any non-negative linear combination of points $\mathbf{X}_1, \mathbf{X}_2$ of $\mathcal{C}$ also belongs to $\mathcal{C}$, *i.e.*, if $(\lambda_1 \mathbf{X}_1 + \lambda_2 \mathbf{X}_2) \in \mathcal{C}$ for all $0 \leq \lambda_i \in \mathbf{R}$ and all $\mathbf{X}_i \in \mathcal{C}$, $i = 1, 2$.

It is important to note that not every cone is convex. However, a hyperplane $\mathcal{H}$ passing through the origin is a convex cone as are the closed half-spaces $[\mathcal{H}^+]$, $[\mathcal{H}^-]$ determined by $\mathcal{H}$. Moreover, the cone $\mathcal{C}=\{\mathbf{Y}|\mathbf{Y}=\lambda\mathbf{X}, \lambda\geq 0, \mathbf{X}\in\mathcal{X}\}$ generated by the set of points $\mathcal{X}=\{\mathbf{X}\}$ is convex if $\mathcal{X}$ itself is a convex set.

For a convex cone $\mathcal{C}=\{\mathbf{Y}\,|\,\mathbf{Y}=\lambda\mathbf{X},\ \lambda\geq 0,\ \mathbf{X}\in\mathbf{R}^n\}$, the ***negative cone*** of $\mathcal{C}$ is the convex cone $-\mathcal{C}=\{-\mathbf{Y}\,|\,-\mathbf{Y}=\lambda(-\mathbf{X}),\ \lambda\geq 0,\ \mathbf{X}\in\mathcal{C}\}$ while the ***orthogonal cone*** of $\mathcal{C}$ is the convex cone $\mathcal{C}^{\perp}=\{\mathbf{V}\,|\,\mathbf{Y}'\mathbf{V}=0 \text{ for all } \mathbf{Y}\in\mathcal{C}\}$. Moreover, $C^{\perp}$ is a subspace of $\mathbf{R}^n$ since if $\mathbf{V}\in\mathcal{C}^{\perp}$, then $\mathbf{Y}'\mathbf{V}=0$ and $\lambda\mathbf{Y}'\mathbf{V}=0$ for *all* λ (not just $\lambda\geq 0$) so that $\lambda\mathbf{Y}'\mathbf{V}\in\mathcal{C}^{\perp}$ for any λ.

The ***dimension*** (or ***rank***) of a convex cone $\mathcal{C}$, $dim(\mathcal{C})$, is the dimension of the *smallest subspace of* $\mathbf{R}^n$ *which contains* $\mathcal{C}$ and thus amounts to the dimension of $\mathcal{C}+(-\mathcal{C})$. Equivalently, $dim(\mathcal{C})$ may be viewed as the maximum number of linearly independent vectors contained in $\mathcal{C}$.

For $\mathcal{C}$ a convex cone in $\mathbf{R}^n$:

(a) The closure $\bar{\mathcal{C}}$ of $\mathcal{C}$ is convex.

(b) $\mathcal{C}$ is termed ***solid*** if $\mathcal{C}^o\neq\phi$.

(c) $\mathcal{C}$ is never a strictly bounded set (although it may be bounded either from above or from below) except if, trivially, $\mathbf{0}$ is its only element.

A convex cone $\mathcal{C}$ in $\mathbf{R}^n$ is said to be ***pointed*** if it contains no subspace other than $\{\mathbf{0}\}$, *i.e.*, given a vector $\boldsymbol{\alpha}\in\mathbf{R}^n$, $\mathcal{C}$ contains no ***line*** $\mathcal{L}=\{\mathbf{X}\,|\,\mathbf{X}=\lambda\boldsymbol{\alpha} \text{ for all } \lambda\in\mathbf{R},\ \mathbf{X}\in\mathbf{R}^n\}$. Based upon this definition it follows that if a convex cone has a vertex, it is unique and must be located at the origin. Hence a convex cone with vertex $\mathbf{0}$ is pointed. It is also true that:

(a) A convex cone $\mathcal{C}$ in $\mathbf{R}^n$ is pointed if and only if $\mathcal{C}\cap(-\mathcal{C})=\{\mathbf{0}\}$.

(b) If $\mathcal{C}$ is a pointed convex cone in $\mathbf{R}^n$, there exists a hyperplane $\mathcal{H}$ passing through $\mathbf{0}$ which separates $\mathcal{C}$, $-\mathcal{C}$ and supports both simultaneously.

(c) A convex cone $\mathcal{C}$ in $\mathbf{R}^n$ which is also a linear subspace has no vertex and consequently is not pointed.

Fundamental operations on convex cones in $\mathbf{R}^n$ are addition, intersection, and duality, *i.e.*,

(a) if $\mathcal{C}_1$, $\mathcal{C}_2$ are convex cones, their ***sum***

$$\mathcal{C}_1 + \mathcal{C}_2 = \{\mathbf{X} \mid \mathbf{X} = \mathbf{X}_1 + \mathbf{X}_2, \mathbf{X}_1 \epsilon \mathcal{C}_1, \mathbf{X}_2 \epsilon \mathcal{C}_2\}$$

is a convex cone.

(b) if $\mathcal{C}_1$, $\mathcal{C}_2$ are convex cones, their ***intersection***

$$\mathcal{C}_1 \cap \mathcal{C}_2 = \{\mathbf{X} \mid \mathbf{X} \epsilon \mathcal{C}_1 \text{ and } \mathbf{X} \epsilon \mathcal{C}_2\}$$

is a convex cone.

(c) if $\mathcal{C}$ is a convex cone, the ***dual cone***

$$\mathcal{C}^* = \{\mathbf{Y} \mid \mathbf{X}'\mathbf{Y} \leq 0 \text{ for all } \mathbf{X} \epsilon \mathcal{C}\}$$

is a convex cone.

Note that $\mathcal{C}^*$ consists of all vectors making a non-acute angle ($\geq \pi/2$) with all vectors of $\mathcal{C}$.

Relative to fundamental operation (c) above we may take the negative of the dual cone $\mathcal{C}^*$ of $\mathcal{C}$ in $\mathbf{R}^n$ so as to form the ***polar cone*** $\mathcal{C}^+ = \{\mathbf{Y} \mid \mathbf{X}'\mathbf{Y} \geq 0$ for all $\mathbf{X} \epsilon \mathcal{C}\}$ consisting of all vectors making a non-obtuse angle ($\leq \pi/2$) with all vectors of $\mathcal{C}$. Clearly $\mathcal{C}^+$ is a convex cone if $\mathcal{C}$ is.

A convex cone $\mathcal{C}$ in $\mathbf{R}^n$ is termed a ***finite cone*** if it consists of the set of all non-negative linear combinations of a finite set of vectors, *i.e.*, for points $\mathbf{X}_j \epsilon \mathbf{R}^n$, $j = 1, \ldots, m$,

$$\mathcal{C} = \left\{\mathbf{X} \mid \mathbf{X} = \sum_{j=1}^{m} \lambda_j \mathbf{X}_j,\ 0 \leq \lambda_j \epsilon \mathbf{R}\right\}. \tag{2.9}$$

Here $\mathcal{C}$ is said to be ***generated*** or ***spanned*** by the points $\mathbf{X}_j$. Hence any vector which can be expressed as a non-negative linear combination of a finite set of vectors $\mathbf{X}_j$, $j = 1, \ldots, m$, lies in the finite cone spanned by those vectors. An alternative way of structuring $\mathcal{C}$ in (2.9) is to state that a convex cone $\mathcal{C}$ in $\mathbf{R}^n$ is a finite cone if for some $(n \times m)$ matrix $\mathbf{A}' = [\mathbf{a}_1, \ldots, \mathbf{a}_m]$, every $\mathbf{X} \epsilon \mathcal{C}$ is a non-negative linear combination of the $(n \times 1)$ column vectors $\mathbf{a}_j$, $j = 1, \ldots, m$, of $\mathbf{A}'$, *i.e.*,

$$\mathcal{C} = \left\{\mathbf{X} \mid \mathbf{X} = \mathbf{A}'\boldsymbol{\lambda} = \sum_{j=1}^{m} \lambda_j \mathbf{a}_j,\ \mathbf{0} \leq \boldsymbol{\lambda} \epsilon \mathbf{R}^m,\ \mathbf{X} \epsilon \mathbf{R}^m\right\}. \tag{2.9.1}$$

A cone such as (2.9.1) is sometimes referred to as a ***closed convex cone.*** As we shall see later on, finite cones (those generated by a finite set of vectors) are synonymous with what will be termed ***polyhedral convex cones***, the latter being formed as the intersection of a finite number of half-planes, each of whose associated hyperplanes pass through the origin. Polyhedral convex cones are typically written as the set $\{\mathbf{X} \mid \mathbf{AX} \leqq \mathbf{0}\}$ in $\mathbf{R}^n$.

A special case of (2.9), (2.9.1) occurs when there is but a single generator. That is, a convex cone $\mathcal{C}$ in $\mathbf{R}^n$ generated by a single vector $\mathbf{a}(\neq\mathbf{0})$ is termed a ***ray*** or ***half-line*** (denoted $(\mathbf{a})$). So for $\mathbf{a} \in \mathbf{R}^n$,

$$(\mathbf{a}) \;=\; \{\mathbf{X} \mid \mathbf{X} = \lambda\mathbf{a},\ 0 \leq \lambda \in \mathbf{R},\ \mathbf{X} \in \mathbf{R}^n\}.$$

Given $(\mathbf{a}) = \{\mathbf{X} \mid \mathbf{X} = \lambda\mathbf{a}, \mathbf{a} \neq \mathbf{0}, \lambda \geq 0, \mathbf{X} \in \mathbf{R}^n\}$, the ***orthogonal cone to (a)*** is the hyperplane $(\mathbf{a})^{\perp} = \mathcal{H} = \{\mathbf{Y} \mid \mathbf{a}'\mathbf{Y} = 0\}$ passing through the origin while the ***dual of (a)*** is the half-space $(\mathbf{a})^* = [\mathcal{H}^-] = \{\mathbf{Y} \mid \mathbf{a}'\mathbf{Y} \leq 0\}$; it consists of all vectors making a non-acute angle with $\mathbf{a}$. Additionally, the ***polar*** of $(\mathbf{a})$ is the half-space $(\mathbf{a})^+ = [\mathcal{H}^+] = \{\mathbf{Y} \mid \mathbf{a}'\mathbf{Y} \geq 0\}$; it consists of all vectors making a non-obtuse angle with $\mathbf{a}$. If we admit to our discussion of cones a finite number of half-lines, then we may arrive at an alternative specification of a finite cone, namely that a convex cone $\mathcal{C}$ in $\mathbf{R}^n$ is termed a finite cone if there exists a finite number of half-lines $(\mathbf{a}_j)$ in $\mathbf{R}^n$, $j=1,\ldots,m$, such that $\mathcal{C} = \Sigma_{j=1}^m (\mathbf{a}_j)$. Here $\mathcal{C}$ is to be interpreted as the sum of the m cones $(\mathbf{a}_j)$ and may be represented as

$$\mathcal{C} \;=\; \left\{\mathbf{X} \mid \mathbf{X} \;=\; \sum_{j=1}^{m} \mathbf{X}_j,\ \mathbf{X}_j \in (\mathbf{a}_j),\ \mathbf{X} \in \mathbf{R}^n\right\}. \tag{2.10}$$

But since $\mathbf{X} = \Sigma_{j=1}^m \mathbf{X}_j = \Sigma_{j=1}^m \lambda_j \mathbf{a}_j = \mathbf{A}'\boldsymbol{\lambda},\ \boldsymbol{\lambda} \geqq \mathbf{0}$, equation (2.10) is just $\mathcal{C} = \{\mathbf{X} \mid \mathbf{X} = \mathbf{A}'\boldsymbol{\lambda},\ \boldsymbol{\lambda} \geqq \mathbf{0},\ \mathbf{X} \in \mathbf{R}^n\}$, thus establishing the equivalence of (2.9), (2.9.1), and (2.10).

In sum, a cone $\mathcal{C}$ in $\mathbf{R}^n$ is termed *finitely generated* if:

(a) each element of $\mathcal{C}$ is expressible as a non-negative linear combination of a finite set of vectors; or equivalently,

(b) $\mathcal{C}$ can be written as the sum of a finite number of half-lines.

In defining a finite cone no mention was made of the linear independence of the spanning set $\{\mathbf{X}_j,\ j=1,\ldots,m\}$. Indeed, the vectors $\mathbf{X}_j$ need not be linearly independent. However, if some subset of the vectors generating a finite cone $\mathcal{C}$ is linearly independent, then we may note that the ***dimension*** of $\mathcal{C}$ is the maximum number of linearly independent vectors in $\mathcal{C}$. So if $\mathcal{C} = \{\mathbf{X} \mid \mathbf{X} = \mathbf{A}'\boldsymbol{\lambda}, \boldsymbol{\lambda} \geqq \mathbf{0}, \mathbf{X} \epsilon \mathbf{R}^n\}$, then $dim(\mathcal{C}) = \rho(\mathbf{A}')$.

It was mentioned earlier that any hyperplane $\mathcal{H}$ through the origin is a convex cone as are $[\mathcal{H}^+]$, $[\mathcal{H}^-]$. In addition, $\mathcal{H}$, $[\mathcal{H}^+]$, and $[\mathcal{H}^-]$ are all finite cones. In this light we note briefly that:

(a) If $\mathbf{A}' = [\mathbf{a}_1, \ldots, \mathbf{a}_m]$ is $(n \times m)$, the solution set $\mathcal{C}$ of the system of homogeneous linear inequalities $\mathbf{A}'\boldsymbol{\lambda} \leqq \mathbf{0}$ is a finite cone. In addition, the solution set $\mathcal{C}$ of $\mathbf{A}'\boldsymbol{\lambda} \geqq \mathbf{0}$ is a finite cone.

(b) If $\mathbf{A}' = [\mathbf{a}_1, \ldots, \mathbf{a}_m]$ is $(n \times m)$, the set of non-negative solutions of the homogeneous linear system $\mathbf{A}'\boldsymbol{\lambda} = \mathbf{0}$ forms a finite cone.

For $\mathcal{C}$ in $\mathbf{R}^n$ a finite cone, the ***dual cone*** $\mathcal{C}^*$ is the finite cone consisting of all vectors making a non-acute angle ($\geq \pi/2$) with the vectors of $\mathcal{C}$, *i.e.*, if $\mathbf{A}' = [\mathbf{a}_1, \ldots, \mathbf{a}_m]$ is $(n \times m)$ and $\mathcal{C} = \{\mathbf{X} \mid \mathbf{X} = \mathbf{A}'\boldsymbol{\lambda}, \boldsymbol{\lambda} \geqq \mathbf{0}, \mathbf{X} \epsilon \mathbf{R}^n\}$, then $\mathcal{C}^* = \{\mathbf{Y} \mid \mathbf{A}\mathbf{Y} \leqq \mathbf{0}, \mathbf{Y} \epsilon \mathbf{R}^n\}$. Moreover, $\mathcal{C}^{**} = \mathcal{C}$.

We may now observe that a cone $\mathcal{C}$ in $\mathbf{R}^n$ is ***closed*** if and only if $\mathcal{C}^{**} = \mathcal{C}$, where $\mathcal{C}^{**}$ is the ***closure*** $\bar{\mathcal{C}}$ of $\mathcal{C}$. Here $\mathcal{C} \subseteq \mathcal{C}^{**} \subseteq \overline{\mathcal{C}^{**}} = \bar{\mathcal{C}}$, which implies $\mathcal{C}^{**} = \bar{\mathcal{C}}$ since $\mathcal{C}^{**}$ is an intersection of half-spaces and consequently closed.

Next, with $\mathcal{C}$ in $\mathbf{R}^n$ a finite cone, the ***polar cone*** $\mathcal{C}^+$ is the finite cone consisting of all vectors making a non-obtuse angle ($\leq \pi/2$) with the vectors of $\mathcal{C}$, *i.e.*, if $\mathbf{A}'$ is $(n \times m)$ and $\mathcal{C} = \{\mathbf{X} \mid \mathbf{X} = \mathbf{A}'\boldsymbol{\lambda}, \boldsymbol{\lambda} \geqq \mathbf{0}, \mathbf{X} \epsilon \mathbf{R}^n\}$, then $\mathcal{C}^+ = \{\mathbf{Y} \mid \mathbf{A}\mathbf{Y} \geqq \mathbf{0}, \mathbf{Y} \epsilon \mathbf{R}^n\}$. Also, $(\mathcal{C}^+)^+ = \mathcal{C}$. The dual cone $\mathcal{C}$ in $\mathbf{R}^n$ is sometimes referred to as the ***negative polar cone*** since $\mathcal{C}^* = -\mathcal{C}^+$.

For a set of vectors $\{\mathbf{a}_1, \ldots, \mathbf{a}_m\}$ in $\mathbf{R}^n$, their ***conical (non-negative linear) combination*** is represented as the point

$$\mathbf{X} = \sum_{j=1}^{m} \lambda_j \mathbf{a}_j, \; 0 \leq \lambda_j \epsilon \mathbf{R} \text{ for all } j.$$

And for $\mathcal{S}$ a set in $\mathbf{R}^n$, the ***conical hull*** of $\mathcal{S}$ is the collection of all conical combinations of vectors from $\mathcal{S}$ or

$$coni(\mathcal{S}) = \left\{ \mathbf{X} \mid \mathbf{X} = \sum_{j=1}^{m} \lambda_j \mathbf{X}_j,\ \mathbf{X}_j \in \mathcal{S},\ 0 \leq \lambda_j \in \mathbf{R}, \text{ and } m \text{ is an arbitrary positive integer} \right\}.$$

Here the conical hull of any set $\mathcal{S}$ in $\mathbf{R}^n$ is a convex cone *spanned* by $\mathcal{S}$ with vertex at the origin; it is the smallest convex cone with vertex $\mathbf{0}$ containing $\mathcal{S}$.

Next, the conical hull *spanned or generated* by a finite collection of points $\{\mathbf{a}_1, \ldots, \mathbf{a}_m\}$ in $\mathbf{R}^n$ is a finite convex cone and is expressed as the sum of a finite number of half-lines or

$$coni\{\mathbf{a}_1, \ldots, \mathbf{a}_m\} = \left\{ \mathbf{X} \mid \mathbf{X} = \sum_{j=1}^{m} (\mathbf{a}_j) = \sum_{j=1}^{m} \lambda_j \mathbf{a}_j,\ 0 \leq \lambda_j \in \mathbf{R} \right\}$$
$$= \{\mathbf{X} \mid \mathbf{X} = \mathbf{A}'\boldsymbol{\lambda},\ \mathbf{0} \leqq \boldsymbol{\lambda} \in \mathbf{R}^m\}.$$

Hence a cone generated by $\{\mathbf{a}_1, \ldots, \mathbf{a}_m\}$ is the set $coni\{\mathbf{a}_1, \ldots, \mathbf{a}_m\}$.

For a cone $\mathcal{C}$ in $\mathbf{R}^n$, the conical hull of $\mathcal{C}$, $coni(\mathcal{C})$, is the set of all conical or non-negative linear combinations of points from $\mathcal{C}$ or

$$coni(\mathcal{C}) = \left\{ \mathbf{X} \mid \mathbf{X} = \sum_{j=1}^{m} \lambda_j \mathbf{X}_j, \mathbf{X}_j \in \mathcal{C},\ 0 \leq \lambda_j \in \mathbf{R}, \text{ and } m \text{ is an arbitrary positive integer} \right\}.$$

Moreover, if $dim(\mathcal{C}) = p$, then any vector $\mathbf{X}(\neq \mathbf{0})$ in $coni(\mathcal{C})$ is expressible as the conical combination of p linearly independent vectors of $\mathcal{C}$.

Next, if $\mathcal{C}$ is a finite cone in $\mathbf{R}^n$ with the vector $\mathbf{a} \in \mathcal{C}^+$, then $(\mathbf{a})^+$ is a ***supporting half-space*** for $\mathcal{C}$ if $\mathcal{C}$ is contained in $(\mathbf{a})^+$ while $(\mathbf{a})^\perp$, the *generator* of $(\mathbf{a})^+$, is called a ***supporting hyperplane*** for $\mathcal{C}$.

We noted above that for an $(n \times m)$ matrix $\mathbf{A}' = [\mathbf{a}_1, \ldots, \mathbf{a}_m]$, the solution set $\mathcal{C}$ of the system of homogeneous linear inequalities $\mathbf{A}'\boldsymbol{\lambda} \leqq \mathbf{0}$ is a finite cone. It is also termed a ***polyhedral convex cone*** since it is the intersection of a finite collection of closed half-spaces whose boundary hyperplanes pass through the origin.

We also noted above that a finite cone in $\mathbf{R}^n$ can be formed as a sum of half-lines. As we shall now see, a given finite cone can also be generated from a set of half-spaces. To develop this point in greater detail let $\{\mathbf{b}_1, \ldots, \mathbf{b}_q\}$ be a set of vectors in $\mathbf{R}^n$. Then a finite cone $\mathcal{C}$ can be written as the intersection of a finite number q of half-spaces $(\mathbf{b}_j)^* = \{\mathbf{X} \mid \mathbf{b}_j'\mathbf{X} \leq 0\}$, $j = 1, \ldots, q$, as

$$\mathcal{C} = \bigcap_{j=1}^{q} (\mathbf{b}_j)^*.$$

Here $\mathcal{C}$ consists of all vectors $\mathbf{X}$ making non-acute angles ($\geq \pi/2$) with each of the vectors $\mathbf{b}_1, \ldots, \mathbf{b}_q$. If $\mathbf{B}' = [\mathbf{b}_1, \ldots, \mathbf{b}_q]$ is of order $(n \times q)$, then the preceding cone may be rewritten as $\mathcal{C} = \{\mathbf{X} \mid \mathbf{B}\mathbf{X} \leqq \mathbf{0},\ \mathbf{X} \in \mathbf{R}^n\}$.

2.10. Basic Solutions to Linear Equalities

Consider the simultaneous linear system $\mathbf{AX} = \mathbf{b}$, where $\mathbf{A}$ is of order $(m \times n)$ and $\rho(\mathbf{A}) = \rho[\mathbf{A}, \mathbf{b}]$ (the system is consistent). If $m < n$, then we have an ***underdetermined equation system*** which, given that it is consistent, possesses an infinite number of particular solutions. Furthermore, let $\rho(\mathbf{A}) = m$ so that $\mathbf{AX} = \mathbf{b}$ is consistent for every $\mathbf{b}$ and none of the linear equations is ***redundant***, *i.e.*, expressible as a linear combination of one or more other equations in the system.

Let us now select m linearly independent columns from $\mathbf{A}$ and express the variables associated with these columns in terms of the remaining $n-m$ variables. If arbitrary values are then assigned to the latter, it is evident that we obtain an infinity of particular solutions. Specifically, let us rearrange the columns of $\mathbf{A}$ so that the first m are linearly independent and constitute the columns of the mth order basis matrix $\mathbf{B} = [\mathbf{b}_1, \ldots, \mathbf{b}_m]$. The remaining $n-m$ columns of $\mathbf{A}$ then form the $(m \times n-m)$ matrix $\mathbf{R} = [\mathbf{r}_1, \ldots, \mathbf{r}_{n-m}]$. Hence

$$\mathbf{AX} = [\mathbf{B}, \mathbf{R}] \begin{bmatrix} \mathbf{X}_B \\ \mathbf{X}_R \end{bmatrix} = \mathbf{B}\mathbf{X}_B + \mathbf{R}\mathbf{X}_R = \mathbf{b},$$

where $\mathbf{X}$ has been partitioned into an $(m \times 1)$ vector $\mathbf{X}_B$, $\mathbf{X}_B' = (x_{B1}, \ldots, x_{Bm})$, containing the m variables corresponding to the m linearly independent columns of $\mathbf{A}$ and an $(n-m \times 1)$ vector $\mathbf{X}_R$, $\mathbf{X}_R' = (x_{R1}, \ldots, x_{R,n-m})$, containing the $n-m$ remaining variables. Since $\rho(\mathbf{B}) = m$, $\mathbf{B}$ is nonsingular. Hence

$$\mathbf{X}_B = \mathbf{B}^{-1}\mathbf{b} - \mathbf{B}^{-1}\mathbf{R}\mathbf{X}_R \tag{2.11}$$

and all solutions to $\mathbf{AX} = \mathbf{b}$ can be generated by assigning arbitrary values to $\mathbf{X}_R$. In particular, if $\mathbf{X}_R = \mathbf{0}$, $\mathbf{X}_B = \mathbf{B}^{-1}\mathbf{b}$ and we obtain what will be called a unique ***basic solution*** to $\mathbf{AX} = \mathbf{b}$, namely

$$\mathbf{X} = \begin{bmatrix} \mathbf{X}_B \\ \mathbf{0} \end{bmatrix} = \begin{bmatrix} \mathbf{B}^{-1}\mathbf{b} \\ \mathbf{0} \end{bmatrix}.^{2} \tag{2.11.1}$$

This type of solution is so named because the columns of $\mathbf{B}$ constitute a basis for $\mathbf{R}^m$, *i.e.*, each column of $\mathbf{R}$ and the vector $\mathbf{b}$ are expressible as a linear combination of the linearly independent columns of $\mathbf{B}$. Hence the variables in $\mathbf{X}_B$ are said to be ***basic variables*** while those within $\mathbf{X}_R$ are termed ***nonbasic variables.*** Moreover, a basic solution to $\mathbf{AX} = \mathbf{b}$ is said to be ***degenerate*** if one or more of the basic variables within $\mathbf{X}_B$ vanishes, *i.e.,* $\mathbf{X}$ contains fewer than m positive variables or, equivalently, $\mathbf{X}$ contains more than $n - m$ zeros.

One of the important aspects of the notion of a basic solution to a linear equation system is its applicability in solving a problem of the form: given an $(m \times n)$ matrix $\mathbf{A}$ and vector $\mathbf{b} \epsilon \mathbf{R}^m$, find an $\mathbf{X} \epsilon \mathbf{R}^n$ such that $\mathbf{AX} = \mathbf{b}$ with $\mathbf{X} \geqq \mathbf{0}$. A problem such as this can serve as system (I) of the following version of Farkas' theorem of the alternative: either

(I) $\mathbf{AX} = \mathbf{b}$, $\mathbf{X} \geqq \mathbf{0}$ has a solution $\mathbf{X} \epsilon \mathbf{R}^n$ or

(II) $\mathbf{A}'\mathbf{Y} \geqq \mathbf{0}$, $\mathbf{b}'\mathbf{Y} < \mathbf{0}$ has a solution $\mathbf{Y} \epsilon \mathbf{R}^m$,

but never both. For system (I), if $\mathbf{X}_B = \mathbf{B}^{-1}\mathbf{b} \geqq \mathbf{0}$, then

$$\mathbf{X} = \begin{bmatrix} \mathbf{B}^{-1}\mathbf{b} \\ \mathbf{0} \end{bmatrix} \geqq \mathbf{0}$$

---

[2]We assumed above that for $\mathbf{AX} = \mathbf{b}$ we have $m < n$. In general, if $\mathbf{A}$ is $(m \times n)$ and $m \gtrless n$, then a solution $\mathbf{X}$ of $\mathbf{AX} = \mathbf{b}$ is a basic solution if the columns of $\mathbf{A}$ corresponding to nonzero components of $\mathbf{X}$ are linearly independent. In this regard, if $\rho(\mathbf{A}) = r$ and $\mathbf{X}$ is a basic solution to $\mathbf{AX} = \mathbf{b}$, then $\mathbf{A}$ has at least one $(m \times r)$ submatrix $\mathbf{N}$ such that

(1) $\rho(\mathbf{N}) = r$; and

(2) the components of $\mathbf{X}$ corresponding to columns of $\mathbf{A}$ not in $\mathbf{N}$ are zero. Conversely, associated with each $(m \times r)$ submatrix $\mathbf{N}$ of $\mathbf{A}$ having rank r there exists a unique basic solution.

is termed a ***basic feasible solution*** to $\mathbf{AX} = \mathbf{b}$. Moreover, a basic feasible solution in ***nondegenerate*** if each basic variable is strictly positive.

Looking to Farkas' system (I) of the preceding paragraph, given any feasible solution $\mathbf{X} \geqq \mathbf{0}$ to $\mathbf{AX} = \mathbf{b}$, under what circumstances can we find at least one basic feasible solution? The answer is provided by the following

2.32. EXISTENCE THEOREM FOR BASIC FEASIBLE SOLUTIONS TO NONHOMOGENEOUS LINEAR EQUALITIES.

For an $(m \times n)$ matrix $\mathbf{A}$ with $\rho(\mathbf{A}) = m < n$ and a vector $\mathbf{b} \in \mathbf{R}^m$, if $\mathbf{AX} = \mathbf{b}$, $\mathbf{X} \geqq \mathbf{0}$ has a feasible solution $\mathbf{X} \in \mathbf{R}^n$, then it also has a basic feasible solution.

PROOF.[3] Assume that there exists a feasible solution to $\mathbf{AX} = \mathbf{b}$ in $k \leq m$ positive variables. (Without loss of generality, let the first k components of $\mathbf{X}$ be strictly positive.) Then $\mathbf{AX} = \mathbf{b}$ becomes

$$\sum_{j=1}^{k} x_j \mathbf{a}_j = \mathbf{b},\ x_j > 0,\ j=1,\ldots,k \leq m; \qquad x_j = 0, j=k+1,\ldots,n. \tag{2.12}$$

What sort of solution does (2.12) represent? Since the vectors $\mathbf{a}_j$, $j=1,\ldots,k \leq m$, may or may not be linearly independent, two cases present themselves.

First, let the vectors $\mathbf{a}_j$, $j=1,\ldots,k \leq m$, be linearly independent. If $k=m$, then associated with this set of vectors is an mth order nonsingular basis matrix $\mathbf{B}$ which provides us with a unique nondegenerate basic feasible solution. If $k < m$, we may add to the vectors $\mathbf{a}_j, j=1,\ldots,k < m$, any $m-k$ additional vectors selected from those remaining in $\mathbf{A}$ to form

$$\sum_{j=1}^{k} x_j \mathbf{a}_j + \sum_{j=k+1}^{m} 0\mathbf{a}_j = \mathbf{b}$$

and thus a basis for $\mathbf{R}^m$. Since the vectors $\mathbf{a}_j$, $j=k+1,\ldots,m$, are admitted to the basis at a zero level, the resulting basic feasible solution is degenerate.

---

[3]The proof of this theorem is *constructive* in that is spells out a computational procedure which actually allows us to obtain a basic feasible solution from any feasible solution to $\mathbf{AX} = \mathbf{b}$.

Next, let the vectors $\mathbf{a}_j, j=1,\ldots,k \leq m$, be linearly dependent. Since our goal is to ultimately attain a basic feasible solution, we shall introduce a procedure which allows us to systematically reduce the number of positive variables step by step until the vectors associated with those that remain are linearly independent. By hypothesis, there exist coefficients y_j not all zero such that

$$\sum_{j=1}^{k} y_j \mathbf{a}_j = \mathbf{0}. \tag{2.13}$$

A multiplication of both sides of this equation by a nonzero scalar θ yields

$$\sum_{j=1}^{k} \theta y_j \mathbf{a}_j = \mathbf{0}. \tag{2.13.1}$$

Upon subtracting this last equation from (2.12) we obtain

$$\sum_{j=1}^{k} (x_j - \theta y_j)\mathbf{a}_j \;=\; \sum_{j=1}^{k} \hat{x}_j \mathbf{a}_j = \mathbf{b}. \tag{2.14}$$

If (2.14) is to yield a feasible solution to $\mathbf{AX} = \mathbf{b}$, then we require that

$$\hat{x}_j \;=\; x_j - \theta y_j \geq 0, j=1,\ldots,k. \tag{2.15}$$

Under what conditions will this inequality hold? If $\theta > 0$, then $\hat{x}_j \geq 0$, $j=1,\ldots,k$, for all $y_j \leq 0$. Hence we need only consider those y_j which are positive. (If all $y_j \neq 0$ are negative, we may multiply (2.13) by -1 so as to insure that there exists at least one $y_j > 0$.) Rearranging (2.15) yields $x_j/y_j \geq \theta$ so that any θ which satisfies

$$\theta \;=\; \hat{\theta} \;=\; \min_j \left\{ \frac{x_j}{y_j}, \; y_j > 0 \right\} \tag{2.16}$$

guarantees that a feasible solution is attained. Suppose the minimum is assumed for $\hat{\theta} = x_h/y_h$. Then if this ratio is substituted into (2.14), the hth coefficient in this equation turns to zero with the result that $\mathbf{a}_h$ is dropped from this linear combination. Hence

$$\sum_{\substack{j=1\\ j\neq h}}^{k} \left[x_j - \left(\frac{x_h}{y_h}\right)y_j\right]\mathbf{a}_j = \mathbf{b} \tag{2.14.1}$$

represents a new feasible solution with no more than $k-1$ positive variables. If the vectors associated with these $k-1$ positive variables are linearly independent, we have actually attained a basic feasible solution; if they are linearly dependent, we may repeat the above procedure and reduce one of the $k-1$ positive variables to zero. If necessary, this elimination technique may be applied continually until the remaining vectors are linearly independent and thus provide us with a basic feasible solution. Q.E.D.

In the next section we shall encounter an extremely important geometric interpretation of this theorem. Specifically, it will be shown that basic feasible solutions to the system $\mathbf{AX}=\mathbf{b}$ are always found at the vertices of the convex polyhedron $\mathcal{S} = \{\mathbf{X} \mid \mathbf{AX}=\mathbf{b},\ \mathbf{X} \geqq \mathbf{0}\ \mathbf{X}\in \mathbf{R}^n\}$.

Let us again assume that $\mathbf{A}$ is of order $(m \times n)$ with $\rho(\mathbf{A}) = m < n$ and $\mathbf{b}\in\mathbf{R}^m$. Then a basic solution to $\mathbf{AX}=\mathbf{b}$ is provided by

$$\mathbf{X} = \begin{bmatrix} B^{-1}\mathbf{b} \\ \mathbf{0} \end{bmatrix} \geqq \mathbf{0}.$$

Our aim is to develop a procedure for replacing a given vector in the basis by another so that the new set of vectors is linearly independent. Here too $\mathbf{A}$ is partitioned as $\mathbf{A} = [\mathbf{B}, \mathbf{R}]$. In this regard, in generating a new basic solution to $\mathbf{AX}=\mathbf{b}$, some vector which currently constitutes a column of the basis matrix $\mathbf{B}$ must be replaced by one of the vectors comprising the columns of $\mathbf{R}$.

Under what conditions can a given nonbasis vector $\mathbf{r}_j$, $j=1,\ldots,n-m$, within $\mathbf{R}$ replace one of the basis vectors in the set $\mathfrak{B} = \{\mathbf{b}_1,\ldots,\mathbf{b}_m\}$ so that the new set of vectors $\bar{\mathfrak{B}}$ forms a basis for $\mathbf{R}^m$? The answer is provided by

2.33. THEOREM. Let the set of vectors $\{\mathbf{b}_1,\ldots,\mathbf{b}_m\}$ constitute a basis for $\mathbf{R}^m$. If a vector $\mathbf{r}\in\mathbf{R}^m$ is expressible as a linear combination of the $\mathbf{b}_i$, $\mathbf{r} = \Sigma_{i=1}^m \lambda_i \mathbf{b}_i$, then any $\mathbf{b}_i$ for which $\lambda_i \neq 0$ can be removed from the basis and replaced by $\mathbf{r}$ with the result that the new collection of vectors is also a basis for $\mathbf{R}^m$.

PROOF. Without loss generality, let us assume that $\lambda_m \neq 0$ so that $\mathbf{r}$ replaces $\mathbf{b}_m$ in the basis. If the new set of vectors $\bar{\mathfrak{B}} = \{\mathbf{b}_1, \ldots, \mathbf{b}_{m-1}, \mathbf{r}\}$ constitutes a basis for $\mathbf{R}^m$, it must be linearly independent and span $\mathbf{R}^m$. Let the null vector be expressible as a linear combination of the elements in $\bar{\mathfrak{B}}$, *i.e.*,

$$\sum_{i=1}^{m-1} \mu_i \mathbf{b}_i + \mu_m \mathbf{r} = \mathbf{0}.$$

A substitution of $\mathbf{r} = \Sigma_{i=1}^m \lambda_i \mathbf{b}_i$ into this expression yields

$$\sum_{i=1}^{m-1} \mu_i \mathbf{b}_i + \sum_{i=1}^{m} \mu_m \lambda_i \mathbf{b}_i = \sum_{i=1}^{m-1} (\mu_i + \mu_m \lambda_i) \mathbf{b}_i + \mu_m \lambda_m \mathbf{b}_m = \mathbf{0}.$$

Since the original set $\mathfrak{B}$ is linearly independent, this last expression equals the null vector if and only if $\mu_i + \mu_m \lambda_i = 0$, $i=1,\ldots,m-1$, $\mu_m = 0$. With $\mu_m = 0$, $\mu_i = -\mu_m \lambda_i = 0$, $i=1,\ldots,m-1$. Hence $\mu_i = 0$, $i=1,\ldots,m$, so that $\bar{\mathfrak{B}}$ is linearly independent and thus forms a basis for $\mathbf{R}^m$. Note that $\lambda_m \neq 0$ is a sufficient condition for $\bar{\mathfrak{B}}$ to be linearly independent; it is also necessary since if $\lambda_m = 0$, then $\mathbf{r} = \Sigma_{i=1}^{m-1} \lambda_i \mathbf{b}_i = \mathbf{0}$ and thus $\bar{\mathfrak{B}}$ would be linearly dependent.

We know that $\mathfrak{B}$ spans $\mathbf{R}^m$. If it can be demonstrated that $\mathfrak{B}$ is expressible in terms of the new set $\bar{\mathfrak{B}}$, then the latter must also span $\mathbf{R}^m$. Since $\lambda_m \neq 0$,

$$\mathbf{b}_m = \lambda_m^{-1} \mathbf{r} - \sum_{i=1}^{m-1} \lambda_i \lambda_m^{-1} \mathbf{b}_i$$

and thus

$$\mathfrak{B} = \{\mathbf{b}_1, \ldots, \mathbf{b}_{m-1}, \lambda_m^{-1}\mathbf{r} - \Sigma_{i=1}^{m-1} \lambda_i \lambda_m^{-1} \mathbf{b}_i\}.$$

And since $\bar{\mathfrak{B}}$ was also found to be linearly independent, it thus represents a basis for $\mathbf{R}^m$. Q.E.D.

Based upon the preceding theorem, it should be evident that since the vectors in $\mathbf{B}$ form a basis for $\mathbf{R}^m$, any vector in $\mathbf{R}$ may be written as a linear combination of the columns of $\mathbf{B}$. If we denote the jth column of $\mathbf{R}$ by $\mathbf{r}_j$, $j=1,\ldots,n-m$, then for all j,

$$\mathbf{r}_j = \sum_{i=1}^{m} \lambda_{ij}\mathbf{b}_i = \mathbf{B}\boldsymbol{\lambda}_j\,. \tag{2.17}$$

Let us assume that we have chosen some vector $\mathbf{r}_j$ to enter the basis and that the $(m \times 1)$ vector λ_j in (2.17) associated with $\mathbf{r}_j$ has at least one positive component λ_{ij}. At the present basic solution $\mathbf{BX}_B = \mathbf{b}$. Upon multiplying both sides of (2.17) by some number θ and subtracting the result from $\mathbf{BX}_B = \mathbf{b}$, we obtain

$$\mathbf{BX}_B - \mathbf{B}(\theta\boldsymbol{\lambda}_j) = \mathbf{b} - \theta\mathbf{r}_j \quad \text{or}$$

$$\mathbf{X}_B - \theta\boldsymbol{\lambda}_j = \mathbf{B}^{-1}(\mathbf{b} - \theta\mathbf{r}_j).$$

Here $(\mathbf{X}_B - \theta\boldsymbol{\lambda}_j, \mathbf{0})'$ is a basic solution to $\mathbf{AX} = \mathbf{b}$ and, if

$$\mathbf{X}_B - \theta\boldsymbol{\lambda}_j \geqq \mathbf{0} \quad \text{or} \quad x_{Bi} - \theta\lambda_{ij} \geq 0,\ i=1,\ldots,m;\ j=1,\ldots,n-m, \tag{2.18}$$

then $(\mathbf{X}_B - \theta\boldsymbol{\lambda}_j, \mathbf{0})'$ represents a basic feasible solution to $\mathbf{AX} = \mathbf{b}$.

Under what circumstances will (2.18) be satisfied? If $\theta > 0$, then (2.18) always holds for all $\lambda_{ij} \leq 0$. Hence we need only concern ourselves with those components of $\boldsymbol{\lambda}_j$ which are positive. In this regard, we need to find a $\theta > 0$ such that (2.18) holds for all $\lambda_{ij} > 0$. Rearranging (2.18) yields $\frac{x_{Bi}}{\lambda_{ij}} \geq \theta$ so that any θ for which

$$0 < \theta \leq \min_i \left\{ \frac{x_{Bi}}{\lambda_{ij}},\ \lambda_{ij} > 0 \right\} \tag{2.19}$$

will do, *i.e.*, will ensure that $(\mathbf{X}_B - \theta\boldsymbol{\lambda}_j, \mathbf{0})'$ is a basic feasible solution to $\mathbf{AX} = \mathbf{b}$. In fact, this latter inequality enables us to select the $\mathbf{b}_i$ to be removed from the basis and the $\mathbf{r}_j$ to be entered into the basis. If

$$\theta = \hat{\theta} = \min_i \left\{ \frac{x_{Bi}}{\lambda_{ij}},\ \lambda_{ij} > 0 \right\} \tag{2.19.1}$$

is substituted into $\mathbf{X}_B - \theta\boldsymbol{\lambda}_j$, then the component of this vector for which the minimum is attained reduces to zero. Hence the column of $\mathbf{B}$ to be replaced by $\mathbf{r}_j$ can always be determined by (2.19.1). In this regard, (2.19.1) is termed an ***exit criterion*** for moving from one basic feasible solution to another.

If $\hat{\mathfrak{B}} = \{\hat{\mathbf{b}}_1, \ldots, \hat{\mathbf{b}}_m\}$ denotes the set of linearly independent vectors obtained by replacing the rth column $\mathbf{b}_r$ within $\mathbf{B}$ by $\mathbf{r}_j$, then the new basic feasible

r^{th} column $\mathbf{b}_r$ within $\mathbf{B}$ by $\mathbf{r}_j$, then the new basic feasible solution can be represented as

$$\hat{\mathbf{X}} = \begin{bmatrix} \mathbf{X}_{\hat{B}} \\ \mathbf{0} \end{bmatrix} \begin{bmatrix} \mathbf{X}_B - \hat{\theta}\boldsymbol{\lambda}_j \\ \mathbf{0} \end{bmatrix} = \begin{bmatrix} \hat{\mathbf{B}}^{-1}\mathbf{b} \\ \mathbf{0} \end{bmatrix}, \tag{2.20}$$

where $\hat{\mathbf{B}} = \left[\hat{\mathbf{b}}_1, \ldots, \hat{\mathbf{b}}_m\right]$ with

$$\hat{\mathbf{b}}_i = \mathbf{b}_i,\ i \neq r;\ \hat{\mathbf{b}}_r = \mathbf{r}_j,$$

and

$$x_{\hat{B}i} = x_{Bi} - \hat{\theta}\lambda_{ij},\ i \neq r;\ x_{\hat{B}r} = \hat{\theta} = \frac{x_{Br}}{\lambda_{rj}}. \tag{2.20.1}$$

In sum, (2.20.1):

(1) determines the vector $\mathbf{b}_r$ to be removed from the present basis $\mathfrak{B}$;

(2) determines the value of the variable $x_{\hat{B}r}$ corresponding to the incoming vector $\mathbf{r}_j$ for the new basis $\hat{\mathfrak{B}}$; and

(3) ensures that $\hat{\mathbf{X}}$, the new basic solution to $\mathbf{AX} = \mathbf{b}$, is feasible, *i.e.*, $\hat{\mathbf{X}} \geqq \mathbf{0}$.

Armed with the concept of a basic feasible solution to $\mathbf{AX} = \mathbf{b}$, let us now address the specific structure of the solution set $\mathcal{S} = \{\mathbf{X} \mid \mathbf{AX} = \mathbf{b},\ \mathbf{X} \geqq \mathbf{0}, \mathbf{X} \in \mathbf{R}^n\}$. As we shall now see, some of the results provided in section 2.3 may be extended to the case where a simultaneous linear equation system comes under a non-negativity restriction $\mathbf{X} \geqq \mathbf{0}$.

Given $\mathcal{S}$, the ***homogeneous system associated with*** $\mathbf{AX} = \mathbf{b},\ \mathbf{X} \geqq \mathbf{0}$ (*i.e.*, associated with $\mathcal{S}$) assumes the form $\mathbf{AY} = \mathbf{0},\ \mathbf{Y} \geqq \mathbf{0}$. Let $\mathcal{T} = \{\mathbf{Y} \mid \mathbf{AY} = \mathbf{0},\ \mathbf{Y} \geqq \mathbf{0}, \mathbf{Y} \in \mathbf{R}^n\}$. As indicated in section 2.9, this set of all $\mathbf{Y}$ yielding homogeneous solutions is a finite or polyhedral convex cone. Furthermore, if $\bar{\mathbf{X}} \in \mathcal{S}$ and $\bar{\mathbf{Y}} \in \mathcal{T}$, then $\bar{\mathbf{X}} + \theta\bar{\mathbf{Y}} \in \mathcal{S}$, $0 \leq \theta \in \mathbf{R}$. Let us now look to the following

2.34. RESOLUTION THEOREM. Let $\{\mathbf{X}_1, \ldots, \mathbf{X}_k\}$ be the set of basic feasible solutions to $\mathbf{AX} = \mathbf{b}$ with $\hat{\mathbf{X}} \in co\{\mathbf{X}_1, \ldots, \mathbf{X}_k\}$ and $\hat{\mathbf{Y}} \in \mathcal{T} = \{\mathbf{Y} \mid \mathbf{AY} = \mathbf{0},\ \mathbf{Y} \geqq \mathbf{0},\ \mathbf{Y} \in \mathbf{R}^n\}$.

If $\mathbf{X} \in \mathcal{S} = \{\mathbf{X} \mid \mathbf{AX} = \mathbf{b}, \mathbf{X} \geqq \mathbf{0}, \mathbf{X} \in \mathbf{R}^n\} \neq \phi$, then $\mathbf{X} = \hat{\mathbf{X}} + \hat{\mathbf{Y}}$.

As this theorem indicates, $\mathbf{X} \in \mathcal{S}$ is expressible as a convex combination of the set of basic feasible solutions to $\mathbf{AX} = \mathbf{b}$ plus a homogeneous solution if $\mathcal{T}$ admits a unique solution $\mathbf{Y}=\mathbf{0}$, *i.e.*, $\mathcal{T}=\{\mathbf{0}\}$. (In the next section a bounded $\mathcal{S}=\{\mathbf{X}|\mathbf{AX}=\mathbf{b},\ \mathbf{X}\geqq\mathbf{0},\ \mathbf{X}\in\mathbf{R}^n\}$ will be referred to as a convex polytope.)

2.11. Faces of Polyhedral Convex Sets: Extreme Points, Facets, and Edges

In this section we shall consider a specialized class of convex sets in $\mathbf{R}^n$, namely those which may be characterized as *polyhedral.* Any such set is generated by intersecting a finite number of closed half-spaces (not all of which pass through the origin) and thus amounts to a closed convex set in $\mathbf{R}^n$. While any closed convex set in $\mathbf{R}^n$ is the intersection of *all* supporting closed half-spaces containing it (see theorem 2.25), a polyhedral convex set can be represented by intersecting only a *finite number m* of closed half-spaces. In this regard, a nonempty set $\mathcal{S}$ in $\mathbf{R}^n$ is a ***polyhedral convex set*** (or simply a ***convex polyhedron***) if $\mathcal{S}=\{\mathbf{X}|\mathbf{a}_i'\mathbf{X}\leq b_i,\ \mathbf{a}_i'\neq\mathbf{0}',\ b_i\in\mathbf{R},\ i=1,\ldots,m,\ \mathbf{X}\in\mathbf{R}^n\}$. For $\mathbf{a}_i'$ the ith row of the $(m\times n)$ matrix $\mathbf{A}$ and $b_i, i=1,\ldots,m$, the ith component of the $(m\times 1)$ vector $\mathbf{b}$, we may alternatively write $\mathcal{S}=\{\mathbf{X}|\mathbf{AX}\leqq\mathbf{b},\mathbf{X}\in\mathbf{R}^n\}$. Here $\mathcal{S}$ serves as a ***representation*** for a convex polyhedron.

If a convex polyhedron in $\mathbf{R}^n$ is bounded, then it is termed a ***convex polytope.*** Equivalently, a convex polytope can be viewed as the convex hull generated or spanned by finitely many points in $\mathbf{R}^n$, *i.e.*, for the finite set $\{\mathbf{X}_1,\ldots,\mathbf{X}_m\}$, $co\{\mathbf{X}_1,\ldots,\mathbf{X}_m\}$ is a convex polytope consisting of all convex combinations of $\{\mathbf{X}_1,\ldots,\mathbf{X}_m\}$.

Generally speaking, a polyhedral convex set $\mathcal{S}$ in $\mathbf{R}^n$ can be represented by a finite number of inequalities and/or equalities (*e.g.*, the equality $a_{11}x_1+a_{12}x_2=b_1$ can be replaced by the two inequalities $a_{11}x_1+a_{12}x_2\leq b_1$, $-a_{11}x_1-a_{12}x_2\leq -b_1$). Hence we may also express a convex polyhedron as $\mathcal{S}=\{\mathbf{X}|\mathbf{AX}=\mathbf{b},\mathbf{X}\in\mathbf{R}^n\}$. Oftentimes the m inequalities indicated above are accompanied by a set of nonnegativity conditions $x_i\geq 0$, $i=1,\ldots,n$. In this instance, $\mathcal{S}=\{\mathbf{X}|\mathbf{AX}\leqq\mathbf{b},\mathbf{X}\geqq\mathbf{0},\mathbf{X}\in\mathbf{R}^n\}$ or possibly $\mathcal{S}=\{\mathbf{X}|\mathbf{AX}=\mathbf{b},\ \mathbf{X}\geqq\mathbf{0},\mathbf{X}\in\mathbf{R}^n\}$.

Moreover any unrestricted variable (one which may be positive, negative, or zero) can be replaced by the difference between two nonnegative variables (*e.g.*, x_k unrestricted in sign is equivalent to $x_k = x'_k - x''_k$, where $x'_k, x''_k \geq 0$).

A vector $\mathbf{X}$ is an ***extreme point*** of a (general) convex set $\mathcal{S}$ in $\mathbf{R}^n$ if $\mathbf{X}$ cannot be written as a positive convex combination of two distinct points in $\mathcal{S}$, *i.e.*, $\mathbf{X}$ is not an interior point of any closed line segment in $\mathcal{S}$. When $\mathcal{S}$ in $\mathbf{R}^n$ is a polyhedral convex set (for concreteness, let $\mathcal{S} = \{\mathbf{X}|\mathbf{AX} \leqq \mathbf{b}, \mathbf{X} \in \mathbf{R}^n\}$), this definition, while still valid, lends itself to further geometric interpretation. To set the stage for what follows let us review the notion of a supporting hyperplane for $\mathcal{S}$. Specifically, let $\mathcal{S}$ be a convex polyhedron in $\mathbf{R}^n$ and, for some vector $\mathbf{C}(\neq \mathbf{0}) \in \mathbf{R}^n$, let $\mathcal{H} = \{\mathbf{X}|\mathbf{C}'\mathbf{X} = \alpha, \alpha \in \mathbf{R}, \mathbf{X} \in \mathbf{R}^n\}$ be a hyperplane such that: (1) $\mathbf{C}'\mathbf{X} \geq \alpha$ (or $\leq \alpha$), $\mathbf{X} \in \mathcal{S}$ ($\mathcal{H}$ has $\mathcal{S}$ on one of its sides); and (2) there is a point $\bar{\mathbf{X}}$ such that $\bar{\mathbf{X}} \in \mathcal{S}$ and $\mathbf{C}'\bar{\mathbf{X}} = \alpha$ (*i.e.*, $\bar{\mathbf{X}} \in \mathcal{S} \cap \mathcal{H}$). Then $\mathcal{H}$ is termed a ***supporting hyperplane*** for $\mathcal{S}$ at the point $\bar{\mathbf{X}}$.

The polyhedral convex set $\mathcal{S}$ specified above is made up of the m supporting closed (defining) half-spaces $\mathbf{a}'_i\mathbf{X} \leq b_i$, each with its associated supporting hyperplane $\mathbf{a}'_i\mathbf{X} = b_i, i=1,\ldots,m$. In this regard, a vector $\bar{\mathbf{X}} \in \mathcal{S}$ is termed an ***extreme point*** (or ***vertex*** or ***corner point***) of $\mathcal{S}$ if and only if $\bar{\mathbf{X}}$ lies on n linearly independent supporting hyperplanes of $\mathcal{S}$, *i.e.*, if at $\bar{\mathbf{X}}$ we have $\mathbf{a}'_i\bar{\mathbf{X}} = b_i, i=1,\ldots,k$, and $\mathbf{a}'_i\bar{\mathbf{X}} < b_i, i=k+1,\ldots,m$ (assume the first k supporting hyperplanes are satisfied at $\bar{\mathbf{X}}$ or the first k inequalities are *tight* there), then $\bar{\mathbf{X}}$ is a unique solution to the system $\mathbf{a}'_i\mathbf{X} = b_i$, $i=1,\ldots,k$, if and only if the set $\{\mathbf{a}_1,\ldots,\mathbf{a}_k\}$ contains n linearly independent vectors or the $(n \times k)$ matrix $[\mathbf{a}_1,\ldots,\mathbf{a}_k]$ has (column) rank n. If more than n supporting hyperplanes pass through $\bar{\mathbf{X}}$, then this extreme point is said to be ***degenerate.*** Hence a nondegenerate extreme point of $\mathcal{S}$ is one which satisfies exactly n supporting hyperplanes of the $\mathbf{AX} \leqq \mathbf{b}$ system.

If polyhedral $\mathcal{S} = \{\mathbf{X}|\mathbf{AX} = \mathbf{b}, \mathbf{X} \geqq \mathbf{0}, \mathbf{X} \in \mathbf{R}^n\}$, with $\mathbf{A}$ of order $(m \times n)$ and $\rho(\mathbf{A}) = m < n$, then an extreme point $\bar{\mathbf{X}} \in \mathcal{S}$ occurs at the intersection of the m linearly independent equations $\mathbf{AX} = \mathbf{b}$ and $n-m$ additional hyperplane equalities taken from the nonnegativity conditions $\mathbf{X} \geqq \mathbf{0}$. And if $\mathcal{S} = \{\mathbf{X}|\mathbf{AX} \leqq \mathbf{b}, \mathbf{X} \geqq \mathbf{0},$

$\mathbf{X} \in \mathbf{R}^n\}$, where again $\mathbf{A}$ is $(m \times n)$ and $\rho(\mathbf{A}) = m < n$, let $\mathbf{D}' = [\mathbf{A}', \mathbf{I}_n]$ represent the coefficient matrix associated with the $m+n$ supporting hyperplane equalities in $\mathcal{S}$. If $\rho(\mathbf{D}) = m+n$ (*i.e.*, $\mathbf{D}$ is of full row rank), then these $m+n$ supports are linearly independent and thus any extreme point $\bar{\mathbf{X}} \in \mathcal{S}$ must lie on some subset of n such supports.

A convex polyhedron $\mathcal{S} = \{\mathbf{X} | \mathbf{A}\mathbf{X} \leqq \mathbf{b}\}$ in $\mathbf{R}^n$ is said to be ***pointed*** if it has extreme points or vertices. A necessary and sufficient condition for a convex polyhedron to be pointed is that it contains no line, *i.e.*, the associated homogeneous system of equations $\mathbf{A}\mathbf{U} = \mathbf{0}$ admits only the trivial solution $\mathbf{U} = \mathbf{0}$.

An inequality $\mathbf{a}_t'\mathbf{X} \leq b_t$ from system $\mathbf{A}\mathbf{X} \leqq \mathbf{b}$ is deemed ***redundant*** (or ***superfluous***) if it can be omitted from the system without changing the structure of the polyhedral convex set $\mathcal{S} = \{\mathbf{X} | \mathbf{A}\mathbf{X} \leqq \mathbf{b},\ \mathbf{X} \in \mathbf{R}^n\}$. Hence any redundant inequality is *implied* by the other inequalities of the $\mathbf{A}\mathbf{X} \leqq \mathbf{b}$ system.

Let $\mathcal{S}$ be a nonempty polyhedral convex set in $\mathbf{R}^n$. The ***dimension of $\mathcal{S}$***, $dim(\mathcal{S})$, is the dimension of the smallest affine space containing $\mathcal{S}$. So given a point $\bar{\mathbf{X}}(\neq \mathbf{0}) \in \mathcal{S}$ and a set of points $\{\mathbf{X}_1, \ldots, \mathbf{X}_r\}$ in $\mathcal{S}$, $dim(\mathcal{S})$ is the maximum number r such that the translated set $\{\mathbf{X}_1 - \bar{\mathbf{X}}, \ldots, \mathbf{X}_r - \bar{\mathbf{X}}\}$ is linearly independent. Looked at in another fashion, $dim(\mathcal{S}) = r$ if the maximum number of affinely independent points in $\mathcal{S}$ is $r+1$. A convex polyhedron $\mathcal{S}$ in $\mathbf{R}^n$ is termed ***full-dimensional*** if $dim(\mathcal{S}) = n$, and this holds if and only if $\mathcal{S}$ has an ***interior point***, *i.e.*, a point $\bar{\mathbf{X}}$ such that $\mathbf{a}_i'\bar{\mathbf{X}} < b_i$ for all i.

Let $\mathcal{S} = \{\mathbf{X} | \mathbf{A}\mathbf{X} \leqq \mathbf{b}, \mathbf{X} \in \mathbf{R}^n\}$ (with $\mathbf{A}\mathbf{X} \leqq \mathbf{b}$ nonredundant) be a nonempty convex polyhedron in $\mathbf{R}^n$ and let $[\mathcal{H}_i^-] = \{\mathbf{X} | \mathbf{a}_i'\mathbf{X} \leq b_i, \mathbf{X} \in \mathbf{R}^n\}$ be a supporting closed half-space with associated supporting hyperplane $\mathcal{H}_i = \{\mathbf{X} | \mathbf{a}_i'\mathbf{X} = \mathbf{b}_i, \mathbf{X} \in \mathbf{R}^n\}$. A ***face*** $\mathcal{F}$ of $\mathcal{S}$ is a subset of points in $\mathcal{S}$ determined by some nonempty collection of defining supports of $\mathcal{S}$. Hence $\mathcal{F}$ is either $\mathcal{S}$ itself, the null set, or the intersection of $\mathcal{S}$ with $\mathcal{H}_i$. In this latter instance $\mathcal{F} = \{\mathbf{X} | \mathbf{a}_i'\mathbf{X} = b_i, \mathbf{X} \in \mathcal{S}\}$. Here the inequality $\mathbf{a}_i'\mathbf{X} \leq b_i$ is said to *define* or *represent* $\mathcal{F}$. In addition, $\mathcal{F}$ is termed ***proper*** if $\mathcal{F} \neq \phi$ and $\mathcal{F} \neq \mathcal{S}$. If $\mathcal{F}$ is a proper face of $\mathcal{S}$, $dim(\mathcal{F}) < dim(\mathcal{S})$. In fact, given a proper face $\mathcal{F}$ for $\mathcal{S}$, if r represents the maximum number of linearly independent support equalities holding at *all* feasible points of $\mathcal{F}$, then $dim(\mathcal{F}) = n-r$. All of the following are types of faces:

(i) ***facet*** – a face whose dimension is one less than $dim(\mathcal{S})$ (since it is the largest face in $\mathcal{S}$ distinct from $\mathcal{S}$).

(ii) ***extreme point*** – a face of dimension zero (since it occurs at the intersection of n linearly independent supporting hyperplanes of $\mathcal{S}$).

(iii) ***edge*** – a face of dimension one (since it is formed by intersecting $n-1$ linearly independent supporting hyperplanes of $\mathcal{S}$).

In this regard, a hyperplane defining a facet corresponds to a supporting closed half-space of $\mathcal{S}$ while an edge is always a line segment joining two extreme points. Here the two extreme points connected by the edge are said to be ***adjacent.*** Hence an extreme point is located at the intersection of two a more distinct edges of $\mathcal{S}$. Furthermore, $\mathcal{S}$ has only finitely many faces, each face being a nonempty convex polyhedron.

Given a polyhedral convex set $\mathcal{S} = \{\mathbf{X} | \mathbf{A}\mathbf{X} \leqq \mathbf{b}, \mathbf{X} \in \mathbf{R}^n\}$, if $\mathbf{b}$ is replaced by the null vector, then the resulting set is termed a ***polyhedral convex cone.*** In this regard, a polyhedral convex cone arises as the intersection of a finite number of closed half-spaces whose defining hyperplanes pass through the origin and appears as $\mathcal{C} = \{\mathbf{X} | \mathbf{A}\mathbf{X} \leqq \mathbf{0}, \mathbf{X} \in \mathbf{R}^n\}$. Here the ith row of $\mathbf{A}, \mathbf{a}_i'$, is the normal to the hyperplane $\mathcal{H}_i = \{\mathbf{X} | \mathbf{a}_i'\mathbf{X} = 0, \mathbf{X} \in \mathbf{R}^n\}$ generating the ith closed half-space $[\mathcal{H}_i^-] = \{\mathbf{X} | \mathbf{a}_i'\mathbf{X} \leq 0, \mathbf{X} \in \mathbf{R}^n\}$. More formally,

2.35. THEOREM. A convex polyhedron $\mathcal{S} = \{\mathbf{X} | \mathbf{A}\mathbf{X} \leqq \mathbf{b}, \mathbf{X} \in \mathbf{R}^n\}$ is a polyhedral convex cone if and only if $\mathbf{b} = \mathbf{0}$, *i.e.*, every polyhedral cone is the solution set of a homogeneous system of linear inequalities.

So if a polyhedral convex set $\mathcal{S}$ in $\mathbf{R}^n$ contains the null vector, then the convex cone generated by $\mathcal{S}$ is polyhedral.

2.12. Extreme Point Representation for Polyhedral Convex Sets

Let us now consider the polyhedral convex set $\mathcal{S} = \{\mathbf{X} | \mathbf{A}\mathbf{X} = \mathbf{b}, \mathbf{X} \geqq \mathbf{0}, \mathbf{X} \in \mathbf{R}^n\}$, where $\mathbf{A}$ is of order $(m \times n)$ and $\rho(\mathbf{A}) = m$. We developed in section 2.10 the concept of a basic feasible solution to a simultaneous linear system of the form $\mathbf{AX} = \mathbf{b}$ given that $\rho(\mathbf{A}) = \rho[\mathbf{A}, \mathbf{b}] = m$. That is, we extracted m linearly independent columns form $\mathbf{A}$ and formed an mth order nonsingular basis matrix $\mathbf{B}$ (hence $\mathbf{B}^{-1}$ exists) and subsequently obtained a unique basic feasible solution

$$\mathbf{X} = \begin{bmatrix} \mathbf{X}_B \\ \mathbf{0} \end{bmatrix} = \begin{bmatrix} \mathbf{B}^{-1}\mathbf{b} \\ \mathbf{0} \end{bmatrix} \geqq \mathbf{0},$$

where the vector of basic variables $\mathbf{X}_B$ has as its components the variables in $\mathbf{X}$ corresponding to the m linearly independent columns of $\mathbf{A}$. As we shall now see, a basic feasible solution is a useful analytical device for locating extreme points of $\mathcal{S}$; it provides us with a *necessary and sufficient characterization* of an extreme point. Specifically, we state the

2.36. EXTREME POINT THEOREM. A vector $\mathbf{X} \epsilon \mathbf{R}^n$ is an extreme point of the convex polyhedron $\mathcal{S} = \{\mathbf{X} | \mathbf{AX} = \mathbf{b}, \mathbf{X} \geqq \mathbf{0}\}$ ($\mathbf{A}$ is $(m \times n)$ and $\rho(\mathbf{A}) = m$) if and only if $\mathbf{X}$ is a basic feasible solution to $\mathbf{AX} = \mathbf{b}$.

PROOF. (sufficiency) What we shall demonstrate first is that if $\bar{\mathbf{X}} \epsilon \mathbf{R}^n$ is a basic feasible solution to $\mathbf{AX} = \mathbf{b}$, then $\bar{\mathbf{X}}$ is an extreme point of $\mathcal{S}$. Given that $\bar{\mathbf{X}}$ is a basic feasible solution, if it is also an extreme point of $\mathcal{S}$, then it cannot be expressed as a convex combination of any other two distinct vectors in $\mathcal{S}$. Hence there do not exist vectors $\mathbf{X}_1, \mathbf{X}_2 \epsilon \mathcal{S}, \mathbf{X}_1 \neq \mathbf{X}_2$, different from $\bar{\mathbf{X}}$ such that

$$\bar{\mathbf{X}} = \theta \mathbf{X}_2 + (1-\theta)\mathbf{X}_1, \quad 0 < \theta < 1. \tag{2.21}$$

If $\mathbf{X}_1' = (\mathbf{Y}_1', \mathbf{0}')$, $\mathbf{X}_2' = (\mathbf{Y}_2', \mathbf{0}')$, where $\mathbf{Y}_1, \mathbf{Y}_2$ are both of order $(m \times 1)$, then (2.21) becomes

$$\begin{bmatrix} \mathbf{X}_B \\ \mathbf{0} \end{bmatrix} = \theta \begin{bmatrix} \mathbf{Y}_2 \\ \mathbf{0} \end{bmatrix} + (1-\theta) \begin{bmatrix} \mathbf{Y}_1 \\ \mathbf{0} \end{bmatrix}, \quad 0 < \theta < 1 .$$

With $\mathbf{X}_B = \theta \mathbf{Y}_2 + (1-\theta)\mathbf{Y}_1 = \mathrm{B}^{-1}\mathbf{b}$ and $\mathbf{AX}_1 = \mathbf{BY}_1 = \mathbf{b}$, $\mathbf{AX}_2 = \mathbf{BY}_2 = \mathbf{b}$, it follows that

$$\theta(\mathbf{Y}_2 - \mathbf{Y}_1) + \mathbf{Y}_1 = \theta(\mathbf{Y}_2 - \mathbf{Y}_1) + \mathbf{B}^{-1}\mathbf{b} = \mathbf{B}^{-1}\mathbf{b}$$

and thus $\theta(\mathbf{Y}_2 - \mathbf{Y}_1) = \mathbf{0}$ or $\mathbf{Y}_1 = \mathbf{Y}_2$. Hence $\mathbf{X}_1 = \mathbf{X}_2 = \bar{\mathbf{X}}$ (*i.e.*, the representation of $\mathbf{b}$ in terms of basis vectors is unique), so that there do not exist distinct vectors

$\mathbf{X}_1, \mathbf{X}_2 \in \mathcal{S}$ such that (2.21) holds. Hence $\bar{\mathbf{X}}$ is an extreme point of $\mathcal{S}$.

(necessity) Next, if $\bar{\mathbf{X}}$ is an extreme point of $\mathcal{S}$, then there are at most m positive components of $\bar{\mathbf{X}}$ which have associated with them a set of linearly independent vectors from $\mathbf{A}$ that provide a basic feasible solution to $\mathbf{AX} = \mathbf{b}$. Without loss of generality, let the first $k \leq m$ components of $\bar{\mathbf{X}}' = (\mathbf{Y}', \mathbf{0}')$ be nonzero, where $\mathbf{Y}$ is of order $(k \times 1)$. Then the vectors associated with these components correspond to the first k columns of $\mathbf{A}$, namely $\bar{\mathbf{a}}_1, \ldots, \bar{\mathbf{a}}_k$, so that

$$\mathbf{A}\bar{\mathbf{X}} = \mathbf{A}_1\mathbf{Y} = \sum_{j=1}^{k} x_j \bar{\mathbf{a}}_j = \mathbf{b},$$

where $\mathbf{A}_1$ is of order $(m \times k)$. If the columns of $\mathbf{A}_1$ yield a basic feasible solution to $\mathbf{AX} = \mathbf{b}$, then they must be linearly independent. Let us assume that the columns of $\mathbf{A}_1$ are linearly dependent, *i.e.*, there exists at least one coefficient $\lambda_j \neq 0$ such that

$$\mathbf{A}_1\boldsymbol{\lambda} = \sum_{j=1}^{k} \lambda_j \bar{\mathbf{a}}_j = \mathbf{0}.$$

or some scalar $\alpha > 0$, $\alpha\mathbf{A}_1\boldsymbol{\lambda} = \mathbf{0}$ so that

$$\mathbf{A}_1\mathbf{Y} + \alpha\mathbf{A}_1\boldsymbol{\lambda} = \mathbf{A}_1(\mathbf{Y}+\alpha\boldsymbol{\lambda}) = \mathbf{A}_1\mathbf{Y}_1 = \mathbf{b},$$
$$\mathbf{A}_1\mathbf{Y} - \alpha\mathbf{A}_1\boldsymbol{\lambda} = \mathbf{A}_1(\mathbf{Y}-\alpha\boldsymbol{\lambda}) = \mathbf{A}_1\mathbf{Y}_2 = \mathbf{b}.$$

With α sufficiently small, both $\mathbf{Y}_1, \mathbf{Y}_2 \geqq \mathbf{0}$ and thus the vectors $\mathbf{X}_1' = (\mathbf{Y}_1', \mathbf{0}')$, $\mathbf{X}_2' = (\mathbf{Y}_2', \mathbf{0}')$ constitute feasible solutions to $\mathbf{AX} = \mathbf{b}$. But $\bar{\mathbf{X}} = \frac{1}{2}\mathbf{X}_1 + \frac{1}{2}\mathbf{X}_2$ so that if the vectors $\mathbf{a}_1, \ldots, \mathbf{a}_k$ are linearly dependent, then $\bar{\mathbf{X}}$ is not an extreme point of $\mathcal{S}$ since it can be written as a convex combination of two distinct vectors $\mathbf{X}_1, \mathbf{X}_2 \in \mathcal{S}$. Since this contradicts the assumption that $\bar{\mathbf{X}}$ is an extreme point, the columns of $\mathbf{A}_1$ must be linearly independent so as to provide a basis for $\mathbf{R}^k$ and thus a basic feasible solution to $\mathbf{AX} = \mathbf{b}$.

If $k = m$, then

$$\sum_{j=1}^{m} x_j \bar{\mathbf{a}}_j = \mathbf{BX}_B = \mathbf{b} \text{ and } \mathbf{X}_B = \mathbf{B}^{-1}\mathbf{b} \tag{2.22}$$

so that a unique basic feasible solution obtains. If $k < m$, we may add to $\mathbf{A}_1$ any

$m-k$ additional columns selected from those remaining in $\mathbf{A}$ to form

$$\sum_{j=1}^{k} x_j \bar{\mathbf{a}}_j + \sum_{j=k+1}^{m} 0\,\bar{\mathbf{a}}_j = \mathbf{b} \tag{2.22.1}$$

and thus a basis for $\mathbf{R}^m$. But since these vectors are admitted at a zero level, it is evident that the resulting basic feasible solution to $\mathbf{AX} = \mathbf{b}$ is not unique, *i.e.*, more than one basic feasible solution may correspond to the same extreme point. So for $x_j = 0, j = k+1,\ldots,m$, it is obvious that such a basic feasible solution is degenerate. Since a basis for $\mathbf{R}^m$ cannot contain more than m linearly independent vectors, there are at most m linearly independent columns in $\mathbf{A}$ and thus at most m positive components in $\bar{\mathbf{X}}$. Q.E.D.

In sum, every extreme point of $\mathcal{S}$ corresponds to a basic feasible solution to $\mathbf{AX} = \mathbf{b}$ (although it may be degenerate) and every basic feasible solution to the same represents an extreme point of $\mathcal{S}$.

It should be intuitively obvious that

2.37. COROLLARY. A convex polyhedron $\mathcal{S} = \{\mathbf{X} | \mathbf{AX} = \mathbf{b}, \mathbf{X} \geqq \mathbf{0},\ \mathbf{X} \epsilon \mathbf{R}^n\}$ has a finite number of extreme points.

In the preceding section we found that by selecting a nonbasis vector $\mathbf{r}_j$ to enter the basis $\mathcal{B} = \{\mathbf{b}_1,\ldots,\mathbf{b}_m\}$ and then applying the exit criterion (2.19.1) to determine the vector $\mathbf{b}_r$ to be removed from $\mathcal{B}$, we could easily move from one basic feasible solution to another. The new basic feasible solution was denoted

$$\hat{\mathbf{X}} = \begin{bmatrix} \mathbf{X}_{\hat{B}} \\ \mathbf{0} \end{bmatrix} = \begin{bmatrix} \hat{\mathbf{B}}^{-1}\mathbf{b} \\ \mathbf{0} \end{bmatrix},$$

where $\hat{\mathbf{B}} = [\hat{\mathbf{b}}_1,\ldots,\hat{\mathbf{b}}_m]$ with

$$\hat{\mathbf{b}}_i = \mathbf{b}_i,\ i \neq r;\ \ \hat{\mathbf{b}}_r = \mathbf{r}_j,$$

and

$$x_{\hat{B}i} = x_{Bi} - \hat{\theta}\lambda_{ij},\ i \neq r;\ x_{\hat{B}r} = \hat{\theta} = x_{Br} \big/ \lambda_{rj}.$$

How can we be sure that this new basic feasible solution is actually an extreme point of $\mathcal{S} = \{\mathbf{X} | \mathbf{AX} = \mathbf{b}, \mathbf{X} \geqq \mathbf{0}, \mathbf{X} \epsilon \mathbf{R}^n\}$? The answer depends upon whether the vectors within $\widehat{\mathbf{B}}$ are linearly independent. If they are, the answer must be yes.

In this regard, for polyhedral convex sets in $\mathbf{R}^n$, we have

2.38. EXISTENCE THEOREM. The convex polyhedron $\mathcal{S} = \{\mathbf{X}|\mathbf{AX} = \mathbf{b}, \mathbf{X} \geqq \mathbf{0}, \mathbf{X} \epsilon \mathbf{R}^n\} \neq \phi$ ($\mathbf{A}$ is $(m \times n)$ and $\rho(\mathbf{A}) = m$) has at least one extreme point.

PROOF. According to theorem 2.32, if $\mathbf{AX} = \mathbf{b}, \mathbf{X} \geqq \mathbf{0}$ has a feasible solution, then it also has a basic feasible solution. And since we know that there exists a one-to-one correspondence between basic feasible and extreme point solutions (by the preceding theorem), the proof is complete. Q.E.D.

We mentioned earlier that the convex hull of a finite number of points is termed the convex polytope spanned by these points. Clearly a convex polytope generated by m points cannot have more than m extreme points (since it is defined as the set of all convex combinations of these m points). Under what circumstances can a point within a convex polytope be represented as a convex combination of its extreme points? To answer this question we have

2.39. REPRESENTATION THEOREM. If $\mathcal{S}$ is a convex polytope in $\mathbf{R}^n$ and $\mathcal{E}$ is its finite set of extreme points, then $co(\mathcal{E}) = \mathcal{S}$.

In sum, every convex polytope is the convex hull of its extreme points. (Conversely, if $\{\mathbf{X}_1, \ldots, \mathbf{X}_m\}$ is a finite set of points, then the convex hull of $\{\mathbf{X}_1, \ldots, \mathbf{X}_m\}$ is a convex polytope $\mathcal{S}$. The set of extreme points $\mathcal{E}$ is a subset of $\{\mathbf{X}_1, \ldots, \mathbf{X}_m\}$.)

2.13. Directions for Polyhedral Convex Sets

If $\mathcal{S}$ in $\mathbf{R}^n$ is the polyhedral convex set $\{\mathbf{X}|\mathbf{AX} \leqq \mathbf{b}, \ \mathbf{X} \geqq \mathbf{0}, \mathbf{X} \epsilon \mathbf{R}^n\} \neq \phi$, where $\mathbf{A}$ is of order $(m \times n)$, then a vector $\mathbf{d}(\neq \mathbf{0})$ is a ***recession direction*** of $\mathcal{S}$ if and only if

$$\mathbf{A}(\mathbf{X} + \lambda \mathbf{d}) \leqq \mathbf{b}$$
$$\mathbf{X} + \lambda \mathbf{d} \geqq \mathbf{0}$$

for each $0 \leq \lambda \epsilon \mathbf{R}$ and $\mathbf{X} \epsilon \mathcal{S}$. Since $\mathbf{X} \epsilon \mathcal{S}$, *i.e.*, $\mathbf{AX} \leqq \mathbf{b}$ and $\mathbf{X} \geqq \mathbf{0}$, the first inequality holds for λ arbitrarily large if and only if $\mathbf{Ad} \leqq \mathbf{0}$. Similarly,

$\mathbf{X}+\lambda\mathbf{d} \geqq \mathbf{0}$ for arbitrarily large λ if and only if $\mathbf{d} \geqq \mathbf{0}$. In sum, $\mathbf{d}(\neq\mathbf{0})$ is a recession direction for polyhedral $\mathcal{S}$ if and only if $\mathbf{Ad} \leqq \mathbf{0}$ and $\mathbf{d} \geqq \mathbf{0}$.

For polyhedral $\mathcal{S} = \{\mathbf{X}|\mathbf{AX} = \mathbf{b}, \mathbf{X} \geqq \mathbf{0}, \mathbf{X} \epsilon \mathbf{R}^n\} \neq \phi$, $\mathbf{d}(\neq\mathbf{0})$ is a recession direction for $\mathcal{S}$ if and only if $\mathbf{Ad} = \mathbf{0}$ and $\mathbf{d} \geqq \mathbf{0}$ (this result obtains by applying the preceding argument to the two sets of linear inequalities $\mathbf{AX} \leqq \mathbf{b}$, $-\mathbf{AX} \leqq -\mathbf{b}$ implied by $\mathbf{AX} = \mathbf{b}$).

Let us now establish the concept of an ***extreme direction*** for a polyhedral convex set $\mathcal{S} = \{\mathbf{X}|\mathbf{AX} = \mathbf{b}, \mathbf{X} \geqq \mathbf{0}, \mathbf{X} \epsilon \mathbf{R}^n\}$, with $\mathbf{A}$ of order $(m \times n)$ and $\rho(\mathbf{A}) = m$. But first we must determine the form of a specialized type of recession direction for $\mathcal{S}$ in $\mathbf{R}^n$. Following the presentation given in section 2.10, we may partition $\mathbf{A}$ as $[\mathbf{B}, \mathbf{R}]$ (with basis matrix $\mathbf{B}$ nonsingular) and $\mathbf{d}' = (\mathbf{d}_1', \mathbf{d}_2')$, where $\mathbf{d}_1$ is of order $(m \times 1)$ and $\mathbf{d}_2$ is of order $((n-m) \times 1)$. Then the requirement $\mathbf{Ad} = \mathbf{0}$ becomes

$$[\mathbf{B}, \mathbf{R}]\begin{bmatrix}\mathbf{d}_1\\ \mathbf{d}_2\end{bmatrix} = \mathbf{0} \quad \text{or} \quad \mathbf{d}_1 = -\mathbf{B}^{-1}\mathbf{R}\mathbf{d}_2.$$

Hence

$$\mathbf{d} = \begin{bmatrix}\mathbf{d}_1\\ \mathbf{d}_2\end{bmatrix} = \begin{bmatrix}-\mathbf{B}^{-1}\mathbf{R}\mathbf{d}_2\\ \mathbf{d}_2\end{bmatrix}.$$

If there exists a nonbasic column $\mathbf{r}_j$ of $\mathbf{R}$ such that $\mathbf{B}^{-1}\mathbf{r}_j \leqq \mathbf{0}$, $j=1,\ldots,n-m$, then set $\mathbf{d}_2 = \bar{d}_j\mathbf{e}_j, \bar{d}_j > 0$ ($\mathbf{e}_j$ is the jth unit column vector), so that

$$\bar{\mathbf{d}}_j = \bar{d}_j\begin{bmatrix}-\mathbf{B}^{-1}\mathbf{r}_j\\ \mathbf{e}_j\end{bmatrix} \geqq \mathbf{0} \tag{2.23}$$

is a recession direction for $\mathcal{S}$. To see this let $\bar{\mathbf{X}}$ be an extreme point of $\mathcal{S}$. Then $\bar{\mathbf{d}}_j(\neq\mathbf{0})$ is a recession direction for $\mathcal{S}$ if $\bar{\mathbf{X}} \epsilon \mathcal{S}$ implies $\mathbf{X} = \bar{\mathbf{X}} + \lambda\bar{\mathbf{d}}_j \epsilon \mathcal{S}$. Since $\mathbf{A}\bar{\mathbf{d}}_j = \mathbf{0}$, $\bar{\mathbf{d}}_j \geqq \mathbf{0}$, and $\mathbf{A}\bar{\mathbf{X}} = \mathbf{b}$, it follows that $\mathbf{AX} = \mathbf{b}$ for $\mathbf{X} = \bar{\mathbf{X}} + \lambda\bar{\mathbf{d}}_j$ and for all λ. In addition, $\mathbf{X}$ is a nonnegative or feasible vector if $\lambda \geq 0$ and $\mathbf{B}^{-1}\mathbf{r}_j \leqq \mathbf{0}$. As we shall now demonstrate, recession directions structured as (2.23) serve as extreme directions of $\mathcal{S}$. Specifically,

2.40. THEOREM. Given the polyhedral convex set $\mathcal{S} = \{\mathbf{X}|\mathbf{AX} = \mathbf{b},\ \mathbf{X} \geqq \mathbf{0}, \mathbf{X} \epsilon \mathbf{R}^n\}$, let $\mathbf{A}$ be of order $(m \times n)$ with

$\rho(\mathbf{A}) = m$. A recession direction $\bar{\mathbf{d}}_j \in \mathcal{S}$ is also an extreme direction if and only if $\bar{\mathbf{d}}_j$ is a positive scalar multiple of the n-vector

$$\begin{bmatrix} -\mathbf{B}^{-1}\mathbf{r}_j \\ \mathbf{e}_j \end{bmatrix} \geqq \mathbf{0},$$

where $\mathbf{B}^{-1}\mathbf{r}_j \leqq \mathbf{0}$ and $\mathbf{r}_j$ is the jth (nonbasic) column of $\mathbf{R}$.

It is easily demonstrated that

2.41. COROLLARY. A convex polyhedron $\mathcal{S} = \{\mathbf{X} | \mathbf{A}\mathbf{X} = \mathbf{b}, \mathbf{X} \geqq \mathbf{0},\ \mathbf{X} \in \mathbf{R}^n\}$ has a finite number of extreme directions.

We next examine the conditions under which extreme directions exist. Specifically, we look to the following

2.42. EXISTENCE THEOREM. The convex polyhedron $\mathcal{S} = \{\mathbf{X} | \mathbf{A}\mathbf{X} = \mathbf{b}, \mathbf{X} \geqq \mathbf{0}, \mathbf{X} \in \mathbf{R}^n\}$, where $\mathbf{A}$ is $(m \times n)$ and $\rho(\mathbf{A}) = m$, has at least one extreme direction if and only if it is unbounded.

In some of the theorems to follow we shall have occasion to invoke the assumption that certain point sets are bounded. To appropriately characterize this concept let us now examine a couple of *unboundedness criteria*. Specifically,

2.43. THEOREM. (a) A convex polyhedron $\mathcal{S} = \{\mathbf{X} | \mathbf{A}\mathbf{X} \leqq \mathbf{b},\ \mathbf{X} \in \mathbf{R}^n\}$ in $\mathbf{R}^n$ is unbounded if and only if it admits a half-line or (b) if and only if the polyhedral convex cone $\mathcal{C} = \{\mathbf{X} | \mathbf{A}\mathbf{X} \leqq \mathbf{0}, \mathbf{X} \in \mathbf{R}^n\}$ contains a non-null vector.

2.14. Combined Extreme Point and Extreme Direction Representation for Polyhedral Convex Sets

In this section we shall see that a point within a convex polyhedron $\mathcal{S}$ in $\mathbf{R}^n$ can be represented as the sum of a convex combination of a finite number of extreme points from $\mathcal{S}$ and a nonnegative linear combination of a finite number of extreme directions for $\mathcal{S}$, *i.e.*, any $\mathbf{X} \in \mathcal{S}$ is representable as the sum of the convex hull of a finite number of extreme points and the conical hull of a finite collection of extreme directions, all taken from $\mathcal{S}$. To this end we have

2.44. REPRESENTATION THEOREM [Minkowski (1910)]. Given a nonempty polyhedral convex set $\mathcal{S} = \{\mathbf{X} | \mathbf{A}\mathbf{X} \leqq \mathbf{b},\ \mathbf{X} \in \mathbf{R}^n\}$, let $\mathbf{A}$ be of order $(m \times n)$ with $\rho(\mathbf{A}) = n$. In addition, let $\{\mathbf{X}_1,\ldots,\mathbf{X}_k\}$, $\{\mathbf{d}_1,\ldots,\mathbf{d}_l\}$ be finite sets of extreme points and extreme directions for $\mathcal{S}$ respectively. Then

$$\mathcal{S} = \left\{\mathbf{X} | \mathbf{X} = \sum_{i=1}^{k} \lambda_i \mathbf{X}_i + \sum_{j=1}^{l} \mu_j \mathbf{d}_j,\ \sum_{i=1}^{k} \lambda_i = 1, \right.$$

$$\left. 0 \leq \lambda_i \in \mathbf{R} \text{ for all } i;\ 0 \leq \mu_j \in \mathbf{R} \text{ for all } j \right\}$$

$$= co\{\mathbf{X}_1,\ldots,\mathbf{X}_k\} + coni\{\mathbf{d}_1,\ldots,\mathbf{d}_l\}.$$

We note briefly that the converse of this theorem is also true, *i.e.*, every $co\{\mathbf{X}_1,\ldots,\mathbf{X}_k\} + coni\{\mathbf{d}_1,\ldots,\mathbf{d}_l\}$ for finite sets $\{\mathbf{X}_1,\ldots,\mathbf{X}_k\}$, $\{\mathbf{d}_1,\ldots,\mathbf{d}_l\}$ is a convex polyhedron.

2.15. Resolution of Convex Polyhedra

We next consider the notion that every polyhedral convex set $\mathcal{S}$ in $\mathbf{R}^n$ is decomposable into the sum of a bounded convex polyhedron and a polyhedral convex cone. To set the stage for this *resolution of polyhedra*, let us first consider a supporting definition. Specifically, a pair of finite point sets $\mathcal{K}_1 = \{\mathbf{X}_1,\ldots,\mathbf{X}_k\}$, $\mathcal{K}_2 = \{\mathbf{X}_{k+1},\ldots,\mathbf{X}_m\}$ is termed a ***finite basis*** for a polyhedron $\mathcal{S}$ in $\mathbf{R}^n$ if $\mathcal{S} = co(\mathcal{K}_1) + coni(\mathcal{K}_2)$, *i.e.*, if $\mathbf{X} \in \mathcal{S}$, then $\mathbf{X}$ has the representation $\mathbf{X} = \Sigma_{i=1}^{k} \lambda_i \mathbf{X}_i + \Sigma_{i=k+1}^{m} \lambda_i \mathbf{X}_i$, $\Sigma_{i=1}^{k} \lambda_i = 1$, $0 \leq \lambda_i \in \mathbf{R}$, $i=1,\ldots,m$.

The resolution theorem for polyhedra to be presented below involves a statement of the so-called

2.45. FINITE BASIS THEOREM. Every convex polyhedron in $\mathbf{R}^n$ has a finite basis.

In view of these considerations we now look to the following

2.46. RESOLUTION THEOREM FOR CONVEX POLYHEDRA [Motzkin (1936); Goldman (1956)]. A polyhedral convex set $\mathcal{S} = \{\mathbf{X} | \mathbf{AX} \leqq \mathbf{b},\ \mathbf{X} \in \mathbf{R}^n\} \neq \phi$ can be expressed as the (vector) sum $co\{\mathbf{X}_1,\ldots,\mathbf{X}_k\} + coni\{\mathbf{X}_{k+1},\ldots,\mathbf{X}_m\} =$

$$\{\mathbf{X}|\mathbf{X}=\mathbf{Y}+\mathbf{Z},\ \mathbf{Y}\in co\{\mathbf{X}_1,\ldots,\mathbf{X}_k\},\ \mathbf{Z}\in coni\{\mathbf{X}_{k+1},\ldots,\mathbf{X}_m\}\} =$$
$$\left\{\mathbf{X}|\mathbf{X}=\Sigma_{i=1}^{k}\lambda_i\mathbf{X}_i+\Sigma_{i=k+1}^{m}\lambda_i\mathbf{X}_i,\ \Sigma_{i=1}^{k}\lambda_i=1, 0\leq\lambda_i\in\mathbf{R}, i=1,\ldots,m\right\} \tag{2.24}$$

of a polytope (bounded convex polyhedron) $co\{\mathbf{X}_1,\ldots,\mathbf{X}_k\}$ and a polyhedral convex cone $coni\{\mathbf{X}_{k+1},\ldots,\mathbf{X}_m\}$. Conversely, any set $\mathcal{S}\neq\phi$ of the form $co\{\mathbf{X}_1,\ldots,\mathbf{X}_k\}+coni\{\mathbf{X}_{k+1},\ldots,\mathbf{X}_m\}$ is a polyhedral convex set.

By virtue of this theorem, a set $\mathcal{S}$ of vectors in $\mathbf{R}^n$ is a convex polyhedron if and only if $\mathcal{S}=\mathcal{S}_1+\mathcal{S}_2$, where $\mathcal{S}_1$ is a convex polytope and $\mathcal{S}_2$ is a polyhedral convex cone.

It is important to note that in any decomposition such as (2.24), $coni\{\mathbf{X}_{k+1},\ldots,\mathbf{X}_m\}=\{\mathbf{X}|\mathbf{AX}\leqq\mathbf{0},\mathbf{X}\in\mathbf{R}^n\}$.

Two immediate consequences of theorem 2.46 are:

2.47. COROLLARY [Goldman (1956)]. Let $\mathcal{S}=\{\mathbf{X}\,|\mathbf{AX}\leqq\mathbf{b},\mathbf{X}\in\mathbf{R}^n\}\neq\phi$ be a polyhedral convex set with $\rho(\mathbf{A})=n$. Then $\mathcal{S}$ has a minimal basis $\{\mathbf{X}_1,\ldots,\mathbf{X}_k;\ \mathbf{X}_{k+1},\ldots,\mathbf{X}_m\}$ which is unique (up to positive multiples of the $\mathbf{X}_i$ for $i=k+1,\ldots,m$). Here $\{\mathbf{X}_1,\ldots,\mathbf{X}_k\}$ is the set of extreme vectors of $\mathcal{S}$.

2.48. COROLLARY [Goldman (1956)].
A polyhedral convex set $\mathcal{S}=\{\mathbf{X}|\mathbf{AX}\leqq\mathbf{b},\mathbf{X}\in\mathbf{R}^n\}\neq\phi$ is bounded if and only if $coni\{\mathbf{X}_{k+1},\ldots,\mathbf{X}_m\}=\{\mathbf{X}\mid\mathbf{AX}\leqq\mathbf{0}\}=\{\mathbf{0}\}$.

Theorem 2.46 also provides the foundation for the assertion of a *finite basis theorem for polytopes.* That is,

2.49. THEOREM. A polyhedral convex set $\mathcal{S}=\{\mathbf{X}|\mathbf{AX}\leqq\mathbf{b},\ \mathbf{X}\in\mathbf{R}^n\}\neq\mathbf{0}$ is a polytope if and only if $\mathcal{S}$ is bounded.

2.16. Simplexes

We noted above that a set $\mathcal{S}$ in $\mathbf{R}^n$ is termed a convex polytope if it is expressible as the convex hull of finitely many points $\{\mathbf{X}_1,\ldots,\mathbf{X}_k\}$ or $\mathcal{S}=\text{co}\{\mathbf{X}_1,\ldots,\mathbf{X}_k\}$. It was also mentioned that a set of $k+1$ vectors $\{\mathbf{X}_0,\mathbf{X}_1,\ldots,\mathbf{X}_k\}$ in $\mathbf{R}^n$ is affinely independent if and only if the translated set

$\{\mathbf{X}_1 - \mathbf{X}_0, \ldots, \mathbf{X}_k - \mathbf{X}_0\}$ is linearly independent or $\rho[\mathbf{X}_1 - \mathbf{X}_0, \ldots, \mathbf{X}_k - \mathbf{X}_0] = k$. Upon combining these definitions we may now state that a convex polytope $\mathcal{I}$ in $\mathbf{R}^n$ is a ***k-dimensional simplex*** (or simply a ***k-simplex***) if and only if it is the convex hull of a set of $k+1$ affinely independent vectors $\{\mathbf{X}_0, \mathbf{X}_1, \cdots, \mathbf{X}_k\}$, $k \leq n+1$. A k-simplex spanned by the points $\{\mathbf{X}_0, \mathbf{X}_1, \cdots, \mathbf{X}_k\}$ in $\mathbf{R}^n$ will be denoted as

$$\begin{aligned}\sigma^k &= co\{\mathbf{X}_0, \mathbf{X}_1, \cdots, \mathbf{X}_k\} \\ &= \left\{\mathbf{X} \,|\, \mathbf{X} = \sum_{i=0}^{k} \lambda_i \mathbf{X}_i, \ \sum_{i=0}^{k} \lambda_i = 1, \ 0 \leq \lambda_i \in \mathbf{R} \text{ for all } i\right\}.^{4} \end{aligned} \tag{2.25}$$

In general, $dim(\sigma^k)$ = number of vertices minus $1 = k$ since the vertices of σ^k determine a ***k-plane*** or affine subspace of dimension k, *i.e.*, the vertices lie on no plane of dimension $k-1$. Clearly the $k+1$ vertices of a k-simplex are extreme points of the same. Hence we may view a k-simplex as a convex polytope having exactly $k+1$ extreme points. We note briefly that a subset of σ^k is said to be a ***subsimplex*** of σ^k if it is itself a simplex.

Based upon the preceding definition of a k-simplex we may now state that if the set of vectors $\{\mathbf{X}_1, \ldots, \mathbf{X}_n\}$ in $\mathbf{R}^n$ is linearly independent, then the convex hull of $\{\mathbf{0}, \mathbf{X}_1, \ldots, \mathbf{X}_n\}$ is an ***n-simplex.*** In particular, the ***standard n-simplex*** in $\mathbf{R}^n$ is the convex hull of $\{\mathbf{0}, \mathbf{e}_1, \ldots, \mathbf{e}_n\}$, where $\mathbf{e}_i, i = 1, \ldots, n$, is the ith unit column vector. Moreover, an n-dimensional ***unit simplex*** in $\mathbf{R}^{n+1}$ is formed as the convex hull of the set of $n+1$ unit column vectors $\{\mathbf{e}_1, \ldots, \mathbf{e}_{n+1}\}$ and appears as

$$\sigma^n = \left\{\mathbf{X} \,\middle|\, \sum_{i=1}^{n+1} x_i = 1, \ \ 0 \leq x_i \in \mathbf{R} \text{ for all } i\right\}.$$

The ***boundary*** of a k-dimensional simplex contains simplexes of lower dimension which are generally called ***simplicial faces*** or just ***faces.*** The number of such faces of dimension ℓ is

[4] Some authors define a k-simplex as the set of all *strictly positive* ($\lambda_i > 0$) convex combinations of a collection of $k+1$ affinely independent points. Clearly any such set must be open. Equation (2.25) is then used to define the *closure of an (open) k-simplex* and subsequently called a *closed* k-simplex.

$$\binom{k+1}{\ell+1} = \frac{(k+1)!}{(\ell+1)!(k-\ell)!}\,.$$

Specifically, an ℓ-simplex σ^ℓ is an ℓ-face of a k-simplex σ^k if all vertices of σ^ℓ are also vertices of σ^k. Here the vertices of σ^ℓ form a subset of the vertices of σ^k and the convex hull of this subset is the resulting face. If $\ell = 0$, $\sigma^\ell = \sigma^0$ is a vertex of σ^k; and if $\ell = k-1$, σ^ℓ is a specialized face which is called a ***facet*** of σ^k. In this latter case there is exactly one vertex in σ^k which is not in σ^ℓ. If this vertex is denoted as $\mathbf{X}_i$ for some $0 \leq i \leq k$, then we may call σ^ℓ the ***face (facet) of σ^k opposite the vertex*** $\mathbf{X}_i$ and refer to it as a $(k-1)$-simplex. For $\mathbf{X} = \Sigma_{i=0}^k \lambda_i \mathbf{X}_i \in \mathrm{co}\{\mathbf{X}_0, \mathbf{X}_1, \ldots, \mathbf{X}_k\}$, let index set $\mathcal{I}(\mathbf{X}) = \{i \mid \lambda_i > 0, \quad 0 \leq i \leq k\}$ $= \{i_0, \ldots, i_\ell\}$. Then $\mathbf{X} \in \sigma^\ell$, where the face of σ^ℓ is termed the ***carrier face*** of $\mathbf{X}$ and is formed as $\mathrm{co}\{\mathbf{X}_{i_0}, \ldots, \mathbf{X}_{i_l}\}$. In fact, $\mathbf{X}$ belongs to exactly one face of σ^k and amounts to the face of lowest dimension of σ^k containing $\mathbf{X}$. The union of all faces of σ^k is its ***closure.***

2.17. Linear Functionals

A ***linear functional*** on $\mathbf{R}^n$ is a function $f: \mathbf{R}^n \rightarrow \mathbf{R}$ which is additive and homogeneous, *i.e.*, for points $\mathbf{X}_1, \mathbf{X}_2 \in \mathbf{R}^n$ and a scalar $\lambda \in \mathbf{R}$, f must satisfy:

(a) $f(\mathbf{X}_1 + \mathbf{X}_2) = f(\mathbf{X}_1) + f(\mathbf{X}_2)$;

(b) $f(\lambda \mathbf{X}_1) = \lambda f(\mathbf{X}_1)$.

The value of a linear functional f at a point $\mathbf{X}$ in $\mathbf{R}^n$ will be represented as the scalar product $f(\mathbf{X}) = \mathbf{C}'\mathbf{X}$. (In fact, if f is a linear functional defined on $\mathbf{R}^n$, then it can be shown that there exists a point $\mathbf{C} \in \mathbf{R}^n$ such that $f(\mathbf{X}) = \mathbf{C}'\mathbf{X}$ for all $\mathbf{X} \in \mathbf{R}^n$.) Hence $f: \mathbf{R}^n \rightarrow \mathbf{R}$ maps $\mathbf{C} \mapsto \mathbf{C}'\mathbf{X}$ so that every linear functional on $\mathbf{R}^n$ can be represented by a vector $\mathbf{C} \in \mathbf{R}^n$. For $f \not\equiv 0$, we may depict the linear functional as hyperplane $\mathcal{H} = \{\mathbf{X} \mid f(\mathbf{X}) = \mathbf{C}'\mathbf{X} = \alpha, \alpha \in \mathbf{R}\}$ in $\mathbf{R}^n$, where the vector $\mathbf{C}(\neq \mathbf{0})$ is a normal to the hyperplane. It can be shown that a linear functional or hyperplane is *continuous and closed.*

We noted above that a set $\mathcal{S}$ in $\mathbf{R}^n$ is compact if it is closed and bounded. The importance of compact sets for our purposes is that *compactness is preserved by continuous mappings*, *i.e.*, if the domain $\mathcal{S}$ of a continuous real-valued function $f:\mathcal{S}\rightarrow\mathbf{R}$ is compact, then so is its range. In this regard, a continuous real-valued function is bounded on a compact set. Moreover, the least upper bound and the greatest lower bound are themselves members of the range of f. That is, there exist points $\mathbf{X}_1, \mathbf{X}_2 \in \mathcal{S}$ such that

$$f(\mathbf{X}_1) = \sup_{\mathbf{X}\in\mathcal{S}} f(\mathbf{X}) = \max_{\mathbf{X}\in\mathcal{S}} f(\mathbf{X});$$

$$f(\mathbf{X}_2) = \inf_{\mathbf{X}\in\mathcal{S}} f(\mathbf{X}) = \min_{\mathbf{X}\in\mathcal{S}} f(\mathbf{X}).$$

These remarks may be summarized by a sufficient condition for the existence of extreme values of a function, namely

2.50. WEIERSTRASS'S THEOREM. Let the real-valued function $y = f(\mathbf{X})$ be continuous over a (nonempty) compact set $\mathcal{S}$ in $\mathbf{R}^n$. Then f assumes a least and a greatest value at least once over $\mathcal{S}$.

A particularization or stronger version of this result is provided by the following

2.51. THEOREM. Let $f:\mathcal{S}\rightarrow\mathbf{R}$ be a linear functional defined on a convex polytope $\mathcal{S}$ in $\mathbf{R}^n$. Then there exists extreme points $\bar{\mathbf{X}}, \bar{\mathbf{Y}} \in \mathcal{S}$ such that

$$f(\mathbf{X}) = \max_{\mathbf{X}\in\mathcal{S}} f(\mathbf{X});\ f(\bar{\mathbf{Y}}) = \min_{\mathbf{X}\in\mathcal{S}} f(\mathbf{X}).$$

PROOF. Since f is continuous and $\mathcal{S}$ is compact, Weierstrass's theorem guarantees the existence of a point $\hat{\mathbf{X}} \in \mathcal{S}$ such that

$$f(\hat{\mathbf{X}}) = \max_{\mathbf{X}\in\mathcal{S}} f(\mathbf{X}).$$

Let $\{\mathbf{X}_1,\ldots,\mathbf{X}_k\}$ depict the set of extreme points of $\mathcal{S}$. Then according to representation theorem 2.39,

$$\hat{\mathbf{X}} = \textstyle\sum_{i=1}^{k}\lambda_i\mathbf{X}_i,\ \sum_{i=1}^{k}\lambda_i = 1,\ 0 \leq \lambda_i \in \mathbf{R},\ i=1,\ldots,k.$$

If none of the extreme points satisfy $f(\mathbf{X}_i) = f(\hat{\mathbf{X}})$, then we must have $f(\mathbf{X}_i) < f(\hat{\mathbf{X}})$ for all i and thus

$$f(\hat{\mathbf{X}}) = \textstyle\sum_{i=1}^{k}\lambda_i f(\mathbf{X}_i) < \sum_{i=1}^{k}\lambda_i f(\hat{\mathbf{X}}) = f(\hat{\mathbf{X}}).$$

This contradiction thus implies that some extreme point $\bar{\mathbf{X}} \in \mathcal{S}$ must satisfy $f(\bar{\mathbf{X}}) = f(\hat{\mathbf{X}})$. A similar demonstration holds for $\bar{\mathbf{Y}}$. Q.E.D.

We shall see in the next chapter that the region of admissible solutions for a linear programming problem is a convex polyhedron $\mathcal{K} = \{\mathbf{X} \mid \mathbf{AX} \leqq \mathbf{b}, \mathbf{X} \geqq \mathbf{0}\}$ in $\mathbf{R}^n$ (if $\mathcal{K}$ is bounded, it is a convex polytope). In this regard we have

2.52. THEOREM. Let $\mathcal{K}$ be a convex polytope (containing no extreme directions) and let the linear functional $f: \mathbf{R}^n \rightarrow \mathbf{R}$ be bounded above on $\mathcal{K}$. Then there exists an extreme point $\bar{\mathbf{X}} \epsilon \mathcal{K}$ such that

$$f(\bar{\mathbf{X}}) = \sup_{\mathbf{X} \epsilon \mathcal{K}} f(\mathbf{X}) = \sup_{\mathbf{X} \epsilon \partial \mathcal{K}} f(\mathbf{X}).$$

The implication of this theorem is that if the linear programming objective function $f(\mathbf{X}) = \mathbf{C}'\mathbf{X}$ (a linear functional) is bounded above on $\mathcal{K}(\neq \phi)$, then the linear program has at least one optimal solution. Furthermore, at least one of the optimal solutions is an extreme point $\bar{\mathbf{X}}$ of $\mathcal{K}$ with $f = \mathbf{C}'\mathbf{X}$ serving as a supporting hyperplane of $\mathcal{K}$ at $\bar{\mathbf{X}}$.

3 INTRODUCTION TO LINEAR PROGRAMMING

3.1. The Canonical Form of a Linear Programming Problem

We may write the basic structure of a linear programming problem in what is called **canonical form** as: maximize a linear functional or hyperplane

$$f(x_1,\ldots,x_p) = c_1x_1+\cdots+c_px_p \tag{3.1}$$

subject to (hereafter abbreviated as "s.t.") a set of m **structural constraints**

$$a_{i1}x_1+\cdots+a_{ip}x_p \leq b_i, i=1,\ldots,m, \tag{3.2}$$

and p **non-negativity conditions** $x_j \geq 0, j=1,\ldots,p$. If $\mathbf{C}, \mathbf{X} \in \mathbf{R}^p$, $\mathbf{b} \in \mathbf{R}^m$, and $\mathbf{A}$ is of order $(m \times p)$, then the linear programming problem may be rewritten in matrix form as

$$max f(\mathbf{X}) = \mathbf{C}'\mathbf{X} \text{ s.t.} \tag{3.1.1}$$

$$\mathbf{AX} \leqq \mathbf{b}, \mathbf{X} \geqq \mathbf{0}, \tag{3.2.1}$$

where

$$\mathbf{C} = \begin{bmatrix} c_1 \\ \vdots \\ c_p \end{bmatrix}, \mathbf{X} = \begin{bmatrix} x_1 \\ \vdots \\ x_p \end{bmatrix}, \mathbf{b} = \begin{bmatrix} b_1 \\ \vdots \\ b_m \end{bmatrix}, \mathbf{A} = \begin{bmatrix} a_{11} & \cdots & a_{1p} \\ \vdots & & \vdots \\ a_{m1} & \cdots & a_{mp} \end{bmatrix}.$$

Here (3.1) (or (3.1.1)) is termed the **objective function** (the expression to be maximized) and, taken altogether, the structural constraints and non-negativity conditions will be referred to as the **constraint system**. (Note that the constraint system is made up of a set of $m+p$ half-planes.) Such a system forms the **feasible region** $\mathcal{K} \subset \mathbf{R}^p$, *i.e.*, the area within which an admissible solution to the problem lies. Thus $\mathcal{K}$ is formed as the intersection of the solution sets of the constraint system or

$$\begin{aligned}\mathcal{K} &= \{\mathbf{X} \mid \mathbf{AX} \leqq \mathbf{b},\ \mathbf{X} \geqq \mathbf{0}\} \\ &= \left(\cap_{i=1}^{m} \left\{ \mathbf{X} \mid \Sigma_{j=1}^{p} a_{ij} x_j \leq b_i \right\} \right) \cap \left(\cap_{j=1}^{p} \{ \mathbf{X} \mid x_j \geq 0 \} \right).\end{aligned}$$

Clearly $\mathcal{K}$ is a convex polyhedron since it is the intersection of a finite number of closed half-spaces. The linear programming problem is termed **feasible** whenever $\mathcal{K} \neq \phi$.

Any $\mathbf{X}$ which satisfies (3.2) (or (3.2.1)) is called a **solution** to the linear programming problem while if such an $\mathbf{X}$ also satisfies $\mathbf{X} \geqq \mathbf{0}$, *i.e.*, $\mathbf{X} \in \mathcal{K}$, then it is termed a **feasible solution.** Moreover, if $\mathbf{X}_0$ is feasible and $f(\mathbf{X}_0) \geq f(\mathbf{X})$, where $\mathbf{X}$ is any other feasible solution, then $\mathbf{X}_0$ is termed an **optimal** (*i.e.*, maximal) **solution.**

Two additional points are in order. First, the linear programming problem has been written in terms of maximizing a hyperplane subject to a set of structural constraints and non-negativity conditions. If we are faced with minimizing the same, then we may utilize a transformation which converts the minimization problem into an equivalent maximization problem, *i.e.*, $min f = -max\{-f\}$ (see Chapter 9 for further details.) Next, in our subsequent analysis we shall assume that each $b_i \geq 0$, i=1,...,m. If it happens that $b_i < 0$ for some value of i, then both sides of the structural constraint to which it corresponds may be multiplied by -1 so as to fulfill this requirement.

3.2. A Graphical Solution to the Linear Programming Problem

The problem that we wish to solve is:

$$\begin{aligned} max f = x_1 + 4x_2 \quad \text{s.t.} \\ \tfrac{1}{2}x_1 + x_2 \leq 50 \\ x_1 - 3x_2 \leq 15 \\ x_1 + x_2 \geq 30 \\ 2x_1 - x_2 \geq 10 \\ x_1,\ x_2 \geq 0. \end{aligned}$$

Let us first illustrate the region of feasible solutions $\mathcal{K}$. Since the structural constraints and non-negativity conditions must all be satisfied simultaneously, it is obvious that $\mathcal{K}$ is simply the intersection of the solution sets of the constraint system (Figure 3.1).

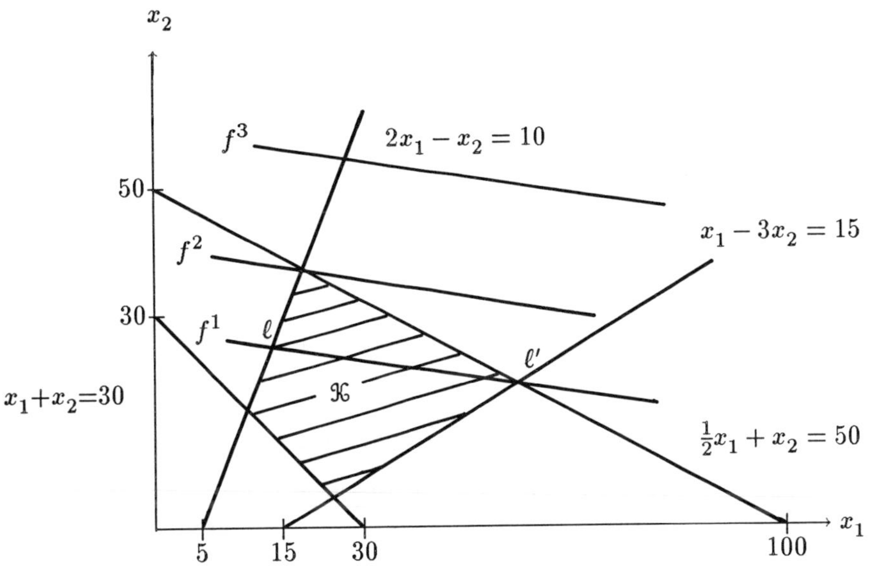

Figure 3.1

A maximal solution to the linear programming problem will be obtained by finding the point (or points) in $\mathcal{K}$ which yields the largest value of the objective function. For any fixed value of f we obtain the straight line $f = x_1 + 4x_2$ and, by choosing different values of f, we generate a series of parallel lines. Such a series is indicated in Figure 3.1. For $f = f^1$ any point on the segment $\ell\ell'$ yields a solution which is feasible but not optimal because, by sliding this line upwards and parallel to itself over $\mathcal{K}$, we obtain a larger value of f with feasibility still maintained. Clearly $f = f^2$ maximizes the objective function since the solution obtained at point A is feasible and f^2 is greater than the value of f at any other point in $\mathcal{K}$. Although $f = f^3$ yields a value of the objective function which is greater than f^2, the requirement that f have at least one point in common with $\mathcal{K}$ is not met. Hence any point on the line corresponding to $f = f^3$ is inadmissible.

Since the lines $2x_1 - x_2 = 10$ and $\frac{1}{2}x_1 + x_2 = 50$ intersect at point A, the simultaneous solution of these two equations yields the optimal values of x_1 and x_2, namely $x_1^0 = 24, x_2^0 = 38$. Hence the maximum value of f is $f^0 = 24 + 4 \cdot 38 = 176$.

In general, at an optimal feasible solution to the linear programming problem, the hyperplane f cannot contain any interior point of $\mathcal{K}$ and must contain at least one boundary point of $\mathcal{K}$, *i.e.*, it is a supporting hyperplane of $\mathcal{K}$. In this light, the optimal solution to a linear program must occur at an extreme point of $\mathcal{K}$ or at some convex combination of extreme points from $\mathcal{K}$.

A formalization of these observations along with a computational technique for generating maximal extreme point solutions is presented below.

3.3. The Standard Form of a Linear Programming Problem

Instead of working with the "$\leq$" structural constraints as a set of inequalities, we may conveniently transform them to a set of equalities so as to obtain the **augmented structural constraint system.** This is accomplished by introducing m new variables into system (3.2). In this regard, if we add a nonnegative variable to the left-hand side of each "$\leq$" structural constraint, then we admit m **slack variables** of the form

$$x_{p+i} = b_i - \sum_{j=1}^{p} a_{ij}x_j \geq 0,\ i=1,\ldots,m. \tag{3.3}$$

In the light of (3.3), the linear programming problem in **standard form** involves maximizing the **augmented objective function**

$$\begin{aligned} &f(x_1,\ldots,x_p,x_{p+1},\ldots,x_{p+m}) \\ &\qquad = c_1x_1 + \cdots + c_px_p + 0x_{p+1} + \cdots + 0x_{p+m} \end{aligned} \tag{3.4}$$

subject to

$$a_{i1}x_1 + \cdots + a_{ip}x_p + x_{p+i} = b_i, i=1,\ldots,m, \tag{3.5}$$

and $x_j \geq 0$, $j=1,\ldots,p,p+1,\ldots,p+m = n$. If $\widehat{\mathbf{C}}$, $\widehat{\mathbf{X}} \epsilon \mathbf{R}^{p+m} = \mathbf{R}^n$ and $\widehat{\mathbf{A}}$ is of order $(m \times n)$, then the preceding problem is expressible in matrix form as

$$max f(\widehat{\mathbf{X}}) = \widehat{\mathbf{C}}'\widehat{\mathbf{X}} \quad \text{s.t.} \tag{3.4.1}$$

$$\widehat{\mathbf{A}}\widehat{\mathbf{X}} = \mathbf{b}, \quad \widehat{\mathbf{X}} \geqq \mathbf{0}, \tag{3.5.1}$$

where

$$\widehat{\mathbf{C}} = \begin{bmatrix} c_1 \\ \vdots \\ c_p \\ c_{p+1} \\ \vdots \\ c_{p+m} \end{bmatrix} = \begin{bmatrix} \mathbf{C} \\ \mathbf{0} \end{bmatrix}, \ \widehat{\mathbf{X}} = \begin{bmatrix} x_1 \\ \vdots \\ x_p \\ x_{p+1} \\ \vdots \\ x_{p+m} \end{bmatrix} = \begin{bmatrix} \mathbf{X} \\ \mathbf{X}_s \end{bmatrix},$$

$$\widehat{\mathbf{A}} = \begin{bmatrix} a_{11} & \cdots & a_{1p} & 1 & 0 & \cdots & 0 \\ a_{21} & \cdots & a_{2p} & 0 & 1 & \cdots & 0 \\ \vdots & & \vdots & \vdots & \vdots & & \vdots \\ a_{m1} & \cdots & a_{mp} & 0 & 0 & \cdots & 1 \end{bmatrix} = \left[\mathbf{A}, \mathbf{I}_m\right],$$

and $\mathbf{X}_s$ is an $(m \times 1)$ vector of slack variables $x_{p+i}, i=1,\ldots,m$.

Given the augmented problem in standard form as depicted by equations (3.4.1) and (3.5.1), the **augmented feasible region** may be represented as

$$\widehat{\mathcal{K}} = \left\{\widehat{\mathbf{X}} \mid \widehat{\mathbf{A}}\widehat{\mathbf{X}} = \mathbf{b}, \ \widehat{\mathbf{X}} \geqq \mathbf{0}\right\}.$$

As was the case with $\mathcal{K}$, region $\widehat{\mathcal{K}}$ is also a convex polyhedron since it can equivalently be expressed as

$$\widehat{\mathcal{K}} = \left\{\widehat{\mathbf{X}} \mid \widehat{\mathbf{A}}\widehat{\mathbf{X}} \leqq \mathbf{b}, -\widehat{\mathbf{A}}\widehat{\mathbf{X}} \leqq -\mathbf{b}, \ \widehat{\mathbf{X}} \geqq \mathbf{0}\right\}.$$

The linear program in standard form is said to be **feasible** if $\widehat{\mathcal{K}} \neq \phi$.

What is the connection between $\mathcal{K}$ and $\widehat{\mathcal{K}}$? Given any feasible solution to $\mathbf{AX} = \mathbf{b}$, the addition of m non-negative slack variables uniquely determines a feasible solution to $\hat{\mathbf{A}}\hat{\mathbf{X}} = \mathbf{b}$. Conversely, for any feasible solution to $\hat{\mathbf{A}}\hat{\mathbf{X}} = \mathbf{b}$, its first p components yield a feasible solution to $\mathbf{AX} = \mathbf{b}$. This is because all vectors in $\hat{\mathcal{K}}$ lie in a p-dimensional hyperplane embedded in $\mathbf{R}^n$ so that the image of any vector $\hat{\mathbf{X}} \in \hat{\mathcal{K}}$ projected onto the hyperplane has as its components the first p components of $\hat{\mathbf{X}}$. Thus there exists a one-to-one correspondence between points in $\mathcal{K}$ and $\hat{\mathcal{K}}$.

To formalize this correspondence let us define a function $\gamma: \mathcal{K} \rightarrow \hat{\mathcal{K}}$ as $\gamma(\mathbf{X}) = \hat{\mathbf{X}}$, where $\mathbf{X} \in \mathcal{K}$ and $\hat{\mathbf{X}} \in \hat{\mathcal{K}}$. Here γ: (a) is one-to-one; (b) preserves convex combinations of vectors; and (c) preserves extreme points. That is:

(a) Choose $\mathbf{X}_1, \mathbf{X}_2 \in \mathcal{K}$ with $\gamma(\mathbf{X}_1) = \gamma(\mathbf{X}_2)$. Then

$$\hat{\mathbf{X}}_1 = \begin{bmatrix} \mathbf{X}_1 \\ \mathbf{b} - \mathbf{AX}_1 \end{bmatrix} = \gamma(\mathbf{X}_1) = \gamma(\mathbf{X}_2) = \begin{bmatrix} \mathbf{X}_2 \\ \mathbf{b} - \mathbf{AX}_2 \end{bmatrix} = \hat{\mathbf{X}}_2.$$

Hence $\mathbf{X}_1 = \mathbf{X}_2$ and γ is one-to-one.

(b) Choose $\mathbf{X}_1, \mathbf{X}_2 \in \mathcal{K}$ with $\mathbf{X}_c = \theta\mathbf{X}_1 + (1-\theta)\mathbf{X}_2$, $0 \leq \theta \leq 1$. Then

$$\gamma(\mathbf{X}_c) = \begin{bmatrix} \mathbf{X}_c \\ \mathbf{b} - \mathbf{AX}_c \end{bmatrix} = \begin{bmatrix} \theta\mathbf{X}_1 + (1-\theta)\mathbf{X}_2 \\ \theta\mathbf{b} + (1-\theta)\mathbf{b} - \theta\mathbf{AX}_1 - (1-\theta)\mathbf{AX}_2 \end{bmatrix}$$

$$= \theta\begin{bmatrix} \mathbf{X}_1 \\ \mathbf{b} - \mathbf{AX}_1 \end{bmatrix} + (1-\theta)\begin{bmatrix} \mathbf{X}_2 \\ \mathbf{b} - \mathbf{AX}_2 \end{bmatrix} = \theta\gamma(\mathbf{X}_1) + (1-\theta)\gamma\mathbf{X}_2).$$

Thus γ preserves convex combinations.

(c) Let $\{\mathbf{X}_1, \ldots, \mathbf{X}_\ell\}$ depict the set of extreme points of $\mathcal{K}$. Then by theorem 2.39,

$$\begin{aligned} \mathcal{K} &= co\{\mathbf{X}_1, \ldots, \mathbf{X}_\ell\} \\ &= \{\mathbf{X} \mid \mathbf{X} = \Sigma_{j=1}^{\ell}\lambda_j\mathbf{X}_j,\ \Sigma_{j=1}^{\ell}\lambda_j = 1,\ 0 \leq \lambda_j \epsilon \mathbf{R} \text{ for all } j\}. \end{aligned}$$

From property (b), γ must preserve convex combinations of extreme points or

$$\gamma(\mathbf{X}) = \Sigma_{j=1}^{\ell}\lambda_j\gamma(\mathbf{X}_j) = \Sigma_{j=1}^{\ell}\lambda_j\hat{\mathbf{X}}_j,$$

where $\hat{\mathbf{X}}_j$ is an extreme point of

$$\begin{aligned} \hat{\mathcal{K}} &= co\{\hat{\mathbf{X}}_1, \ldots, \hat{\mathbf{X}}_\ell\} \\ &= \{\hat{\mathbf{X}} \mid \hat{\mathbf{X}} = \Sigma_{j=1}^{\ell}\lambda_j\hat{\mathbf{X}}_j,\ \Sigma_{j=1}^{\ell}\lambda_j = 1,\ 0 \leq \lambda_j \in \mathbf{R} \text{ for all } j\}. \end{aligned}$$

Thus extreme points in $\mathcal{K}$ are in one-to-one correspondence with extreme points in $\widehat{\mathcal{K}}$.

What is the relationship between the original objective function (3.1) and the augmented one (3.4)? Since the coefficients on the slack variables in (3.4) are all zero, the original objective function is unchanged, *i.e.*, $f(\mathbf{X}) = f(\widehat{\mathbf{X}})$. In this regard, if $\widehat{\mathbf{X}}_0' = (\mathbf{X}_0', \mathbf{0}')$ optimizes $f\widehat{\mathbf{X}})$ subject to (3.5.1), then the first p components of $\widehat{\mathbf{X}}_0$ optimize $f(\mathbf{X})$ subject to (3.2.1) with the result that $f(\widehat{\mathbf{X}}_0) = f(\mathbf{X}_0)$.

By virtue of the aforementioned one-to-one correspondence between points in $\mathcal{K}$ and $\widehat{\mathcal{K}}$ and the fact that $f(\mathbf{X}) = f(\widehat{\mathbf{X}})$, we will, for the sake of notational efficiently, adopt the convention of expressing a linear program as either

$$\left.\begin{array}{l} max\ f(\mathbf{X}) = \mathbf{C}'\mathbf{X} \quad \text{s.t.} \\ \mathbf{X} \in \mathcal{K} = \{\mathbf{X} \mid \mathbf{A}\mathbf{X} \leqq \mathbf{b}, \mathbf{X} \geqq \mathbf{0}\} \subset \mathbf{R}^p \end{array}\right\} \text{ or } \left\{\begin{array}{l} max f(\mathbf{X}) = \mathbf{C}'\mathbf{X} \quad \text{s.t.} \\ \mathbf{X} \in \mathcal{K} = \{\mathbf{X} \mid \mathbf{A}\mathbf{X} = \mathbf{b}, \mathbf{X} \geqq \mathbf{0}\} \subset R^n. \end{array}\right.$$

In the former case $\mathbf{C}$, $\mathbf{X} \in \mathbf{R}^p$ and $\mathbf{A}$ is of order $(m \times p)$ while in the latter instance it is to be understood that $\mathbf{C}$, $\mathbf{X} \in \mathbf{R}^n$ and $\mathbf{A}$ is of order $(m \times n)$. The form actually selected will be determined by the requirements of the problem at hand.

3.4. Properties of the Feasible Region

We noted in Chapter 2 that a closed half-plane is a closed convex set. It was also indicated there that the intersection of a finite number of closed (convex) sets is itself closed (convex). Since $\mathcal{K} = \{\mathbf{X} \mid \mathbf{A}\mathbf{X} = \mathbf{b}, \mathbf{X} \geqq \mathbf{0}\} \subset \mathbf{R}^n$ is the intersection of the $2m + p$ solution sets of the constraint system, it is obvious that $\mathcal{K}$ is a convex polyhedron, *i.e.*, $\mathcal{K}$ is a closed convex set whose boundary consists of, at most, a finite number of edges and thus a finite number of extreme points.

In addition, since the non-negativity conditions span only the positive orthant of $\mathbf{R}^n$, it is evident that $\mathcal{K}$ is bounded from below. If it is also bounded from above (*i.e.*, there exists a non-negative vector $\mathbf{V} \in \mathbf{R}^n$ with finite components such that $\mathbf{X} \leqq \mathbf{V}$ for all $\mathbf{X} \in \mathcal{K}$), then $\mathcal{K}$ is strictly bounded and thus a convex polytope, *i.e.*, $\mathcal{K}$ is a strictly-bounded, closed, and convex set with a finite number of extreme points (*e.g.*, see Figure 3.1). Hence $\mathcal{K}$ can be written as the convex hull of its set of extreme points. In this regard, if the set of extreme points of $\mathcal{K}$ appears as $\{\mathbf{X}_1,\ldots,\mathbf{X}_k\}$, then by virtue of theorem 2.39,

$$\mathcal{K} = co\{\mathbf{X}_1,\ldots,\mathbf{X}_k\} = \{\mathbf{X} \mid \mathbf{X} = \textstyle\sum_{j=1}^{k} \theta_j \mathbf{X}_j, \\ \textstyle\sum_{j=1}^{k} \theta_j = 1,\ 0 \leq \theta_j \in \mathbf{R}, j=1,\ldots,k\}.$$

That is, each $\mathbf{X} \in \mathcal{K}$ is expressible as a convex combination of the set of extreme points of $\mathcal{K}$.

However, if $\mathcal{K}$ is not strictly bounded, then $\mathcal{K}$ is not a convex polytope and thus, as determined by representation theorem 2.44, $\mathcal{K}$ has a finite basis, *i.e.*, $\mathcal{K}$ is expressible as the sum of the convex hull of a finite number of extreme points and the conical hull of a finite collection of extreme directions, all taken from $\mathcal{K}$. If the set of extreme points of $\mathcal{K}$ again appears as $\{\mathbf{X}_1,\ldots,\mathbf{X}_k\}$ and $\{\mathbf{d}_1,\ldots,\mathbf{d}_\ell\}$ depicts the set of extreme directions in $\mathcal{K}$, then if $\mathbf{A}$ is of order $(m \times n)$ with $\rho(\mathbf{A}) = n$,

$$\begin{aligned} \mathcal{K} &= co\{\mathbf{X}_1,\ldots,\mathbf{X}_k\} + coni\{\mathbf{d}_1,\ldots,\mathbf{d}_\ell\} \\ &= \Big\{\mathbf{X} \mid \mathbf{X} = \textstyle\sum_{j=1}^{k} \theta_j \mathbf{X}_j + \sum_{r=1}^{\ell} \mu_r \mathbf{d}_r,\ \sum_{j=1}^{k} \theta_j = 1, \\ &\qquad 0 \leq \theta_j \in \mathbf{R}, j=1,\ldots,k;\ 0 \leq \mu_r \in \mathbf{R},\ r=1,\ldots,\ell\Big\}. \end{aligned}$$

Thus any $\mathbf{X} \in \mathcal{K}$ can be represented as the sum of a convex combination of a finite number of extreme points from $\mathcal{K}$ and a non-negative linear combination of a finite number of extreme directions for $\mathcal{K}$.

3.5. Existence and Location of Finite Optimal Solutions

In this section we shall exhibit the significance of the notion that $\mathcal{K}$ is a convex polyhedron along with a collection of important theorems pertaining to the existence and location of optimal solutions. We first look to a necessary and sufficient condition for the existence of a finite optimal (maximal) solution. To this end we state a general

3.1. EXISTENCE THEOREM. For the linear program $max f = \mathbf{C}'\mathbf{X}$ s.t. $\mathbf{X} \epsilon \mathcal{X} = \{\mathbf{X} \mid \mathbf{AX} = \mathbf{b}, \mathbf{X} \geqq \mathbf{0}, \mathbf{X} \epsilon \mathbf{R}^n\} \neq \phi$, let $\rho(\mathbf{A}) = m$ and let $\{\mathbf{X}_1, \ldots, \mathbf{X}_k\}$, $\{\mathbf{d}_1, \ldots, \mathbf{d}_\ell\}$ depict sets of extreme points and extreme directions for $\mathcal{X}$ respectively. Then f attains a finite maximum at an extreme point $\mathbf{X}_r$ if and only if $\mathbf{C}'\mathbf{d}_j \leq 0$, $j=1,\ldots,\ell$. (The angle between $\mathbf{C}$ and each $\mathbf{d}_j$ must be nonobtuse.) If f assumes its finite maximum at more than one extreme point, then it takes on the same value at each convex combination of those extreme points.

PROOF. By representation theorem 2.44, $\mathbf{X} \epsilon \mathcal{X}$ if and only if

$$\mathbf{X} = \sum_{i=1}^{k} \theta_i \mathbf{X}_i + \sum_{j=1}^{\ell} \mu_j \mathbf{d}_j, \ \sum_{i=1}^{k} \theta_i = 1, \ 0 \leq \theta_i \epsilon \mathbf{R}, \\ i=1,\ldots,k; \ 0 \leq \mu_j \ \epsilon \mathbf{R}, \ j=1,\ldots,\ell.$$

Hence we may rewrite the linear program as

$$max f = \mathbf{C}'\mathbf{X} = \sum_{i=1}^{k} \theta_i \mathbf{C}'\mathbf{X}_i + \sum_{j=1}^{\ell} \mu_j \mathbf{C}'\mathbf{d}_j \ \text{ s.t.} \\ \sum_{i=1}^{k} \theta_i = 1 \\ \theta_i \geq 0, \ i=1,\ldots,k \\ \mu_j \geq 0, \ j=1,\ldots,\ell.$$

If $\mathbf{C}'\mathbf{d}_j > 0$ for some $1 \leq j \leq \ell$, then $f \rightarrow +\infty$ as $\mu_j \rightarrow +\infty$ so that f is unbounded from above. Hence f attains a finite maximal solution if and only if $\mathbf{C}'\mathbf{d}_j \leq 0$, $j=1,\ldots,\ell$. Let $\mathbf{C}'\mathbf{d}_j \leq 0$ for all j and set $\mu_j = 0$ so that the above problem becomes

$$max f = \sum_{i=1}^{k} \theta_i \mathbf{C}'\mathbf{X}_i \ \text{ s.t. } \ \sum_{i=1}^{k} \theta_i = 1, \theta_i \geq 0, i=1,\ldots,k.$$

If $\mathbf{C}'\mathbf{X}_r = \max_{1 \leq i \leq k} \mathbf{C}'\mathbf{X}_i$, then replacing each $\mathbf{C}'\mathbf{X}_i$ by $\mathbf{C}'\mathbf{X}_r$ in the preceding $1 \leq i \leq k$ objective function does not decrease this sum so that

$$\mathbf{C}'\mathbf{X} = \sum_{i=1}^{k} \theta_i \mathbf{C}'\mathbf{X}_i \leq \mathbf{C}'\mathbf{X}_r \sum_{i=1}^{k} \theta_i = \mathbf{C}'\mathbf{X}_r.$$

Hence f takes on its finite maximum at an extreme point of $\mathcal{X}$.

Next, let f attain its finite maximum at more than one extreme point, *e.g.*, for extreme points $\mathbf{X}_1, \ldots, \mathbf{X}_s$, $s < k$, assume $\mathbf{C}'\mathbf{X}_1 = \cdots = \mathbf{C}'\mathbf{X}_s = \mathbf{C}'\mathbf{X}_r$. Let $\bar{\mathbf{X}}$ be any convex combination of $\mathbf{X}_1, \ldots, \mathbf{X}_s$, *i.e.*,

$$\bar{\mathbf{X}} = \sum_{i=1}^{s} \bar{\theta}_i \mathbf{X}_i,\ \sum_{i=1}^{s} \bar{\theta}_i = 1,\ 0 \leq \bar{\theta}_i \in \mathbf{R},\ i=1,\ldots,s.$$

Then

$$\mathbf{C}'\bar{\mathbf{X}} = \sum_{i=1}^{s} \bar{\theta}_i \mathbf{C}'\mathbf{X}_i = \mathbf{C}'\mathbf{X}_r \sum_{i=1}^{s} \bar{\theta}_i = \mathbf{C}'\mathbf{X}_r$$

since $\mathbf{C}'\mathbf{X}_r = \mathbf{C}'\mathbf{X}_i, i=1,\ldots,s$, and thus the second part of the theorem is verified. Q.E.D.

This theorem informs us that if $\mathcal{K}$ is a convex polyhedron and $\mathbf{C}'\mathbf{d}_j \leq 0$ for each extreme direction $\mathbf{d}_j, j=1,\ldots,\ell$, for $\mathcal{K}$, then f takes on its finite maximum at an extreme point of $\mathcal{K}$ or at some convex combination of extreme points taken from $\mathcal{K}$. However, if $\mathcal{K}$ is a convex polytope (a bounded convex polyhedron), then f is indeed bounded from above on $\mathcal{K}$ and thus the existence of a finite maximum of f at an extreme point of $\mathcal{K}$ follows immediately from theorem 2.52 . As Figure 3.2.a reveals, the maximum of f occurs at a unique extreme point A while in Figure 3.2.b, f attains the same maximal value at extreme points A, B, and at every convex combination C of A, B.

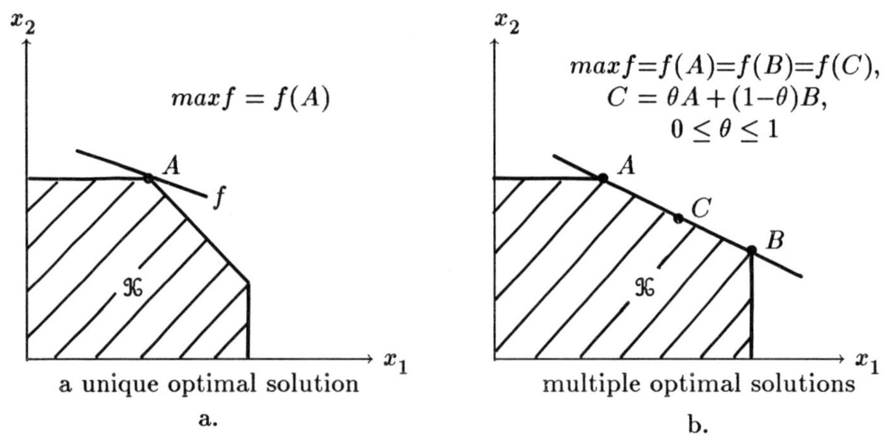

Figure 3.2

The above example problem (Figure 3.1 of section 3.2) dealt with the case where $\mathcal{K}$ was a convex polytope. However, if $\mathcal{K}$ is not bounded, then the notion of whether or not the linear program attains a finite maximum depends upon the characteristics of the f hyperplane itself. For instance, let us

$$\begin{aligned} max f &= -\tfrac{3}{2}x_1 + 3x_2 \quad \text{s.t.} \\ -x_1 + x_2 &\leq 1 \\ -\tfrac{1}{2}x_1 + x_2 &\leq 3 \\ -\tfrac{1}{3}x_1 + x_2 &\leq 5 \\ x_1, x_2 &\geq 0. \end{aligned}$$

Here $\mathcal{K}$ is a convex polyhedron that is unbounded (Figure 3.3). Moreover, $\mathcal{K}$ has four extreme points

$$\mathbf{X}_1 = \begin{bmatrix} 0 \\ 0 \end{bmatrix}, \mathbf{X}_2 = \begin{bmatrix} 0 \\ 1 \end{bmatrix}, \mathbf{X}_3 = \begin{bmatrix} 4 \\ 5 \end{bmatrix}, \text{ and } \mathbf{X}_4 = \begin{bmatrix} 12 \\ 9 \end{bmatrix}$$

and two extreme directions that may be constructed as

$$\mathbf{d}_1 = \mathbf{X}_5 - \mathbf{X}_1 = \begin{bmatrix} 1 \\ 0 \end{bmatrix} - \begin{bmatrix} 0 \\ 0 \end{bmatrix} = \begin{bmatrix} 1 \\ 0 \end{bmatrix},$$

$$\mathbf{d}_2 = \mathbf{X}_6 - \mathbf{X}_4 = \begin{bmatrix} 15 \\ 10 \end{bmatrix} - \begin{bmatrix} 12 \\ 9 \end{bmatrix} = \begin{bmatrix} 3 \\ 1 \end{bmatrix}.$$

Hence a point $\mathbf{X} \in \mathcal{K}$ can be represented generally as

$$\mathbf{X} = co\{\mathbf{X}_1, \mathbf{X}_2, \mathbf{X}_3, \mathbf{X}_4\} + coni\{\mathbf{d}_1, \mathbf{d}_2\}.$$

In particular, $\mathbf{X}_7 \in \mathcal{K}$ can be expressed as

$$\mathbf{X}_7 = \theta_1\mathbf{X}_1 + \theta_4\mathbf{X}_4 + \mu_2\mathbf{d}_2, \theta_1 + \theta_4 = 1, \ \ \theta_1, \theta_4 \geq 0; \ \mu_2 \geq 0.$$

Since $\mathbf{C}'\mathbf{X}_1 = 0, \mathbf{C}'\mathbf{X}_2 = 3, \mathbf{C}'\mathbf{X}_3 = 9, \mathbf{C}'\mathbf{X}_4 = 9, \mathbf{C}'\mathbf{d}_1 = -\frac{15}{2}$, and $\mathbf{C}'\mathbf{d}_2 = -\frac{3}{2}$, our problem is equivalent to

$$\begin{aligned} max f &= 3\theta_2 + 9\theta_3 + 9\theta_4 - \tfrac{15}{2}\mu_1 - \tfrac{3}{2}\mu_2 \quad \text{s.t.} \\ \theta_1 + \theta_2 &+ \theta_3 + \theta_4 = 1 \\ \theta_1, \ldots, \theta_4 &\geq 0; \ \mu_1, \mu_2 \geq 0. \end{aligned}$$

With both $\mathbf{C}'\mathbf{d}_1$, $\mathbf{C}'\mathbf{d}_2 \leq 0$, set $\mu_1 = \mu_2 = 0$. Then to maximize $f = 3\theta_2 + 9\theta_3 + 9\theta_4$ subject to the indicated constraints we may, for instance, set $\theta_1 = \theta_2 = \theta_3 = 0$ and let $\theta_4 = 1$. Hence the optimal extreme point is $\mathbf{X}_4$. Are there other options? The answer is yes since any point along the entire edge from $\mathbf{X}_3$ to $\mathbf{X}_4$ serves as a maximal point with the maximum of f being 9. However, if the objective hyperplane were changed to $f = x_1 + x_2$, it is easily demonstrated that no finite maximum exists since the value of f may be increased without bound with feasibility always maintained.

To see this we first find $\mathbf{C}'\mathbf{X}_1 = 0$, $\mathbf{C}'\mathbf{X}_2 = 1$, $\mathbf{C}'\mathbf{X}_3 = 9$, $\mathbf{C}'\mathbf{X}_4 = 21$, $\mathbf{C}'\mathbf{d}_1 = 5$, and $\mathbf{C}'\mathbf{d}_2 = 4$. Hence an equivalent version of the original problem is

$$max f = \theta_2 + 9\theta_3 + 21\theta_4 + 5\mu_1 + 4\mu_2 \text{ s.t.}$$
$$\theta_1 + \theta_2 + \theta_3 + \theta_4 = 1$$
$$\theta_1, \ldots, \theta_4 \geq 0; \ \mu_1, \mu_2 \geq 0.$$

Since, say, $\mathbf{C}'\mathbf{d}_2 > 0$, μ_2 can be made arbitrarily large without violating the feasibility of this problem. Hence $f \rightarrow \infty$ as $\mu_2 \rightarrow \infty$ (with $\mu_1 = 0$, $\theta_1 = 1$, and $\theta_2 = \theta_3 = \theta_4 = 0$) so that no finite maximum exists.

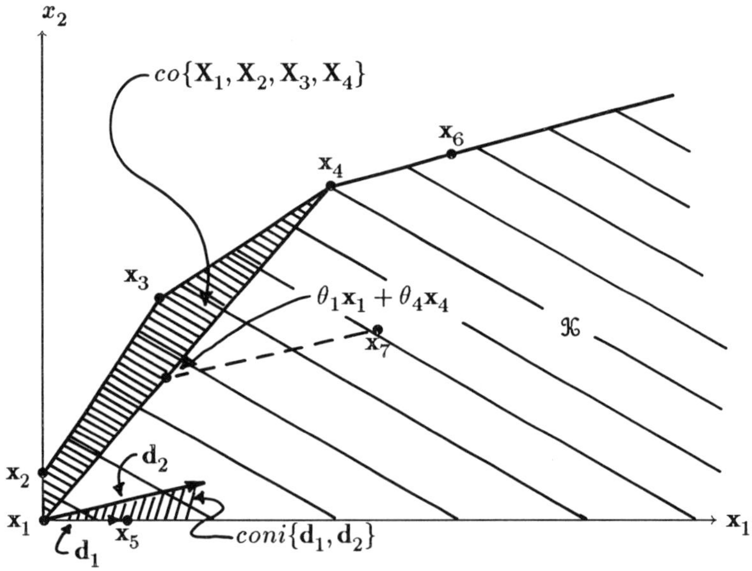

Figure 3.3

The importance of existence theorem 3.1 is that it drastically limits the size of the set of points within $\mathcal{K}$ which are possible candidates for an optimal solution. That is to say, instead of examining all points along the boundary of $\mathcal{K}$ in our search for a finite maximal solution, we need only look to the set of extreme points for it is among these points that the optimal solution must be found. And since there are at most a finite number of such points, we may be assured that only a finite number of extreme point solutions need be examined in our search for a maximum of f. An iterative method of search which does in fact proceed to the maximum of f in a finite number of steps is the *simplex method,* which will be discussed in subsequent chapters.

We close this section by noting briefly that:

(a) if $\mathcal{K} = \phi$, the linear programming problem has no solution;

(b) if $\mathcal{K}$ is bounded ($\mathcal{K}$ is a convex polytope), the linear program has a finite optimal solution which occurs at an extreme point or at some convex combination of extreme points; and

(c) if $\mathcal{K}$ is unbounded, the linear programming problem has a solution that may be unbounded.

3.6. Basic Feasible and Extreme Point Solutions to the Linear Programming Problem

Because of the importance of the connection between extreme points and basic feasible solutions to a linear program, let us review our earlier discussion of *extreme point solutions as basic solutions to linear equalities* which was presented in Chapter 2. In what follows we shall work with the linear programming problem in standard form or

$$max f = \mathbf{C}'\mathbf{X} \text{ s.t.}$$
$$\mathbf{AX} = \mathbf{b},\ \mathbf{X} \geqq \mathbf{0},$$

where $\mathbf{X} \in \mathbf{R}^n$ and $\mathbf{A}$ is of order $(m \times n)$.

As noted in section 2.10, if the (augmented structural constraint) system $\mathbf{AX} = \mathbf{b}$, $\rho(\mathbf{A}) = m$, has a feasible solution $\mathbf{X} \geqq \mathbf{0}$, then it also possesses a basic feasible solution. Moreover, it was also mentioned that a vector $\mathbf{X} \in \mathbf{R}^n$ is a basic feasible solution to $\mathbf{AX} = \mathbf{b}$ if and only if $\mathbf{X}$ is an extreme point of a convex polyhedron (taken herein as $\mathcal{K}$). These observations are of paramount importance since they inform us that we may undertake a search for an optimal extreme point solution to a linear programming problem by seeking only basic feasible solutions. In this regard, if we can find a set of m linearly independent vectors within $\mathbf{A}$ and arrive at a basic feasible solution to $\mathbf{AX} = \mathbf{b}$, it must yield an extreme point of $\mathcal{K}$, *i.e.*, extreme point solutions are always generated by m linearly independent columns from $\mathbf{A}$.

At the outset of our search for basic feasible solutions we shall impose the restriction that $\rho(\mathbf{A}) = \rho[\mathbf{A}, \mathbf{b}]$. If such an assumption is not made, then $\mathbf{AX} = \mathbf{b}$ represents an inconsistent system of equations and we are thus precluded from obtaining a solution to the linear programming problem. Since $\mathbf{AX} = \mathbf{b}$ is comprised of m equations in n unknowns with $m < n$, it is clear that the augmented structural constraints represent an *underdetermined system* of simultaneous linear equations. For such a system an infinite number of particular solutions exists if it is consistent. In this regard, if we further assume that $\rho(\mathbf{A}) = m$, so that none of the augmented structural constraints is *redundant* (*i.e.*, a particular structural constraint cannot be expressed as a linear combination of any others), we may select m linearly independent columns from $\mathbf{A}$ and uniquely express the variables associated with these columns in terms of the remaining $n-m$ variables. If arbitrary values are then assigned to the latter, it is evident that an infinity of solutions ensues. More formally, let us rearrange the columns of $\mathbf{A}$ so that the first m are linearly independent and constitute the columns of the $(m \times m)$ matrix $\mathbf{B} = [\mathbf{b}_1, \ldots, \mathbf{b}_m]$. The remaining $n-m$ columns of $\mathbf{A}$ then form the $(m \times n-m)$ matrix $\mathbf{R} = [\mathbf{r}_1, \ldots, \mathbf{r}_{n-m}]$. Hence $\mathbf{AX} = \mathbf{b}$ may be written as

$$
\begin{aligned}
\mathbf{AX} &= [\mathbf{B}, \mathbf{R}] \begin{bmatrix} \mathbf{X}_B \\ \mathbf{X}_R \end{bmatrix} \\
&= \mathbf{BX}_B + \mathbf{RX}_R = \mathbf{b},
\end{aligned}
$$

where $\mathbf{X}$ has been partitioned into an $(m \times 1)$ matrix $\mathbf{X}_B, \mathbf{X}'_B = (x_{B1}, \ldots, x_{Bm})$, containing the m variables corresponding to the m linearly independent columns of $\mathbf{A}$ and an $(n-m\times 1)$ matrix $\mathbf{X}_R, \mathbf{X}'_R = (x_{R1},\ldots,x_{R,n-m})$, containing the $n-m$ remaining variables. Since $\rho(\mathbf{A})=m$, $\mathbf{B}$ is nonsingular. Hence

$$\mathbf{X}_B=\mathbf{B}^{-1}\mathbf{b}-\mathbf{B}^{-1}\mathbf{R}\mathbf{X}_R \tag{3.6}$$

and all solutions to the augmented structural constraint system can be generated by assigning arbitrary values to $\mathbf{X}_R$. In particular, if $\mathbf{X}_R=\mathbf{0}$,

$$\mathbf{X}_B = \begin{bmatrix} x_{B1} \\ x_{B2} \\ \vdots \\ x_{Bm} \end{bmatrix} = \mathbf{B}^{-1}\mathbf{b}$$

and we obtain a **basic solution** to $\mathbf{AX}=\mathbf{b}$, namely $\mathbf{X}'=(\mathbf{X}'_B,\mathbf{0}')$, since the columns of $\mathbf{B}$ constitute a basis for $\mathbf{R}^m$, *i.e.*, each column of $\mathbf{R}$ and the vector $\mathbf{b}$ are expressible as a linear combination of the linearly independent columns of $\mathbf{B}$. Hence the variables within $\mathbf{X}_B$ are said to be **basic variables** while those within $\mathbf{X}_R$ are termed **nonbasic variables.**

To summarize: given the augmented structural constraint system $\mathbf{AX}=\mathbf{b}$ with $\rho(\mathbf{A})=\rho[\mathbf{A},\mathbf{b}]=m$, if any m linearly independent columns are chosen from $\mathbf{A}$ and if all the remaining $n-m$ variables not associated with these columns are set equal to zero, the resulting solution $\mathbf{X}'=(\mathbf{X}'_B,\mathbf{0}')$ contains no more than m nonzero variables and is a basic solution to $\mathbf{AX}=\mathbf{b}$. Furthermore, if $\mathbf{X}_B \geqq \mathbf{0}$ so that $\mathbf{X}\geqq\mathbf{0}$, then $\mathbf{X}$ is termed a **basic feasible solution** to $\mathbf{AX}=\mathbf{b}$ and corresponds to an extreme point of $\mathcal{K}$. That is, $\mathbf{X}$ is an extreme point of $\mathcal{K}$ if and only if $\mathbf{X}'=(\mathbf{B}^{-1}\mathbf{b},\mathbf{0})'\geqq\mathbf{0}'$. Finally, a basic solution (feasible or otherwise) to $\mathbf{AX}=\mathbf{b}$ is said to be **degenerate** if one or more of the basic variables within $\mathbf{X}_B$ vanishes, *i.e.*, $\mathbf{X}$ contains less than m positive components.

EXAMPLE 3.1. Find a basic solution to

$$\begin{aligned} x_1+2x_2+x_3+x_4&=1\\ 2x_1-2x_2-x_3+x_4&=1. \end{aligned}$$

Here

$$\mathbf{A}=\begin{bmatrix} 1 & 2 & 1 & 1 \\ 2 & -2 & -1 & 1 \end{bmatrix},\ \mathbf{X}=\begin{bmatrix} x_1 \\ x_2 \\ x_3 \\ x_4 \end{bmatrix},\quad \text{and}\quad \mathbf{b}=\begin{bmatrix} 1 \\ 1 \end{bmatrix}.$$

Since $\rho(\mathbf{A}) = \rho[\mathbf{A},\mathbf{b}] = 2$, the equations are consistent and nonredundant. Hence we may extract from $\mathbf{A}$ two linearly independent columns so as to form the nonsingular basis matrix $\mathbf{B}$. Let us try columns two and three. Since

$$\begin{vmatrix} 2 & 1 \\ -2 & -1 \end{vmatrix} = 0,$$

these columns are obviously not linearly independent so that $\mathbf{B}^{-1}$ does not exist. Let us now try columns one and three. Since

$$\begin{vmatrix} 1 & 1 \\ 2 & -1 \end{vmatrix} = -3 \neq 0,$$

we may now form

$$\mathbf{B}=\begin{bmatrix} 1 & 1 \\ 2 & -1 \end{bmatrix},\ \mathbf{B}^{-1} = \begin{bmatrix} \frac{1}{3} & \frac{1}{3} \\ \frac{2}{3} & -\frac{1}{3} \end{bmatrix},$$

$$\mathbf{R}=\begin{bmatrix} 2 & 1 \\ -2 & 1 \end{bmatrix},\ \mathbf{X}_B=\begin{bmatrix} x_1 \\ x_3 \end{bmatrix},\ \text{and}\quad \mathbf{X}_R=\begin{bmatrix} x_2 \\ x_4 \end{bmatrix}.$$

We noted above that to obtain a basic solution to $\mathbf{AX}=\mathbf{b}$ we need only solve for the basic variables in terms of the nonbasic variables, and then set the nonbasic variables equal to zero. Proceeding in this fashion yields, from (3.6),

$$\mathbf{X}_B = \mathbf{B}^{-1}\mathbf{b} - \mathbf{B}^{-1}\mathbf{R}\mathbf{X}_R \quad \text{or}$$

$$\begin{bmatrix} x_1 \\ x_3 \end{bmatrix} = \begin{bmatrix} \frac{2}{3} & & -\frac{2}{3}x_4 \\ \frac{1}{3} & -2x_2 & -\frac{1}{3}x_4 \end{bmatrix}.$$

If $\mathbf{X}_R=\mathbf{0}$, $\mathbf{X}'_B=\left(\frac{2}{3},\ \frac{1}{3}\right)$ so that a basic solution to the above system is $\mathbf{X}'=\left(\frac{2}{3},\frac{1}{3},0,0\right)$. Note that since no component of $\mathbf{X}_B$ is zero and $\mathbf{X}\geqq\mathbf{0}$, we actually have obtained a nondegenerate basic feasible solution.

3.7. Solutions and Requirements Spaces

We may now examine two types of Euclidian spaces commonly connected with a linear programming problem. In doing so we shall shed some additional light on the significance of the nonredundancy assumption that $\rho(\mathbf{A}) = m$. It was noted above that the augmented feasible region $\mathcal{K}$ lies within $\mathbf{R}^n$. That part of $\mathbf{R}^n$ which contains $\mathcal{K}$ will be called the **solutions space** of the constrained equation system $\mathbf{AX} = \mathbf{b}$, $\mathbf{X} \geqq \mathbf{0}$, $\mathbf{X} \epsilon \mathbf{R}^n$. If a vector $\mathbf{X}$ satisfies $\mathbf{AX} = \mathbf{b}$, then $\mathbf{X}$ lies within the intersection of a set of hyperplanes which comprise the rows of $[\mathbf{A}, \mathbf{b}]$. If $\rho(\mathbf{A}) = m$, then the said intersection cannot be formed by less than m hyperplanes, *i.e.*, any one hyperplane cannot be expressed as a linear combination of the others.

The second space in question will be referred to as **requirements space.** The elements of this space are the columns of $[\mathbf{A}, \mathbf{b}]$, where $\mathbf{b}$ is termed the **requirements vector.** Since the columns of $[\mathbf{A}, \mathbf{b}]$ represent a set of $(m \times 1)$ vectors, it is obvious that requirements space is a subset of $\mathbf{R}^m$. If $\mathbf{a}_j$ denotes the jth column of $\mathbf{A}$, then

$$\mathbf{AX} = \sum_{j=1}^{n} \mathbf{a}_j x_j = \mathbf{b}.$$

Since a feasible solution to $\mathbf{AX} = \mathbf{b}$ obtains when $\mathbf{b}$ is expressible as a nonnegative linear combination of the columns of $\mathbf{A}$, $\mathbf{b}$ must be an element of the convex polyhedral cone generated by the columns of $\mathbf{A}$. Hence that part of $\mathbf{R}^m$ which contains the convex polyhedral cone

$$\mathcal{C} = \left\{ \mathbf{Y} \mid \mathbf{Y} = \mathbf{AX} = \sum_{j=1}^{n} \mathbf{a}_j x_j, \mathbf{X} \geqq \mathbf{0} \right\}$$

is requirements space. Now if $\rho(\mathbf{A}) = m$, the columns of $\mathbf{A}$ yield a basis for $\mathbf{R}^m$ so that $\mathbf{b}$ cannot be expressible as a nonnegative linear combination of fewer than m of the $\mathbf{a}_j$. Thus the number of $x_j > 0$ is never greater than m. In this regard, if the m linearly independent columns of $\mathbf{A}$ constitute the $(m \times m)$ matrix $\mathbf{B}$, then a basic feasible solution to $\mathbf{AX} = \mathbf{b}$ results when $\mathbf{b}$ is expressible as a nonnegative linear combination of the columns of **B** so that **b** is an element of the convex polyhedral cone

$$\mathcal{C}_B = \left\{ \mathbf{Y} \mid \mathbf{Y} = [\mathbf{B}, \mathbf{R}]\begin{bmatrix} \mathbf{X}_B \\ \mathbf{0} \end{bmatrix} = \mathbf{B}\mathbf{X}_B = \sum_{i=1}^{m} \mathbf{b}_i x_{Bi}, \mathbf{X}_B \geqq \mathbf{0} \right\},$$

where $\mathbf{b}_i$ is the ith column of **B**.

For instance, let $\mathbf{A}\mathbf{X} = [\mathbf{a}_1, \ldots, \mathbf{a}_5]\mathbf{X} = \mathbf{b} \in \mathbf{R}^n$ (Figure 3.4.a). Since $\mathbf{b} \in \mathcal{C}$, *i.e.*, $\mathbf{b} = \Sigma_{i=1}^5 \mathbf{a}_i x_i$, system $\mathbf{A}\mathbf{X} = \mathbf{b}$ has a feasible solution in requirements space or $\mathcal{K} \neq \phi$. Moreover, **b** is expressible as

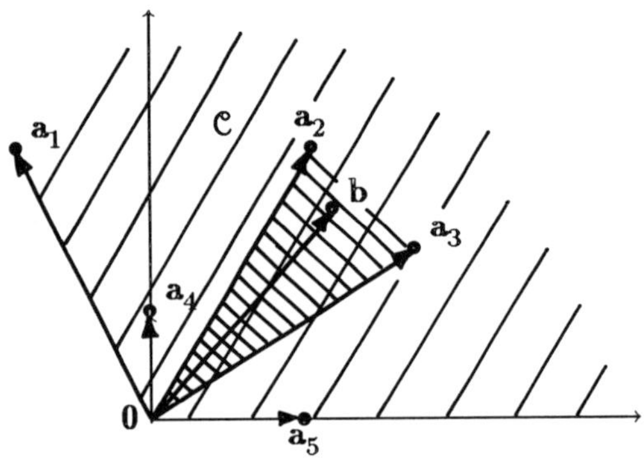

a.

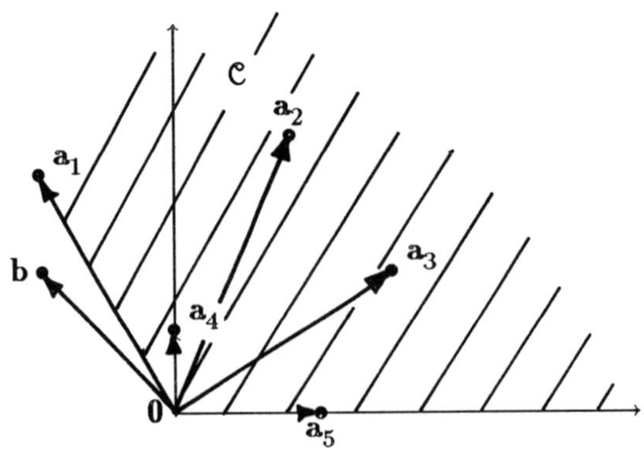

b.
Figure 3.4

a nonnegative linear combination of $\mathbf{a}_2$, $\mathbf{a}_3$ alone so that $\mathcal{C}_B = \{\mathbf{Y}|\mathbf{B}\mathbf{X}_B = \mathbf{b}, \mathbf{X}_B \geqq \mathbf{0}\}$, where $\mathbf{B} = [\mathbf{a}_2, \mathbf{a}_3]$, $\mathbf{X}'_B = (x_{B1}, x_{B2}) = (x_2, x_3)$. However, if $\mathbf{b}$ is redrawn as in Figure 3.4.b, then $\mathcal{K} = \phi$ — now $\mathbf{AX} = \mathbf{b}$ is inconsistent in requirements space.

Figure 3.1 depicted the attainment of an optimal solution to a linear program in solutions space. Let us now illustrate the notion of optimality in requirements space. Specifically, our problem is to: $max\, f(\mathbf{X}) = \mathbf{C}'\mathbf{X}$ s.t. $\mathbf{AX} = \mathbf{b}$, $\mathbf{X} \geqq \mathbf{0}$. If we treat the objective function as a $m+1^{st}$ equation in the system

$$\begin{bmatrix} \mathbf{A} \\ \mathbf{C}' \end{bmatrix} \mathbf{X} = \begin{bmatrix} \mathbf{b} \\ f \end{bmatrix}$$

or

$$\sum_{i=1}^{n} \begin{bmatrix} \mathbf{a}_i \\ c_i \end{bmatrix} x_i = \sum_{i=1}^{n} \bar{\mathbf{a}}_i x_i = \begin{bmatrix} \mathbf{b} \\ f \end{bmatrix}, \tag{3.7}$$

then we desire to find a solution to (3.7) for which each $x_i \geq 0$, i=1,...,n, and f is as large as possible. Clearly the vector $\begin{bmatrix} \mathbf{b} \\ f \end{bmatrix}$, f maximal, must be an element of the finite cone $\bar{\mathcal{C}}$ spanned by the vectors $\bar{\mathbf{a}}'_i = (\mathbf{a}'_i, c_i)$, i=1,...,n.

For example, let us

$$\begin{aligned} max\, f(\mathbf{X}) &= -x_1 + 2x_2 \quad \text{s.t.} \\ x_1 + x_2 + x_3 &= 5 \\ x_1, x_2, x_3 &\geq 0, \end{aligned}$$

where x_3 is a slack variable. As indicated above our problem can be rewritten as: choose $x_1, x_2, x_3 \geq 0$ so that

$$\begin{bmatrix} 1 \\ -1 \end{bmatrix} x_1 + \begin{bmatrix} 1 \\ 2 \end{bmatrix} x_2 + \begin{bmatrix} 1 \\ 0 \end{bmatrix} x_3 = \sum_{i=1}^{3} \bar{\mathbf{a}}_i x_i = \begin{bmatrix} 5 \\ f \end{bmatrix}$$

and f is maximal. From Figure 3.5, it is evident that the maximal solution is a multiple of the vector $\bar{\mathbf{a}}'_2 = (1,2)$ or

$$\bar{\mathbf{a}}_1 \cdot 0 + \bar{\mathbf{a}}_2 x_2 + \bar{\mathbf{a}}_3 \cdot 0 = \begin{bmatrix} 1 \\ 2 \end{bmatrix} x_2 = \begin{bmatrix} 5 \\ f \end{bmatrix}.$$

Hence $x_2 = 5$ and $f = f^0 = 2x_2 = 10$.

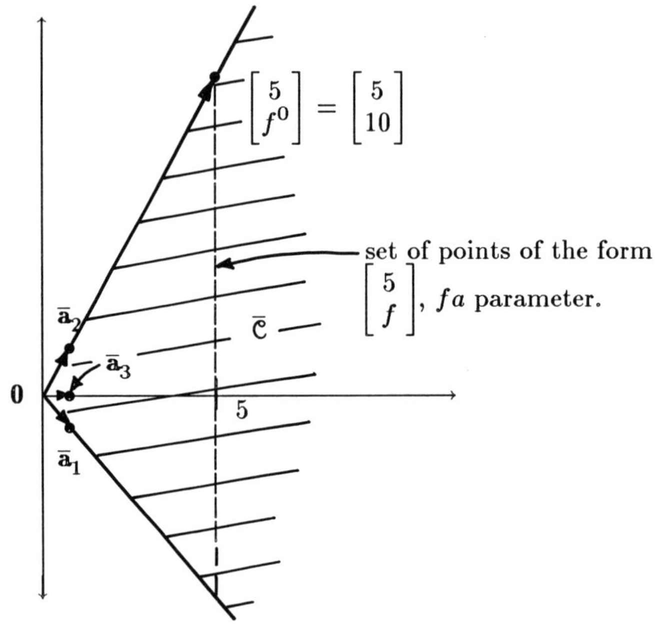
$\begin{bmatrix}5\\f^0\end{bmatrix} = \begin{bmatrix}5\\10\end{bmatrix}$
set of points of the form $\begin{bmatrix}5\\f\end{bmatrix}$, f a parameter.
$\bar{\mathbf{a}}_2$
$\bar{\mathbf{a}}_3$
$\bar{\mathbf{a}}_1$
$\bar{c}$
0
5

Figure 3.5

4 DUALITY THEORY

4.1. The Symmetric Dual

Associated with every linear programming problem is another such problem called its dual. Quite generally, duality theory addresses itself to the study of the connection between two related linear programming problems, where one of them, the **primal**, is a maximization (minimization) problem and the other, the **dual**, is a minimization (maximization) problem. Moreover, this association is such that the existence of an optimal solution to one of these problems guarantees an optimal solution to the other with the result that their extreme values are equal.

We now turn to the expression of the primal problem in canonical form and then look to the construction of its associated dual. Specifically, if the primal problem appears as

$$\begin{gathered} max\ f = c_1x_1 + \cdots + c_px_p \quad \text{s.t.} \\ a_{11}x_1 + \cdots + a_{1p}x_p \le b_1 \\ a_{21}x_1 + \cdots + a_{2p}x_p \le b_2 \\ \cdots\cdots\cdots\cdots\cdots\cdots \\ a_{m1}x_1 + \cdots + a_{mp}x_p \le b_m \\ x_1, \ldots, x_p \ge 0, \end{gathered}$$

then its dual is

$$\begin{aligned} min\, g = b_1u_1 + \cdots + b_mu_m \quad \text{s.t.} \\ a_{11}u_1 + \cdots + a_{m1}u_m \geq c_1 \\ a_{12}u_1 + \cdots + a_{m2}u_m \geq c_2 \\ \cdots\cdots\cdots\cdots\cdots\cdots \\ a_{1p}u_1 + \cdots + a_{mp}u_m \geq c_p \\ u_1, \ldots, u_m \geq 0. \end{aligned}$$

Henceforth this form of the dual will be termed the **symmetric dual**, *i.e.*, in this instance, the primal and dual problems are said to be mutually symmetric. (A rationalization of the structure assumed by the symmetric dual program will be offered in the next chapter wherein we develop a necessary and sufficient condition for the attainment of primal optimality.) In matrix form, the symmetric primal-dual pair of problems appears as

Primal	Symmetric Dual
$max\ f(\mathbf{X}) = \mathbf{C}'\mathbf{X}$ s.t.	$min\ g(\mathbf{U}) = \mathbf{b}'\mathbf{U}$ s.t.
$\mathbf{AX} \leqq \mathbf{b},\ \mathbf{X} \geqq \mathbf{0},\ \mathbf{X} \in \mathbf{R}^p$ or	$\mathbf{A}'\mathbf{U} \geqq \mathbf{C},\ \mathbf{U} \geqq \mathbf{0},\ \mathbf{U} \in \mathbf{R}^m$ or
$\mathbf{X} \in \mathcal{K}_P = \{\mathbf{X} \mid \mathbf{AX} \leqq \mathbf{b}, \mathbf{X} \geqq \mathbf{0}, \mathbf{X} \in \mathbf{R}^p\}$	$\mathbf{U} \in \mathcal{K}_D = \{\mathbf{U} \mid \mathbf{A}'\mathbf{U} \geqq \mathbf{C}, \mathbf{U} \geqq \mathbf{0}, \mathbf{U} \in \mathbf{R}^m\}$

where $\mathbf{X}$ and $\mathbf{C}$ are of order $(p \times 1)$, $\mathbf{U}$ and $\mathbf{b}$ are of order $(m \times 1)$, $\mathbf{A}$ is of order $(m \times p)$, and $\mathcal{K}_P, \mathcal{K}_D$ denote the regions of admissible solutions for the primal and dual problems respectively. In standard form, the primal and its symmetric dual may be written as

Primal	Symmetric Dual
$max\ f(\mathbf{X}, \mathbf{X}_s) = \mathbf{C}'\mathbf{X} + \mathbf{0}'\mathbf{X}_s$ s.t.	$min\ g(\mathbf{U}, \mathbf{U}_s) = \mathbf{b}'\mathbf{U} + \mathbf{0}'\mathbf{U}_s$ s.t.
$\mathbf{AX} + \mathbf{I}_m\mathbf{X}_s = \mathbf{b}$	$\mathbf{A}'\mathbf{U} - \mathbf{I}_p\mathbf{U}_s = \mathbf{C}$
$\mathbf{X}, \mathbf{X}_s \geqq \mathbf{0}$	$\mathbf{U}, \mathbf{U}_s \geqq \mathbf{0}$

where $\mathbf{X}_s, \mathbf{U}_s$ denote $(m \times 1)$ and $(p \times 1)$ vectors of primal slack and dual surplus variables respectively. Here $\mathbf{U}_s$ has as its jth component the **dual surplus variable**

$$u_{m+j} = \sum_{i=1}^{m} a_{ij}u_i - c_j \geq 0, j=1,\ldots,p. \tag{4.1}$$

An examination of the structure of the primal and symmetric dual problems reveals the following salient features:

(a) if the primal problem involves maximization, then the dual problem involves minimization and conversely;

(b) a new set of variables appears in the dual; in this regard,

(c) if there are p variables and m structural constraints in the primal problem, there will be m variables and p structural constraints in the dual problem and conversely. In particular, there exists a one-to-one correspondence between the jth primal variable and the jth dual structural constraint and between the ith dual variable and the ith primal structural constraint;

(d) the primal objective function coefficients become the right-hand sides of the dual structural constraints and conversely;

(e) the coefficient matrix of the dual structural constraint system is the transpose of the coefficient matrix of the primal structural constraint system and conversely. Thus the coefficients on the x_j in the primal problem become the coefficients of the jth dual structural constraint; and

(f) the inequality signs in the structural constraints of the primal problem are reversed in the dual problem and conversely.

EXAMPLE 4.1. Given that the primal problem appears as

$$\begin{aligned} max\ f &= 2x_1 + 3x_2 + 5x_3 \text{ s.t.} \\ x_1 + x_2 + 3x_3 &\leq 5 \\ 2x_1 - x_2 + x_3 &\leq 6 \\ x_1 + 3x_2 - 4x_3 &\leq 8 \\ x_2 + x_3 &\leq 7 \\ x_1, x_2, x_3 &\geq 0, \end{aligned}$$

find its dual. In light of the preceding discussion it is obvious that the symmetric dual problem is

$$\begin{aligned} \min g &= 5u_1 + 6u_2 + 8u_3 + 7u_4 \quad \text{s.t.} \\ u_1 + 2u_2 + u_3 \qquad &\geq 2 \\ u_1 - u_2 + 3u_3 + u_4 &\geq 3 \\ 3u_1 + u_2 - 4u_3 + u_4 &\geq 5 \\ u_1, u_2, u_3, u_4 &\geq 0. \end{aligned}$$

4.2. Unsymmetric Duals

In this section we shall develop a variety of duality relationships and forms which obtain when the primal maximum problem involves structural constraints other than the " $\leq$ " variety and variables which are unrestricted in sign. In what follows our standard operating procedure will be to initially write the given primal problem in canonical form and then from it obtain the symmetric dual problem. To this end let us first assume that the primal problem appears as

$$\max f(\mathbf{X}) = \mathbf{C}'\mathbf{X} \quad \text{s.t.}$$
$$\mathbf{AX} = \mathbf{b}, \quad \mathbf{X} \geqq \mathbf{0}.$$

In this instance two sets of inequality structural constraints are implied by $\mathbf{AX} = \mathbf{b}$, *i.e.*,

$$\left.\begin{array}{l} \mathbf{AX} \leqq \mathbf{b} \\ \\ \mathbf{AX} \geqq \mathbf{b} \end{array}\right\} \text{ or } \left\{\begin{array}{l} \mathbf{AX} \leqq \mathbf{b} \\ \\ -\mathbf{AX} \leqq -\mathbf{b} \end{array}\right.$$

must hold simultaneously. Then the above primal problem has the canonical form

$$\max f(\mathbf{X}) = \mathbf{C}'\mathbf{X} \quad \text{s.t.}$$
$$\begin{bmatrix} \mathbf{A} \\ -\mathbf{A} \end{bmatrix} \mathbf{X} \leqq \begin{bmatrix} \mathbf{b} \\ -\mathbf{b} \end{bmatrix}, \quad \mathbf{X} \geqq \mathbf{0}.$$

The symmetric dual associated with this problem becomes

$$\min g(\mathbf{W}, \mathbf{V}) = \mathbf{b}'\mathbf{W} - \mathbf{b}'\mathbf{V} \quad \text{s.t.}$$
$$(\mathbf{A}', -\mathbf{A}') \begin{bmatrix} \mathbf{W} \\ \mathbf{V} \end{bmatrix} \geqq \mathbf{C} \text{ or } \mathbf{A}'\mathbf{W} - \mathbf{A}'\mathbf{V} \geqq \mathbf{C}; \ \mathbf{W}, \mathbf{V} \geqq \mathbf{0},$$

where the dual variables $\mathbf{W}, \mathbf{V}$ are associated with the primal structural constraints $\mathbf{AX} \leqq \mathbf{b}$, $-\mathbf{AX} \leqq -\mathbf{b}$ respectively. Upon combining terms this dual problem appears as

$$\begin{aligned} &min\, g(\mathbf{W},\mathbf{V}) = \mathbf{b}'(\mathbf{W}-\mathbf{V}) \;\text{ s.t.} \\ &\mathbf{A}'(\mathbf{W}-\mathbf{V}) \geqq \mathbf{C};\;\; \mathbf{W},\mathbf{V} \geqq \mathbf{0}. \end{aligned}$$

If $\mathbf{U} = \mathbf{W}-\mathbf{V}$, then we ultimately obtain

$$\begin{aligned} &min\, g(\mathbf{U}) = \mathbf{b}'\mathbf{U} \;\text{ s.t.} \\ &\mathbf{A}'\mathbf{U} \geqq \mathbf{C},\; \mathbf{U} \text{ unrestricted} \end{aligned}$$

since if both $\mathbf{W},\mathbf{V} \geqq \mathbf{0}$, their difference may be greater than or equal to, less than or equal to, or exactly equal to the null vector.

Next, suppose the primal problem appears as

$$\begin{aligned} &max\; f(\mathbf{X}) = \mathbf{C}'\mathbf{X} \;\text{ s.t.} \\ &\mathbf{A}\mathbf{X} \geqq \mathbf{b},\;\; \mathbf{X} \geqq \mathbf{0}. \end{aligned}$$

In canonical form we seek to

$$\begin{aligned} &max\; f(\mathbf{X}) = \mathbf{C}'\mathbf{X} \;\text{ s.t.} \\ &-\mathbf{A}\mathbf{X} \leqq -\mathbf{b},\;\; \mathbf{X} \geqq \mathbf{0}. \end{aligned}$$

The symmetric dual associated with this problem is

$$\begin{aligned} &min\, g(\mathbf{V}) = -\mathbf{b}'\mathbf{V} \;\text{ s.t.} \\ &-\mathbf{A}'\mathbf{V} \geqq \mathbf{C},\;\; \mathbf{V} \geqq \mathbf{0}. \end{aligned}$$

If $\mathbf{U} = -\mathbf{V}$, then we finally

$$\begin{aligned} &min\; g(\mathbf{U}) = \mathbf{b}'\mathbf{U} \;\text{ s.t.} \\ &\mathbf{A}'\mathbf{U} \geqq \mathbf{C},\;\; \mathbf{U} \leqq \mathbf{0}. \end{aligned}$$

Lastly, if the primal problem is of the form

$$\begin{aligned} &max\; f(\mathbf{X}) = \mathbf{C}'\mathbf{X} \;\text{ s.t.} \\ &\mathbf{A}\mathbf{X} \leqq \mathbf{b},\;\; \mathbf{X} \text{ unrestricted}, \end{aligned}$$

then $\mathbf{X}$ may by expressed as the difference between two vectors $\mathbf{X}_1,\mathbf{X}_2$ whose components are all nonnegative. Hence the primal problem becomes

$$\begin{aligned} &max\; f(\mathbf{X}_1,\mathbf{X}_2) = \mathbf{C}'(\mathbf{X}_1-\mathbf{X}_2) \;\text{ s.t.} \\ &\mathbf{A}(\mathbf{X}_1-\mathbf{X}_2) \leqq \mathbf{b};\;\; \mathbf{X}_1,\mathbf{X}_2 \geqq \mathbf{0} \end{aligned}$$

or

$$max\ f(\mathbf{X}_1, \mathbf{X}_2) = (\mathbf{C}', -\mathbf{C}')\begin{bmatrix}\mathbf{X}_1\\ \mathbf{X}_2\end{bmatrix} \quad \text{s.t.}$$

$$(\mathbf{A}, -\mathbf{A})\begin{bmatrix}\mathbf{X}_1\\ \mathbf{X}_2\end{bmatrix} \leqq \mathbf{b}; \quad \mathbf{X}_1, \mathbf{X}_2 \geqq \mathbf{0}.$$

The symmetric dual of this latter problem is

$$min\, g(\mathbf{U}) = \mathbf{b}'\mathbf{U} \quad \text{s.t.}$$

$$\begin{bmatrix}\mathbf{A}'\\ -\mathbf{A}'\end{bmatrix}\mathbf{U} \geqq \begin{bmatrix}\mathbf{C}\\ -\mathbf{C}\end{bmatrix} \quad \text{or} \quad \mathbf{A}'\mathbf{U} = \mathbf{C}, \quad \mathbf{U} \geqq \mathbf{0}.$$

To summarize, let us assume that the primal maximum problem contains a mixture of structural constraints. Then:

(1) If the primal variables are all nonnegative, the dual minimum problem has the following properties:

(a) the form of the dual structural constraint system is indepen dent of the form of the primal structural constraint system, *i.e.*, the former always appears as $\mathbf{A}'\mathbf{U} \geqq \mathbf{C}$.

Moreover, only the dual variables are influenced by the primal structural constraints. Specifically:

(b) the dual variable u_r corresponding to the rth "$\leq$" primal structural constraint is nonnegative;

(c) the dual variable u_s corresponding to the sth "$\geq$" primal structural constraint is nonpositive;

(d) the dual variable u_t corresponding to the tth "$=$" primal structural constraint is unrestricted in sign.

(2) If some variable x_j in the primal problem is unrestricted in sign, then the jth dual structural constraint will hold as an equality.

EXAMPLE 4.2. The dual of the primal problem

$$\begin{aligned} max\ f &= 3x_1 - x_2 + x_3 - x_4 \quad \text{s.t.}\\ x_1 + x_2 + 2x_3 + 3x_4 &\leq 5\\ x_3 - x_4 &\geq 4\\ x_1 - x_2 &= -1\\ x_1, x_2, x_4 \geq 0,\ & x_3 \text{ unrestricted} \end{aligned}$$

is

$$\begin{array}{llll} min\, g = 5u_1 + 4u_2 - u_3 & \text{s.t.} \\ u_1 \quad + u_3 \geq 3 \\ u_1 \quad - u_3 \geq -1 \\ 2u_1 + u_2 \quad = 1 \\ 3u_1 - u_2 \quad \geq -1 \\ u_1 \geq 0,\ u_2 \leq 0, \quad u_3 \text{ unrestricted.} \end{array}$$

4.3. Duality Theorems

To further enhance our understanding of the relationship between the primal and dual linear programming problems, let us consider the following sequence of duality theorems. First,

4.1. THEOREM. The dual of the dual is the primal problem itself.

PROOF. An equivalent representation of the dual minimum problem is

$$max\,\{-g(\mathbf{U}) = -\mathbf{b}'\mathbf{U}\} \quad \text{s.t.}$$
$$-\mathbf{A}'\mathbf{U} \leqq -\mathbf{C}, \quad \mathbf{U} \geqq \mathbf{0}.$$

The symmetric dual to this problem is

$$min\ h(\mathbf{X}) = -\mathbf{C}'\mathbf{X} \quad \text{s.t.}$$
$$-\mathbf{A}\mathbf{X} \geqq -\mathbf{b}, \quad \mathbf{X} \geqq \mathbf{0}$$

or

$$max\ \{-h(\mathbf{X}) = \mathbf{C}'\mathbf{X}\} \quad \text{s.t.}$$
$$\mathbf{A}\mathbf{X} \leqq \mathbf{b}, \quad \mathbf{X} \geqq \mathbf{0}. \qquad \text{Q.E.D.}$$

The implication of this theorem is that it is immaterial which problem is called the dual.

Our next two theorems address themselves to the relationship between the objective function values of the primal-dual pair of problems. First,

4.2. WEAK DUALITY THEOREM. If $\mathbf{X}_0$ is a feasible solution to the primal maximum problem and $\mathbf{U}_0$ is a feasible solution to the dual minimum problem, then $\mathbf{C}'\mathbf{X}_0 \leq \mathbf{b}'\mathbf{U}_0$.

PROOF. Since $\mathbf{A}\mathbf{X}_0 \leqq \mathbf{b}$, it follows that $\mathbf{U}_0'\mathbf{A}\mathbf{X}_0 \leq \mathbf{U}_0'\mathbf{b} = \mathbf{b}'\mathbf{U}_0$. Furthermore, with $\mathbf{A}'\mathbf{U}_0 \geqq \mathbf{C}$, $\mathbf{U}_0'\mathbf{A} \geqq \mathbf{C}'$ and thus $\mathbf{U}_0'\mathbf{A}\mathbf{X}_0 \geq \mathbf{C}'\mathbf{X}_0$. Combining the two results yields $\mathbf{b}'\mathbf{U}_0 \geq \mathbf{U}_0'\mathbf{A}\mathbf{X}_0 \geq \mathbf{C}'\mathbf{X}_0$. Q.E.D.

This theorem shows that a feasible solution to either problem yields a bound on the objective value of the other problem. Next,

4.3. SUFFICIENT OPTIMALITY CRITERION. If $\mathbf{X}_0, \mathbf{U}_0$ are feasible solu-tions to the primal maximum and dual minimum problems respectively and $\mathbf{C}'\mathbf{X}_0 = \mathbf{b}'\mathbf{U}_0$, then both $\mathbf{X}_0, \mathbf{U}_0$ are optimal solutions.

PROOF. Let $\bar{\mathbf{X}}$ be any other feasible solution to the primal problem. Then $\mathbf{C}'\bar{\mathbf{X}} \leq \mathbf{b}'\mathbf{U}_0 = \mathbf{C}'\mathbf{X}_0$ and thus $\mathbf{X}_0$ is an optimal solution to the primal problem. Similarly, if $\bar{\mathbf{U}}$ is any other feasible solution to the dual problem, then $\mathbf{b}'\bar{\mathbf{U}} \geq \mathbf{C}'\mathbf{X}_0 = \mathbf{b}'\mathbf{U}_0$ and thus $\mathbf{U}_0$ is an optimal solution to the dual problem. Q.E.D.

We now turn to the development of some background material which will serve to set the stage for our discussion of what are commonly called fundamental theorems of linear programming, namely the duality, existence, and complementary slackness theorems. Following Goldman and Tucker (1956), let us begin by considering the $m+p+1$ inequalities

$$\begin{aligned} -\mathbf{AX} + \mathbf{b} &\geqq \mathbf{0} \\ \mathbf{A}'\mathbf{U} \qquad\quad -\mathbf{C} &\geqq \mathbf{0} \\ -\mathbf{b}'\mathbf{U} + \mathbf{C}'\mathbf{X} \qquad &\geq 0, \end{aligned} \tag{4.2}$$

where the last inequality in this system has been obtained by reversing the sense of the inequality in $\mathbf{b}'\mathbf{U} \geq \mathbf{C}'\mathbf{X}$ (theorem 4.2). In matrix form, system (4.2) may be written as

$$\mathbf{KY} \geqq \mathbf{0}, \tag{4.3}$$

where

$$\mathbf{K} = \begin{bmatrix} \mathbf{0} & -\mathbf{A} & \mathbf{b} \\ \mathbf{A}' & \mathbf{0} & -\mathbf{C} \\ -\mathbf{b}' & \mathbf{C}' & 0 \end{bmatrix}, \quad \mathbf{Y} = \begin{bmatrix} \mathbf{U} \\ \mathbf{X} \\ 1 \end{bmatrix},$$

and $\mathbf{K}$ is skew-symmetric. If we replace the $m+p+1^{st}$ component of $\mathbf{Y}$ by the variable t, $\mathbf{KY} \geqq \mathbf{0}$ is transformed to the homogeneous system of inequalities

$$\mathbf{K}\begin{bmatrix}\mathbf{U}\\ \mathbf{X}\\ t\end{bmatrix} \geq \mathbf{0}. \tag{4.4}$$

Under what conditions will this modified combined primal-dual linear programming problem possess a nonnegative solution? According to the skew-symmetric theorem 2.7.a, there exists a vector $\overline{\mathbf{Y}}' = (\overline{\mathbf{U}}', \overline{\mathbf{X}}', \overline{t}) \geqq \mathbf{0}'$ such that

$$\left.\begin{array}{r}\mathbf{K}\overline{\mathbf{Y}} \geqq \mathbf{0}\\ \overline{\mathbf{Y}} + \mathbf{K}\overline{\mathbf{Y}} > \mathbf{0}\end{array}\right\} \quad \text{or}$$

$$-\mathbf{A}\overline{\mathbf{X}} + \mathbf{b}\overline{t} \geqq \mathbf{0} \tag{4.5}$$

$$\mathbf{A}'\overline{\mathbf{U}} - \mathbf{C}\overline{t} \geqq \mathbf{0} \tag{4.6}$$

$$-\mathbf{b}'\overline{\mathbf{U}} + \mathbf{C}'\overline{\mathbf{X}} \geq 0 \tag{4.7}$$

$$\overline{\mathbf{U}} - \mathbf{A}\overline{\mathbf{X}} + \mathbf{b}\overline{t} > \mathbf{0} \tag{4.8}$$

$$\mathbf{A}'\overline{\mathbf{U}} + \overline{\mathbf{X}} - \mathbf{C}\overline{t} > \mathbf{0} \tag{4.9}$$

$$-\mathbf{b}'\overline{\mathbf{U}} + \mathbf{C}'\overline{\mathbf{X}} + \overline{t} > 0. \tag{4.10}$$

Then, with $\overline{t} \geq 0$, there are two cases to consider: $\overline{t} > 0$ and $\overline{t} = 0$.

First,

4.4. THEOREM [Goldman, Tucker (1956)]. If $\overline{t} > 0$, then there exist optimal vectors $\widehat{\mathbf{X}}, \widehat{\mathbf{U}}$ for the primal maximum and dual minimum problems respectively such that $\mathbf{C}'\widehat{\mathbf{X}} = \mathbf{b}'\widehat{\mathbf{U}}$.

PROOF. With $\overline{t} > 0$, the homogeneous inequalities (4.5) - (4.10) may be transformed to

$$\left.\begin{array}{r}\mathbf{K}\widehat{\mathbf{Y}} \geqq \mathbf{0}\\ \widehat{\mathbf{Y}} + \mathbf{K}\widehat{\mathbf{Y}} > \mathbf{0}\end{array}\right\} \quad \text{or}$$

$$-\mathbf{A}\widehat{\mathbf{X}} + \mathbf{b} \geqq \mathbf{0} \tag{4.5.1}$$

$$\mathbf{A}'\widehat{\mathbf{U}} - \mathbf{C} \geqq \mathbf{0} \tag{4.6.1}$$

$$-\mathbf{b}'\widehat{\mathbf{U}} + \mathbf{C}'\widehat{\mathbf{X}} \geq 0 \tag{4.7.1}$$

$$\widehat{\mathbf{U}} - \mathbf{A}\widehat{\mathbf{X}} + \mathbf{b} > \mathbf{0} \tag{4.8.1}$$

$$\mathbf{A}'\widehat{\mathbf{U}} + \widehat{\mathbf{X}} - \mathbf{C} > \mathbf{0} \tag{4.9.1}$$

$$-\mathbf{b}'\widehat{\mathbf{U}} + \mathbf{C}'\widehat{\mathbf{X}} + 1 > 0, \tag{4.10.1}$$

where

$$\hat{\mathbf{Y}} = \begin{bmatrix} \hat{\mathbf{U}} \\ \hat{\mathbf{X}} \\ 1 \end{bmatrix} = \begin{bmatrix} \bar{\mathbf{U}}/\bar{t} \\ \bar{\mathbf{X}}/\bar{t} \\ 1 \end{bmatrix}.$$

Then (4.5.1), (4.6.1) indicate that $\hat{\mathbf{X}}$, $\hat{\mathbf{U}}$ are feasible, (4.7.1) and theorem 4.2 imply that $\mathbf{C}'\hat{\mathbf{X}} = \mathbf{b}'\hat{\mathbf{U}}$, and thus, by virtue of theorem 4.3, the optimality of $\hat{\mathbf{X}}$ and $\hat{\mathbf{U}}$. Q.E.D.

Next,

4.5. THEOREM [Goldman, Tucker (1956)]. If $\bar{t}=0$, then one of the following mutually exclusive cases holds: (a) at least one of the dual problems has no feasible solution; (b) if the primal maximum problem has a feasible solution, then its set of feasible solutions is unbounded and $\mathbf{C}'\mathbf{X}$ is unbounded from above. Dually, if the minimum problem has a feasible solution, then its set of feasible solutions is unbounded and $\mathbf{b}'\mathbf{U}$ is unbounded from below; and (c) neither the primal nor the dual problem has an optimal solution.

PROOF. To prove (a), let $\bar{\mathbf{X}}$ represent a feasible solution to the primal maximum problem. From (4.6) (with $\bar{t}=0$) $\mathbf{A}'\bar{\mathbf{U}} \geqq \mathbf{0}$ and thus, with $\bar{\mathbf{X}} \geqq \mathbf{0}$, $\bar{\mathbf{X}}'\mathbf{A}'\bar{\mathbf{U}} \geq 0$. This last inequality, along with (4.10) ($\bar{t}=0$) and $\bar{\mathbf{X}} \geqq \mathbf{0}$, yields

$$\mathbf{C}'\bar{\mathbf{X}} > \mathbf{b}'\bar{\mathbf{U}} \geq \bar{\mathbf{X}}'\mathbf{A}'\bar{\mathbf{U}} \geq 0 \tag{4.11}$$

(since when $\mathbf{b} \geqq \mathbf{A}\bar{\mathbf{X}}, \mathbf{b}' \geqq \bar{\mathbf{X}}'\mathbf{A}'$ and thus $\mathbf{b}'\bar{\mathbf{U}} \geq \bar{\mathbf{X}}'\mathbf{A}'\bar{\mathbf{U}}$). If the dual minimum problem also has a feasible solution $\bar{\mathbf{U}}$, (4.5) ($\bar{t}=0$) and $\bar{\mathbf{U}} \geqq \mathbf{0}$ imply that $\mathbf{C}'\bar{\mathbf{X}} \leq \bar{\mathbf{U}}'\mathbf{A}\bar{\mathbf{X}} \leq 0$ (since if $\mathbf{A}'\bar{\mathbf{U}} \geqq \mathbf{C}, \bar{\mathbf{U}}'\mathbf{A} \geqq \mathbf{C}'$ and thus $\bar{\mathbf{U}}'\mathbf{A}\bar{\mathbf{X}} \geq \mathbf{C}'\bar{\mathbf{X}}$). This last set of inequalities clearly contradicts (4.11). In sum, it is not possible for both the primal and dual problems to have a feasible solution when $\bar{t}=0$.

To prove (b) let us construct the infinite ray $\mathbf{X}_* + \lambda\bar{\mathbf{X}} \geq \mathbf{0}$, $0 \leq \lambda \in \mathbf{R}$, where $\mathbf{X}_*$ represents a feasible solution to the primal maximum problem and $\bar{\mathbf{X}}$ satisfies (4.5)-(4.10). From (4.5) (with $\bar{t}=0$) we obtain $\mathbf{A}\bar{\mathbf{X}} \leqq \mathbf{0}$ and thus, for $\lambda > 0$, $\mathbf{A}(\lambda\bar{\mathbf{X}}) \leqq \mathbf{0}$. With $\mathbf{X}_*$ feasible, $\mathbf{A}\mathbf{X}_* \leqq \mathbf{b}$. Upon combining these last two inequalities it is easily seen that $\mathbf{A}(\mathbf{X}_*+\lambda\bar{\mathbf{X}}) \leqq \mathbf{A}\mathbf{X}_* \leqq \mathbf{b}$. Thus the entire infinite ray consists of feasible solutions, indicating that as λ increases without bound, $\mathbf{X}_* + \lambda\bar{\mathbf{X}}$ remains feasible. Turning now to the latter part of the first assertion in (b), with $\mathbf{C}'\bar{\mathbf{X}} > 0$ (by (4.11)), it follows from the feasibility of $\mathbf{X}_*$ that $\mathbf{C}'(\mathbf{X}_*+\lambda\bar{\mathbf{X}}) = \mathbf{C}'\mathbf{X}_* + \lambda\mathbf{C}'\bar{\mathbf{X}}$ can be made arbitrarily large by choosing an appropriate value of λ. A symmetric argument holds for the dual feasible region and objective function. So for $\bar{t} = 0$, whether the primal or dual problem has a feasible solution, its objective function is unbounded.

That (c) is true follows as a direct consequence of (b). Q.E.D.

At this point two immediate corollaries to theorems 4.4, 4.5 emerge. First,

4.6. COROLLARY [Goldman, Tucker (1956)]. Either boththe primal maximum and dual minimum problems have optimal vectors $\widehat{\mathbf{X}}$ and $\widehat{\mathbf{U}}$ respectively with $\mathbf{C}'\widehat{\mathbf{X}} = \mathbf{b}'\widehat{\mathbf{U}}$, or neither has an optimal vector.

PROOF. If one of these problems has an optimal vector, then part (c) of theorem 4.4 implies that $\bar{t} > 0$, to wit theorem 4.3 states that both problems have optimal vectors $\widehat{\mathbf{X}}, \widehat{\mathbf{U}}$ with $\mathbf{C}'\widehat{\mathbf{X}} = \mathbf{b}'\widehat{\mathbf{U}}$. Q.E.D.

In addition,

4.7. COROLLARY [Goldman, Tucker (1956)]. One (and thus both) of the primal maximum and dual minimum problems has an optimal solution if and only if either $\mathbf{C}'\mathbf{X}$ or $\mathbf{b}'\mathbf{U}$ is bounded on its associated nonempty feasible region.

PROOF. (necessity) If $\widehat{\mathbf{X}}$ ($\widehat{\mathbf{U}}$) represents an optimal solution to the primal maximum (dual minimum) problem, then clearly $\mathbf{C}'\mathbf{X}$ ($\mathbf{b}'\mathbf{U}$) cannot be unbounded over $\mathcal{K}_P \neq \phi$ ($\mathcal{K}_D \neq \phi$). (sufficiency) If the primal maximum (dual minimum) problem is such that $\mathbf{C}'\mathbf{X}$ ($\mathbf{b}'\mathbf{U}$) is bounded over $\mathcal{K}_P \neq \phi$ ($\mathcal{K}_D \neq \phi$), then part (a) of theorem 4.5 implies that $\bar{t} > 0$ so that by theorem 4.4, both problems have optimal vectors. Q.E.D.

Having determined the conditions under which (4.4) possesses a nonnegative solution, let us now look to a discussion of the first two aforementioned fundamental theorems of linear programming. First,

4.8. STRONG DUALITY THEOREM [Goldman, Tucker (1956)]. A feasible solution $\widehat{\mathbf{X}}$ to the primal maximum problem is optimal if and only if the dual minimum problem has a feasible solution $\widehat{\mathbf{U}}$ with $\mathbf{b}'\widehat{\mathbf{U}} = \mathbf{C}'\widehat{\mathbf{X}}$. (Likewise, a feasible solution $\widehat{\mathbf{U}}$ is optimal if and only if a feasible $\widehat{\mathbf{X}}$ exists for which $\mathbf{C}'\widehat{\mathbf{X}} = \mathbf{b}'\widehat{\mathbf{U}}$.)

PROOF. (necessity) If $\widehat{\mathbf{X}}$ is optimal, then by part (c) of theorem 4.5, $\bar{t} > 0$. With $\bar{t} > 0$, theorem 4.4 insures us that the dual problem also has an optimal solution $\widehat{\mathbf{U}}$ such that $\mathbf{b}'\widehat{\mathbf{U}} = \mathbf{C}'\widehat{\mathbf{X}}$. (suffi ciency) If $\widehat{\mathbf{U}}$ is a feasible solution to the dual problem and $\mathbf{b}'\widehat{\mathbf{U}} = \mathbf{C}'\widehat{\mathbf{X}}$, then theorem 4.3 indicates that $\widehat{\mathbf{U}}$, $\widehat{\mathbf{X}}$ are optimal. The proof of the parenthetical statement parallels that of the first. Q.E.D.

The importance of this theorem is that it allows us to conclude, for instance, that:

> if $\widehat{\mathbf{X}}$ is an optimal solution for the primal problem, then there exists an optimal solution $\widehat{\mathbf{U}}$ for the dual problem and $\widehat{f} = \mathbf{C}'\widehat{\mathbf{X}} = \mathbf{b}'\widehat{\mathbf{U}} = \widehat{g}$.

Conversely,

> if $\widehat{\mathbf{U}}$ is an optimal solution for the dual problem, then there exists an optimal solution $\widehat{\mathbf{X}}$ for the primal problem and again $\widehat{f} = \widehat{g}$.

Looked at in another fashion, the sufficient optimality condition 4.3 established the result that if a pair of feasible solutions to the primal and dual problems render equal objective function values, then these solutions are also optimal. The strong duality theorem 4.8 posits that the converse of the sufficient optimality condition is also true, *i.e.*, if either the primal or dual program has a finite optimal solution, then so does the other and their optimal objective function values are the same. This latter point is easily illustrated in Figure 4.1. Here the primal and dual objective functions approach the same optimal value A but from

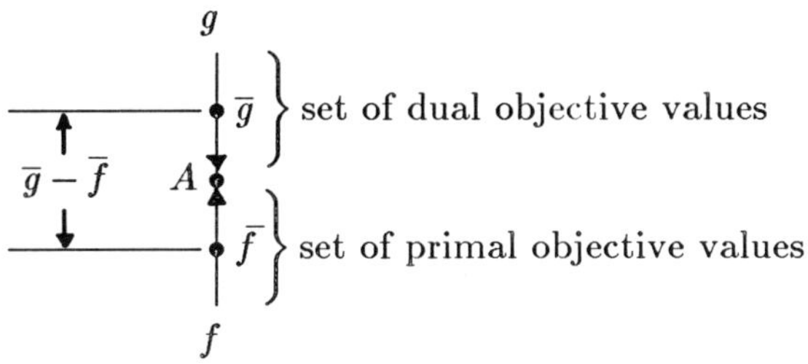

Figure 4.1

different directions (f approaches optimality in a monotonically increasing fashion while g approaches the same along a monotonically worsening path). Thus given primal and dual feasible points $\bar{\mathbf{X}} \epsilon \mathcal{K}_P$ and $\bar{\mathbf{U}} \epsilon \mathcal{K}_D$ respectively with $\bar{f}, \bar{g}$ their corresponding objective values, we know that $\bar{g}-\bar{f} > 0$ by virtue of the weak duality theorem. If $\bar{g}-\bar{f} = 0$ ($\bar{f}=\bar{g}=A$), then $\bar{\mathbf{X}}$ and $\bar{\mathbf{U}}$ are also optimal according to the sufficient optimality condition. Conversely, according to the strong duality theorem, if $\bar{\mathbf{X}}$ yields $\bar{f}=A$, then $\bar{\mathbf{U}}$ must give $\bar{g}=A$ also and $\bar{g}-\bar{f} = 0$ — there is no **duality gap.**

The above proof of the strong duality theorem 4.8 relied upon so-called *classical results of linear algebra*, *i.e.*, it utilized a fundamental result (the skew-symmetric theorem) characterizing the existence of a nonnegative solution for certain dual systems of linear inequalities. Other classical proofs of the strong duality theorem are provided by Shapiro (1979) (Farkas' theorem is applied to a system of primal-dual nonhomogeneous equations) and Luenberger (1984) (a strong separation theorem is used to establish the existence of a hyperplane separating a closed convex set and a point not in the set).

Since the strong duality theorem 4.8 will be of considerable importance in our subsequent discussions of linear programming, a variety of alternative proofs will follow as appropriate. For instance, a constructive proof which also defines the vector of dual variables will be offered in the next chapter wherein we obtain an optimality criterion for the primal problem based upon the simplex method. In addition, another proof based upon saddle point optimality conditions will also be presented.

Up to now in our discussion of the connection between the primal and dual problems we have primarily addressed situations in which both of these problems have feasible as well as optimal solutions. However, it is important to specify exactly what happens to the relationship between the primal-dual pair of problems if an infeasible or unbounded solution exists in either of these programs. To this end we state

4.9. EXISTENCE THEOREM [Goldman, Tucker (1956)].

(a) The primal maximum and dual minimum problems possess finite optimal solutions if and only if both have feasible solutions.

(b) If the primal (dual) objective is unbounded from above (below), then the dual minimum (primal maximum) problem does not possess a feasible solution.

(c) If the dual minimum (primal maximum) problem is infeasible but the primal maximum (dual minimum) problem has a feasible solution, then the primal (dual) objective is unbounded from above (below).

(Here (a) can be thought of as a corollary to the strong duality theorem while (b), (c) can be considered as corollaries to the weak duality theorem.)

PROOF. (a) (necessity) If the primal problem has an optimal feasible solution, then the dual problem has a feasible solution by virtue of the strong duality theorem. (sufficiency) If $\widehat{\mathbf{X}}, \widehat{\mathbf{U}}$ are feasible solutions to the primal and dual problems respectively, then both $\mathbf{C}'\widehat{\mathbf{X}}, \mathbf{b}'\widehat{\mathbf{U}}$ are finite with $\mathbf{b}'\widehat{\mathbf{U}} \geq \mathbf{C}'\widehat{\mathbf{X}}$ (the weak duality theorem). And since $\mathbf{b}'\widehat{\mathbf{U}}$ serves as an upper bound on $\mathbf{C}'\mathbf{X}$ for any feasible $\mathbf{X}$, the primal problem must possess a finite maximum.

(b) If the primal objective is unbounded, then the primal problem has a feasible

solution. If the dual problem also has a feasible solution, then according to part (a) above, the primal problem must have a finite optimum — clearly a contradiction. (A similar proof holds if it is assumed that the dual objective is unbounded.)
(c) A feasible solution $\hat{\mathbf{X}}$ to the primal problem cannot also be optimal since, if it were, a feasible solution $\hat{\mathbf{U}}$ to the dual problem would have to exist by virtue of the strong duality theorem — again a contradiction. But if no feasible solution to the primal problem can be maximal, then clearly the primal objective must be unbounded from above. (A similar proof can be offered if it is assumed that the primal problem is infeasible.) Q.E.D.

The contents of the duality and existence theorems may be conveniently summarized by Table 4.1. As indicated therein, either both the primal maximum and dual minimum problems have optimal vectors $\hat{\mathbf{X}}, \hat{\mathbf{U}}$ respectively,

Table 4.1

		Dual Minimum Problem	
		Feasible	Infeasible
Primal Maximum Problem	Feasible	$\hat{\mathbf{X}}, \hat{\mathbf{U}}$ are optimal with $\mathbf{C}'\hat{\mathbf{X}} = \mathbf{b}'\hat{\mathbf{U}}$	$\mathbf{C}'\mathbf{X}$ unbounded from above
	Infeasible	$\mathbf{b}'\mathbf{U}$ unbounded from below	No solution exists

with $\mathbf{C}'\hat{\mathbf{X}} = \mathbf{b}'\hat{\mathbf{U}}$, or neither has an optimal vector. In addition, one (and thus both) of the primal maximum and dual minimum problems has an optimal solution if and only if either the primal or dual objective is bounded on its associated nonempty feasible region.

EXAMPLE 4.3. The optimal solution to the primal problem

$$max\ f = 24x_1 + 25x_2 \quad \text{s.t.}$$

$$\begin{aligned} x_1 + 5x_2 &\leq 1 \\ 4x_1 + \ \ x_2 &\leq 3 \\ x_1, x_2 &\geq 0 \end{aligned}$$

is $\mathbf{X}'_0 = \left(\frac{14}{19}, \frac{1}{19}\right)$ and $f^0 = 19$ (Figure 4.2.a) while the same for the dual problem

$$
\begin{aligned}
min\, g &= u_1 + 3u_2 \quad \text{s.t.} \\
u_1 + 4u_2 &\geq 24 \\
5u_1 + \ u_2 &\geq 25 \\
u_1, u_2 &\geq 0
\end{aligned}
$$

is $\mathbf{U}'_0 = (4,5)$ and $g^0 = 19$ (Figure 4.2.b). Here both the primal and dual problems possess optimal vectors $\mathbf{X}_0, \mathbf{U}_0$ respectively with $f^0 = g^0$, as required by theorem 4.8.

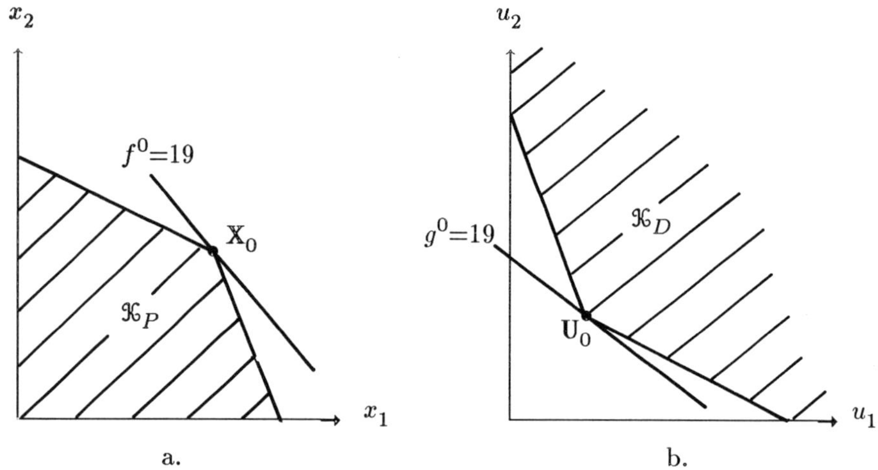

Figure 4.2

EXAMPLE 4.4. As Figure 4.3.a reveals, the primal problem

$$
\begin{aligned}
max\ f &= 2x_1 + 3x_2 \quad \text{s.t.} \\
-2x_1 + x_2 &\leq 3 \\
-\tfrac{1}{2}x_1 + x_2 &\leq 6 \\
x_1, x_2 &\geq 0
\end{aligned}
$$

has an unbounded objective function. From our preceding discussion we should anticipate that its dual,

$$\begin{aligned} \min g = 3u_1 + 6u_2 \quad & \text{s.t.} \\ -2u_1 - \tfrac{1}{2}u_2 &\geq 2 \\ u_1 + \ \ u_2 &\geq 3 \\ u_1, u_2 &\geq 0, \end{aligned}$$

does not possess a feasible solution, as Figure 4.3.b clearly indicates.

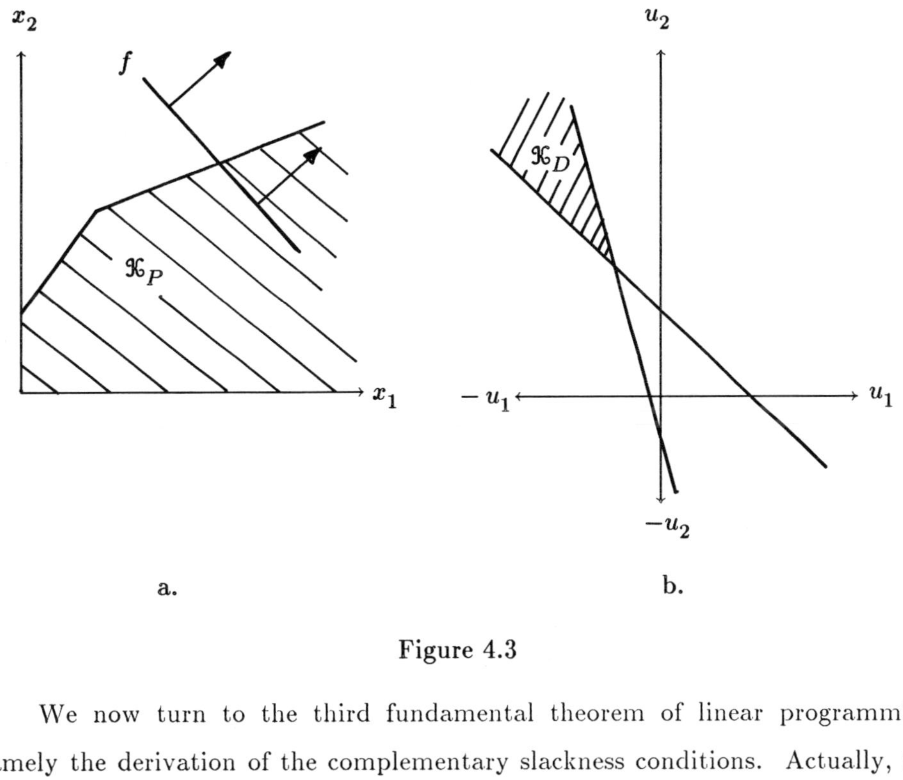

a. b.

Figure 4.3

We now turn to the third fundamental theorem of linear programming, namely the derivation of the complementary slackness conditions. Actually, two such theorems will be presented — the strong and weak cases. First,

4.10. WEAK COMPLEMENTARY SLACKNESS THEOREM.

The solutions $(\mathbf{X}_0, (\mathbf{X}_s)_0), (\mathbf{U}_0, (\mathbf{U}_s)_0)$ to the primal maximum and dual minimum problems respectively are optimal if and only if they satisfy the complementary slackness conditions

$$(\mathbf{U}_s)_0'\mathbf{X}_0 + \mathbf{U}_0'(\mathbf{X}_s)_0 = 0 \text{ or } \begin{cases} u_i^0\, x_{p+i}^0 = 0, & i=1,\ldots,m, \\ x_j^0\, u_{m+j}^0 = 0, & j=1,\ldots,p, \text{i.e.,} \end{cases} \tag{4.12}$$

for optimal solutions $(\mathbf{X}_0, (\mathbf{X}_s)_0), (\mathbf{U}_0, (\mathbf{U}_s)_0)$ to the primal maximum and dual minimum problems respec tively:

(a) if a primal slack variable x_{p+i}^0, $i=1,\ldots,m$, is positive, then the ith dual variable $u_i^0, i=1,\ldots,m$, must equal zero. Conversely, if u_i^0, $i=1,\ldots,m$, is positive, then x_{p+i}^0, $i=1,\ldots,m$, equals zero;

(b) if the jth primal variable $x_j^0, j=1,\ldots,p$, is positive, then the dual surplus variable u_{m+j}^0, $j=1,\ldots,p$, equals zero. Conversely, if u_{m+j}^0, $j=1,\ldots,p$, is positive, then x_j^0, $j=1,\ldots,p$, must equal zero.

PROOF. (necessity) The augmented structural con straint systems of the primal and dual problems were written earlier as

$$\mathbf{AX} + \mathbf{I}_m\mathbf{X}_s = \mathbf{b},\ \mathbf{A}'\mathbf{U} - \mathbf{I}_p\mathbf{U}_s = \mathbf{C}$$

respectively. If the first of these equalities is multiplied by $\mathbf{U}'$, we obtain $\mathbf{U}'(\mathbf{AX} + \mathbf{I}_m\mathbf{X}_s) = \mathbf{U}'\mathbf{b}$. Subtracting this expression from the primal objective function yields

$$\begin{aligned} \mathbf{C}'\mathbf{X} - \mathbf{U}'(\mathbf{AX} + \mathbf{I}_m\mathbf{X}_s) &= f(\mathbf{X}) - \mathbf{U}'\mathbf{b} \qquad \text{or} \\ \mathbf{C}'\mathbf{X} - \mathbf{U}'\mathbf{AX} - \mathbf{U}'\mathbf{X}_s &= (\mathbf{C}' - \mathbf{U}'\mathbf{A})\mathbf{X} - \mathbf{U}'\mathbf{X}_s \\ &= -\mathbf{U}_s'\mathbf{X} - \mathbf{U}'\mathbf{X}_s = f(\mathbf{X}) - \mathbf{U}'\mathbf{b}. \end{aligned}$$

Since $\mathbf{U}'\mathbf{b}$ is a scalar, this last equation may be rewritten as

$$\mathbf{U}_s'\mathbf{X} + \mathbf{U}'\mathbf{X}_s = g(\mathbf{U}) - f(\mathbf{X}).$$

Given that both the primal and dual problems have optimal solutions, the preceding expression becomes

$$(\mathbf{U}_s)_0'\mathbf{X}_0 + \mathbf{U}_0'(\mathbf{X}_s)_0 = g(\mathbf{U}_0) - f(\mathbf{X}_0)$$

and thus, by virtue of theorem 4.8,

$$\begin{aligned} (\mathbf{U}_s)_0'\mathbf{X}_0 + \mathbf{U}_0'(\mathbf{X}_s)_0 &= 0 \qquad \text{or} \\ u_i^0\, x_{p+i}^0 &= 0,\ i=1,\ldots,m, \\ x_j^0\, u_{m+j}^0 &= 0, j=1,\ldots,p. \end{aligned}$$

(sufficiency) If both sides of the primal and dual augmented structural constraint equalities are multiplied by $\mathbf{U}_0', \mathbf{X}_0'$ respectively, with $(\mathbf{U}_s)_0'\mathbf{X}_0 = \mathbf{U}_0'(\mathbf{X}_s)_0 = 0$, then $\mathbf{U}_0'(\mathbf{A}\mathbf{X}_0) = \mathbf{U}_0'\mathbf{b}$ and $\mathbf{X}_0'(\mathbf{A}'\mathbf{U}_0) = \mathbf{X}_0'\mathbf{C}$. Transposing this latter equality thus yields $\mathbf{U}_0'\mathbf{A}\mathbf{X}_0 = \mathbf{C}'\mathbf{X}_0 = \mathbf{b}'\mathbf{U}_0$. Hence $\mathbf{X}_0, \mathbf{U}_0$ represent optimal solutions. Q.E.D.

How may we interpret (4.12)? Specifically, $u_i^0 x_{p+i}^0 = 0,\ i=1,\ldots,m$, implies that either $x_{p+i}^0 > 0$ and $u_i^0 = 0$ or $u_i^0 > 0$ and $x_{p+i}^0 = 0$ or both $u_i^0,\ x_{p+i}^0 = 0$. Similarly, $x_j^0 u_{m+j}^0 = 0, j=1,\ldots,p$, implies that either $x_j^0 > 0$ and $u_{m+j}^0 = 0$ or $u_{m+j}^0 > 0$ and $x_j^0 = 0$ or both $x_j^0,\ u_{m+j}^0 = 0$. Alternatively, if the constraint system for the primal-dual pair of problems is written as

$$
\text{Primal Constraints}\left\{
\begin{array}{l}
\left.\begin{array}{c} x_1 \geq 0 \\ \vdots \\ x_p \geq 0 \end{array}\right\}
\underset{\longleftarrow}{\overset{\longrightarrow}{\text{Complementary Slack}}}
\left\{\begin{array}{c} a_{11}u_1+\cdots+a_{m1}u_m \geq c_1 \\ \cdots\cdots\cdots\cdots \\ a_{1p}u_1+\cdots+a_{mp}u_m \geq c_p \end{array}\right. \\[2ex]
\left.\begin{array}{c} a_{11}x_1+\cdots+a_{1p}x_p \leq b_1 \\ \cdots\cdots\cdots\cdots \\ a_{m1}x_1+\cdots+a_{mp}x_p \leq b_m \end{array}\right\}
\underset{\longleftarrow}{\overset{\longrightarrow}{\text{Complementary Slack}}}
\left\{\begin{array}{c} u_1 \geq 0 \\ \vdots \\ u_m \geq 0 \end{array}\right.
\end{array}
\right\}\text{Dual Constraints}
$$

and (4.12) is rewritten as

$$
\begin{aligned}
u_i^0\left(b_i-\sum_{j=1}^{p} a_{ij}x_j^0\right) &= 0, \quad i=1,\ldots,m, \\
x_j^0\left(\sum_{i=1}^{m} a_{ij}u_i^0-c_j\right) &= 0, \quad j=1,\ldots,p,
\end{aligned}
\tag{4.12.1}
$$

it is easily seen that if the kth dual variable is positive, then its complementary primal structural constraint is binding (*i.e.*, holds as an equality). Conversely, if the kth primal structural constraint holds with strict inequality, then its complementary dual variable is zero. Likewise, if the ℓth primal variable is positive, then its complementary dual structural constraint is binding. Conversely, if the ℓth dual structural constraint holds as a strict inequality, its complementary primal variable is zero. In short, $x_j^0 \geq 0$ and $b_i - \sum_{j=1}^{p} a_{ij}x_j^0 \geq 0$ as well as $u_i^0 \geq 0$ and $\sum_{i=1}^{m} a_{ij}u_i^0 - c_j \geq 0$ are complementary slack since within each pair of inequalities at least one of them must hold as an equality.

As a final observation, the geometric interpretation of the weak complementary slackness conditions is that at an optimal solution to the primal-dual pair of problems, the vector of primal variables $\mathbf{X}_0$ is orthogonal to the vector of dual surplus variables $(\mathbf{U}_s)_0$ while the vector of dual variables $\mathbf{U}_0$ is orthogonal to the vector of primal slack variables $(\mathbf{X}_s)_0$.

It is important to note that the weak complementary slackness conditions (4.12), (4.12.1) must hold for every pair of optimal solutions to the primal and dual problems. However, it may happen that, say, for some particular value of $j(i)$, both $x_j^0(u_i^0)$ and u_{m+j}^0 (x_{p+i}^0) vanish at the said solutions. In this instance, can we be sure that there exists at least one pair of optimal solutions to the primal and dual problems for which this cannot occur? The answer is in the affirmative, as the following theorem demonstrates.

4.11. STRONG COMPLEMENTARY SLACKNESS THEOREM. For optimal solutions $(\mathbf{X}_0,(\mathbf{X}_s)_0)$, $(\mathbf{U}_0,(\mathbf{U}_s)_0)$ to the primal maximum and dual minimum problems respectively: (a) if the primal slack variable $x_{p+i}^0, i=1,\ldots,m$, equals zero, then the ith dual variable $u_i^0, i=1,\ldots,m$, is positive. Conversely, if $u_i^0, i=1,\ldots,m$, equals zero, then $x_{p+i}^0, i=1,\ldots,m$, is positive; (b) if the jth primal variable $x_j^0, j=1,\ldots,p$, equals zero, then the dual surplus variable $u_{m+j}^0, j=1,\ldots,p$, is positive. Conversely, if $u_{m+j}^0, j=1,\ldots,p$, equals zero, then $x_j^0, j=1,\ldots,p$, is positive. Parts (a), (b) taken together may be summarized as:

$$\left.\begin{matrix}(\mathbf{X}_s)_0+\mathbf{U}_0>\mathbf{0}\\(\mathbf{U}_s)_0+\mathbf{X}_0>\mathbf{0}\end{matrix}\right\}\quad\text{or}\quad\left\{\begin{matrix}u_i^0+x_{p+i}^0>0,\ i=1,\ldots,m,\\x_j^0+u_{m+j}^0>0,\ j=1,\ldots,p.\end{matrix}\right.\tag{4.13}$$

PROOF. Given that both the primal and dual problems have optimal solutions, part (c) of theorem 4.5 implies that $\bar{t}>0$. Then from (4.8.1), (4.9.1),

$$\mathbf{U}_0+\mathbf{b}-\mathbf{A}\mathbf{X}_0>\mathbf{0}$$
$$\mathbf{X}_0+\mathbf{A}'\mathbf{U}_0-\mathbf{C}>\mathbf{0}$$

and thus

$$\mathbf{U}_0 + (\mathbf{X}_s)_0 > \mathbf{0},$$
$$\mathbf{X}_0 + (\mathbf{U}_s)_0 > \mathbf{0} \quad \text{or}$$

$$u_i^0 + x_{p+i}^0 > 0,\ i=1,\ldots,m,$$
$$x_j^0 + u_{m+j}^0 > 0,\ j=1,\ldots,p. \quad \text{Q.E.D.}$$

By way of interpretation: $u_i^0 + x_{p+i}^0 > 0$, $i=1,\ldots,m$, implies that either $u_i^0 = 0$ and $x_{p+i}^0 > 0$ or $x_{p+i}^0 = 0$ and $u_i^0 > 0$, but not both $u_i^0, x_{p+i}^0 = 0$; while $x_j^0 + u_{m+j}^0 > 0$, $j=1,\ldots,p$, implies that either $x_j^0 > 0$ and $u_{m+j}^0 = 0$ or $u_{m+j}^0 > 0$ and $x_j^0 = 0$, but not both $x_j^0, u_{m+j}^0 = 0$.

Let us now establish the connection between the weak and strong complementary slackness conditions by first rewriting (4.13) as

$$\begin{aligned} u_i^0 + \Big(b_i - \sum_{j=1}^{p} a_{ij}x_j^0\Big) &> 0,\ i=1,\ldots,m, \\ x_j^0 + \Big(\sum_{i=1}^{m} a_{ij}u_i^0 - c_j\Big) &> 0,\ j=1,\ldots,p. \end{aligned} \tag{4.13.1}$$

Then, for the strong case, if the kth dual variable is zero, then its complementary primal structural constraint holds as a strict inequality. Conversely, if the kth primal structural constraint is binding, then its complementary dual variable is positive. Note that this situation cannot prevail under the weak case since if $u_i^0 = 0\Big(b_i - \sum_{j=1}^{p} a_{ij}x_j^0 = 0\Big)$, it does not follow that $b_i - \sum_{j=1}^{p} a_{ij}x_j^0 > 0 (u_i^0 > 0)$, *i.e.*, under the weak conditions, $u_i^0 = 0\Big(b_i - \sum_{j=1}^{p} a_{ij}x_j^0 = 0\Big)$ implies that $b_i - \sum_{j=1}^{p} a_{ij}x_j^0 \geq 0 (u_i^0 \geq 0)$. Similarly (again considering the strong case), if the ℓth primal variable is zero, it follows that its complementary dual structural constraint holds with strict inequality. Conversely, if the ℓth dual structural constraint is binding, it follows that its complementary primal variable is positive. Here too a comparison of the strong and weak cases indicates that, for the latter, if $x_j^0 = 0\Big(\sum_{i=1}^{m} a_{ij}u_j^0 - c_j = 0\Big)$, we may conclude that $\sum_{i=1}^{m} a_{ij}u_j^0 - c_j \geq 0 (x_j^0 \geq 0)$ but not that $\sum_{i=1}^{m} a_{ij}u_j^0 - c_j > 0\ (x_j^0 > 0)$ as in the former.

To summarize:

<u>Weak Complementary Slackness</u>

$u_i^0 > 0$ implies $b_i - \sum_{j=1}^{p} a_{ij} x_j^0 = 0$

and

$b_i - \sum_{j=1}^{p} a_{ij} x_j^0 > 0$ implies $u_i^0 = 0$,

$i = 1, \ldots, m$; while

$x_j^0 > 0$ implies $\sum_{i=1}^{m} a_{ij} u_i^0 - c_j = 0$

and

$\sum_{i=1}^{m} a_{ij} u_i^0 - c_j > 0$ implies $x_j^0 = 0$,

$j = 1, \ldots, p$.

<u>Strong Complementary Slackness</u>

$u_i^0 = 0$ implies $b_i - \sum_{j=1}^{p} a_{ij} x_j^0 > 0$

and

$b_i - \sum_{j=1}^{p} a_{ij} x_j^0 = 0$ implies $u_i^0 > 0$,

$i = 1, \ldots, m$; while

$x_j^0 = 0$ implies $\sum_{i=1}^{m} a_{ij} u_i^0 - c_j > 0$

and

$\sum_{i=1}^{m} a_{ij} u_i^0 - c_j = 0$ implies $x_j^0 > 0$,

$j = 1, \ldots, p$.

5 THE THEORY OF LINEAR PROGRAMMING

5.1. Finding Primal Basic Feasible Solutions

We noted in Chapter 3 that the primal feasible region $\mathcal{K}_P$ has a finite number of extreme points. Since each such point has associated with it a basic feasible solution (unique or otherwise), it follows that there exists a finite number of basic feasible solutions. Hence an optimal solution to the primal linear programming problem will be contained within the set of basic feasible solutions to $\mathbf{AX} = \mathbf{b}$. How many elements does this set possess? Since a basic feasible solution has at most m of n variables different from zero, an upper bound to the number of basic feasible solutions is

$$\binom{n}{m} = \frac{n!}{m!(n-m)!},$$

i.e., we are interested in the total number of ways in which m basic variables can be selected (without regard to their order within the vector of primal basic variables $\mathbf{X}_B$) from a group of n variables. Clearly for large n and m it becomes an exceedingly tedious task to examine each and every basic feasible solution. What is needed is a computational scheme which examines, in a selective fashion, only some small subset of the set of basic feasible solutions. Such a scheme is the **simplex method** (Dantzig (1951)). Starting from an initial basic feasible solution this technique systematically proceeds to alternative basic feasible solutions and, in a finite number of steps or iterations, ultimately arrives at an optimal basic feasible solution. The path taken to the optimum is one for which the value of the objective function at any extreme point is at least as great as at an adjacent extreme point (two extreme points are said to be **adjacent** if they are joined by an edge of a convex polyhedron). For instance, if in Figure 5.1 the extreme point A represents our initial basic feasible solution, the first iteration slides f upwards parallel to itself over $\mathcal{K}_P$ until it passes through its adjacent

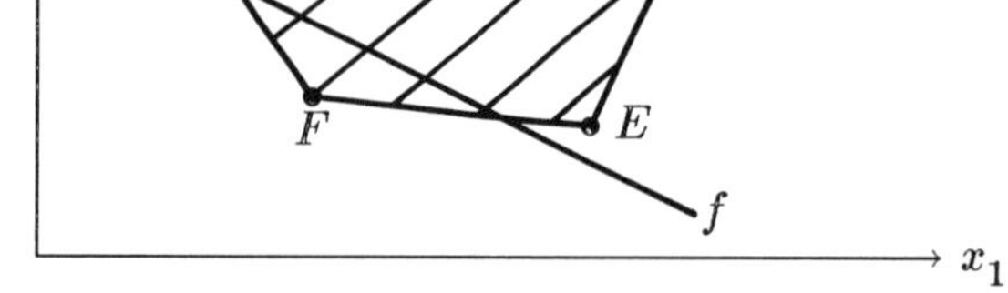

Figure 5.1

extreme point B. The next iteration advances f in a similar fashion to its optimal (maximal) basic feasible solution at C. So with f nondecreasing between successive basic feasible solutions, this search technique does not examine all basic feasible solutions but only those which yield a value of f at least as large as at the preceding basic feasible solution.

To set the stage for the development of the simplex method, let us

$$max\ f = \mathbf{C}'\mathbf{X} \quad \text{s.t.}$$
$$\mathbf{AX} = \mathbf{b},\ \mathbf{X} \geqq \mathbf{0},$$

where $\mathbf{X} \in \mathbf{R}^n$. To insure that the simplex method terminates in a finite number of steps at an optimal basic feasible solution (provided one exists), let us invoke the following **nondegeneracy assumption:** every set of m columns from $\mathbf{A}$ is linearly independent. Under this restriction, we are guaranteed that:

(1) every basic feasible solution has exactly m nonzero components; and

(2) f increases in value between successive basic feasible solutions.

With $\rho(\mathbf{A}) = m$, we may select from $\mathbf{A}$ a set of m linearly independent vectors which constitute a basis for $\mathbf{R}^m$ and thus the columns of the $(m \times m)$ nonsingular basis matrix $\mathbf{B}$. Then

$$\begin{aligned} \mathbf{AX} &= [\mathbf{B}, \mathbf{R}] \begin{bmatrix} \mathbf{X}_B \\ \mathbf{X}_R \end{bmatrix} \\ &= \mathbf{BX}_B + \mathbf{RX}_R = \mathbf{b}. \end{aligned}$$

Since $|\mathbf{B}| \neq 0$, $\mathbf{B}^{-1}$ exists and we thus obtain

$$\mathbf{X}_B = \mathbf{B}^{-1}\mathbf{b} - \mathbf{B}^{-1}\mathbf{RX}_R. \tag{5.1}$$

Let us next partition the $(n \times 1)$ coefficient matrix $\mathbf{C}$ into $(m \times 1)$ and $(n-m \times 1)$ matrices $\mathbf{C}_B$ and $\mathbf{C}_R$ respectively, where $\mathbf{C}_B$ contains the objective function coefficients corresponding to the basic variables in $\mathbf{X}_B$ and $\mathbf{C}_R$ contains the coefficients on the nonbasic variables in $\mathbf{X}_R$. Then

$$\begin{aligned} f(\mathbf{X}) &= \mathbf{C}'\mathbf{X} \\ &= [\mathbf{C}'_B, \mathbf{C}'_R] \begin{bmatrix} \mathbf{X}_B \\ \mathbf{X}_R \end{bmatrix} \\ &= \mathbf{C}'_B\mathbf{X}_B + \mathbf{C}'_R\mathbf{X}_R. \end{aligned} \tag{5.2}$$

Substituting $\mathbf{X}_B$ from above into (5.2) yields

$$f(\mathbf{X}) = \mathbf{C}'_B\mathbf{B}^{-1}\mathbf{b} + (\mathbf{C}'_R - \mathbf{C}'_B\mathbf{B}^{-1}\mathbf{R})\mathbf{X}_R. \tag{5.2.1}$$

At a basic feasible solution to the linear programming problem, $\mathbf{X}_R = \mathbf{0}$ so that

$$\begin{aligned} \mathbf{X} &= \begin{bmatrix} \mathbf{X}_B \\ \mathbf{0} \end{bmatrix} = \begin{bmatrix} \mathbf{B}^{-1}\mathbf{b} \\ \mathbf{0} \end{bmatrix} \geqq \mathbf{0}, \\ f(\mathbf{X}) &= \mathbf{C}'_B\mathbf{B}^{-1}\mathbf{b}. \end{aligned} \tag{5.3}$$

If $\bar{\mathbf{b}} = \mathbf{B}^{-1}\mathbf{b} \geqq \mathbf{0}$, then $\mathbf{B}$ will be termed a **primal feasible basis.**

5.2. The Reduced Primal Problem

In what follows we shall find it convenient to express the primal linear programming problem along with equations (5.1), (5.2.1) in a slightly different form. In particular, since the vectors in $\mathbf{B}$ form a basis for $\mathbf{R}^m$, any vector in $\mathbf{R}$ may be written as a linear combination of the columns of $\mathbf{B}$. If we denote the jth column of $\mathbf{R}$ by $\mathbf{r}_j$, $j = 1,\ldots,n-m$, then for all j,

$$\mathbf{r}_j = \sum_{i=1}^{m} y_{ij}\mathbf{b}_i = \mathbf{B}\mathbf{Y}_j$$

so that $\mathbf{Y}_j = \mathbf{B}^{-1}\mathbf{r}_j$, $j = 1,\ldots,n-m$, *i.e.*, $\mathbf{Y}_j$ is the jth column of $\mathbf{B}^{-1}\mathbf{R}$. Hence, from (5.1),

$$\begin{aligned} \mathbf{X}_B &= \mathbf{B}^{-1}\mathbf{b} - \left[\mathbf{Y}_1,\ldots,\mathbf{Y}_{n-m}\right]\mathbf{X}_R \\ &= \bar{\mathbf{b}} - \sum_{j=1}^{n-m}\mathbf{Y}_j x_{Rj}. \end{aligned} \tag{5.1.1}$$

From (5.2.1), let $\bar{\mathbf{C}}'_R = \mathbf{C}'_R - \mathbf{C}'_B\mathbf{B}^{-1}\mathbf{R}$ have as its jth component $\bar{c}_j = c_{Rj} - \mathbf{C}'_B\mathbf{Y}_j = c_{Rj} - f_j$, $j = 1,\ldots,n-m$. Then

$$\begin{aligned} f &= \mathbf{C}'_B\bar{\mathbf{b}} + \bar{\mathbf{C}}'_R\mathbf{X}_R \\ &= \bar{f} + \sum_{j=1}^{n-m}\bar{c}_j x_{Rj}. \end{aligned} \tag{5.2.2}$$

On the basis of equations (5.1.1), (5.2.2) the primal linear programming problem can be expressed alternatively as

$$\begin{aligned} f &= \bar{f} + max\ \bar{\mathbf{C}}'_R\mathbf{X}_R \quad \text{s.t.} \\ &\sum_{j=1}^{n-m}\mathbf{Y}_j x_{Rj} + \mathbf{X}_B = \bar{\mathbf{b}} \\ &\mathbf{X}_B \geqq \mathbf{0},\ x_{Rj} \geq 0,\ j = 1,\ldots,n-m, \end{aligned} \tag{5.4}$$

or

$$\begin{aligned} f &= \bar{f} + max\ \sum_{j=1}^{n-m}\bar{c}_j x_{Rj} \quad \text{s.t.} \\ &\sum_{j=1}^{n-m} y_{ij}x_{Rj} + x_{Bi} = \bar{b}_i,\ i = 1,\ldots,m \\ &x_{Bi} \geq 0,\ i = 1,\ldots,m \\ &x_{Rj} \geq 0,\ j = 1,\ldots,n-m. \end{aligned} \tag{5.4.1}$$

These primal programs are said to be written in **reduced form** in nonbasic variable space since the number of dimensions involved has been reduced to $n - m = p$, the number of nonbasic variables. In this regard the new set of objective function coefficients $\bar{c}_j = c_{Rj} - f_j$, $j = 1,\ldots,n-m$, are called **reduced benefit (profit) coefficients** (or **reduced cost coefficients** in the case of a minimization problem). Moreover, the current set of basic variables x_{Bi}, $i = 1,\ldots,m$, serves as a set of slack variables for the structural constraints of (5.4)(or (5.4.1)).

5.3. The Primal Optimality Criterion

At this point in our discussion of primal basic feasible solutions an important question emerges: how do we know whether or not a particular basic feasible solution is optimal? The answer is provided by the following

5.1. OPTIMALITY THEOREM. A primal feasible basis $\mathbf{B}$ with $\bar{\mathbf{b}} = \mathbf{B}^{-1}\mathbf{b} > \mathbf{0}$ is optimal if and only if $\bar{\mathbf{C}}'_R = \mathbf{C}'_R - \mathbf{C}'_B\mathbf{B}^{-1}\mathbf{R} \leqq \mathbf{0}'$ or $\bar{c}_j = c_{Rj} - f_j \leq 0$, $j = 1,\ldots,n-m$. Moreover, an optimal primal basic feasible solution is unique if $\bar{c}_j < 0$, $j = 1,\ldots,n-m$.

PROOF. (sufficiency) From (5.2.2), $f(\mathbf{X}) = \bar{f} + \bar{\mathbf{C}}'_R\mathbf{X}_R$. Since $\bar{\mathbf{C}}'_R \leqq \mathbf{0}'$ and $\mathbf{X}_R \geqq \mathbf{0}$, we have $f(\mathbf{X}) \leq \bar{f}$. An since $\bar{f}$ is obtained from the basic feasible solution $\bar{\mathbf{X}}' = (\bar{\mathbf{X}}'_B, \mathbf{0}') = (\bar{\mathbf{b}}', \mathbf{0}')$, $\bar{f} = f(\bar{\mathbf{X}})$ is the maximum value of the primal objective function.

(necessity) For some nonbasic variable x_{Rj} let $\bar{c}_j > 0$. Hence we can obtain a new basic feasible solution $\mathbf{X}$ with an objective function value f larger than $\bar{f}$. That is, starting from $\bar{\mathbf{X}}$, we move to $\mathbf{X}$ in a fashion such that only x_{Rj} increases from its current value of zero to the level $\theta > 0$. Hence $\mathbf{X}' = (\mathbf{X}'_B, \mathbf{X}'_R) = (\mathbf{X}'_B, \theta\mathbf{e}'_j)$. Since $\mathbf{X}$ must be feasible, we require that

$$\begin{aligned} \mathbf{X}_B &= \bar{\mathbf{b}} - \mathbf{B}^{-1}\mathbf{R}\mathbf{X}_R = \bar{\mathbf{b}} - \theta\mathbf{B}^{-1}\mathbf{R}\mathbf{e}_j \\ &= \bar{\mathbf{b}} - \theta\mathbf{B}^{-1}\mathbf{r}_j = \bar{\mathbf{b}} - \theta\mathbf{Y}_j \geqq \mathbf{0}. \end{aligned} \tag{5.5}$$

Since $\bar{\mathbf{b}} \geqq \mathbf{0}$, θ can be made arbitrarily small in order for $\mathbf{X}_B$ to remain nonnegative as required. Then

$$\begin{aligned} f(\mathbf{X}) &= \bar{f} + \bar{\mathbf{C}}'_R\mathbf{X}_R = \bar{f} + \theta\bar{\mathbf{C}}'_R\mathbf{e}_j \\ &= \bar{f} + \theta\bar{c}_j > \bar{f} . \end{aligned} \tag{5.6}$$

Thus $\bar{\mathbf{X}}$ cannot represent a maximal solution and thus $\mathbf{B}$ is not an optimal primal feasible basis.

That $\bar{c}_j < 0$ for all j implies a unique optimal basic feasible solution for the primal problem follows from the fact that if $\bar{c}_j < 0$ for all $j \neq k$ and $\bar{c}_k = 0$, then, by setting $j = k$ in (5.5) and (5.6), we can move to an alternative optimal feasible basis that renders the same value of f, namely $\bar{f}$. Q.E.D.

Given that each $\bar{c}_j \leq 0$, $j = 1,\ldots,n-m$, at an optimal primal basic feasible solution, these coefficients are often referred to as **optimality evaluators.**

5.4. Constructing the Dual Solution

We noted in the previous chapter (corollary 4.6) that either both of the primal-dual pair of problems have optimal solutions, with the values of their objective functions being the same, or neither has an optimal solution. And as evidenced by the strong duality theorem, if one of these problems possesses an optimal solution, then the other does likewise. In this light, let us now undertake the derivation of the optimal dual solution from the optimal primal solution. In fact, this exercise can be thought of as an alternative and constructive proof of the strong duality theorem 4.8.

To this end let us assume that the primal maximum problem

$$\begin{aligned} max\, f &= \mathbf{C}'\mathbf{X} + \mathbf{0}'\mathbf{X}_s \text{ s.t.} \\ \mathbf{A}\mathbf{X} + \mathbf{I}_m\mathbf{X}_s &= \mathbf{b} \\ \mathbf{X}, \mathbf{X}_s &\geqq \mathbf{0} \end{aligned}$$

has an optimal basic feasible solution consisting of

$$\begin{aligned} \mathbf{X}_B &= \mathbf{B}^{-1}\mathbf{b}, \\ f(\mathbf{X}) &= \mathbf{C}_B'\mathbf{X}_B = \mathbf{C}_B'\mathbf{B}^{-1}\mathbf{b}. \end{aligned}$$

Note that in the last equation the optimal value of the primal objective function is written in terms of the optimal values of the primal basic variables x_{Bi}, $i = 1,\ldots,m$. As an alternative, let us respecify the optimal value of the primal objective function in terms of the optimal dual variables by expressing the latter as

$$
\begin{aligned}
\mathbf{C}_B'\mathbf{B}^{-1} &= (c_{B1},\ldots,c_{Bm})\begin{bmatrix} b_{11} & \cdots & b_{1m} \\ \vdots & & \vdots \\ b_{m1} & \cdots & b_{mm} \end{bmatrix} \\
&= \left(\sum_{i=1}^{m} c_{Bi}b_{i1},\ldots,\sum_{i=1}^{m} c_{Bi}b_{im}\right) \\
&= (u_1,\ldots,u_m),
\end{aligned}
$$

where $c_{Bi}, i = 1,\ldots,m$, is the ith component of $\mathbf{C}_B$ and $b_{ij}, i,j = 1,\ldots,m$, is the element in the ith row and j^{th} column of $\mathbf{B}^{-1}$. Then $\mathbf{C}_B'\mathbf{B}^{-1} = \mathbf{U}'$ and thus $f(\mathbf{X}) = \mathbf{U}'\mathbf{b}$. Thus the value of f corresponding to an optimal basic feasible solution may be expressed in terms of the optimal values of either the primal or dual variables.

Given that the optimality criterion is satisfied, we must have

$$\mathbf{C}_R' - \mathbf{C}_B'\mathbf{B}^{-1}\mathbf{R} \leqq \mathbf{0}' \quad \text{or} \quad \bar{c}_j = c_{Rj} - \mathbf{C}_B'\mathbf{B}^{-1}\mathbf{r}_j \leq 0,\ j = 1,\ldots,n-m,$$

where $\mathbf{r}_j$ is the jth nonbasic vector. Moreover, this inequality must hold for basic vectors as well, *i.e.*, relative to the ith basic variable $x_{Bi}, i = 1,\ldots,m$,

$$\bar{c}_i = c_{Bi} - \mathbf{C}_B'\mathbf{e}_i \;=\; c_{Bi} - c_{Bi} \;=\; 0.$$

Hence the optimality criterion holds for all columns of $[\mathbf{A},\mathbf{I}_m]$ (or $[\mathbf{B},\mathbf{R}]$). In this regard:

(a) for the columns of $\mathbf{A}$,

$$\mathbf{C}_B'\mathbf{B}^{-1}\mathbf{a}_j \geq c_j,\ j = 1,\ldots,p, \quad \text{or}$$

$$\mathbf{C}_B'\mathbf{B}^{-1}\mathbf{A} \geqq \mathbf{C}'$$

and thus

$$\mathbf{U}'\mathbf{A} \geqq \mathbf{C}' \quad \text{or} \quad \mathbf{A}'\mathbf{U} \geqq \mathbf{C},$$

i.e., $\mathbf{U} = (\mathbf{B}^{-1})'\mathbf{C}_B$ represents a solution to the dual structural constraint system;

(b) for the columns in $\mathbf{I}_m$,

$$\mathbf{C}_B'\mathbf{B}^{-1}\mathbf{e}_i \geq 0,\ i = 1,\ldots,m, \quad \text{or}$$

$$\mathbf{C}_B'\mathbf{B}^{-1}\mathbf{I}_m \geqq \mathbf{0}'$$

and thus

$$\mathbf{U}'\mathbf{I}_m \geqq \mathbf{0}' \quad \text{or} \quad \mathbf{U} \geqq \mathbf{0},$$

i.e.,**U** is a feasible solution to the dual structural constraints. urthermore, since

$$g(\mathbf{U}) \;=\; \mathbf{b}'\mathbf{U} \;=\; \mathbf{U}'\mathbf{b} \;=\; \mathbf{C}_B'\mathbf{B}^{-1}\mathbf{b} \;=\; \mathbf{C}_B'\mathbf{X}_B \;=\; f(\mathbf{X}),$$

U is an optimal solution to the dual minimum problem given that **X** represents an optimal basic feasible solution to the primal maximum problem.

An important observation which emerges from this construction is that *the concept of primal optimality is equivalent to the notion of dual feasibility.* Hence if $\mathbf{C}_R' - \mathbf{C}_B'\mathbf{B}^{-1}\mathbf{R} \leqq \mathbf{0}'$, **B** is termed a **dual feasible basis.** In this regard, at an optimal dual feasible solution **U**, the optimality evaluators $\bar{c}_j$, $j = 1,\ldots,n-m$, within $\bar{\mathbf{C}}_R$ correspond to the dual surplus variables in the dual structural constraint system $\mathbf{A}'\mathbf{U} \geqq \mathbf{C}$, *i.e.*, forming $\mathbf{A}'\mathbf{U} - \bar{\mathbf{C}}_R = \mathbf{C}$, it follows that $-\bar{\mathbf{C}}_R = \mathbf{C} - \mathbf{A}'\mathbf{U} \leqq \mathbf{0}$ under primal optimality and thus $\bar{\mathbf{C}}_R \geqq \mathbf{0}$. So if all $\bar{c}_j$'s ≤ 0, the associated dual solution is dual feasible and hence **B** is a dual feasible basis. If any of the $\bar{c}_j$'s determined from the basis **B** are positive, then the associated dual solution is infeasible and hence **B** is dual infeasible.

Having developed a criterion for primal optimality (equivalently, dual feasibility) and a definition of the vector of dual variables $\mathbf{U}' = \mathbf{C}_B'\mathbf{B}^{-1}$, let us see how we can rationalize the structure of the symmetric dual problem. Again let the primal problem correspond to

$$\begin{aligned} \max f \;&=\; \mathbf{C}'\mathbf{X} \text{ s.t.} \\ \mathbf{A}\mathbf{X} &\leqq \mathbf{b},\ \mathbf{X} \geqq \mathbf{0}. \end{aligned}$$

Let us form the primal Lagrangian function

$$\begin{aligned} \mathcal{L}(\mathbf{U}) \;&=\; \max\{\mathbf{C}'\mathbf{X} + \mathbf{U}'(\mathbf{b} - \mathbf{A}\mathbf{X})\} \\ &= \mathbf{U}'\mathbf{b} + \max_{\mathbf{X} \geqq \mathbf{0}} (\mathbf{C}' - \mathbf{U}'\mathbf{A})\mathbf{X} \end{aligned} \tag{5.7}$$

by weighting or *pricing out* the structural constraints by the vector of dual variables $\mathbf{U} \geqq \mathbf{0}$. Two cases can now be addressed.

First, if $\mathbf{C}' - \mathbf{U}'\mathbf{A} \leqq \mathbf{0}'$ (note that for $\mathbf{U}' = \mathbf{C}_B'\mathbf{B}^{-1}$, this inequality is simply the primal optimality criterion), then the maximum in (5.7) is attained for $\mathbf{X} = \mathbf{0}$ and thus $\mathcal{L}(\mathbf{U}) = \mathbf{U}'\mathbf{b}$. Here $\mathcal{L}(\mathbf{U})$ is a finite upper bound on the maximal primal objective value (by the weak duality theorem 4.2). Second, if $\mathbf{C}' = \mathbf{U}'\mathbf{A} \not\leqq \mathbf{0}'$, say $c_k = \mathbf{U}'\mathbf{a}_k > 0$, then for $x_j = 0,\ j \neq k$,

$$\mathcal{L}(\mathbf{U}) \;=\; \mathbf{U}'\mathbf{b} + \max_{x_k \geq 0}\ (c_k - \mathbf{U}'\mathbf{a}_k)x_k.$$

Clearly $\mathcal{L}(\mathbf{U}) \rightarrow \infty$ as $x_k \rightarrow \infty$, *i.e.*, according to existence theorem 4.9, if the dual problem is infeasible but the primal problem has a feasible solution, then the primal objective value is unbounded from above.

To construct the dual problem let us assume that primal optimality holds. Hence we many determine, via the weak duality result, the least upper bound of $\mathcal{L}(\mathbf{U})$ or

$$\min \mathcal{L}(\mathbf{U}) \ \text{ s.t.}$$
$$\mathbf{C}' - \mathbf{U}'\mathbf{A} \leqq \mathbf{0}',\ \mathbf{U} \geqq \mathbf{0}$$

or

$$\min g = \mathbf{U}'\mathbf{b} \ \text{ s.t.}$$
$$\mathbf{A}'\mathbf{U} \geqq \mathbf{C},\ \mathbf{U} \geqq \mathbf{0},$$

the symmetric dual problem introduced at the outset of Chapter 4.

Looking to the dual variables themselves, let us note that for any particular requirements vector $\mathbf{b}$, $f(\mathbf{b})$ depicts the maximum value of f when $\mathbf{AX} \leqq \mathbf{b}$, $\mathbf{X} \geqq \mathbf{0}$, and $\mathbf{b}$ is fixed. Then at optimal feasible solutions $\overline{\mathbf{X}}, \overline{\mathbf{U}}$ to the primal-dual pair of problems, the strong duality theorem 4.8 informs us that $\bar{f} = \mathbf{C}'\overline{\mathbf{X}} = \mathbf{b}'\overline{\mathbf{U}} = \Sigma_{i=1}^{m} b_i\bar{u}_i$ and thus $\partial\bar{f}/\partial b_i = \bar{u}_i$ (generally). (Later on in the text we shall elaborate on the circumstances under which this strict inequality might not hold.) That is, when the dual optimal feasible solution $\overline{\mathbf{U}}$ is unique, the ith dual variable $\bar{u}_i$ depicts the rate of change of the optimal primal objective value with respect to small (per unit) perturbations in the right-hand side value b_i of the ith primal structural constraint, given that all other requirements b_ℓ, $\ell \neq i$, remain constant. Here $\bar{u}_i \geq 0$ for all i since increasing slightly the size of the primal feasible region cannot decrease the maximal value of f.

Sometimes u_i is referred to as the **marginal imputed value** of the ith primal structural constraint right-hand side value b_i. It is also referred to as a **simplex multiplier** since it serves as a Lagrange multiplier in (5.7) above. In this regard, since its role in (5.7) is to *price out* the ith primal structural constraint, u_i is often called a **shadow (fictitious) price.** We shall return to our interpretation of the dual variables later on when we explore the geometry of the primal-dual pair of problems.

Having discussed the nature of the dual variables $u_i, i = 1,\ldots,m$, we are now equipped to offer two distinct interpretations of the optimality evaluators $\bar{c}_j, j = 1,\ldots,n-m$. But first, however, let us examine the significance of the columns within the matrix product $\mathbf{B}^{-1}\mathbf{R}$ in

$$\mathbf{X}_B = \bar{\mathbf{b}} - \mathbf{B}^{-1}\mathbf{R}\mathbf{X}_R = \bar{\mathbf{b}} - [\mathbf{Y}_1,\ldots,\mathbf{Y}_{n-m}]\mathbf{X}_R.$$

For j fixed at k, the preceding expression becomes

$$\mathbf{X}_B = \bar{\mathbf{b}} - \mathbf{Y}_k x_{Rk} \text{ or } x_{Bi} = \bar{b}_i - y_{ik}x_{Rk}, i = 1,\ldots,m.$$

Thus y_{ik}, $i = 1,\ldots,m$, represents the amount by which the ith basic variable would change in any primal basic solution given that the kth nonbasic variable increases from zero (its current nonbasic value) to one, with all other nonbasic variables remaining at their zero level.

Looking now to the optimality evaluators, again let $j = k$ so that

$$\bar{c}_k = c_{Rk} - f_k = c_{Rk} - \sum_{i=1}^{m} c_{Bi}y_{ik}.$$

We may interpret this expression as the *net change* in f given that x_{Rk} increases from zero to one. In what sense shall we interpret the notion of net change? As x_{Rk} increases by a unit, it *directly* impacts f through its objective function coefficient c_{Rk}. Next, x_{Rk} affects f *indirectly* through its impact on the m basic variables x_{Bi}, which in turn precipitate a change in f via their objective function coefficients c_{Bi}. Thus the combined indirect effect of a one unit increase in x_{Rk} on f is $f_k = \sum_{i=1}^{m} c_{Bi}y_{ik}$. Hence $\bar{c}_k = c_{Rk} - f_k$ is the net effect on f of increasing x_{Rk} by a single unit. In sum,

$$
\begin{aligned}
\bar{c}_k &= \text{marginal net effect on } f \text{ of an increase in } x_{Rk} \\
&= c_{Rk} - f_k \\
&= \text{marginal } \textit{direct effect} \text{ on } f \text{ of an increase in } x_{Rk} \\
&- \text{marginal } \textit{indirect effect} \text{ on } f \text{ of an increase in } x_{Rk}.
\end{aligned}
$$

So at an optimal feasible solution to the primal linear program we must have $\bar{c}_k \leq 0$, *i.e.*, the marginal indirect effect on f of increasing x_{Rk} cannot fall short of its direct counterpart.

An alternative approach to interpreting the $\bar{c}_j$'s, $j = 1,\ldots,n-m$, is to let $\bar{\mathbf{C}}'_R = \mathbf{C}'_R - \mathbf{C}'_B\mathbf{B}^{-1}\mathbf{R} = \mathbf{C}'_R - \mathbf{U}'\mathbf{R}$. Then for $j = k$ we have

$$
\begin{aligned}
\bar{c}_k &= c_{Rk} - \mathbf{U}'\mathbf{r}_k = c_{Rk} - \mathbf{U}'\mathbf{a}_k = c_{Rk} - f_k \\
&= c_{Rk} - \sum_{i=1}^{m} u_i a_{ik}\,.
\end{aligned}
$$

(Here $\mathbf{r}_k = \mathbf{a}_k$. If $\mathbf{r}_k = \mathbf{e}_k$, then $\bar{c}_k = -u_k$.) While the direct effect on f of a unit increase in x_{Rk} is the same as before (f changes marginally by c_{Rk}), the indirect impact on f when $x_{Rk} = 1$ is now given by $f_k = \Sigma_{i=1}^{m} u_i a_{ik}$. Given that a_{ik} represents the contribution of the ith constraint in supporting x_{Rk} at the unit level and u_i depicts the imputed contribution to f of relaxing the ith constraint by one unit (since typically $u_i = \partial f/\partial b_i \geq 0$), we see that f_k amounts to the total imputed change in f resulting from a per unit increase in x_{Rk} given that at least one primal structural constraint is binding (*i.e.*, hence at least one u_i can be positive by virtue of the weak complementary slackness conditions) and the per unit increase in x_{Rk} is brought about by increasing the right-hand side of at least one primal structural constraint.

Equivalently, $\bar{c}_k$ is the **relative cost** of making x_{Rk} basic, *i.e.*, it is c_{Rk}, the direct effect on f of increasing x_{Rk} by a unit, less the value, in terms of f, of the **real opportunity cost** of doing so. This real opportunity cost of making x_{Rk} basic, in terms of a reduction in f, is $\mathbf{U}'\mathbf{r}_k = \Sigma_{i=1}^{m} u_i a_{ik}$, the **shadow value of the contributions of all constraints** in supporting x_{Rk} at the unit level.

EXAMPLE 5.1. To see exactly how (5.3) is obtained in actual practice, let us

$$\begin{aligned} max\, f = x_1 + 3x_2 + 4x_3 \ \text{ s.t.} \\ 2x_1 + \ x_2 + 4x_3 \ &\leq 100 \\ x_1 + 3x_2 + \ x_3 \ &\leq \ 40 \\ x_1, x_2, x_3 \geq 0. \end{aligned}$$

Since our starting point is the primal linear programming problem in standard form, we introduce two nonnegative slack variables x_4, x_5 and ultimately

$$\begin{aligned} max\, f = x_1 + 3x_2 + 4x_3 + 0x_4 + 0x_5 \quad \text{s.t.} \\ 2x_1 + \ x_2 + 4x_3 + \ x_4 \qquad &= 100 \\ x_1 + 3x_2 + \ x_3 \qquad + x_5 \ &= \ 40 \\ x_1, \ldots, x_5 \geq 0. \end{aligned}$$

In terms or our previous notation,

$$\mathbf{A} = \begin{bmatrix} 2 & 1 & 4 & 1 & 0 \\ 1 & 3 & 1 & 0 & 1 \end{bmatrix}, \ \mathbf{C} = \begin{bmatrix} 1 \\ 3 \\ 4 \\ 0 \\ 0 \end{bmatrix}, \ \mathbf{X} = \begin{bmatrix} x_1 \\ x_2 \\ \vdots \\ x_5 \end{bmatrix}, \text{ and } \mathbf{b} = \begin{bmatrix} 100 \\ 40 \end{bmatrix}.$$

Since the determinant of the (2×2) submatrix formed from columns three and five of $\mathbf{A}$ is different from zero, $\rho(\mathbf{A}) = 2$. With these two columns linearly independent, we may consequently select x_3 and x_5 as basic variables with x_1, x_2, and x_4 nonbasic and proceed to generate our first basic feasible solution. In this regard,

$$\mathbf{B} = \begin{bmatrix} 4 & 0 \\ 1 & 1 \end{bmatrix}, \ \mathbf{B}^{-1} = \begin{bmatrix} \frac{1}{4} & 0 \\ -\frac{1}{4} & 1 \end{bmatrix}, \ \mathbf{R} = \begin{bmatrix} 2 & 1 & 1 \\ 1 & 3 & 0 \end{bmatrix}, \ \mathbf{X}_B = \begin{bmatrix} x_3 \\ x_5 \end{bmatrix},$$

$$\mathbf{X}_R = \begin{bmatrix} x_1 \\ x_2 \\ x_4 \end{bmatrix}, \ \mathbf{C}_B = \begin{bmatrix} 4 \\ 0 \end{bmatrix}, \text{ and } \mathbf{C}_R = \begin{bmatrix} 1 \\ 3 \\ 0 \end{bmatrix}.$$

From (5.1),

$$\mathbf{X}_B = \mathbf{B}^{-1}\mathbf{b} - \mathbf{B}^{-1}\mathbf{R}\mathbf{X}_R \quad \text{or}$$

$$\begin{bmatrix} x_{B1} \\ x_{B2} \end{bmatrix} = \begin{bmatrix} 25 \\ 15 \end{bmatrix} - \begin{bmatrix} \frac{1}{2} & \frac{1}{4} & \frac{1}{4} \\ \frac{1}{2} & \frac{11}{4} & -\frac{1}{4} \end{bmatrix} \begin{bmatrix} x_{R1} \\ x_{R2} \\ x_{R3} \end{bmatrix} \quad \text{or}$$

$$\begin{bmatrix} x_3 \\ x_5 \end{bmatrix} = \begin{bmatrix} 25 - \frac{1}{2}x_1 - \frac{1}{4}x_2 - \frac{1}{4}x_4 \\ 15 - \frac{1}{2}x_1 - \frac{11}{4}x_2 + \frac{1}{4}x_4 \end{bmatrix}. \tag{5.8}$$

If $\mathbf{X}_R = \mathbf{0}$, we obtain the nondegenerate primal basic feasible solution $\mathbf{X}' = (0, 0, 25, 0, 15)$.

We previously denoted the jth column of $\mathbf{B}^{-1}\mathbf{R}$ by $\mathbf{Y}_j$. Hence y_{ij}, the ith component of $\mathbf{Y}_j$, represents the amount by which the ith basic variable would change in any primal basic solution given that the jth nonbasic variable increases from zero (its current nonbasic value) to one. For instance, in our example, y_{13} indicates that if x_4 increases from zero to one, x_3 decreases by $\frac{1}{4}$ to $x_3 = 25 - \frac{1}{4}(1) = \frac{99}{4}$.

From (5.2.1),

$$\begin{aligned} f &= \mathbf{C}'_B\mathbf{B}^{-1}\mathbf{b} - (\mathbf{C}'_R - \mathbf{C}'_B\mathbf{B}^{-1}\mathbf{R})\mathbf{X}_R \\ &= 100 + (-1, 2, -1) \begin{bmatrix} x_{R1} \\ x_{R2} \\ x_{R3} \end{bmatrix} \\ &= 100 - x_1 + 2x_2 - x_4. \end{aligned} \tag{5.9}$$

Here $\bar{c}_1 = -1$, $\bar{c}_2 = 2$, and $\bar{c}_3 = -1$. Upon setting $\mathbf{X}_R = \mathbf{0}$ it is easily seen that the value of the objective function corresponding to this basic feasible solution is 100. We mentioned earlier that $\bar{c}_j$ depicts the marginal net effect on f of an increase in x_{Rj}. For example, let us assume that x_2 increases from zero to one with the remaining nonbasic variables still equal to zero. The effect of this change is two-fold. First, there is the direct influence of x_2 on f through its corresponding coefficient in $\mathbf{C}_R$, *i.e.*, since

$$\mathbf{C}_R'\mathbf{X}_R = (c_{R1}, c_{R2}, c_{R3})\begin{bmatrix} 0 \\ x_{R2} \\ 0 \end{bmatrix}$$

$$= (1,3,0)\begin{bmatrix} 0 \\ x_2 \\ 0 \end{bmatrix} = 3x_2,$$

it is easily seen that when $x_2 = 1$, $f = 3(1) = 3$. Hence f directly increases by three units. Next, x_2 affects f indirectly through its influence on the basic variables x_3 and x_5, which in turn produce a change in f. Moreover, the said effect is transmitted through the term $\mathbf{C}_B'\mathbf{Y}_2$. That is to say, given

$$\mathbf{Y}_2 x_{R2} = \begin{bmatrix} \frac{1}{4} \\ \frac{11}{4} \end{bmatrix} x_2,$$

we see that if $x_2 = 1$, x_3 changes by $\frac{1}{4}$ unit and x_5 changes by $\frac{11}{4}$ units since, from (5.8),

$$\mathbf{X}_B = \bar{\mathbf{b}} - \mathbf{Y}_2 x_{R2} \quad \text{or}$$

$$\begin{bmatrix} x_3 \\ x_5 \end{bmatrix} = \begin{bmatrix} 25 \\ 15 \end{bmatrix} - \begin{bmatrix} \frac{1}{4} \\ \frac{11}{4} \end{bmatrix} x_2 .$$

But the effect of a change in x_3 and x_5 upon f is issued through their corresponding coefficients in $\mathbf{C}_B$, *i.e.*,

$$f_2 x_{R2} = \mathbf{C}_B'\mathbf{Y}_2 x_{R2} = (4,0)\begin{bmatrix} \frac{1}{4} \\ \frac{11}{4} \end{bmatrix} x_2 = x_2.$$

For $x_2 = 1$, the indirect effect of x_2 on f is thus $f_2 = 1$ unit. Hence the net change if f, being the difference between the direct and indirect effects of x_2, is given by $\bar{c}_2 = c_{R2} - f_2 = 3 - 1 = 2$ units, which is just the coefficient on x_2 in (5.9).

We noted above that since $f = 100 - x_1 + 2x_2 - x_4$, $f = 100$ when $\mathbf{X}_R = \mathbf{0}$ is definitely not maximal since f can increase in value if x_2 does. Now at the moment x_2 is equal to zero so that if we increase its value from the zero level by some positive amount, it is obvious that, with $\rho(\mathbf{B})=2$, we must change the status of some variable which is currently basic to nonbasic since there are at most two linearly independent vectors in a basis for $\mathbf{R}^2$ and thus at most two basic variables. Hence we must determine which basic variable, x_3 or x_5, should now become nonbasic (*i.e.*, decrease in value to zero). The answer to this question is the topic of the next section wherein we consider the generation of a succession of extreme point solutions to a linear programming problem using the primal simplex method.

5.5. The Primal Simplex Method

Let us assume that we have obtained a basic feasible solution to a linear programming problem but that the solution is not optimal, *i.e.*, at least one $\bar{c}_j = c_{Rj} - f_j > 0$, $j = 1,\ldots,n-m$, in

$$\begin{aligned} f &= \mathbf{C}_B'\mathbf{B}^{-1}\mathbf{b} + (\mathbf{C}_R' - \mathbf{C}_B'\mathbf{B}^{-1}\mathbf{R})\mathbf{X}_R \\ &= \mathbf{C}_B'\mathbf{B}^{-1}\mathbf{b} + \bar{c}_1 x_{R1} + \bar{c}_2 x_{R2} + \cdots + \bar{c}_{n-m} x_{R,n-m}. \end{aligned} \tag{5.10}$$

How may we obtain a new basic feasible solution with an objective function value which is at least as great as at the old basic feasible solution? That is to say, given a basic feasible solution $\mathbf{X}' = (\mathbf{X}_B', \mathbf{0}')$ with $f(\mathbf{X}) = \mathbf{C}_B'\mathbf{X}_B$, we week to determine an alternative basic feasible solution $\hat{\mathbf{X}}' = (\mathbf{X}_{\hat{B}}', \mathbf{0}') \in \mathbf{R}^n$ such that $f(\hat{\mathbf{X}}) = \mathbf{C}_{\hat{B}}'\mathbf{X}_{\hat{B}} \geq \mathbf{C}_B'\mathbf{X}_B$, where $\hat{\mathbf{B}}$ denotes the new m^{th} order nonsingular basis matrix. If more than one $\bar{c}_j > 0$, $j = 1,\ldots,n-m$, we must first determine which nonbasic variable is to become basic (*i.e.*, increase in value from zero). Let us adopt the convention of picking the nonbasic variable in (5.10) with the largest positive $\bar{c}_j$ coefficient since such a choice involves the largest increment in f per unit change in x_{Rj}, $j = 1,\ldots,n-m$. In this regard, if $\mathfrak{J}$ denotes the index set consisting of all j for which $\bar{c}_j > 0$, *i.e.*, $\mathfrak{J} = \{j \mid \bar{c}_j > 0\}$, and

$$\text{ENTRY CRITERION:} \quad \bar{c}_k = max\{\bar{c}_j, j \in \mathfrak{J}\}, \tag{5.11}$$

then x_{Rk}, the nonbasic variable corresponding to the kth column of $\mathbf{R}$, $\mathbf{r}_k$, becomes basic. Next, how do we go about selecting a column of the current basis matrix $\mathbf{B}$ to be replaced by $\mathbf{r}_k$? The answer is provided by the following corollary to optimality theorem 5.1. As we shall now see, the proof of this corollary is constructive in that it outlines a process called the **primal simplex method** for actually obtaining a new basic feasible solution which renders an increase in the value of f.

5.2. COROLLARY (THE PRIMAL SIMPLEX METHOD). Let $\mathbf{B}$ represent a primal feasible basis with $\bar{\mathbf{b}} = \mathbf{B}^{-1}\mathbf{b} > \mathbf{0}$ (nondegeneracy) and $\bar{c}_j > 0$ for some nonbasic variable x_{Rj}, $j = 1,\ldots,n-m$. Then either:

(a) there exists an extreme direction $\mathbf{d}_j$ with $\mathbf{C}'\mathbf{d}_j > \mathbf{0}$, in which case x_{Rj} can be increased indefinitely without violating feasibility so that the primal objective function value is unbounded; or

(b) we obtain a new primal feasible basis and a new (possibly degenerate) basic feasible solution that yields an improved value of f.

PROOF. Let the current basic feasible solution be depicted as $\bar{\mathbf{X}}' = (\bar{\mathbf{X}}'_B, \mathbf{X}'_R) = (\bar{\mathbf{b}}', \mathbf{0}')$ with $\mathbf{C}'\bar{\mathbf{X}} = \mathbf{C}'_B\bar{\mathbf{b}}$. Also, let $\mathbf{C}'_R \not\leq \mathbf{0}'$ so that primal optimality does not hold. In particular, let $\bar{c}_k = c_{Rk} - \mathbf{C}'_B\mathbf{B}^{-1}\mathbf{r}_k > 0$. Hence we may increase x_{Rk} above its current nonbasic value of zero to the level $\theta > 0$ so as to obtain a larger value of f. To this end, using (5.1.1), $\bar{\mathbf{X}}_B = \bar{\mathbf{b}}$ changes to $\mathbf{X}_B = \bar{\mathbf{b}} - \theta\mathbf{Y}_k$ while $\mathbf{X}_R = \mathbf{0}$ becomes $\theta\mathbf{e}_k$. Thus our new solution point is

$$\begin{aligned}\mathbf{X} &= \begin{bmatrix}\mathbf{X}_B\\ \mathbf{X}_R\end{bmatrix} = \begin{bmatrix}\bar{\mathbf{b}} - \theta\mathbf{Y}_k\\ \theta\mathbf{e}_k\end{bmatrix} = \begin{bmatrix}\bar{\mathbf{b}}\\ \mathbf{0}\end{bmatrix} + \theta\begin{bmatrix}-\mathbf{Y}_k\\ \mathbf{e}_k\end{bmatrix}\\ &= \bar{\mathbf{X}} + \theta\mathbf{d}_k. \end{aligned} \tag{5.12}$$

Looking to the primal objective value,

$$\begin{aligned}\mathbf{C}'\mathbf{X} &= \mathbf{C}'\bar{\mathbf{X}} + \theta\mathbf{C}'\mathbf{d}_k\\ &= \mathbf{C}'_B\bar{\mathbf{b}} + \theta(-\mathbf{C}'_B\mathbf{Y}_k + c_{Rk})\\ &= \mathbf{C}'_B\bar{\mathbf{b}} + \theta\bar{c}_k. \end{aligned} \tag{5.13}$$

At this point two cases must now be considered.

1. $\mathbf{Y}_k \leqq \mathbf{0}$. In this instance $\mathbf{d}_k' = (-\mathbf{Y}_k', e_k') \geqq \mathbf{0}'$ serves as an extreme direction since the requirement $\mathbf{A}\mathbf{d}_k = \mathbf{0}$ renders $\mathbf{X}$ feasible, *i.e.*, $\mathbf{AX} = \mathbf{A}(\overline{\mathbf{X}} + \theta \mathbf{d}_k) = \mathbf{A}\overline{\mathbf{X}} + \theta \mathbf{A}\mathbf{d}_k = \mathbf{A}\overline{\mathbf{X}} = \mathbf{b}$. Then $\mathbf{C}'\mathbf{d}_k = -\mathbf{C}_B'\mathbf{Y}_k + c_{Rk} = \overline{c}_k > 0$, *i.e.*, we have found an extreme direction $\mathbf{d}_k$ along which f can be increased without bound since, from (5.13), $\mathbf{C}'\mathbf{X} \rightarrow \infty$ as $\theta \rightarrow \infty$ when $\overline{c}_k > 0$.

2. $y_{ik} > 0$ for some i. Then θ cannot be increased indefinitely without violating primal feasibility. To see this let us employ (5.4.1) and write

$$x_{Bi} = \overline{b}_i - y_{ik}\,\theta \geq 0,\ i = 1,\ldots,m. \tag{5.14}$$

In this regard, we need to find a $\theta > 0$ such that (5.14) holds for all $y_{ik} > 0$. Rearranging (5.14) yields $x_{Bi}/y_{ik} \geq \theta$ so that any θ for which

$$0 < \theta \leq \min_i \left\{ \frac{x_{Bi}}{y_{ik}}, y_{ik} > 0 \right\}$$

will do, *i.e.*, will insure that primal feasibility is not violated.

In fact, this latter inequality enables us to select the $\mathbf{b}_i$ to be removed from the basis when $\mathbf{r}_k$ is entered into the basis. If

$$\text{EXIT CRITERION: } \widehat{\theta} = \min_i \left\{ \frac{x_{Bi}}{y_{ik}}, y_{ik} > 0 \right\} = \frac{x_{Br}}{y_{rk}} \tag{5.15}$$

is substituted into $\overline{\mathbf{b}} - \theta \mathbf{Y}_k$, then the component(s) of this vector for which the minimum (assumed unique) is attained reduces to zero or becomes nonbasic, *i.e.*,

$$\begin{aligned} x_{Br} &= \overline{b}_r - \widehat{\theta}\, y_{rk} = \overline{b}_r - \frac{x_{Br}}{y_{rk}}\, y_{rk} \\ &= \overline{b}_r - \frac{\overline{b}_r}{y_{rk}}\, y_{rk} = 0. \end{aligned}$$

If the minimum in (5.15.1) is unique (nondegeneracy) then the column of $\mathbf{B}$ to be replaced by $\mathbf{r}_k$ can always be uniquely determined by the exit criterion. In this circumstance, if $\{\widehat{\mathbf{b}}_1, \ldots, \widehat{\mathbf{b}}_m\}$ denotes the set of linearly independent vectors obtained by replacing the rth column $\mathbf{b}_r$ within $\{\mathbf{b}_1, \ldots, \mathbf{b}_m\}$ by $\mathbf{r}_k$, then the new basic feasible and thus extreme point solution can be represented as

$$\hat{\mathbf{X}} = \begin{bmatrix} \mathbf{X}_{\hat{B}} \\ \mathbf{0} \end{bmatrix} = \begin{bmatrix} \hat{\mathbf{B}}^{-1}\mathbf{b} \\ \mathbf{0} \end{bmatrix}, \tag{5.16}$$

where $\hat{\mathbf{B}} = [\hat{\mathbf{b}}_1, \ldots, \hat{\mathbf{b}}_m]$ with

$$\hat{\mathbf{b}}_i = \mathbf{b}_i,\ i \neq r;\ \ \hat{\mathbf{b}}_r = \mathbf{r}_k$$

and

$$x_{\hat{B}i} = \bar{b}_i - \hat{\theta} y_{ik},\ i \neq r;\ x_{\hat{B}r} = \hat{\theta} = x_{Br}/y_{rk}. \tag{5.16.1}$$

Clearly $\hat{\mathbf{X}}$ must have exactly m positive components.

In addition, since $c_{\hat{B}i} = c_{Bi},\ i \neq r;\ c_{\hat{B}r} = c_{Rk}$; and $c_{Br}(x_{Br} - \hat{\theta} y_{rk}) = 0$, it follows that

$$\begin{aligned} f(\hat{\mathbf{X}}) &= \mathbf{C}'_{\hat{B}}\mathbf{X}_{\hat{B}} = \mathbf{C}'_B(\bar{\mathbf{b}} - \theta\mathbf{Y}_k) + c_{Rk}x_{\hat{B}r} \\ &= \mathbf{C}'_B\bar{\mathbf{b}} + (c_{Rk} - \mathbf{C}'_B\mathbf{Y}_k)\theta \\ &= \mathbf{C}'_B\bar{\mathbf{b}} + \bar{c}_k\theta. \end{aligned} \tag{5.17}$$

So with $\theta > 0$, $f(\hat{\mathbf{X}}) > \mathbf{C}'_B\bar{\mathbf{b}}$ if $\bar{c}_k > 0$. Hence we obtain an improved solution to the primal problem. Q.E.D.

Based upon the preceding discussion it is evident that the exit criterion (5.15.1): (1) determines the vector $\mathbf{b}_r$ to be removed from the present primal feasible basis $\mathbf{B}$; (2) determines the value of the variable $x_{\hat{B}r}$ corresponding to the incoming vector $\mathbf{r}_k$ for the new primal feasible basis $\hat{\mathbf{B}}$; and (3) in the absence of degeneracy, insures that $\hat{\mathbf{X}}$ has exactly m positive components and thus constitutes a basic feasible solution to $\mathbf{AX} = \mathbf{b}$. And since the columns of $\hat{\mathbf{B}}$ are linearly independent, $\hat{\mathbf{X}}$ represents an extreme point of the primal feasible region.

We may summarize the salient features of the preceding construction by outlining the following stepwise procedure for implementing the simplex method:

5.3. PRIMAL SIMPLEX ALGORITHM.

Step 0. Initialize.

Obtain an initial primal feasible basis $\mathbf{B}$. This provides us with our first basic feasible and thus extreme point solution

$\mathbf{X}' = (\mathbf{X}'_B, \mathbf{X}'_R) = (\bar{\mathbf{b}}', \mathbf{0}')$ and beginning objective value $\mathbf{C}'\mathbf{X} = \mathbf{C}'_B\bar{\mathbf{b}}$, where $\bar{\mathbf{b}} = \mathbf{B}^{-1}\mathbf{b}$.

Step 1. Check the optimality evaluators $\bar{\mathbf{C}}'_R = \mathbf{C}'_R - \mathbf{C}'_B\mathbf{B}^{-1}\mathbf{R}$.

(a) If $\bar{\mathbf{C}}'_R \geqq \mathbf{0}'$, stop. We have obtained an optimal basic feasible solution.

(b) If $\bar{\mathbf{C}}'_R \ngeqq \mathbf{0}'$, apply the entry criterion (5.11).

Step 2 If x_{Rk} is chosen to become basic, check $\mathbf{Y}_k$.

(a) If $\mathbf{Y}_k \leqq \mathbf{0}$, stop. The objective value is unbounded along the infinite ray $\mathbf{X} + \theta\mathbf{d}_k$, where $x_{Rk} = \theta \geq 0$, $\mathbf{d}'_k = (-\mathbf{Y}'_k, \mathbf{e}'_k)$.

(b) If $\mathbf{Y}_k \nleqq \mathbf{0}$, apply the exit criterion (5.15.1) to determine the variable x_{Br} to become nonbasic.

Step 3. Form a new primal feasible basis $\widehat{\mathbf{B}}$ by replacing $\mathbf{b}_r$ within $\mathbf{B}$ by $\mathbf{r}_k$ and return to step 1.

The entire process involving steps (1) - (3) is repeated until either primal optimality is attained or we detect that the objective is unbounded. Since the simplex routine examines (adjacent) extreme points, which are finite in number, the simplex process must terminate in a finite number of iterations. And as (5.17) reveals, since $\bar{c}_k \geq 0$, the objective function will be nondecreasing between successive basic feasible or extreme point solutions.[1]

Let us now return to the example set forth at the outset of this chapter and proceed to generate an optimal basic feasible solution given the initial one. We previously found, using (5.1.1) and (5.2.2), that

$$\mathbf{X}_B = \mathbf{B}^{-1}\mathbf{b} - \mathbf{B}^{-1}\mathbf{R}\mathbf{X}_R = \mathbf{B}^{-1}\mathbf{b} - [\mathbf{Y}_1, \mathbf{Y}_2, \mathbf{Y}_3]\begin{bmatrix} x_{R1} \\ x_{R2} \\ x_{R3} \end{bmatrix} \quad \text{or}$$

$$\begin{bmatrix} x_3 \\ x_5 \end{bmatrix} = \begin{bmatrix} 25 - \frac{1}{2}x_1 - \frac{1}{4}x_2 - \frac{1}{4}x_4 \\ 15 - \frac{1}{2}x_1 - \frac{11}{4}x_2 + \frac{1}{4}x_4 \end{bmatrix};$$

[1] For details on an interesting variant of the standard simplex routine see the *steepest edge simplex method* presented in Appendix B.

$$\begin{aligned} f &= \mathbf{C}'_B\mathbf{B}^{-1}\mathbf{b} + (\mathbf{C}'_R - \mathbf{C}'_B\mathbf{B}^{-1}\mathbf{R})\mathbf{X}_R \\ &= \mathbf{C}'_B\mathbf{B}^{-1}\mathbf{b} + (\bar{c}_1, \bar{c}_2, \bar{c}_3)\begin{bmatrix} x_{R1} \\ x_{R2} \\ x_{R3} \end{bmatrix} \\ &= 100 - x_1 + 2x_2 - x_4. \end{aligned}$$

Setting $\mathbf{X}_R = \mathbf{0}$, our initial nondegenerate basic feasible solution was found to be $\mathbf{X}' = (0, 0, 25, 0, 15)$ with $f = 100$.

Since $\bar{c}_2(=2) > 0$, it is obvious that the current basic feasible solution is not optimal. Hence we must change the status of x_2 from nonbasic to basic. To this end we have

$$\mathbf{X}_B = \mathbf{B}^{-1}\mathbf{b} - x_{R2}\mathbf{Y}_2 \quad \text{or}$$

$$\begin{bmatrix} x_3 \\ x_5 \end{bmatrix} = \begin{bmatrix} x_{B1} - y_{12}x_{R2} \\ x_{B2} - y_{22}x_{R2} \end{bmatrix} = \begin{bmatrix} 25 - \frac{1}{4}x_2 \\ 15 - \frac{11}{4}x_2 \end{bmatrix}.$$

Setting this expression equal to the null vector renders the values of $x_{R2} = x_2$ which respectively reduce x_3, x_5 to zero, namely $x_2 = \frac{x_{B1}}{y_{12}} = 100$, $x_2 = \frac{x_{B2}}{y_{22}} = \frac{60}{11}$ and thus

$$\hat{\theta} = \frac{60}{11} = min\left\{\frac{x_{B1}}{y_{12}} = 100, \frac{x_{B2}}{y_{22}} = \frac{60}{11}\right\},$$

i.e., the largest increment in x_2 consistent with primal feasibility is $\frac{60}{11}$. Hence x_5 becomes nonbasic and $\mathbf{r}_2$ replaces $\mathbf{b}_2$ in the previous basis matrix so that our new basis matrix is

$$\hat{\mathbf{B}} = \begin{bmatrix} 4 & 1 \\ 1 & 3 \end{bmatrix}$$

with

$$\hat{\mathbf{B}}^{-1} = \begin{bmatrix} \frac{3}{11} & -\frac{1}{11} \\ -\frac{1}{11} & \frac{4}{11} \end{bmatrix}$$

(see Appendix A for a method of transforming $\mathbf{B}$ to $\hat{\mathbf{B}}$). In this instance,

$$\hat{\mathbf{R}} = \begin{bmatrix} 2 & 0 & 1 \\ 1 & 1 & 0 \end{bmatrix}, \quad \mathbf{X}_{\hat{B}} = \begin{bmatrix} x_3 \\ x_5 \end{bmatrix}, \quad \mathbf{X}_{\hat{R}} = \begin{bmatrix} x_1 \\ x_5 \\ x_4 \end{bmatrix}, \quad \mathbf{C}_{\hat{B}} = \begin{bmatrix} 4 \\ 3 \end{bmatrix}, \text{ and } \mathbf{C}_{\hat{R}} = \begin{bmatrix} 1 \\ 0 \\ 0 \end{bmatrix}$$

and thus an iteration produces

$$\mathbf{X}_{\hat{B}} = \hat{\mathbf{B}}^{-1}\mathbf{b} - \hat{\mathbf{B}}^{-1}\hat{\mathbf{R}}\mathbf{X}_{\hat{R}} \quad \text{or}$$

$$\begin{bmatrix} x_3 \\ x_2 \end{bmatrix} = \begin{bmatrix} 23.63 - \frac{5}{11}x_1 + \frac{1}{11}x_5 - \frac{1}{11}x_4 \\ 5.46 - \frac{2}{11}x_1 - \frac{4}{11}x_5 - \frac{1}{11}x_4 \end{bmatrix};$$

$$\begin{aligned} f &= \mathbf{C}'_{\hat{B}}\hat{\mathbf{B}}^{-1}\mathbf{b} + (\bar{c}_1, \bar{c}_2, \bar{c}_3)\mathbf{X}_{\hat{R}} \\ &= 110.91 - \tfrac{15}{11}x_1 - \tfrac{8}{11}x_5 - \tfrac{9}{11}x_4. \end{aligned}$$

Upon setting $\mathbf{X}_{\hat{R}} = \mathbf{0}$, $\hat{\mathbf{X}}' = (0, 5.46, 23.63, 0, 0)$ and $f = 110.91$. And since all the $\bar{c}_j$'s are negative, this second nondegenerate basic feasible solution is optimal.

What are the optimal values of the dual variables associated with this primal solution? To answer this we need only compute

$$\hat{\mathbf{U}} = \mathbf{C}'_{\hat{B}}\hat{\mathbf{B}}^{-1} = (4,3)\begin{bmatrix} \frac{3}{11} & -\frac{1}{11} \\ -\frac{1}{11} & \frac{4}{11} \end{bmatrix} = \left(\tfrac{9}{11}, \tfrac{8}{11}\right).$$

In addition,

$$f = g = \mathbf{b}'\hat{\mathbf{U}} = (100, 40)\begin{bmatrix} \frac{9}{11} \\ \frac{8}{11} \end{bmatrix} = 110.91$$

are required.

5.6. Degenerate Basic Feasible Solutions

Earlier in our discussion of the simplex method we instituted a nondegeneracy assumption, the purpose of which was to insure that every basic feasible solution to the augmented structural constraint system $\mathbf{AX} = \mathbf{b}$ was nondegenerate. In this section we shall examine the implications of the existence of degenerate basic feasible solutions to $\mathbf{AX} = \mathbf{b}$. We noted in Chapter 3 that if

$\rho(\mathbf{A}) = m$, we may select m linearly independent columns from $\mathbf{A}$ and form an $(m \times m)$ nonsingular matrix $\mathbf{B}$. We then obtain a basic solution to $\mathbf{AX} = \mathbf{b}$ by solving $\mathbf{BX}_B = \mathbf{b}$. Hence such a solution occurs when $\mathbf{b}$ is expressible as a linear combination of the columns of $\mathbf{B}$. Moreover, the said solution is feasible if $\mathbf{b}$ can be represented as a nonnegative linear combination of the same, *i.e.*, $\mathbf{b}$ is an element of the convex polyhedral cone $\mathcal{C}_B = \{\mathbf{Y} \mid \mathbf{Y} = \mathbf{BX}_B, \mathbf{X}_B \geqq \mathbf{0}\}$ generated by the columns of $\mathbf{B}$. Let us now get a bit more restrictive. If $\mathbf{b}$ is expressible as a strictly positive linear combination of all the basis vectors in $\mathbf{B}$, then the basic feasible solution to $\mathbf{AX} = \mathbf{b}$ nondegenerate. In this instance, none of the basic variables within $\mathbf{X}_B$ vanishes. In sum, the point being made is that degeneracy occurs if $\mathbf{b}$ lies within a convex polyhedral cone spanned by only some subset of vectors of $\mathbf{B}$ *i.e.*, $\mathbf{b}$ is a positive linear combination of fewer than m basis vectors. Hence $\mathbf{b}$ lies along an edge of $\mathcal{C}_B$ and not within $\mathcal{C}_B$ proper so that not all components of $\mathbf{X}_B$ are positive.

To relate these observations to our discussion in the previous section pertaining to the generation of new or improved basic feasible solutions, let us address ourselves to a couple of remarks concerning the value of $\widehat{\theta}$ which constitutes the minimum in (5.15). As we shall now see, the presence or absence of degeneracy is reflected by the behavior of $\widehat{\theta}$. What exceptional situation may arise? Let us assume that $\widehat{\theta}$ is not unique. Hence we may remove from the basis any one of the columns of $\mathbf{B}$ for which the same minimum value of x_{Rk} $(\widehat{\theta})$ is assumed and obtain a new (call it the ℓth) basic feasible solutions. In this instance, the ℓth basic feasible solution is degenerate since $x_{\widehat{B}i} = 0$ for all i corresponding to those columns that have the same $\theta = \widehat{\theta}$ as the eliminated column. To see this, let us assume that the status of x_{Ru} is to be changed from nonbasic to basic and that $\widehat{\theta} = \frac{x_{Br}}{y_{ru}}$, $y_{ru} > 0$, so that column $\mathbf{b}_r$ of $\mathbf{B}$ is to be removed from the basis. In addition, for some other column $\mathbf{b}_s$ in $\mathbf{B}$, let $\frac{x_{Br}}{y_{ru}} = \frac{x_{Bs}}{y_{su}}$, $y_{su} > 0$. Then from (5.16.1),

$$\begin{aligned} x_{\hat{B}i} &= x_{Bi} - \hat{\theta} y_{iu} \\ &= x_{Bi} - \frac{x_{Br}}{y_{ru}} y_{iu} \quad \text{or} \end{aligned}$$

$$\frac{x_{\hat{B}i}}{y_{iu}} = \frac{x_{Bi}}{y_{iu}} - \frac{x_{Br}}{y_{ru}}$$

and, for $i = s$,

$$\frac{x_{\hat{B}s}}{y_{su}} = \frac{x_{Bs}}{y_{su}} - \frac{x_{Br}}{y_{ru}} = 0 \quad \text{or} \quad x_{\hat{B}s} = 0.$$

So when the variable x_{Ru} turns basic (becoming $x_{\hat{B}r}$) and thus increases from zero to a level coincident with $\hat{\theta}$, it reduces to zero not only the variable which it is displacing, namely x_{Br}, but also any other variable $x_{\hat{B}s}$ whose original value x_{Bs} was used in determining $\hat{\theta}$. In this regard, if another iteration is performed and we obtain the $\ell+1^{st}$ basic feasible solution, the sth component $x_{\hat{B}s}$ within $\mathbf{X}_{\hat{B}} = \hat{\mathbf{B}}^{-1}\mathbf{b}$ is zero. It is at this point in our discussion of degeneracy that a major complication introduced by this notion emerges. For if we now determine that the vth column of $\hat{\mathbf{R}}$ is to be introduced into the basis, then $\hat{\theta} = \frac{x_{\hat{B}s}}{y_{sv}} = \frac{0}{y_{sv}} = 0$, $y_{sv} > 0$, *i.e.*, $\hat{\theta}$ vanishes since the current or $\ell+1^{st}$ basic feasible solution is degenerate. With $\hat{\theta} = 0$, (5.16.1) informs us that if the next or $\ell+2^{nd}$ iteration is executed, the values of the variables common to the $\ell+1^{st}$ and $\ell+2^{nd}$ basic feasible solutions are unchanged while the value of the variable $x_{\hat{R}v}$ whose status has changed from nonbasic to basic is zero. Hence the $\ell+2^{nd}$ basic feasible solution is also degenerate. In this situation, it is clear that no improvement in the value of the objective function occurs, *i.e.*, although we have actually obtained a new degenerate basic feasible solution from the old one, we are still at the same extreme point. On this latter point pertaining to the absence of a one-to-one correspondence between basic feasible solutions and extreme points in the presence of degeneracy, see the argument underlying (2.22.1).

An additional complication precipitated by degeneracy is that we may not be able to unambiguously interpret the dual variable u_i as the marginal change in the optimal primal objective value per unit change in the ith requirement b_i, *i.e.*, $u_i = \partial f/\partial b_i$, $i = 1,\ldots,m$, may not be uniquely defined in the presence of degeneracy, a point to which we shall return later on.

It must be mentioned that a new basic feasible solution will not always be degenerate given that the previous one was. For example, if for some $x_{Br} = 0$ we have $y_{rj} < 0$, then, since negative y_{ij}'s are not used in computing $\widehat{\theta}$, $\frac{x_{Br}}{y_{rj}} = 0$ is not a candidate for the minimum in (5.15) so that $x_{\hat{B}r} = \widehat{\theta} \neq 0$ in the next iteration. Thus the new basic feasible solution will be nondegenerate.

To summarize, when a degenerate basic feasible solution occurs at some particular round of the simplex method, the value of the objective function is unchanged when the next iteration is carried out. In fact, it may remain constant for some large number of subsequent iterations. In addition, there also exists the possibility that all succeeding basic feasible solutions will be degenerate. In this latter instance, the simplex process is said to **cycle** forever in an endless loop, *i.e.*, a given sequence of bases is repeated again and again so that convergence to an optimal solution never occurs.[2] In the absence of degeneracy, the value of the objective function increases at each iteration so that we never return to a previous basis. And as mentioned earlier, since there are a finite number of bases, we are guaranteed that the simplex method will terminate in a finite number of iterations.

When degeneracy presents itself, how may it be resolved? The answer is quite simple. In light of the previously observed fact that the threat of cycling is, for all practical purposes, virtually nonexistent, no antidegeneracy procedure is really needed since the effect of degeneracy is simply to slow down the simplex routine along its path to the optimum rather than to preclude the optimum from ever being attained. All that one really needs to do is to make an arbitrary

[2]While cycling is indeed theoretically possible, it is essentially an academic matter since, in actual practice, no ordinary linear programming problem has ever been known to cycle (although the sort of degeneracy discussed above is quite common). Artificial problems which exhibit cycling have been discussed by Beal (1955), Hoffman (1953), Marshall and Suurballe (1969), Balinski and Tucker (1969), and Yudin and Gol'shtien (1965).

choice from among the tied ratios which constitute the minimum in (5.15). If a specific anticycling procedure is desired, we may employ:

(1) the perturbation method of Charnes (1952) which perturbs or moves the vector **b** a suitably restricted distance and in a fashion such that **b** becomes a strictly positive linear combination of the columns of **B**.

(2) a lexicographic rule for selecting the exiting variable proposed by Dantzig, Orden, and Wolfe (1954); and

(3) the rule of Bland (1977) for selecting both entering and exiting variables.

Additional considerations regarding anticycling routines are provided by Dantizig (1951a), Wolfe (1963), Perold (1980), Azpeitia and Dickinson (1964,), and Gass (1985).

5.7. Unbounded Solutions Reexamined

We noted earlier that any vector $\mathbf{r}_j$ with $\overline{c}_j > 0$ may be introduced into the basis provided that at least one component of $\mathbf{Y}_j$ is positive, *i.e.*, there must exist at least one $y_{ij} > 0$, $i = 1,\ldots,m$. Given that the present basic feasible solution is nondegenerate, this restriction not only ensures that the new solution will be basic as well as feasible, it also precludes the existence of feasible solutions which are nonbasic (contain more than m nonnegative variables) and for which the objective function possesses no finite upper bound. To verify this let us amplify our earlier comments on the existence of an extreme direction $\mathbf{d}_j$ along which f can be made to increase without bound (corollary 5.2) and assume that we are to introduce into the basis some vector $\mathbf{r}_j$ for which $\overline{c}_j > 0$ and all $y_{ij} \leq 0, i = 1,\ldots,m$. In this circumstance, an examination of the discussion underlying (5.15) indicates that for the new basic solution to be feasible, $\mathbf{r}_j$ must enter the basis at the zero level. But to determine the effect of $\mathbf{r}_j$ on the objective function, inserting this vector at a zero level will definitely not do. Hence the following trade-off will be made. Instead of trying to obtain a basic solution which is feasible, let us derive a feasible solution which is nonbasic, *i.e.*, which expresses **b** as a nonnegative linear combination of $m+1$ vectors.

With all $y_{ij} \leq 0$, $i = 1,\ldots,m$, $\mathbf{r}_j$ cannot replace any vector currently in the basis since $x_{Bi} - \hat{\theta} y_{ij} > 0$, $i = 1,\ldots,m$, given that $\hat{\theta} > 0$, *i.e.*, in this situation, it is impossible to change the status of any variable from basic to nonbasic. So if the current basic feasible solution is nondegenerate, $\mathbf{b}$ is expressible as a nonnegative linear combination of the $m+1$ vectors $\mathbf{b}_1,\ldots,\mathbf{b}_m, \mathbf{b}_r$ and thus $m+1$ variables or components of $\hat{\mathbf{X}}$ are different from zero. To this end let us add and subtract $\hat{\theta}\mathbf{r}_j$ in the current basic feasible solution to obtain

$$\mathbf{B}\mathbf{X}_B + \hat{\theta}\mathbf{r}_j - \hat{\theta}\mathbf{r}_j = \mathbf{b}.$$

Since $\mathbf{r}_j = \mathbf{B}\mathbf{Y}_j$,

$$\mathbf{B}(\mathbf{X}_B - \hat{\theta}\mathbf{Y}_j) + \hat{\theta}\mathbf{r}_j = \mathbf{b}.$$

With $y_{ij} \leq 0, i = 1,\ldots,m,$ and $\hat{\theta} > 0$, $\mathbf{X}_B - \hat{\theta}\mathbf{Y}_j \geqq \mathbf{0}$ so that while this solution to the augmented structural constraint system $\mathbf{AX} = \mathbf{b}$ is feasible, it is nonbasic since $\mathbf{b}$ is expressible as a nonnegative linear combination of $m+1$ vectors, *i.e.*, $\hat{\mathbf{X}}$ has $m+1$ nonnegative components. The value of the objective function corresponding to this nonbasic feasible solution is, from (5.17),

$$\begin{aligned} f(\hat{\mathbf{X}}) &= \mathbf{C}'_B(\mathbf{X}_B - \hat{\theta}\mathbf{Y}_j) + c_{Rj}\hat{\theta} \\ &= f(\mathbf{X}) + \bar{c}_j\hat{\theta}, \end{aligned}$$

where $c_{Bi}(x_{Bi} - \hat{\theta} y_{ij}) \neq 0$, $i = 1,\ldots,m$. So with $\bar{c}_j > 0$, $f(\hat{\mathbf{X}})$ can be made arbitrarily large simply by making an appropriate selection of $\hat{\theta}$. In this instance f is said to be **unbounded** since there is, theoretically, no finite upper limit to our choice of $\hat{\theta}$ and thus to the value of f. In sum, if all $y_{ij} \leq 0$, $i = 1,\ldots,m$, for some vector $\mathbf{r}_j$ in $\mathbf{R}$ with $\bar{c}_j > 0$, then there exists a feasible solution in the $m+1$ variables $x_{B1},\ldots,x_{Bm}$, x_{Rj} with the result that the values of the objective function and at least two of these variables may be increased without bounded. If this case obtains, the simplex process should be terminated.

5.8. Multiple Optimal Solutions

We noted in the previous chapter that, by virtue of the extreme point theorem, the objective function assumes its optimal value at an extreme point of the feasible region $\mathcal{K}_P$. Moreover, if the optimum occurs at more than one extreme point, the objective function takes on that same value at every convex combination of those points. Now, extreme point solutions correspond to basic feasible solutions and conversely so that in this latter instance involving multiple optimal solutions, there exists only a finite number of optimal basic feasible solutions but an infinite number of optimal nonbasic feasible solutions. In short, while the optimal value of the objective function is unique, the set of basic variables which yield the optimum will not be unique in the case of multiple optimal solutions.

How may we determine whether two or more different (nondegenerate) basic feasible solutions yield the same value of the objective function? To answer this question let us assume that we have obtained an optimal basic feasible solution to a linear programming problem and that $\bar{c}_j = 0$ for some vector $\mathbf{r}_j$ in $\mathbf{R}$ with at least one $y_{ij} > 0$, $i = 1,\ldots,m$. In this instance it is evident that when $\mathbf{r}_j$ replaces one of the vectors currently in the basis, we obtain a new basic feasible solution $\hat{\mathbf{X}}$ with $f(\hat{\mathbf{X}}) = f(\mathbf{X}) + \bar{c}_j\hat{\theta} = f(\mathbf{X})$. Hence $\hat{\mathbf{X}}$ must also be optimal. So when the coefficient $\bar{c}_j$ on the nonbasic variable x_{Rj} is zero for some particular optimal basic feasible solution, the set of basic variables which yield the optimum is not unique, *i.e.*, x_{Rj} can become basic without changing the maximal value of the objective function so that we obtain an alternative optimal basic feasible solution. Such a situation is illustrated in Figure 5.2.a. Here the extreme points A, B yield optimal basic feasible solutions while any nonextreme point $C = \lambda A + (1-\lambda)B, 0 < \lambda < 1$, yields an optimal though nonbasic feasible solution with $f(A) = f(B) = f(C)$.

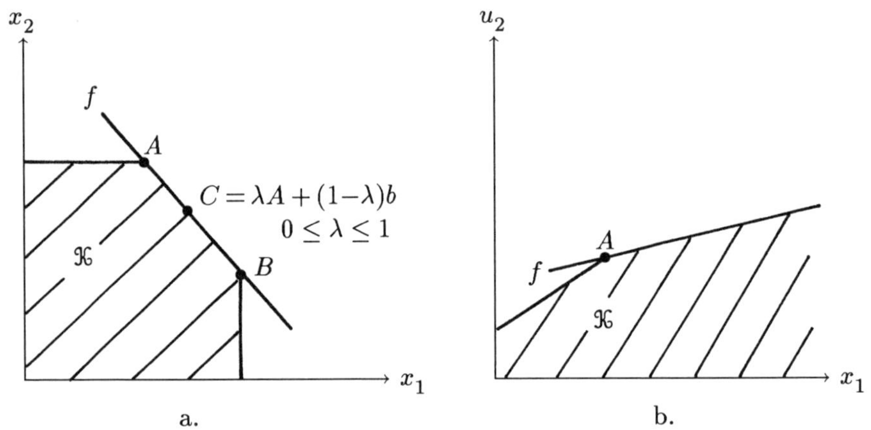

Figure 5.2

The above discussion has indicated that if there exists at least two different optimal basic feasible solutions to a linear programming problem, then there also exists an infinite number of optimal nonbasic solutions. However, it must be mentioned that there may occur a situation in which we have an infinite number of optimal nonbasic feasible solutions but only a single optimal basic feasible solution. To verify this let us assume that we have obtained an optimal basic feasible solution and that $\bar{c}_j = 0$ for some vector $\mathbf{r}_j$ in $\mathbf{R}$ with all $y_{ij} \geq 0$, $i = 1, \ldots, m$. If $\mathbf{r}_j$ is introduced into the basis, we obtain a new feasible nonbasic solution

$$\mathbf{B}(\mathbf{X}_B - \hat{\theta}\mathbf{Y}_j) + \hat{\theta}\mathbf{r}_j = \mathbf{b}$$

in the $m+1$ vectors $\mathbf{b}_1, \ldots, \mathbf{b}_m, \mathbf{r}_j$. In this instance $f(\hat{\mathbf{X}}) = f(\mathbf{X}) + \bar{c}_j\hat{\theta} = f(\mathbf{X})$ so that $\hat{\mathbf{X}}$ is also an optimal solution for any arbitrary $\hat{\theta} > 0$. With $\mathbf{r}_j \neq \mathbf{0}$, at least one $y_{ij} < 0$ so that there exist optimal nonbasic solutions for which at least two of the $m+1$ variables can be made arbitrarily large since the value of $\hat{x}_{Rj}$ can be increased without bound. This case is illustrated in Figure 5.2.b. Here the optimal hyperplane is tangent to the feasible region along an edge which extends to infinity.

How does a situation involving the existence of multiple optimal basic feasible solutions to a linear programming problem formally compare with one in which a degenerate basic feasible solution occurs. A definite similarity emerges since, quite generally, the value of the objective function remains unchanged in each case as we move from one basic feasible solution to another. Specifically, in a problem involving multiple optimal solutions, we obtain the same value of the objective function as at the preceding solution since $\bar{c}_j = 0$ for some vector $\mathbf{r}_j$ not in the basis with $\widehat{\theta} > 0$, while a degenerate solution yields, after an additional iteration, the same value of the objective function because $\widehat{\theta} = 0$. In the first instance we move from one extreme point and thus basic feasible solution to another while in the latter we remain at the same extreme point although we obtain a new degenerate basic feasible solution.

6 DUALITY THEORY REVISITED

6.1. The Geometry of Duality and Optimality

Starting with the primal problem, let us

$$\begin{aligned} max\, f(\mathbf{X}) &= \mathbf{C}'\mathbf{X} \quad \text{s.t.} \\ \mathbf{AX} \leqq \mathbf{b},\ \mathbf{X} &\geqq \mathbf{0}, \quad \mathbf{X} \in R^n, \end{aligned}$$

where $\mathbf{A}$ is of order $(m \times n)$ with rows $\boldsymbol{\alpha}_1, \ldots, \boldsymbol{\alpha}_m$ and $\mathbf{b}$ is an $(m \times 1)$ vector with components $b_1, \ldots, b_m$. Alternatively, if

$$\bar{\mathbf{A}} = \begin{bmatrix} \mathbf{A} \\ \cdots \\ -\mathbf{I}_n \end{bmatrix}, \quad \bar{\mathbf{b}} = \begin{bmatrix} \mathbf{b} \\ \cdots \\ \mathbf{0} \end{bmatrix}$$

are respectively of order $(m+n \times n)$ and $(m+n \times 1)$, then we may

$$\begin{aligned} max\, f(\mathbf{X}) &= \mathbf{C}'\mathbf{X} \quad \text{s.t.} \\ \bar{\mathbf{b}} - \bar{\mathbf{A}}\mathbf{X} &\geqq \mathbf{0}, \end{aligned}$$

where $\mathbf{X} \in R^n$ is now unrestricted. In this formulation $x_j \geq 0,\ j = 1, \ldots, n,$ is treated as a structural constraint so that $\bar{\mathbf{b}} - \bar{\mathbf{A}}\mathbf{X} \geq \mathbf{0}$ defines a **region of feasible or admissible solutions** $\mathcal{K} \subset R^n$.

If $\mathbf{X}_0$ yields an optimal solution to the primal maximum problem, then no permissible change in $\mathbf{X}$ (*i.e.*, one which does not violate any of the constraints specifying $\mathcal{K}$) can improve the optimal value of the objective function. How may we characterize such admissible changes? It is evident that, if starting at some feasible $\mathbf{X}$, we can find a vector $\mathbf{h}$ such that a small movement along it violates no constraint, then $\mathbf{h}$ specifies a feasible direction at $\mathbf{X}_0$. More formally, let $\delta(\mathbf{X}_0)$ be a suitably restricted spherical δ-neighborhood about the point $\mathbf{X}_0 \in \mathbf{R}^n$. Then the $(n \times 1)$ vector $\mathbf{h}$ is a **feasible direction** at $\mathbf{X}_0$ if there exists a scalar $t \geq 0$, $0 \leq t < \delta$, such that $\mathbf{X}_0 + t\mathbf{h}$ is feasible. In this light, for $\delta(\mathbf{X}_0)$ again a spherical δ-neighborhood about the feasible point $\mathbf{X}_0 \in R^n$, the **set of feasible directions** at $\mathbf{X}_0, \mathfrak{D}(\mathbf{X}_0)$, is the collection of all $(n \times 1)$ vectors $\mathbf{h}$ such that $\mathbf{X}_0 + t\mathbf{h}$ is feasible for t sufficiently small, *i.e.*,

$$\mathfrak{D}(\mathbf{X}_0) = \{\mathbf{h} \mid \text{if } \mathbf{X}_0 \in \mathfrak{K}, \text{ then } \mathbf{X}_0 + t\mathbf{h} \in \mathfrak{K}, 0 \leq t < \delta\}.$$

Here $\mathfrak{D}(\mathbf{X}_0)$ is the **tangent support cone** consisting of all feasible directions at $\mathbf{X}_0$ and is generated by the tangent or supporting hyperplanes to $\mathfrak{K}$ at $\mathbf{X}_0$ (Figure 6.1). And since each such hyperplane

$$\begin{aligned} -\bar{\mathbf{a}}_i\mathbf{h} &= -\bar{\mathbf{a}}_i(\mathbf{X}-\mathbf{X}_0) \\ &= \bar{b}_i - \bar{\mathbf{a}}_i\mathbf{X} = 0, \ i = 1,\ldots,m+n,^1 \end{aligned}$$

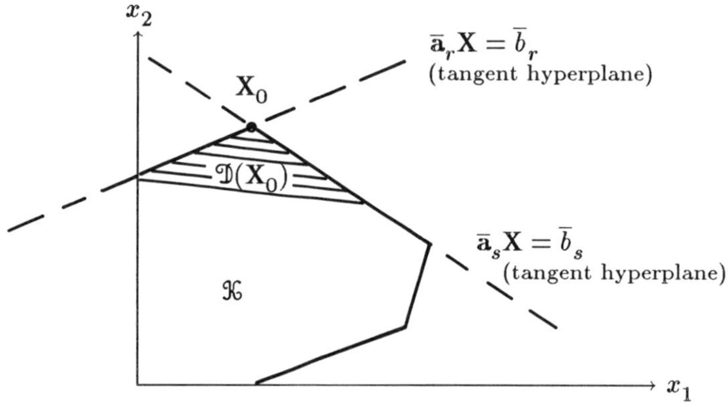

Figure 6.1

specifies a closed half-space $\bar{\mathbf{a}}_i\mathbf{X} \leq \bar{b}_i$, $i = 1,\ldots,m+n$, the tangent support cone $\mathfrak{D}(\mathbf{X}_0)$ represents that portion of the intersection of these closed half-places in the immediate vicinity of $\mathbf{X}_0$.

---

[1]Here $\bar{\mathbf{a}}_i$ is the ith row of $\bar{\mathbf{A}}$ while $\bar{b}_i$ is the ith component of $\bar{\mathbf{b}}$. Specifically: $\bar{\mathbf{a}}_1 = \boldsymbol{\alpha}_1,\ldots,\bar{\mathbf{a}}_m = \boldsymbol{\alpha}_m$; $\bar{\mathbf{a}}_{m+j} = \mathbf{e}'_j$, $j = 1,\ldots,n$; $\bar{b}_1 = b_1,\ldots,\bar{b}_m = b_m$; and $\bar{b}_{m+j} = 0$, $j = 1,\ldots,n$.

To discern the relationship between $\mathcal{D}(\mathbf{X}_0)$ ($\mathbf{X}_0$ optimal) and: (a) $f(\mathbf{X}_0)$; (b) the constraints $\bar{\mathbf{b}} - \bar{\mathbf{A}}\mathbf{X}_0 \geqq \mathbf{0}$, let us consider the following two theorems. First,

6.1. THEOREM. If the primal objective function attains its maximum at $\mathbf{X}_0$, then any movement from $\mathbf{X}_0$ along a feasible direction $\mathbf{h}$ cannot increase the value of f, *i.e.*,

$$df^0 = \mathbf{C}'\mathbf{h} = \mathbf{C}'(\mathbf{X}-\mathbf{X}_0) \leq 0 \quad \text{or } \mathbf{C}'\mathbf{X} \leq \mathbf{C}'\mathbf{X}_0.$$

PROOF. If $f(\mathbf{X}_0) = \mathbf{C}'\mathbf{X}_0$ is the maximum of f over $\mathcal{K}$, then for $t > 0$

$$\mathbf{C}'(\mathbf{X}_0+t\mathbf{h}) - \mathbf{C}'\mathbf{X}_0 = t\mathbf{C}'\mathbf{h} \leq 0$$

and thus $\mathbf{C}'\mathbf{h} \leq 0$ for $t > 0$. Q.E.D.

Hence f must decrease along any feasible direction $\mathbf{h}$. Geometrically, no feasible direction may form an acute angle ($< \pi/2$) between itself and $\mathbf{C}$.

To set the stage for our next theorem, we note that for any optimal $\mathbf{X} \epsilon \mathcal{K}$, it is usually the case that that not all of the inequality constraints are binding or hold as an equality. To incorporate this observation into our discussion, let us divide all of the constraints $\bar{\mathbf{b}} - \bar{\mathbf{A}}\mathbf{X} \geqq \mathbf{0}$ into two mutually exclusive classes — those which are binding at $\mathbf{X}$, $\bar{b}_i - \bar{\mathbf{a}}_i\mathbf{X} = 0$, and those which are not, $\bar{b}_i - \bar{\mathbf{a}}_i\mathbf{X} > 0$, $i = 1,\ldots, m+n$. So if

$$\mathcal{J} = \left\{ i \mid \bar{b}_i - \bar{\mathbf{a}}_i\mathbf{X} = 0, \ \mathbf{X} \epsilon R^n \right\}$$

depicts the index set of binding constraints, then $\bar{b}_i - \bar{\mathbf{a}}_i\mathbf{X} > 0$ for $i \notin \mathcal{J}$. In this regard,

6.2. THEOREM. The vector $\mathbf{h} \epsilon \mathcal{D}(\mathbf{X}_0)$ ($\mathbf{X}_0$ optimal) if and only if $-\bar{\mathbf{a}}_i\mathbf{h} \geq 0$, $i \epsilon \mathcal{J}$.

PROOF. (necessity) Let $\mathbf{X}_0 \epsilon \mathcal{K}$ with $\bar{\mathbf{a}}_i\mathbf{X}_0 = \bar{b}_i$, $i \epsilon \mathcal{J}$. For $\mathbf{h} \epsilon \mathcal{D}(\mathbf{X}_0)$, $\bar{\mathbf{a}}_i(\mathbf{X}_0 + t\mathbf{h}) \leq \bar{b}_i$, $0 \leq t < \delta$, for some i. If $i \epsilon \mathcal{J}$, $\bar{\mathbf{a}}_i\mathbf{X}_0 = \bar{b}_i$ and thus $t\bar{\mathbf{a}}_i\mathbf{h} \leq 0$ or $-\bar{\mathbf{a}}_i\mathbf{h} \geq 0$, $t > 0$, for all $i \epsilon \mathcal{J}$. (sufficiency). Let $-\bar{\mathbf{a}}_i\mathbf{h} \geq 0$ for all $i \epsilon \mathcal{J}$. Then for some $i \notin \mathcal{J}$, $\bar{\mathbf{a}}_i\mathbf{X}_0 < \bar{b}_i$ and thus

$$\bar{\mathbf{a}}_i(\mathbf{X}_0 + t\mathbf{h}) = \bar{\mathbf{a}}_i\mathbf{X}_0 + t\bar{\mathbf{a}}_i\mathbf{h} < \bar{b}_i$$

for all $t \geq 0$. If $i \epsilon \mathcal{J}$, $\bar{\mathbf{a}}_i\mathbf{X}_0 = \bar{b}_i$ so that

$$\bar{\mathbf{a}}_i(\mathbf{X}_0+t\mathbf{h}) = \bar{b}_i + t\bar{\mathbf{a}}_i\mathbf{h} \leq \bar{b}_i$$

for each $t \geq 0$. Q.E.D.

To interpret this theorem we note that $-\bar{\mathbf{a}}_i'$ may be characterized as an inward pointing or interior normal to $-\bar{\mathbf{a}}_i\mathbf{h}=0$ (or $\bar{b}_i-\bar{\mathbf{a}}_i\mathbf{X}=0$) at $\mathbf{X}_0$. So if $\mathbf{h}$ is a feasible direction, it makes a nonobtuse angle ($\leq \pi/2$) with all of the interior normals to the boundary of $\mathcal{K}$ at $\mathbf{X}_0$. Geometrically, these interior normals or gradients of the binding constraints form a finite cone containing all feasible directions making nonobtuse angles with the supporting hyperplanes to $\mathcal{K}$ at $\mathbf{X}_0$. Such a cone is polar to the tangent support cone $\mathcal{D}(\mathbf{X}_0)$ and will be termed the **polar support cone** $\mathcal{D}(\mathbf{X}_0)^+$ (Figure 6.2.a). Thus $\mathcal{D}(\mathbf{X}_0)^+$ is the cone spanned by the gradients $-\bar{\mathbf{a}}_i'$ such that for $\mathbf{h}\in\mathcal{D}(\mathbf{X}_0)$, $-\bar{\mathbf{a}}_i\mathbf{h}>0$, $i\in\mathcal{I}$. Looked at from another perspective, $\bar{\mathbf{a}}_i'$ may be considered an outward pointing or exterior normal to the boundary of $\mathcal{K}$ at $\mathbf{X}_0$. In this instance, if $\mathbf{h}$ is a feasible direction, it must now make a nonacute angle ($\geq \pi/2$) with all of the outward pointing normals to the boundary of $\mathcal{K}$ at $\mathbf{X}_0$. Again looking to geometry, the exterior normals or negative gradients of the binding constraints form a finite cone containing all feasible directions making nonacute angles with the hyperplanes tangent to $\mathcal{K}$ at $\mathbf{X}_0$. This cone is the dual of the tangent support cone $\mathcal{D}(\mathbf{X}_0)$ and is termed the **dual support cone** $\mathcal{D}(\mathbf{X}_0)^*$ (Figure 6.2.b). Thus $\mathcal{D}(\mathbf{X}_0)^*$ is the cone spanned by the negative gradients $\bar{\mathbf{a}}_i'$ such that for all $\mathbf{h}\in\mathcal{D}(\mathbf{X}_0)$, $\bar{\mathbf{a}}_i\mathbf{h}\leq 0$, $i\in\mathcal{I}$.

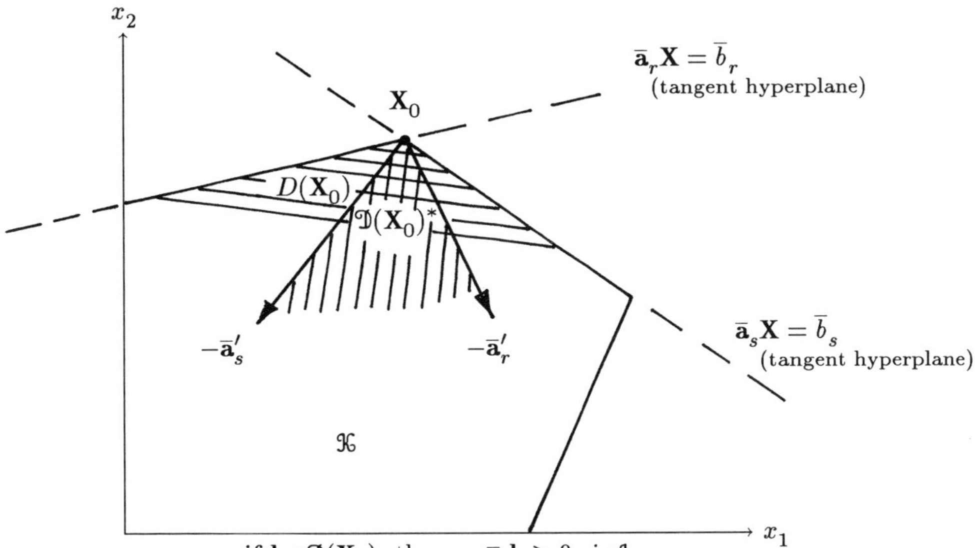

if $\mathbf{h} \in \mathfrak{D}(\mathbf{X}_0)$, then $-\bar{\mathbf{a}}_i\mathbf{h} \geq 0,\ i \in \mathfrak{J}$

a.

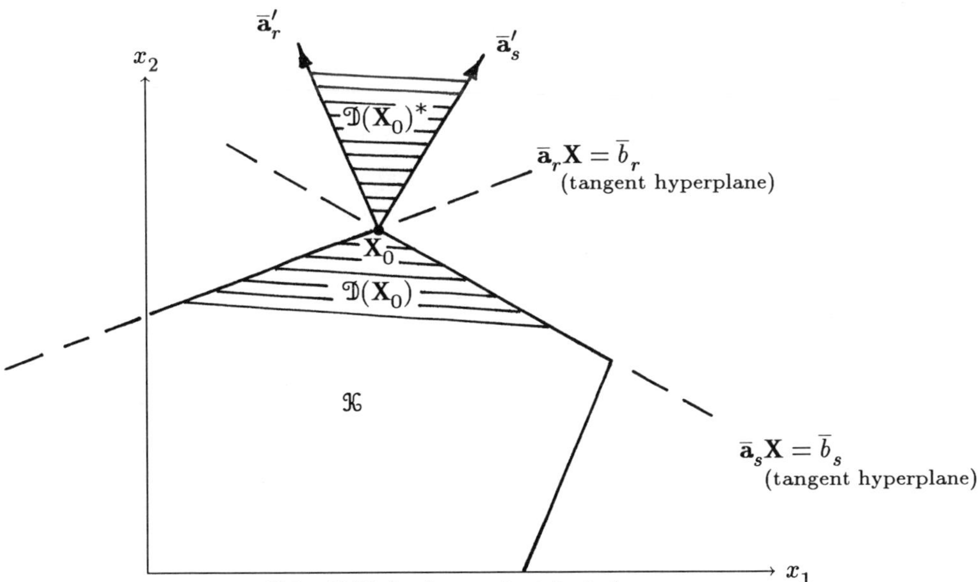

if $\mathbf{h} \in \mathfrak{D}(\mathbf{X}_0)$, then $\bar{\mathbf{a}}_i\mathbf{h} \leq 0,\ i \in \mathfrak{J}$

b.

Figure 6.2

At this point let us collect our major results. We found that if $f(\mathbf{X})$ subject to $\bar{b}_i - \bar{\mathbf{a}}_i\mathbf{X} = 0$, $i \in \mathcal{I}$, attains its maximum at $\mathbf{X}_0$, then

$$\mathbf{C}'\mathbf{h} \leq 0 \text{ for all } \mathbf{h} \text{ satisfying } \bar{\mathbf{a}}_i\mathbf{h} \leq 0,\ i \in \mathcal{I},\ \mathbf{h} \in \mathcal{D}(\mathbf{X}_0). \tag{6.1}$$

How may we interpret this condition? Given any $\mathbf{h} \in \mathcal{D}(\mathbf{X}_0)$, (6.1) holds if the gradient of f or $\mathbf{C}$ lies within the finite cone spanned by the exterior normals $\bar{\mathbf{a}}_i'$, $i \in \mathcal{I}$, *i.e.*, $\mathbf{C} \in \mathcal{D}(\mathbf{X}_0)^*$ (or if $-\mathbf{C}$ is contained within the finite cone generated by the interior normals $-\bar{\mathbf{a}}_i'$, $i \in \mathcal{I}$, *i.e.*, $-\mathbf{C} \in \mathcal{D}(\mathbf{X}_0)^+$). Hence (6.1) requires that the gradient of f be a nonnegative linear combination of the negative gradients of the binding constraints at $\mathbf{X}_0$ (Figure 6.3). In this regard, there must exist real numbers $\bar{u}_i^0 \geq 0$ such that

$$\mathbf{C} = \sum_i \bar{u}_i^0\, \bar{\mathbf{a}}_i',\ i \in \mathcal{I}. \tag{6.2}$$

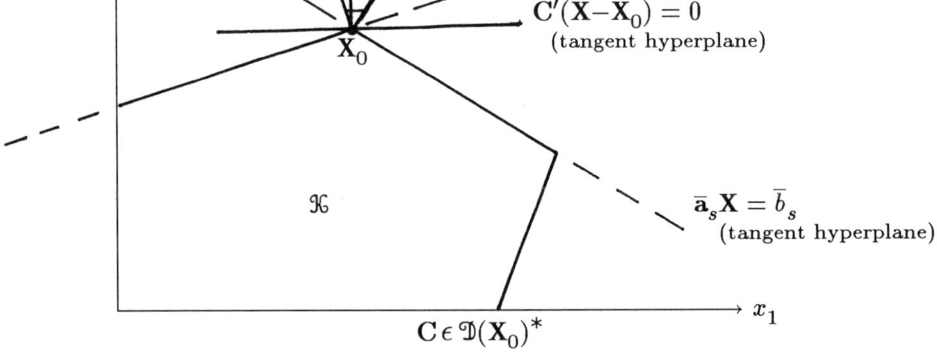

Figure 6.3

Under what conditions will the numbers $\bar{u}_i^0 \geq 0$, $i \in \mathcal{I}$, exist? To answer this question let us employ the

6.3. MINKOWSKI-FARKAS THEOREM. A necessary and sufficient condition for the n-component vector $\mathbf{V}$ to lie within the finite cone spanned by the columns of the $(m \times n)$ matrix $\mathbf{B}$ is that $\mathbf{V}'\mathbf{Y} \leq 0$ for all $\mathbf{Y}$ satisfying $\mathbf{B}'\mathbf{Y} \leqq \mathbf{0}$, *i.e.*, there exists an n-component vector $\boldsymbol{\lambda} \geqq \mathbf{0}$ such that $\mathbf{B}\boldsymbol{\lambda} = \mathbf{V}$ if and only if $\mathbf{V}'\mathbf{Y} \leq 0$ for all $\mathbf{Y}$ satisfying $\mathbf{B}'\mathbf{Y} \leqq \mathbf{0}$.

PROOF. (sufficiency) if $\mathbf{B}\boldsymbol{\lambda} = \mathbf{V}$, $\boldsymbol{\lambda} \geqq \mathbf{0}$, then $\boldsymbol{\lambda}'\mathbf{B}' = \mathbf{V}'$ and $\boldsymbol{\lambda}'\mathbf{B}'\mathbf{Y} = \mathbf{V}'\mathbf{Y} \leq 0$ for all $\mathbf{Y}$ for which $\mathbf{B}'\mathbf{Y} \leqq \mathbf{0}$. (necessity) If $\mathbf{V}'\mathbf{Y} \leq 0$ for each $\mathbf{Y}$ satisfying $\mathbf{B}'\mathbf{Y} \leqq \mathbf{0}$, then the (primal) linear programming problem

$$\begin{aligned} &max\ f(\mathbf{Y}) = \mathbf{V}'\mathbf{Y} \quad \text{s.t.} \\ &\mathbf{B}'\mathbf{Y} \leqq \mathbf{0},\ \mathbf{Y} \text{ unrestricted} \end{aligned}$$

has as its optimal solution $\mathbf{Y} = \mathbf{0}$. Hence the dual problem

$$\begin{aligned} &min\ g(\boldsymbol{\lambda}) = \mathbf{0}'\boldsymbol{\lambda} \quad \text{s.t.} \\ &\mathbf{B}\boldsymbol{\lambda} = \mathbf{V},\ \boldsymbol{\lambda} \geqq \mathbf{0} \end{aligned}$$

also has an optimal solution by the duality theorem presented above. Hence there exists at least one $\boldsymbol{\lambda} \geqq \mathbf{0}$ satisfying $\mathbf{B}\boldsymbol{\lambda} = \mathbf{V}$. Q.E.D.

Returning now to our previous question, in terms of the notation used above, if: $\mathbf{C} = \mathbf{V}$; the vectors $\bar{\mathbf{a}}_i'$, $i \in \mathfrak{I}$, are taken to be the columns of $\mathbf{B}$ (*i.e.*, $\mathbf{B} = -\bar{\mathbf{A}}'$); and the $\bar{u}_i^0 s$, $i \in \mathfrak{I}$, are the elements of $\boldsymbol{\lambda} \geqq \mathbf{0}$, then a necessary and sufficient condition for $\mathbf{C}$ to lie within the finite cone generated by the vectors $\bar{\mathbf{a}}_i'$ is that $\bar{\mathbf{C}}'\mathbf{h} \leq 0$ for all $\mathbf{h}$ satisfying $\bar{\mathbf{a}}_i\mathbf{h} \leq 0$, $i \in \mathfrak{I}$. Hence there exist real numbers $\bar{u}_i^0 \geq 0$ such that (6.2) holds.

We may incorporate all of the constraints (active or otherwise) $\bar{b}_i - \bar{\mathbf{a}}_i\mathbf{X} = 0$, $i=1,\ldots,m+n$, into our discussion by defining the scalar $\bar{u}_i^0$ as zero whenever $\bar{b}_i - \bar{\mathbf{a}}_i\mathbf{X} > 0$ or $i \notin \mathfrak{I}$. Then (6.2) is equivalent to the system

$$\left.\begin{aligned} &\mathbf{C} - \sum_{i=1}^{m+n} \bar{u}_i^0\, \bar{\mathbf{a}}_i' = \mathbf{0} \\ &\bar{u}_i^0(\bar{b}_i - \bar{\mathbf{a}}_i\mathbf{X}_0) = 0 \\ &\bar{b}_i - \bar{\mathbf{a}}_i\mathbf{X}_0 \geq \mathbf{0} \\ &\bar{u}_i^0 \geq 0 \\ &\mathbf{X}_0 \text{ unrestricted} \end{aligned}\right\} i=1,\ldots,m+n \quad \text{or} \quad \begin{aligned} &\mathbf{C} - \bar{\mathbf{A}}'\bar{\mathbf{U}}_0 = \mathbf{0} \\ &\bar{\mathbf{U}}_0'(\bar{\mathbf{b}} - \bar{\mathbf{A}}\mathbf{X}_0) = 0 \\ &\bar{\mathbf{b}} - \bar{\mathbf{A}}\mathbf{X}_0 \geqq \mathbf{0} \\ &\bar{\mathbf{U}}_0 \geqq \mathbf{0} \\ &\mathbf{X}_0 \text{ unrestricted} \end{aligned} \tag{6.3}$$

where $\bar{\mathbf{U}}_0' = (\bar{u}_1^0, \ldots, \bar{u}_{m+n}^0)$. Note that $\bar{u}_i^0(\bar{b}_i - \bar{\mathbf{a}}_i\mathbf{X}) = 0$ for all i values since if $\bar{b}_i - \bar{\mathbf{a}}_i\mathbf{X} > 0$, then $\bar{u}_i^0 = 0$ while if $\bar{b}_i - \bar{\mathbf{a}}_i\mathbf{X} = 0$, then $\bar{u}_i^0 \geq 0$.

When we formed the structural constraint inequality $\bar{\mathbf{b}} - \bar{\mathbf{A}}\mathbf{X} \geqq \mathbf{0}$ the n nonnegativity conditions $\mathbf{X} \geqq \mathbf{0}$ were treated as structural constraints. That is, $x_j \geq 0$ was converted to $\bar{b}_{m+j} - \bar{\mathbf{a}}_{m+j}\mathbf{X} = \mathbf{e}_j'\mathbf{X} \geq 0$, $j=1,\ldots,n$. However, if the nonnegativity conditions are not written in this fashion but appear explicitly as $\mathbf{X} \geqq \mathbf{0}$, then (6.3) may be rewritten in an equivalent form provided by the

6.4. KARUSH-KUHN-TUCKER THEOREM FOR LINEAR PROGRAMS. The point $\mathbf{X}_0 \in \mathbf{R}^n$ solves the primal problem

$$max f(\mathbf{X}) = \mathbf{C}'\mathbf{X} \quad \text{s.t.}$$
$$\mathbf{b} - \mathbf{A}\mathbf{X} \geqq \mathbf{0},\ \mathbf{X} \geqq \mathbf{0}$$

if and only if

$$\begin{array}{llll}
 & & \text{(a.1) } \mathbf{C} - \sum_{i=1}^{m} \bar{u}_i^0 \boldsymbol{\alpha}_i' \leqq \mathbf{0} \text{ or} & \\
\text{(a) } \mathbf{C} - \mathbf{A}'\mathbf{U}_0 \leqq \mathbf{0} & & c_j - \sum_{i=1}^{m} \bar{u}_i^0 a_{ij} \leq 0, j=1,\ldots,n & \\
\text{(b) } \mathbf{X}_0'(\mathbf{C} - \mathbf{A}'\mathbf{U}_0) = 0 & & \text{(b.1) } \sum_{j=1}^{n} x_j^0 (c_j - \sum_{i=1}^{m} \bar{u}_i^0 a_{ij}) = 0 & \\
\text{(c) } \mathbf{U}_0' (\mathbf{b} - \mathbf{A}\mathbf{X}_0) = 0 & \text{or} & \text{(c.1) } \sum_{i=1}^{m} \bar{u}_i^0 (b_i - \boldsymbol{\alpha}_i \mathbf{X}_0) = 0 & \\
\text{(d) } \mathbf{b} - \mathbf{A}\mathbf{X}_0 \geqq \mathbf{0} & & \left.\begin{array}{l}\text{(d.1) } b_i - \boldsymbol{\alpha}_i \mathbf{X}_0 \geq 0 \\ \text{(e.1) } u_i^0 \geq 0\end{array}\right\} i=1,\ldots,m & (6.4) \\
\text{(e) } \mathbf{U}_0 \geqq \mathbf{0} & & & \\
\text{(f) } \mathbf{X}_0 \geqq \mathbf{0} & & \text{(f.1) } x_j^0 \geq 0, j=1,\ldots,n &
\end{array}$$

where $\mathbf{U}_0$ is an $(m \times 1)$ vector with components $u_1^0, \ldots, u_m^0$.[2]

PROOF. (necessity). To solve the primal problem let us employ the technique of Lagrange. In this regard, we first convert each of the inequality constraints to an equality by subtracting from its left-hand side a nonnegative slack variable. (Actually, since we require that each slack variable be nonnegative, its square will be subtracted from the left-hand side to insure its nonnegativity.) That is, $b_i - \boldsymbol{\alpha}_i\mathbf{X} \geq 0$ is transformed to $b_i - \boldsymbol{\alpha}_i\mathbf{X} - x_{n+i}^2 = 0$, $i=1,\ldots,m$, while $x_j \geq 0$ is converted to $x_j - x_{n+m+j}^2 = 0$, $j=1,\ldots,n$, where x_{n+i}^2,

[2]Here (6.4) (as well as (6.3)) represents the Karush-Kuhn-Tucker conditions for a maximum of a function subject to inequality constraints. On all this see Panik (1976).

$i=1,\ldots,m$; x^2_{n+m+j}, $j=1,\ldots,n$, are all squares of slack variables. Hence the primal Lagrangian appears as

$$\begin{aligned}\mathcal{L}(x_1,\ldots,x_n,\ x_{n+1},\ldots,\ x_{n+m},\ x_{n+m+1},\ldots,x_{2n+m},u_1,\ldots,u_m,v_1,\ldots,v_n)\\ = \sum_{j=1}^{n} c_j x_j + \sum_{i=1}^{m} u_i(b_i-\boldsymbol{\alpha}_i\mathbf{X}-x^2_{n+i}) + \sum_{j=1}^{n} v_j(x_j-x^2_{n+m+j}),\end{aligned}$$

where $u_1,\ldots,u_m,v_1,\ldots,v_n$ are Lagrange multipliers. In matrix terms, the primal Lagrangian may be written as

$$\mathcal{L}(\mathbf{X},\mathbf{X}^2_{s1},\mathbf{X}^2_{s2},\ \mathbf{U},\mathbf{V}) = \mathbf{C}'\mathbf{X}+\mathbf{U}'(\mathbf{b}-\mathbf{A}\mathbf{X}-\mathbf{X}^2_{s1})+\mathbf{V}'(\mathbf{X}-\mathbf{X}^2_{s2}),$$

where

$$\mathbf{X}^2_{s1} = \begin{bmatrix} x^2_{n+1} \\ \vdots \\ x^2_{n+m}\end{bmatrix},\quad \mathbf{X}^2_{s2} = \begin{bmatrix} x^2_{n+m+1} \\ \vdots \\ x^2_{2n+m}\end{bmatrix},\ \mathbf{U} = \begin{bmatrix} u_1 \\ \vdots \\ u_m\end{bmatrix},\quad \mathbf{V} = \begin{bmatrix} v_1 \\ \vdots \\ v_m\end{bmatrix}.$$

Here $\mathbf{X}^2_{s1}$ and $\mathbf{U}$ are of order $(m\times 1)$ while $\mathbf{X}^2_{s2}$ and $\mathbf{V}$ have order $(n\times 1)$. Then

$$\begin{aligned}
&\text{(a)}\ \nabla_{\mathbf{X}}\mathcal{L} = \mathbf{C}-\mathbf{A}'\mathbf{U}+\mathbf{V} = \mathbf{0}\\
&\text{(b)}\ \mathcal{L}_{x_{n+i}} = 2u_i x_{n+i} = 0,\ i=1,\ldots,m\\
&\text{(c)}\ \mathcal{L}_{x_{n+m+j}} = 2v_j x_{n+m+j} = 0,\ j=1,\ldots,n\\
&\text{(d)}\ \nabla_{\mathbf{U}}\mathcal{L} = \mathbf{b}-\mathbf{A}\mathbf{X}-\mathbf{X}^2_{s1} = \mathbf{0}\\
&\text{(e)}\ \nabla_{\mathbf{V}}\mathcal{L} = \mathbf{X}-\mathbf{X}^2_{s2} = \mathbf{0}
\end{aligned} \tag{6.5}$$

Let us first transform (6.5.b, c) to $\mathbf{U}'\mathbf{X}^2_{s1}=0$, $\mathbf{V}'X^2_{s2}=0$ respectively. Then from $\mathbf{U}'\mathbf{X}^2_{s1}=0$ and (6.5.d) we obtain, at $\mathbf{X}_0$ and $\mathbf{U}_0$, $\mathbf{U}_0'(\mathbf{b}-\mathbf{A}\mathbf{X}_0)=0$. That is, if the constraint $b_i-\boldsymbol{\alpha}_i\mathbf{X}\geq 0$ is not binding at $\mathbf{X}_0$, then $x^0_{n+i}>0$, $u^0_i=0$ while if it is binding, $x^0_{n+i}=0$, $u^0_i\geq 0$, $i=1,\ldots,m$. Hence, at least one of each pair $(u^0_i, b_i-\boldsymbol{\alpha}_i\mathbf{X}_0)$ vanishes, thus guaranteeing that $\sum_{i=1}^{m} u^0_i(b_i-\boldsymbol{\alpha}_i\mathbf{X}_0)=0$. Next, combining $\mathbf{V}'\mathbf{X}^2_{s2}=0$ and (6.5) yields $\mathbf{X}'\mathbf{V}=0$. From (6.5.a) we have $-\mathbf{V}=\mathbf{C}-\mathbf{A}'\mathbf{U}$. Then, at $\mathbf{X}_0$ and $\mathbf{U}_0$, these latter two expressions yield $\mathbf{X}_0'(\mathbf{C}-\mathbf{A}'\mathbf{U}_0)=0$ or $\sum_{j=1}^{n} x^0_j\left(c_j-\sum_{i=1}^{m} u^0_i a_{ij}\right)=0$. In this regard, either $x^0_j=0$, allowing c_j to be less than $\sum_{j=1}^{m} u^0_i a_{ij}$, or $c_j-\sum_{i=1}^{m} u^0_i a_{ij}=0$, in which case x^0_j may be positive. To see this let us assume, without loss of generality, that the first $k < m$ structural constraints are binding at $\mathbf{X}_0$. Then $b_i - \boldsymbol{\alpha}_i\mathbf{X}_0 = 0(x^0_{n+i} = 0,\ u^0_i \geq 0)$, $i=1,\ldots,k$, while $b_i - \boldsymbol{\alpha}_i\mathbf{X}_0 > 0$ $(x^0_{n+i} > 0,\ u^0_i = 0)$, $i=k+1,\ldots,m$. Now, if the primal problem possesses a solution at $\mathbf{X}_0$, $\mathbf{C}'\mathbf{h} \leq 0$ for vectors $\mathbf{h}$ which do not violate $\boldsymbol{\alpha}_i\mathbf{h} = 0$, $i=1,\ldots,k$. If we multiply each equality $\boldsymbol{\alpha}_i\mathbf{h} = 0$. by some constraint u^0_i and form the sum $\mathbf{C}'\mathbf{h} - \sum_{i=1}^{k} u^0_i\boldsymbol{\alpha}_i\mathbf{h} \leq 0$, we ultimately obtain

$$\left(\mathbf{C} - \sum_{i=1}^{m} u^0_i\boldsymbol{\alpha}'_i\right)'\mathbf{h} \leq 0. \tag{6.6}$$

Let us further assume that at least one component of $\mathbf{X}_0$, say x^0_ℓ, is strictly positive (and thus $x^0_{n+m+\ell} > 0$, $v^0_\ell = 0$). Then there exists a set of sufficiently small positive and negative deviations $h_\ell = \delta, -\delta$ such that, for $h_\ell = \delta > 0$, (6.6) becomes

$$c_\ell - \sum_{i=1}^{k} u^0_i a_{i\ell} \leq 0 \tag{6.7}$$

while for $h_\ell = -\delta$, (6.6) may be written as

$$c_\ell - \sum_{i=1}^{k} u^0_i a_{i\ell} \geq 0 \tag{6.8}$$

whence, upon combining (6.7), (6.8),

$$c_\ell - \sum_{i=1}^{k} u^0_i a_{i\ell} = 0.$$

If $x^0_\ell = 0$ (so that $x_{n+m+\ell} = 0$, $v^0_\ell > 0$), h_ℓ can only be positive. Hence $h_\ell = \delta > 0$ in (6.6) yields

$$c_\ell - \sum_{i=1}^{k} u^0_i a_{i\ell} \leq 0.$$

In general then,

$$\mathbf{C} - \sum_{i=1}^{m} u^0_i\boldsymbol{\alpha}'_i \leq \mathbf{0}.$$

(sufficiency) For $\mathbf{X} \in \mathcal{K}$, let us express the Lagrangian of f as

$$\mathbf{C}'\mathbf{X} \leq \mathbf{C}'\mathbf{X} + \mathbf{U}'_0(\mathbf{b}-\mathbf{A}\mathbf{X}) + \mathbf{V}'_0\mathbf{X} = \mathcal{L}(\mathbf{X}, \mathbf{U}_0, \mathbf{V}_0).$$

With $\mathcal{L}(\mathbf{X}, \mathbf{U}_0, \mathbf{V}_0)$ concave in $\mathbf{X}$ over $\mathcal{K}$, the linear approximation to $\mathcal{L}$ at $\mathbf{X}_0$ is

$$
\begin{aligned}
\mathcal{L}(\mathbf{X}, \mathbf{U}_0, \mathbf{V}_0) \leq \mathcal{L}(\mathbf{X}, \mathbf{U}_0, \mathbf{V}_0) &+ \nabla_{\mathbf{X}} \mathcal{L}(\mathbf{X}, \mathbf{U}_0, \mathbf{V}_0)'(\mathbf{X}-\mathbf{X}_0) \\
&= \mathbf{C}'\mathbf{X}_0 + (\mathbf{C}-\mathbf{A}'\mathbf{U}_0+\mathbf{V}_0)'(\mathbf{X}-\mathbf{X}_0) \\
&= \mathbf{C}'\mathbf{X}_0.^3 \qquad (6.9)
\end{aligned}
$$

Hence $\mathbf{C}'\mathbf{X}_0 \geq \mathcal{L}(\mathbf{X}, \mathbf{U}_0, \mathbf{V}_0) \geq \mathbf{C}'\mathbf{X}, \mathbf{X} \in \mathcal{K}$, and thus $\mathbf{X}_0$ solves the primal problem. Q.E.D.

We note briefly that if the primal problem appears as

$$
\min f(\mathbf{X}) = \mathbf{C}'\mathbf{X} \quad \text{s.t.}
$$
$$
\mathbf{AX} \geqq \mathbf{b}, \ \mathbf{X} \geqq \mathbf{0}
$$

then (6.4) becomes

$$
\begin{aligned}
&\text{(a)}\ \mathbf{C} - \mathbf{A}'\mathbf{U} \geqq \mathbf{0} \\
&\text{(b)}\ \mathbf{X}_0'(\mathbf{C}-\mathbf{A}'\mathbf{U}) = 0 \\
&\text{(c)}\ \mathbf{U}_0'(\mathbf{b}-\mathbf{AX}_0) = 0 \\
&\text{(d)}\ \mathbf{b} - \mathbf{AX}_0 \geqq \mathbf{0} \qquad (6.4.1) \\
&\text{(e)}\ \mathbf{U}_0 \geqq \mathbf{0} \\
&\text{(f)}\ \mathbf{X}_0 \geqq \mathbf{0}
\end{aligned}
$$

It is certainly not coincidental that the Lagrange multipliers appearing in (6.2), (6.3), and (6.4) ((6.4.1)) are denoted as "u's," for they are the dual variables in the dual linear programming problem. To formalize this observation, let us examine the following corollaries to the preceding theorem.

6.5. COROLLARY. Let (6.4) hold so that the primal maximum problem possess an optimal solution. Then the dual problem

$$
\min g(\mathbf{U}) = \mathbf{b}'\mathbf{U} \quad \text{s.t.}
$$
$$
\mathbf{A}'\mathbf{U} \geqq \mathbf{C}, \ \mathbf{U} \geqq \mathbf{0}
$$

also has an optimal solution $\mathbf{U}_0$ with $\mathbf{C}'\mathbf{X}_0 = \mathbf{b}'\mathbf{U}_0$

PROOF. From (6.4.a, e) $\mathbf{A}'\mathbf{U}_0 \geqq \mathbf{C}$ and $\mathbf{U}_0 \geqq \mathbf{0}$ so that $\mathbf{U}_0$ is a feasible solution to the dual problem. With $\mathbf{AX}_0 \leqq \mathbf{b}$, $\mathbf{X}_0'\mathbf{A}' \leqq \mathbf{b}'$ or $\mathbf{X}_0'\mathbf{A}'\mathbf{U}_0 \leq \mathbf{b}'\mathbf{U}_0$.

[3]Since $\mathcal{L}(\mathbf{X}, \mathbf{U}_0, \mathbf{V}_0)$ is linear in $\mathbf{X}$, it is concave over $\mathcal{K}$. Moreover, with $\mathcal{L}$ concave, it lies everywhere on or beneath its tangent hyperplane at $\mathbf{X}_0$ as (6.9) indicates.

Since $\mathbf{A}'\mathbf{U}_0 \geqq \mathbf{C}$, $\mathbf{X}_0'\mathbf{C} = \mathbf{C}'\mathbf{X}_0 \leq \mathbf{b}'\mathbf{U}_0$ for $\mathbf{U}_0$ feasible. Moreover, from (6.4.b, c), $\mathbf{X}_0'\mathbf{C} - \mathbf{X}_0'\mathbf{A}'\mathbf{U}_0 = \mathbf{U}_0'\mathbf{b} - \mathbf{U}_0'\mathbf{A}\mathbf{X}_0$ or $\mathbf{C}'\mathbf{X}_0 = \mathbf{b}'\mathbf{U}_0$ since $\mathbf{X}_0'\mathbf{A}'\mathbf{U}_0 = \mathbf{U}_0'\mathbf{A}\mathbf{X}_0$. Hence $\mathbf{U}_0$ must be an optimal solution to the dual problem. Q.E.D.

It should be evident that corollary 6.5 is a restatement of the strong duality theorem 4.8. Based upon this result we may now state

6.6. COROLLARY. The optimal value of the ith dual variable u_i^0 is generally a measure of the amount by which the optimal value of the primal objective function changes given a small (incremental) change in the amount of the ith requirement b_i with all other requirements $b_\ell, \ell \neq i$, i=1,...,m, remaining constant, *i.e.*,

$$\frac{\partial f^0}{\partial b_i} = u_i^0. \tag{6.10}$$

PROOF. Let $\mathbf{X}_0, \mathbf{U}_0$ be feasible solutions to the primal maximum and dual minimum problems respectively with $f^0 = \mathbf{C}'\mathbf{X}_0$, $g^0 = \mathbf{b}'\mathbf{U}_0$. If (6.4) holds, then (from corollary 6.5) $f^0 = g^0$, *i.e.*, $\mathbf{X}_0$ and $\mathbf{U}_0$ are optimal solutions and thus $f^0 = \sum_{i=1}^{m} b_i u_i^0$ so that $\dfrac{\partial f^0}{\partial b_i} = u_i^0$. Q.E.D.

So if the right-hand side b_i of the ith primal structural constraint were changed by a sufficiently small amount $\in_i$, then the corresponding change in the optimal value of f, f^0, would be $df^0 = u_i^0 \in_i$. Moreover, since $df^0 = \sum_{i=1}^{m} \dfrac{\partial f^0}{\partial b_i} db_i = \sum_{i=1}^{m} u_i^0 db_i$, we may determine the amount by which the optimal value of the primal objective function changes given small (marginal) changes in the right-hand sides of any specific subset of structural constraints. For instance, let us assume, without loss of generality, that the right-hand sides of the first $t < m$ primal structural constraints are increased by sufficiently small increments $\in_1, \ldots, \in_t$ respectively. Then the optimal value of the primal objective function would be increased by $df^0 = \sum_{i=1}^{t} u_i^0 \in_i$.

Throughout the preceding discussion the reader may have been wondering why the qualification *generally* was included in the statement of corollary 6.6. The fact of the matter is that there exist values of, say b_k, for which $\dfrac{\partial f^0}{\partial b_k}$ is not defined, *i.e.*, points were $\frac{\partial f^0}{\partial b_k}$ is discontinuous so that (6.10) does not hold. To develop this argument[4] let us express f in terms of b_k as

$$f(b_k) = \underset{\mathbf{X}}{max}\{\mathbf{C}'\mathbf{X}|\boldsymbol{\alpha}_i\mathbf{X} \leq b_i^0,\ i \neq k,\ i{=}1,\ldots,m;\ \boldsymbol{\alpha}_k\mathbf{X} \leq b_k,\ \mathbf{X} \geqq \mathbf{0}\},$$

where the b_i^0's depict fixed levels of the requirements $b_i, i \neq k$, i=1,...,m. Clearly f must be non-decreasing in b_k since increasing the value of b_k expands the size of the feasible region.

Moreover, the function $f(b_k)$ is concave. In fact, this assertion is part of a more general result pertaining to the **optimal value function**

$$f(\mathbf{b}) = max\{\mathbf{C}'\mathbf{X}|\mathbf{A}\mathbf{X} \leqq \mathbf{b},\ \mathbf{X} \geqq \mathbf{0}\}.$$

(Here $f(\mathbf{b})$ denotes the maximum of the primal objective as a function of the requirements vector $\mathbf{b}$.) Specifically,

6.7. THEOREM. The optimal value function $f(\mathbf{b})$ is concave.

PROOF. For $\mathbf{b}_1, \mathbf{b}_2 \in \mathbf{R}^m$, let $\mathbf{b}_c = \theta\mathbf{b}_1 + (1-\theta)\mathbf{b}_2$, $0 \leq \theta \leq 1$. If either $f(\mathbf{b}_1)$ or $f(\mathbf{b}_2)$ equals $-\infty$, then trivially $f(\mathbf{b}_c) \geq \theta f(\mathbf{b}_1) + (1-\theta)f(\mathbf{b}_2)$. Hence we shall assume that both $f(\mathbf{b}_1)$, $f(\mathbf{b}_2)$ are greater than $-\infty$. If either $f(\mathbf{b}_1)$ or $f(\mathbf{b}_2)$ equals $+\infty$, then the dual problem is infeasible so that $f(\mathbf{b}) = +\infty$ for all $\mathbf{b} \in \mathbf{R}^m$.

Next, let both $f(\mathbf{b}_1) = \mathbf{C}'\mathbf{X}_1$, $f(\mathbf{b}_2) = \mathbf{C}'\mathbf{X}_2$ be finite for primal feasible $\mathbf{X}_1$, $\mathbf{X}_2$. Then for $\mathbf{X}_c = \theta\mathbf{X}_1 + (1-\theta)\mathbf{X}_2$, $0 \leq \theta \leq 1$, we have $\mathbf{A}\mathbf{X}_c \leqq \mathbf{b}_c$ and thus, from the definition of the optimal value function, $f(\mathbf{b}_c) \geq \mathbf{C}'\mathbf{X}_c = \theta f(\mathbf{b}_1) + (1-\theta)f(\mathbf{b}_2)$ as required. Q.E.D.

If in Figure 6.4 we take area OABCG to be the initial feasible region when $b_k = b_k'$ (here the line segment CG consists of all feasible points $\mathbf{X}$ which satisfy $\boldsymbol{\alpha}_k\mathbf{X} = b_k'$), then increasing the kth requirement from b_k' to b_k'' changes the optimal extreme point from C to D. The result of this increase in b_k is a concomitant increase in f^0 proportionate to the distance $\in_k = b_k'' - b_k'$, where the constant of proportionality is u_k^0. Thus, over this range of b_k values, $\frac{\partial f^0}{\partial b_k}$ exists (is

[4]The discussion which follows has, in part, been adapted from the developments presented in Balinski, Baumol (1968) and Williams (1963).

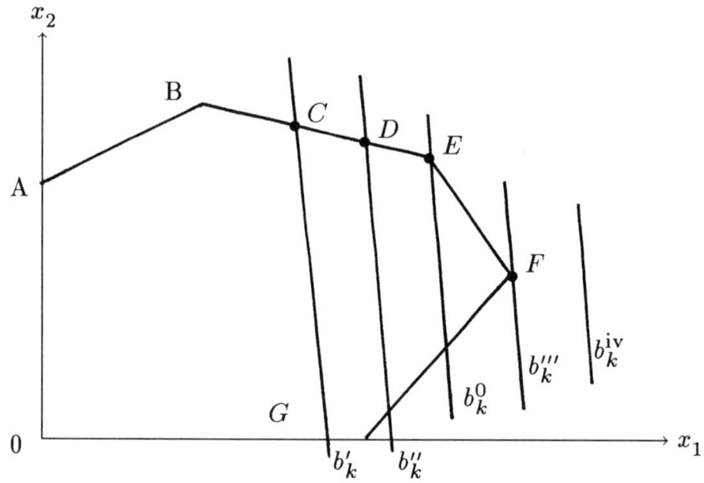

Figure 6.4

continuous). If we again augment the level of b_k by the same amount as before, this time shifting the kth structural constraint so that it passes through point E, $\frac{\partial f^0}{\partial b_k}$ becomes discontinuous since now the maximal point moves along a more steeply sloping portion (segment EF) of the boundary of the feasible region so that any further equal parallel shift in the kth structural constraint would produce a much larger increment in the optimal value of f than before. Thus each time the constraint line $\boldsymbol{\alpha}_k\mathbf{X} = b_k$ passes through an extreme point of the feasible region, there occurs a discontinuity in $\frac{\partial f^0}{\partial b_k}$.

To relate this argument directly to $f(b_k)$ above, let us describe the behavior of $f(b_k)$ as b_k assumes the values b_k', b_k'', b_k^0, b_k''' and b_k^{iv}. In Figure 6.5 the piecewise continuous curve $f(b_k)$ indicates the effect on $f(b_k)$ precipitated by changes in b_k. (Note that each time the constraint $\boldsymbol{\alpha}_k\mathbf{X} = b_k$ passes through an extreme point of the feasible region in Figure 6.4, the $f(b_k)$ curve assumes a kink, *i.e.*, a point of discontinuity in its derivative. Note also that for a requirements level of $b_k > b_k'''$ (say b_k^{iv}) f cannot increase any further so that beyond F', $f(b_k)$ is horizontal.) As far as the dual objective function $g(\mathbf{U}) = \mathbf{b}'\mathbf{U}$ is concerned, g may be expressed as $g(b_k) = u_0 + b_k u_k$, where $u_0 = \sum_{i \neq k}^{m} b_i u_i =$ constant is independent of b_k. For $b_k = b_k^0$ (b_k^0 optimal), $f(b_k^0) = g(b_k^0)$. Moreover, since the dual structural constraints are unaffected by changes in b_k, dual feasibility is preserved for variations in b_k away from b_k^0 so that, in general, for any pair of feasible solutions to the primal and dual problems, $f(b_k) \leq g(b_k)$, *i.e.*, $g(b_k)$ is tangent to $f(b_k)$ for $b_k = b_k^0$ (point E' in Figure 6.5) while $f(b_k^0 + \in_k) < g(b_k^0 + \in_k)$ for $\in_k(\lesseqgtr 0)$ sufficiently small. In this regard, for $\in_k \to 0+$, $\partial g(b_k^0)/\partial b_k = u_k^0 > \overset{+}{\partial} f(b_k^0)/\partial b_k$ while, for $\in_k \to 0-$, $u_k^0 < \bar{\partial} f(b_k^0)/\partial b_k$ whence $\overset{+}{\partial} f(b_k^0)/\partial b_k < u_k^0 < \bar{\partial} f(b_k^0)/\partial b_k$.

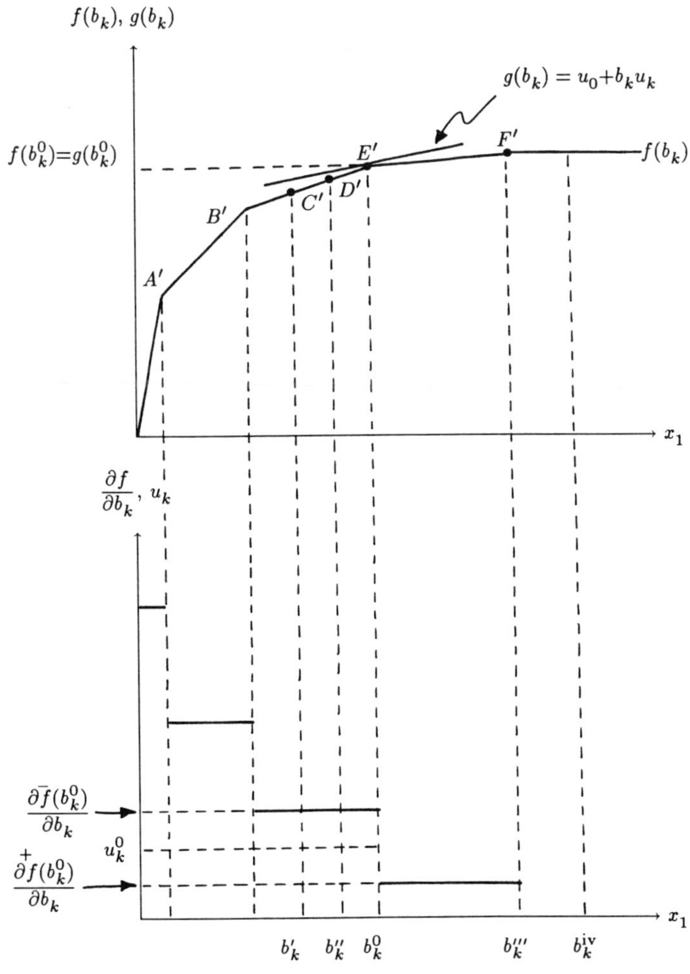

Figure 6.5

So while both the left- and right-hand derivatives of $f(b_k)$ exist at b_k^0, $\partial f(b_k^0)/\partial b_k$ itself does not exist since $\overset{+}{\partial} f(b_k^0)/\partial b_k \neq \overset{-}{\partial} f(b_k^0)/\partial b_k$ *i.e.*, $\partial f/\partial b_k$ possesses a finite discontinuity at b_k^0. In fact, $\partial f/\partial b_k$ is discontinuous at each of the points A', B', E' and F' in Figure 6.5. For $b_k = b_k''$, it is evident that $\overset{+}{\partial} f(b_k'')/\partial b_k = u_k'' = \overset{-}{\partial} f(b_k'')/\partial b_k$ since $f(b_k)$, $g(b_k)$ coincide all along the segment $B'E'$ of Figure 6.5. In this instance, we many unequivocally write $\partial f(b_k'')/\partial b_k = u_k''$. Moreover, for $b_k = b_k^{iv}$ (see Figures 6.4, 6.5), $\partial f(b_k^{iv})/\partial b_k = u_k^{iv} = 0$ since the constraint $\boldsymbol{\alpha}_k \mathbf{X} = b_k^{iv}$ is **superfluous**, i.e., it does not form any part of the boundary of the feasible region. In general then, $\overset{+}{\partial} f/\partial b_k \leq u_k \leq \overset{-}{\partial} f/\partial b_k$.

To formalize the preceding discussion we state

6.8. THEOREM. Let the requirements b_i be held fixed at b_i^0, $i=1,\ldots,m$. If $f(b_k) = \underset{\mathbf{X}}{max}\{\mathbf{C}'\mathbf{X} \mid \boldsymbol{\alpha}_i \mathbf{X} \leq b_i^0,\ i \neq k,\ i=1,\ldots,m;\ \boldsymbol{\alpha}_k \mathbf{X} \leq b_k,\ \mathbf{X} \geqq \mathbf{0}\}$ possesses finite right- and left-hand derivatives at an optimal point $b_k = b_k^0$, then u_k^0, the optimal value of the corresponding dual variable, lies between these derivatives, *i.e.*,

$$\frac{\overset{+}{\partial} f(b_k^0)}{\partial b_k} \leq u_k^0 \leq \frac{\overset{-}{\partial} f(b_k^0)}{\partial b_k} \quad \text{for all } k.$$

Moreover, if $\overset{+}{\partial} f(b_k^0)/\partial b_k = \overset{-}{\partial} f(b_k^0)/\partial b_k$ so that

$\partial f(b_k^0)/\partial b_k$ exists, then $\dfrac{\partial f(b_k^0)}{\partial b_k} = u_k^0$.

PROOF. Since $g(b_k) = u_0 + b_k u_k$ is linear, $g(b_k^0) = f(b_k^0) = \underset{\mathbf{X}}{max}\{\mathbf{C}'\mathbf{X} \mid \boldsymbol{\alpha}_i \mathbf{X} \leq b_i^0, i \neq k, i = 1,\ldots,m; \boldsymbol{\alpha}_k \mathbf{X} \leq b_k^0, \mathbf{X} \geqq \mathbf{0}\}$ for b_k^0 optimal (see, for instance, point E of Figure 6.4). And since $f(b_k^0 + \in_k) < g(b_k^0 + \in_k)$, $\in_k$ unrestricted in sign and sufficiently small, it follows that, for $\in_k > 0$,

$$\frac{g(b_k^0 + \in_k) - g(b_k^0)}{\in_k} = u_k^0 \geq \frac{f(b_k^0 + \in_k) - f(b_k^0)}{\in_k}$$

and thus

$$u_k^0 \geq \lim_{\in_k \to 0+} \frac{f(b_k^0 + \in_k) - f(b_k^0)}{\in_k} = \frac{\overset{+}{\partial} f(b_k^0)}{\partial b_k};$$

while for $\in_k < 0$,

$$u_k^0 \leq \frac{f(b_k^0 + \in_k) - f(b_k^0)}{\in_k}$$

so that

$$u_k^0 \leq \lim_{\in_k \to 0-} \frac{f(b_k^0 + \in_k) - f(b_k^0)}{\in_k} = \frac{\bar{\partial} f(b_k^0)}{\partial b_k} .$$

Hence $\overset{+}{\partial} f(b_k^0)/\partial b_k \leq u_k^0 \leq \bar{\partial} f(b_k^0)/\partial b_k$ for all k. The *moreover* portion of the theorem holds by definition. Q.E.D.

To summarize: if both the left- and right-hand derivatives $\bar{\partial} f/\partial b_i$, $\overset{+}{\partial} f/\partial b_i$ exist and their common value is u_i^0, then $\partial f^0/\partial b_i = u_i^0$; otherwise $\overset{+}{\partial} f/\partial b_i \leq u_i^0 \leq \bar{\partial} f/\partial b_i$, i=1,...,m.

An additional source of difficulty in interpreting u_k^0 as $\partial f^0/\partial b_k$ emerges when the optimal basic feasible solution to the primal problem is degenerate. In this instance the dual possesses multiple optimal basic feasible solutions, with $f(\mathbf{b})$ nondifferentiable at such points. More specifically, alternative optimal dual basic feasible solutions all correspond to alternative bases associated with the primal degenerate basic feasible solution $\mathbf{X}_0$. Hence u_k^0 is not uniquely determined.

If $\mathbf{X}_0$ represents a nondegenerate optimal basic feasible solution to the primal problem, then, according to the (weak) complementary slackness conditions, the dual structural constraints associated with the primal basic variable columns of $\mathbf{A}$ must hold as a set of equalities or $\mathbf{U}'\mathbf{B} = \mathbf{C}_B'$. Hence we obtain, as the *unique* optimal dual basic feasible solution, $\mathbf{U}' = \mathbf{C}_B' \mathbf{B}^{-1}$ with $u_k^0 = \partial f^0/\partial b_k$ $(=\overset{+}{\partial} f^0/\partial b_k = \bar{\partial} f^0/\partial b_k)$.

While $f(\mathbf{b})$ is nondifferentiable at points where there exist multiple optimal dual solutions, it does have directional derivatives which exist everywhere. Such derivatives can be derived from the set of optimal dual solutions and are defined as: the **directional derivative of f at b in the direction h** $(\neq \mathbf{0}) \epsilon \mathbf{R}^m$, if it exists, is

$$D_{\mathbf{h}} f(\mathbf{b}) = \lim_{t \to 0+} \frac{f(\mathbf{b}+t\mathbf{h}) - f(\mathbf{b})}{t} .$$

To see this let us assume that $\mathcal{K}_D = \{\mathbf{U} \mid \mathbf{A}'\mathbf{U} \geqq \mathbf{C},\ \mathbf{U} \geqq \mathbf{0}\} \neq \phi$. Hence the dual problem is assumed to posses a feasible solution so that the primal objective is not unbounded for any $\mathbf{b}$. By the strong duality theorem, $f(\mathbf{b}) = \min_{\mathbf{U} \epsilon \mathcal{K}_D} \{\mathbf{b}'\mathbf{U}\}$

(possibly $f(\mathbf{b}) = -\infty$). By a suitable modification of existence theorem 3.1 to address the dual problem, $f(\mathbf{b})$ attains a finite minimum if $\mathbf{b}'\mathbf{d}_j \geq 0$ for all extreme directions $\mathbf{d}_j, j=1,\ldots,\ell$, of $\mathcal{K}_D$. Moreover, $f(\mathbf{b}) = \mathbf{b}'\mathbf{U}_r$ for any extreme points of $\mathcal{K}_D$ such that $\mathbf{b}'\mathbf{U}_r = min\{\mathbf{b}'\mathbf{U}_s\}$, where the index s varies over all extreme points of $\mathcal{K}_D$.

In this regard, consider a $\mathbf{b} \in \mathbf{R}^m$ such that $f(\mathbf{b})$ is finite. Then $\mathbf{h}(\neq \mathbf{0}) \in \mathbf{R}^m$ is a feasible direction of f at $\mathbf{b}$ if $f(\mathbf{b}+t\mathbf{h})$ is finite for all $0 \leq t \leq \delta$. Since $f(\mathbf{b})$ is finite and concave, a feasible direction $\mathbf{h}$ will always exist at a point $\mathbf{b}$ if we can find at least one other point in $\mathbf{R}^m$ which yields a finite value of f. Hence

6.9. THEOREM [Shapiro (1979)]. Let $f(\mathbf{b}) = \underset{\mathbf{U} \in \mathcal{K}_D}{min}\{\mathbf{b}'\mathbf{U}\}$ be finite with $\mathbf{h}(\neq \mathbf{0})$ a feasible direction of f at $\mathbf{b}$. Then the directional derivative of f at $\mathbf{b}$ in the direction $\mathbf{h}$ exists and is given by $D_{\mathbf{h}}f(\mathbf{b}) = min\{\mathbf{h}'\mathbf{U}, \mathbf{U} \in \mathcal{E}\}$, where $\mathcal{E}$ is the set of optimal dual extreme points.

When $\mathcal{E}$ is the singleton $\{\mathbf{U}_0\} \in \mathbf{R}^m$ (the optimal dual extreme point is unique), then $D_{\mathbf{h}}f(\mathbf{b}) = \mathbf{h}'\mathbf{U}_0$ for all feasible directions $\mathbf{h}$. In addition, if the set of feasible directions is all of $\mathbf{R}^m\backslash\mathbf{0}$, then f is differentiable and $\partial f^0/\partial b_i = u_i^0, i=1,\ldots,m$ (Shapiro (1979)).

EXAMPLE 6.1. Using the Karush-Kuhn-Tucker conditions specified above, let us determine an optimal solution to the linear program

$$
\begin{aligned}
max\ f &= x_1 + 3x_2 + 4x_3 \quad \text{s.t.}\\
2x_1 + 2x_1 + x_2 + 4x_3 &\leq 100\\
x_1 + 3x_2 + x_3 &\leq 40\\
x_1, x_2, x_3 &\geq 0.
\end{aligned}
$$

For

$$
\bar{\mathbf{A}} = \begin{bmatrix} \mathbf{A} \\ \cdots \\ -\mathbf{I}_3 \end{bmatrix} = \begin{bmatrix} 2 & 1 & 4 \\ 1 & 3 & 1 \\ -1 & 0 & 0 \\ 0 & -1 & 0 \\ 0 & 0 & -1 \end{bmatrix}, \quad \bar{\mathbf{b}} = \begin{bmatrix} 100 \\ 40 \\ 0 \\ 0 \\ 0 \end{bmatrix},
$$

we have, from (6.3),

$$\mathbf{C}-\bar{\mathbf{A}}'\bar{\mathbf{U}}_0 = \begin{bmatrix} 1-2\bar{u}_1^0-\bar{u}_2^0+\bar{u}_3^0 & & \\ 3-\bar{u}_1^0-3\bar{u}_2^0 & +\bar{u}_4^0 & \\ 4-4\bar{u}_1^0-\bar{u}_2^0 & & +\bar{u}_5^0 \end{bmatrix} = \mathbf{0},$$

$$\begin{aligned}\bar{\mathbf{U}}_0'(\bar{\mathbf{b}}-\bar{\mathbf{A}}\mathbf{X}_0) &= \bar{u}_1^0(100-2x_1^0-x_2^0-4x_3^0)+\bar{u}_2^0(40-x_1^0-3x_2^0-x_3^0)\\ &\quad +\bar{u}_3^0x_1^0+\bar{u}_4^0x_2^0+\bar{u}_5^0x_3^0 = 0,\end{aligned} \tag{6.3.1}$$

$$\bar{\mathbf{b}}-\bar{\mathbf{A}}\mathbf{X}_0 = \begin{bmatrix} 100-2x_1^0-x_2^0-4x_3^0 \\ 40-x_1^0-3x_2^0-x_3^0 \\ x_1^0 \\ x_2^0 \\ x_3^0 \end{bmatrix} \geqq \mathbf{0},\ \bar{\mathbf{U}}_0 = \begin{bmatrix} \bar{u}_1^0 \\ \bar{u}_2^0 \\ \bar{u}_3^0 \\ \bar{u}_4^0 \\ \bar{u}_5^0 \end{bmatrix} \geqq \mathbf{0},$$

If $i = 2, 3, 5 \in \mathfrak{I}$, (6.3.1) may be rewritten as

$$\begin{aligned}
&\text{(a)}\ 1-\bar{u}_2^0+\bar{u}_3^0 = 0\\
&\text{(b)}\ 3-3\bar{u}_2^0 = 0\\
&\text{(c)}\ 4-\bar{u}_2^0+\bar{u}_5^0 = 0\\
&\quad 0(100-2x_1^0-x_2^0-4x_3^0)+\bar{u}_2^0(0)+\bar{u}_3^0(0)+0(x_2^0)+\bar{u}_5^0(0) = 0 \qquad (6.3.2)\\
&\text{(d)}\ 100-2x_1^0-x_2^0-4x_3^0>0,\ \bar{u}_1^0 = 0\\
&\text{(e)}\ 40-x_1^0-3x_2^0-x_3^0 = 0,\ \geq \bar{u}_2^0 \geq 0\\
&\text{(f)}\ x_1^0 = 0,\ \bar{u}_3^0 \geq 0\\
&\text{(g)}\ x_2^0 > 0,\ \bar{u}_4^0 = 0\\
&\text{(h)}\ x_3^0 = 0,\ \bar{u}_5^0 \geq 0.
\end{aligned}$$

Using (6.3.2.a, b, c) we obtain $\bar{u}_2^0 = 1$, $\bar{u}_3^0 = 0$ and $\bar{u}_5^0 = -3$. Moreover, (6.3.2.e, f, h) yield $x_2^0 = \frac{40}{3}$ while (6.3.2.d, f, h) inform us that $x_2^0 < 100$. Since (6.3.1) is violated (*i.e.*, $\bar{u}_5^0 = -3 < 0$), we are prompted to test additional points for optimality.

For $i = 1, 2, 3 \in \mathfrak{I}$, (6.3.1) becomes

$$
\begin{aligned}
&\text{(a) } 1-2\bar{u}_1^0+\bar{u}_2^0+\bar{u}_3^0 = 0\\
&\text{(b) } 3-\bar{u}_1^0-3\bar{u}_2^0 = 0\\
&\text{(c) } 4-4\bar{u}_1^0-\bar{u}_2^0 = 0\\
&\qquad \bar{u}_1^0(0)+\bar{u}_2^0(0)+\bar{u}_3^0(0)+0(x_2^0)+0(x_3^0) = 0 \qquad (6.3.3)\\
&\text{(d) } 100-2x_1^0-x_2^0-4x_3^0 = 0,\quad \bar{u}_1^0 \geq 0\\
&\text{(e) } 40-x_1^0-3x_2^0-x_3^0 = 0,\ \bar{u}_2^0 \geq 0\\
&\text{(f) } x_1^0 > 0,\quad \bar{u}_3^0 \geq 0\\
&\text{(g) } x_2^0 > 0,\quad \bar{u}_4^0 = 0\\
&\text{(h) } x_3^0 > 0,\quad \bar{u}_5^0 = 0\,.
\end{aligned}
$$

Using (6.3.3. b, c) $\bar{u}_1^0 = \frac{9}{11}$, $\bar{u}_2^0 = \frac{8}{11}$. A substitution of these values into (6.3.3.a) renders $\bar{u}_3^0 = \frac{15}{11}$. Additionally, from (6.3.3.d, e, f), $x_2^0 = \frac{60}{11}$, $x_3^0 = \frac{260}{11}$. Finally, a substitution of this latter set of values into f yields $f^0 = \frac{1220}{11}$. By systematically choosing other combinations of values of $i \in \mathcal{I}$ one can generate additional solutions to the above problem. However, the one exhibited by (6.3.3) maximizes f subject to the constraints.

6.2. Lagrangian Saddle Points and Primal Optimality

If we structure the primal-dual pair of problems as

PRIMAL PROBLEM	DUAL PROBLEM
$max\ f(\mathbf{X}) = \mathbf{C}'\mathbf{X}$ s.t	$min\ g(\mathbf{U}) = \mathbf{b}'\mathbf{U}$ s.t.
$\mathbf{AX} \leqq \mathbf{b},\ \mathbf{X} \geqq \mathbf{0}$	$\mathbf{A}'\mathbf{U} \geqq \mathbf{C},\ \mathbf{U} \geqq \mathbf{0}$

then the Lagrangian associated with the primal problem is $\mathcal{L}(\mathbf{X},\ \mathbf{U}) = \mathbf{C}'\mathbf{X} + \mathbf{U}'(\mathbf{b}-\mathbf{AX})$ with $\mathbf{X} \geqq \mathbf{0},\ \mathbf{U} \geqq \mathbf{0}$. And if we now express the dual problem as maximize $\{-\mathbf{b}'\mathbf{U}\}$ subject to $-\mathbf{A}'\mathbf{U} \leqq \mathbf{C},\ \mathbf{U} \geqq \mathbf{0}$, then the Lagrangian of the dual problem is $\mathcal{M}(\mathbf{U},\ \mathbf{X}) = -\mathbf{b}'\mathbf{U} + \mathbf{X}'(-\mathbf{C}+\mathbf{A}'\mathbf{U})$, where again $\mathbf{U} \geqq \mathbf{0},\ \mathbf{X} \geqq \mathbf{0}$. Moreover , since $\mathcal{M}(\mathbf{U},\mathbf{X}) = -\mathbf{b}'\mathbf{U} - \mathbf{X}'\mathbf{C} + \mathbf{X}'\mathbf{A}'\mathbf{U} = -\mathbf{C}'\mathbf{X} - \mathbf{U}'\mathbf{b} + \mathbf{U}'\mathbf{AX} = -\mathcal{L}(\mathbf{X},\ \mathbf{U})$, we see that the Lagrangian of the dual problem is actually the negative of the Lagrangian of the primal problem. Alternatively, if $x_j \geq 0$ is converted to $\bar{b}_{m+j} - \bar{a}_{m+j}\mathbf{X} = \mathbf{e}_j'\mathbf{X} \geq 0$, j=1,..., n, then the revised primal problem appears as

$$\begin{aligned} \max f(\mathbf{X}) &= \mathbf{C}'\mathbf{X} \quad \text{s.t.} \\ \bar{\mathbf{A}}\mathbf{X} &\leqq \bar{\mathbf{b}}, \ \mathbf{X} \text{ unrestricted} \end{aligned}$$

with associated Lagrangian $\mathcal{L}(\mathbf{X}, \bar{\mathbf{U}}) = \mathbf{C}'\mathbf{X} + \bar{\mathbf{U}}(\bar{\mathbf{b}} - \bar{\mathbf{A}}\mathbf{X})$, where $\bar{\mathbf{U}}' = (u_1, \ldots, u_{n+m}) \geqq \mathbf{0}'$. As far as the dual of this revised problem is concerned, we seek to

$$\begin{aligned} \min g(\bar{\mathbf{U}}) &= \bar{\mathbf{b}}'\bar{\mathbf{U}} \quad \text{s.t.} \\ \bar{\mathbf{A}}'\bar{\mathbf{U}} &= \mathbf{C}, \ \bar{\mathbf{U}} \geqq \mathbf{0}. \end{aligned}$$

In this instance $\mathcal{M}(\bar{\mathbf{U}}, \mathbf{X}) = -\bar{\mathbf{b}}'\bar{\mathbf{U}} + \mathbf{X}'(-\mathbf{C} + \bar{\mathbf{A}}'\bar{\mathbf{U}}) = -\mathcal{L}(\mathbf{X}, \bar{\mathbf{U}})$ also.

We next turn to the notion of a saddle point of a Lagrangian. Specifically, a point $(\mathbf{X}_0, \mathbf{U}_0) \in R^{n+m}$ is termed a **saddle point** of the Lagrangian $\mathcal{L}(\mathbf{X}, \mathbf{U}) = \mathbf{C}'\mathbf{X} + \mathbf{U}'(\mathbf{b} - \mathbf{A}\mathbf{X})$ if

$$\mathcal{L}(\mathbf{X}, \mathbf{U}_0) \leq \mathcal{L}(\mathbf{X}_0, \mathbf{U}_0) \leq \mathcal{L}(\mathbf{X}_0, \mathbf{U}) \tag{6.11}$$

for all $\mathbf{X} \geqq \mathbf{0}$, $\mathbf{U} \geqq \mathbf{0}$. Alternatively, $(\mathbf{X}_0, \bar{\mathbf{U}}_0) \in R^{2n+m}$ is a saddle point of $\mathcal{L}(\mathbf{X}, \bar{\mathbf{U}}) = \mathbf{C}'\mathbf{X} + \bar{\mathbf{U}}'(\bar{\mathbf{b}} - \bar{\mathbf{A}}\mathbf{X})$ if

$$\mathcal{L}(\mathbf{X}, \bar{\mathbf{U}}_0) \leq \mathcal{L}(\mathbf{X}_0, \bar{\mathbf{U}}_0) \leq \mathcal{L}(\mathbf{X}_0, \bar{\mathbf{U}}) \tag{6.12}$$

for all $\mathbf{X}$ unrestricted and $\bar{\mathbf{U}} \geqq \mathbf{0}$. What these definitions imply is that $\mathcal{L}(\mathbf{X}, \mathbf{U})$ $\big(\mathcal{L}(\mathbf{X}, \bar{\mathbf{U}})\big)$ simultaneously attains a maximum with respect to $\mathbf{X}$ and a minimum with respect to $\mathbf{U}$ $(\bar{\mathbf{U}})$. Hence (6.11), (6.12) appear respectively as

$$\mathcal{L}(\mathbf{X}_0, \mathbf{U}_0) = \max_{\mathbf{X} \geqq \mathbf{0}}\{\min_{\mathbf{U} \geqq \mathbf{0}} \mathcal{L}(\mathbf{X}, \mathbf{U})\} = \min_{\mathbf{U} \geqq \mathbf{0}}\{\max_{\mathbf{X} \geqq \mathbf{0}} \mathcal{L}(\mathbf{X}, \mathbf{U})\}, \tag{6.11.1}$$

$$\mathcal{L}(\mathbf{X}_0, \bar{\mathbf{U}}_0) = \max_{\mathbf{X}}\{\min_{\bar{\mathbf{U}} \geqq \mathbf{0}} \mathcal{L}(\mathbf{X}, \bar{\mathbf{U}})\} = \min_{\bar{\mathbf{U}} \geqq \mathbf{0}}\{\max_{\mathbf{X}} \mathcal{L}(\mathbf{X}, \bar{\mathbf{U}})\}, \ \mathbf{X} \text{ unrestricted.} \tag{6.12.1}$$

Under what conditions will $\mathcal{L}(\mathbf{X}, \mathbf{U})$ $\big(\mathcal{L}(\mathbf{X}, \bar{\mathbf{U}})\big)$ possess a saddle point at $(\mathbf{X}_0, \mathbf{U}_0)$ $\big((\mathbf{X}_0, \bar{\mathbf{U}}_0)\big)$? To answer these questions we must look to the solution of what may be called the

SADDLE POINT PROBLEM. To find a point $(\mathbf{X}_0, \mathbf{U}_0)$ such that (6.11) holds for all $(\mathbf{X}, \mathbf{U}) \in R^{n+m}$, $\mathbf{X} \geqq \mathbf{0}$, $\mathbf{U} \geqq \mathbf{0}$; or, to find a point $(\mathbf{X}_0, \bar{\mathbf{U}}_0)$ such that (6.12) holds for all $(\mathbf{X}, \bar{\mathbf{U}}) \in R^{2n+m}$, $\mathbf{X}$ unrestricted, $\bar{\mathbf{U}} \geqq \mathbf{0}$. In the light of (6.11.1), (6.12.1) we shall ultimately see that determining a saddle point of a

Lagrangian corresponds to maximinimizing or minimaximizing it. To this end we look to

6.10. THEOREM (A Necessary and Sufficient Condition). Let $\mathcal{L}(\mathbf{X}, \bar{\mathbf{U}})$ be defined for all $(\mathbf{X}, \bar{\mathbf{U}})\epsilon\mathbf{R}^{2n+m}$, where $\mathbf{X}$ is unrestricted and $\bar{\mathbf{U}} \geqq \mathbf{0}$. Then $\mathcal{L}(\mathbf{X}, \bar{\mathbf{U}})$ has a saddle point at $(\mathbf{X}_0, \bar{\mathbf{U}}_0)$ if and only if

(a) $\mathcal{L}(\mathbf{X}, \bar{\mathbf{U}}_0)$ attains a maximum at $\mathbf{X}_0$,

(b) $\bar{\mathbf{b}} - \bar{\mathbf{A}}\mathbf{X}_0 \geqq \mathbf{0}$ (6.13)

(c) $\bar{\mathbf{U}}_0'(\bar{\mathbf{b}}-\bar{\mathbf{A}}\mathbf{X}_0) = 0.$

PROOF. (necessity) The left-hand inequality of (6.12) is equivalent to (6.13.a) while it is right-hand inequality implies that $\mathbf{C}'\mathbf{X}_0 + \bar{\mathbf{U}}'(\bar{\mathbf{b}}-\bar{\mathbf{A}}\mathbf{X}_0) \geq \mathbf{C}'\mathbf{X}_0 + \bar{\mathbf{U}}_0'(\bar{\mathbf{b}}-\bar{\mathbf{A}}\mathbf{X}_0)$ or $(\bar{\mathbf{U}} - \bar{\mathbf{U}}_0)'(\bar{\mathbf{b}} - \bar{\mathbf{A}}\mathbf{X}_0) \geq 0$. If $\bar{\mathbf{b}} - \bar{\mathbf{A}}\mathbf{X}_0 \leqq \mathbf{0}$, a $\bar{\mathbf{U}}$ may be chosen sufficiently large so that $\bar{\mathbf{U}} - \bar{\mathbf{U}}_0 \geqq \mathbf{0}$ or $(\bar{\mathbf{U}}-\bar{\mathbf{U}}_0)'(\bar{\mathbf{b}}-\bar{\mathbf{A}}\mathbf{X}_0) \leq 0$ whence $\mathcal{L}(\mathbf{X}_0, \bar{\mathbf{U}}) \geq \mathcal{L}(\mathbf{X}_0, \bar{\mathbf{U}}_0)$ is violated. Hence (6.13.b) must hold. If $\bar{\mathbf{U}} = \mathbf{0}$, then $(\bar{\mathbf{U}}-\bar{\mathbf{U}}_0)'(\bar{\mathbf{b}}-\bar{\mathbf{A}}\mathbf{X}_0) = \bar{\mathbf{U}}_0'(\bar{\mathbf{b}}-\bar{\mathbf{A}}\mathbf{X}_0) \leq 0$. But since $\bar{\mathbf{U}}_0 \geqq \mathbf{0}$, $\bar{\mathbf{b}} - \bar{\mathbf{A}}\mathbf{X}_0 \geqq \mathbf{0}$ together imply $\bar{\mathbf{U}}_0'(\bar{\mathbf{b}}-\bar{\mathbf{A}}\mathbf{X}_0) \geq 0$, we see that (6.13.c) must hold as well. So if $\mathcal{L}(\mathbf{X}, \bar{\mathbf{U}})$ assumes a saddle point at $(\mathbf{X}_0, \bar{\mathbf{U}}_0)$, (6.13) holds.

(sufficiency) If $\mathbf{X}_0$ maximizes $\mathcal{L}(\mathbf{X}, \bar{\mathbf{U}}_0)$, it then follows that $\mathcal{L}(\mathbf{X}_0, \bar{\mathbf{U}}_0) \geq \mathcal{L}(\mathbf{X}, \bar{\mathbf{U}}_0)$. If $\bar{\mathbf{U}}_0'(\bar{\mathbf{b}}-\bar{\mathbf{A}}\mathbf{X}_0) = 0$, then $\mathcal{L}(\mathbf{X}_0, \bar{\mathbf{U}}_0) = \mathbf{C}'\mathbf{X}_0$. Hence $\mathcal{L}(\mathbf{X}_0, \bar{\mathbf{U}}) = \mathbf{C}'\mathbf{X}_0 + \bar{\mathbf{U}}'(\bar{\mathbf{b}}-\bar{\mathbf{A}}\mathbf{X}_0) = \mathcal{L}(\mathbf{X}_0, \bar{\mathbf{U}}_0) + \bar{\mathbf{U}}'(\bar{\mathbf{b}}-\bar{\mathbf{A}}\mathbf{X}_0)$ or $\mathcal{L}(\mathbf{X}_0, \bar{\mathbf{U}}) \geq \mathcal{L}(\mathbf{X}_0, \bar{\mathbf{U}}_0)$ since $\bar{\mathbf{U}}'(\bar{\mathbf{b}}-\bar{\mathbf{A}}\mathbf{X}_0) \geq 0$ if $\bar{\mathbf{b}} - \bar{\mathbf{A}}\mathbf{X}_0 \geqq \mathbf{0}$. So if (6.13) is satisfied, $\mathcal{L}(\mathbf{X}, \bar{\mathbf{U}})$ has a saddle point at $(\mathbf{X}_0, \bar{\mathbf{U}}_0)$. Q.E.D.

Our discussion in this chapter has centered around the solution of two important types of problems – the primal linear programming problem and the saddle point problem. As we shall now see, there exists an important connection between them. Specifically, what does the attainment of a saddle point of the Lagrangian of the primal problem imply about the existence of a solution to the primal problem? To answer this we cite

6.11. THEOREM (A Sufficient Condition). Let $\mathcal{L}(\mathbf{X}, \bar{\mathbf{U}})$ be defined for all $(\mathbf{X}, \bar{\mathbf{U}})\epsilon\,\mathbf{R}^{2n+m}$, where $\mathbf{X}$ is unrestricted and $\bar{\mathbf{U}} \geqq \mathbf{0}$. If $(\mathbf{X}_0, \bar{\mathbf{U}}_0)$ is a saddle point for $\mathcal{L}(\mathbf{X}, \bar{\mathbf{U}})$, then $\mathbf{X}_0$ solves the primal problem.

PROOF. Let $(\mathbf{X}_0, \bar{\mathbf{U}}_0)$ be a saddle point for $\mathcal{L}(\mathbf{X}, \bar{\mathbf{U}})$. Then by the previous theorem (6.13) holds and thus $\mathcal{L}(\mathbf{X}_0, \bar{\mathbf{U}}_0) \geq \mathcal{L}(\mathbf{X}, \bar{\mathbf{U}}_0)$ or $\mathbf{C}'\mathbf{X}_0 + \bar{\mathbf{U}}_0'(\bar{\mathbf{b}} - \bar{\mathbf{A}}\mathbf{X}_0) \geq \mathbf{C}'\mathbf{X} + \bar{\mathbf{U}}_0'(\bar{\mathbf{b}} - \bar{\mathbf{A}}\mathbf{X})$. Since $\bar{\mathbf{U}}_0'(\bar{\mathbf{b}} - \bar{\mathbf{A}}\mathbf{X}_0) = 0$ and $\bar{\mathbf{U}}_0'(\bar{\mathbf{b}} - \bar{\mathbf{A}}\mathbf{X}) \geq 0$, $\mathbf{C}'\mathbf{X}_0 \geq \mathbf{C}'\mathbf{X}$ for all $\mathbf{X}$. Q.E.D.

The importance of the developments in this section is that they set the stage for an analysis of the Karush-Kuhn-Tucker equivalence theorem. This theorem (presented below) establishes the notion that the existence of an optimal solution to the primal problem is equivalent to the existence of a saddle point of its associated Lagrangian, *i.e.*, solving the primal problem is equivalent to maximinimizing (minimaximizing) its Lagrangian. In this regard we look to the

6.12. KARUSH-KUHN-TUCKER EQUIVALENCE THEOREM (A Necessary and Sufficient Condition). A vector $\mathbf{X}_0$ is an optimal solution to the primal linear programming problem

$$\begin{aligned} max\, f(\mathbf{X}) &= \mathbf{C}'\mathbf{X} \quad \text{s.t.} \\ \bar{\mathbf{b}} - \bar{\mathbf{A}}\mathbf{X} &\geqq \mathbf{0}, \mathbf{X} \text{ unrestricted} \end{aligned}$$

if and only if there exists a vector $\bar{\mathbf{U}}_0 \geqq \mathbf{0}$ such that $(\mathbf{X}_0, \bar{\mathbf{U}}_0)$ is a saddle point of the Lagrangian $\mathcal{L}(\mathbf{X}, \bar{\mathbf{U}})$, in which case

$$\begin{aligned} \mathbf{C} - \bar{\mathbf{A}}'\bar{\mathbf{U}}_0 &= \mathbf{0} \\ \bar{\mathbf{U}}_0'(\bar{\mathbf{b}} - \bar{\mathbf{A}}\mathbf{X}_0) &= 0 \\ \bar{\mathbf{b}} - \bar{\mathbf{A}}\mathbf{X}_0 &\geqq \mathbf{0}. \end{aligned}$$

PROOF. (sufficiency) Let $\mathcal{L}(\mathbf{X}_0, \bar{\mathbf{U}}_0)$ solve the saddle point problem. Then by theorem (6.10), system (6.13) obtains. In this regard, if $\mathbf{X}_0$ maximizes $\mathcal{L}(\mathbf{X}, \bar{\mathbf{U}}_0) = \mathbf{C}'\mathbf{X} + \bar{\mathbf{U}}_0'(\bar{\mathbf{b}} - \bar{\mathbf{A}}\mathbf{X})$, then

$$\nabla_{\mathbf{X}}\mathcal{L}(\mathbf{X}_0, \bar{\mathbf{U}}_0) = \mathbf{C} - \bar{\mathbf{A}}'\mathbf{X}_0 = \mathbf{0}.$$

Hence this latter expression replaces (6.13.a) so that (6.13) is equivalent to (6.3), *i.e.*, if $\mathcal{L}(\mathbf{X}, \bar{\mathbf{U}})$ has a saddle point at $(\mathbf{X}_0, \bar{\mathbf{U}}_0)$, then (6.3) holds. Thus $\mathbf{X}_0$ solves the primal problem.

(necessity) Let $\mathbf{X}_0$ solve the primal problem. Then by corollary 6.5, the optimality of $\mathbf{X}_0$ implies the existence of a $\bar{\mathbf{U}}_0 \geqq \mathbf{0}$ such that $\mathbf{C}'\mathbf{X}_0 = \bar{\mathbf{b}}'\bar{\mathbf{U}}_0$. And since $\bar{\mathbf{U}}_0$ represents an optimal solution to the dual problem, $\bar{\mathbf{U}}_0'(\bar{\mathbf{b}} - \bar{\mathbf{A}}\mathbf{X}_0) = 0$.

Since $\mathbf{X}_0$ satisfies $\bar{\mathbf{b}}-\bar{\mathbf{A}}\mathbf{X}_0 \geqq \mathbf{0}$, then for any $\bar{\mathbf{U}} \geqq \mathbf{0}$, it follows that $\bar{\mathbf{U}}'(\bar{\mathbf{b}}-\bar{\mathbf{A}}\mathbf{X}_0) \geq 0$. Thus

$$\mathcal{L}(\mathbf{X}_0, \bar{\mathbf{U}}_0) = \mathbf{C}'\mathbf{X}_0 + \bar{\mathbf{U}}_0'(\bar{\mathbf{b}}-\bar{\mathbf{A}}\mathbf{X}_0) \leq \mathbf{C}'\mathbf{X}_0 + \bar{\mathbf{U}}'(\bar{\mathbf{b}}-\bar{\mathbf{A}}\mathbf{X}_0).$$

Hence the right-hand side of (6.12) is established. To verify its left-hand side we note that since $\mathbf{C}-\bar{\mathbf{A}}'\bar{\mathbf{U}}_0 = \mathbf{0}$ is equivalent to (6.13.a), $\mathcal{L}(\mathbf{X}_0, \bar{\mathbf{U}}_0) \geq \mathcal{L}(\mathbf{X}, \bar{\mathbf{U}}_0)$ by the sufficiency portion of theorem 6.10. Q.E.D.

It is instructive to analyze this equivalence from an alternative viewpoint. We noted above that solving the primal linear programming problem amounts to maximinimizing (minimaximizing) its associated Lagrangian. In this regard, if we express the primal problem as

$$\begin{aligned} max\, f(\mathbf{X}) &= \mathbf{C}'\mathbf{X} \quad \text{s.t.} \\ \mathbf{b}-\mathbf{A}\mathbf{X} &\geqq \mathbf{0},\ \mathbf{X} \geqq \mathbf{0}, \end{aligned}$$

then if $\mathcal{L}(\mathbf{X}, \mathbf{U}_0)$ has a maximum in the $\mathbf{X}$-direction at $\mathbf{X}_0$, it follows from (6.4.a, b, f) that

$$\left.\begin{array}{l} \nabla_{\mathbf{X}}\mathcal{L}^0 \leqq \mathbf{0} \\ (\nabla_{\mathbf{X}}\mathcal{L}^0)'\mathbf{X}_0 = 0 \\ \mathbf{X}_0 \geqq \mathbf{0} \end{array}\right\} \text{ or } \left\{\begin{array}{l} \mathbf{C}-\mathbf{A}'\mathbf{U}_0 \leqq \mathbf{0} \\ \mathbf{X}_0'(\mathbf{C}-\mathbf{A}'\mathbf{U}_0) = 0 \\ \mathbf{X}_0 \geqq \mathbf{0} \end{array}\right.$$

while if $L(\mathbf{X}_0, \mathbf{U})$ attains a minimum in the $\mathbf{U}$-direction at $\mathbf{U}_0$, then from (6.4.c, d, e)

$$\left.\begin{array}{l} -\nabla_{\mathbf{U}}\mathcal{L}^0 \leqq \mathbf{0} \text{ or } \nabla_{\mathbf{U}}\mathcal{L}^0 \geqq \mathbf{0} \\ (\nabla_{\mathbf{U}}\mathcal{L}^0)'\mathbf{U}_0 = 0 \\ \mathbf{U}_0 \geqq \mathbf{0} \end{array}\right\} \text{ or } \left\{\begin{array}{l} \mathbf{b}-\mathbf{A}\mathbf{X}_0 \geqq \mathbf{0} \\ \mathbf{U}_0'(\mathbf{b}-\mathbf{A}\mathbf{X}_0) = 0 \\ \mathbf{U}_0 \geqq \mathbf{0}\,. \end{array}\right.$$

Thus $\mathbf{X}_0$ is an optimal solution to the preceding primal problem if and only if there exists a vector $\mathbf{U}_0 \geqq \mathbf{0}$ such that $(\mathbf{X}_0, \mathbf{U}_0)$ is a saddle point of $\mathcal{L}(\mathbf{X}, \mathbf{U})$, in which case system (6.4) is satisfied.

As a result of the discussion underlying the Karush-Kuhn-Tucker equivalence theorem we have yet another version of the strong duality theorem, namely

6.13. COROLLARY. Let (6.3) hold so that the primal maximum problem has an optimal solution $\mathbf{X}_0$. Then the dual minimum problem

$$\begin{aligned} \min g(\overline{\mathbf{U}}) &= \overline{\mathbf{b}}'\overline{\mathbf{U}} \quad \text{s.t.} \\ \overline{\mathbf{A}}'\overline{\mathbf{U}} &= \mathbf{C},\ \overline{\mathbf{U}} \geqq \mathbf{0} \end{aligned}$$

also has an optimal solution $\overline{\mathbf{U}}_0 \geqq \mathbf{0}$ with $\mathbf{C}'\mathbf{X}_0 = \overline{\mathbf{b}}'\overline{\mathbf{U}}_0$.

PROOF. If $\mathbf{X}_0$ solves the primal problem, then by the preceding theorem there exists a vector $\overline{\mathbf{U}}_0 \geqq \mathbf{0}$ such that $(\mathbf{X}_0, \overline{\mathbf{U}}_0)$ is a saddle point of $\mathcal{L}(\mathbf{X},\overline{\mathbf{U}}) = -\mathcal{M}(\overline{\mathbf{U}},\mathbf{X})$, whence $\mathcal{M}(\overline{\mathbf{U}}_0,\mathbf{X}) \geq \mathcal{M}(\overline{\mathbf{U}}_0,\mathbf{X}_0) \geq \mathcal{M}(\overline{\mathbf{U}},\mathbf{X}_0)$, *i.e.*, $\mathcal{M}(\overline{\mathbf{U}},\mathbf{X})$ attains a maximum with respect to $\overline{\mathbf{U}}$ and a minimum with respect to $\mathbf{X}$ and thus possesses a saddle point at $(\overline{\mathbf{U}}_0, \mathbf{X}_0)$. And since $\overline{\mathbf{U}}_0'(\overline{\mathbf{b}}-\overline{\mathbf{A}}\mathbf{X}_0) = \mathbf{X}_0'(\mathbf{C}-\overline{\mathbf{A}}'\overline{\mathbf{U}}_0) = 0$, it follows that $\overline{\mathbf{U}}_0'\overline{\mathbf{b}} - \overline{\mathbf{U}}_0'\overline{\mathbf{A}}\mathbf{X}_0 = \mathbf{X}_0'\mathbf{C} - \mathbf{X}_0'\overline{\mathbf{A}}'\overline{\mathbf{U}}_0 = 0$ or $\mathbf{C}'\mathbf{X}_0 = \overline{\mathbf{b}}'\overline{\mathbf{U}}_0$. Q.E.D.

A final point is in order. We may relate the preceding material to the development of the (weak) complementary slackness theorem by expressing the primal-dual pair of problems as

PRIMAL PROBLEM	DUAL PROBLEM
$max\ f(\mathbf{X}) = \mathbf{C}'\mathbf{X}$ s.t.	$min\ g(\overline{\mathbf{U}}) = \overline{\mathbf{b}}'\overline{\mathbf{U}}$ s.t.
$\overline{\mathbf{A}}\mathbf{X} \leqq \overline{\mathbf{b}}$, $\mathbf{X}$ unrestricted.	$\overline{\mathbf{A}}'\overline{\mathbf{U}} = \mathbf{C},\ \overline{\mathbf{U}} \geqq \mathbf{0}$.

Then their associated Lagrangians may be written respectively as

$$\begin{aligned} \mathcal{L}(\mathbf{X},\overline{\mathbf{U}},\overline{\mathbf{X}}_s) &= \mathbf{C}'\mathbf{X} + (\overline{\mathbf{b}}-\overline{\mathbf{A}}\mathbf{X}-\overline{\mathbf{X}}_s), \\ \mathcal{M}(\overline{\mathbf{U}},\mathbf{X}) &= -\overline{\mathbf{b}}'\overline{\mathbf{U}} - \mathbf{X}'(\mathbf{C}-\overline{\mathbf{A}}'\overline{\mathbf{U}}), \end{aligned}$$

where $\overline{\mathbf{X}}_s$ is an $(m+n \times 1)$ vector of nonnegative primal slack variables, *i.e.*,

$$\overline{\mathbf{X}}_s = \begin{bmatrix} \overline{x}_{n+1} \\ \vdots \\ \overline{x}_{n+m} \\ \overline{x}_{n+m+1} \\ \vdots \\ \overline{x}_{2n+m} \end{bmatrix} \geqq \mathbf{0} \text{ with } \begin{cases} \overline{x}_{n+i} = \overline{b}_i - \overline{\mathbf{a}}_i\mathbf{X} \geq 0, \quad i=1,\ldots,m; \\ \overline{x}_{n+m+j} = \overline{b}_{m+j} - \overline{\mathbf{a}}_{m+j}\,\mathbf{X} \\ \qquad = \mathbf{e}_j'\,\mathbf{X} \\ \qquad = x_j \text{ unrestricted}, \quad j=1,\ldots,n. \end{cases}$$

In this regard we now look to the

6.14. WEAK COMPLEMENTARY SLACKNESS THEOREM.

The vectors $\mathbf{X}_0, \bar{\mathbf{U}}_0$ solve the primal and dual problems

$$\begin{array}{ll} max\, f(\mathbf{X}) \;=\; \mathbf{C}'\mathbf{X} \text{ s.t.} & min\, g(\bar{\mathbf{U}}) \;=\; \mathbf{b}'\bar{\mathbf{U}} \text{ s.t.} \\ \bar{\mathbf{A}}\mathbf{X} \leqq \bar{\mathbf{b}},\ \mathbf{X} \text{ unrestricted} & \bar{\mathbf{A}}'\bar{\mathbf{U}} \;=\; \mathbf{C},\ \bar{\mathbf{U}} \geqq \mathbf{0} \end{array}$$

respectively if and only if $\mathbf{X}_0, \bar{\mathbf{U}}_0$ satisfy the (weak) complementary slackness conditions

$$\bar{\mathbf{U}}_0'(\bar{\mathbf{X}}_s)_0 = 0 \text{ or } \begin{cases} \text{(a) } \bar{u}_i^0\, \bar{x}_{n+i}^0 = 0,\ i=1,\ldots,m; \\ \text{(b) } \bar{u}_{m+j}^0\, \bar{x}_{n+m+j}^0 = 0,\ j=1,\ldots,n. \end{cases} \tag{6.14}$$

PROOF. (necessity) If $\mathbf{X}_0$ solves the primal problem, then there exists a vector $\bar{\mathbf{U}}_0 \geqq \mathbf{0}$ such that $(\mathbf{X}_0, \bar{\mathbf{U}}_0)$ is a saddle point of $\mathcal{L}(\mathbf{X}, \bar{\mathbf{U}}) = -\mathcal{M}(\bar{\mathbf{U}}, \mathbf{X})$, whence

$$\begin{aligned} \mathcal{L}(\mathbf{X}_0, \bar{\mathbf{U}}_0, (\bar{\mathbf{X}}_s)_0) + \mathcal{M}(\bar{\mathbf{U}}_0, \mathbf{X}_0) &= \mathbf{C}'\mathbf{X}_0 + \bar{\mathbf{U}}_0'(\bar{\mathbf{b}} - \bar{\mathbf{A}}\mathbf{X}_0 - (\bar{\mathbf{X}}_s)_0) \\ &\quad - \bar{\mathbf{b}}'\bar{\mathbf{U}}_0 - \mathbf{X}_0'(\mathbf{C} - \bar{\mathbf{A}}'\bar{\mathbf{U}}_0) \\ &= -\bar{\mathbf{U}}_0'(\bar{\mathbf{X}}_s)_0 \;=\; 0 \end{aligned}$$

or $\bar{\mathbf{U}}_0'(\bar{\mathbf{X}}_s)_0 = 0$. With $\bar{\mathbf{U}}_0' = (\bar{u}_1^0, \ldots, \bar{u}_m^0, \bar{u}_{m+1}^0, \ldots, \bar{u}_{m+n}^0)$, it follows that

$$\begin{aligned} \bar{\mathbf{U}}_0'(\bar{\mathbf{X}}_s)_0 &= \sum_{i=1}^{m} \bar{u}_i^0\, \bar{x}_{n+i}^0 + \sum_{j=1}^{n} \bar{u}_{m+j}^0\, \bar{x}_{n+m+j}^0 \\ &= \sum_{i=1}^{m} \bar{u}_i^0\, \bar{x}_{n+i}^0 + \sum_{j=1}^{n} \bar{u}_{m+j}^0\, x_j^0 \;=\; 0. \end{aligned}$$

(sufficiency) If $\bar{\mathbf{U}}_0'(\bar{\mathbf{X}}_s)_0 = \bar{\mathbf{U}}_0'(\bar{\mathbf{b}} - \bar{\mathbf{A}}\mathbf{X}) = 0$, then $\mathcal{L}(\mathbf{X}_0, \bar{\mathbf{U}}_0) = \mathbf{C}'\mathbf{X}_0$, $\mathcal{M}(\bar{\mathbf{U}}_0, \mathbf{X}_0) = -\bar{\mathbf{b}}'\bar{\mathbf{U}}_0 = -\mathcal{L}(\mathbf{X}_0, \bar{\mathbf{U}}_0)$, and thus $\mathbf{C}'\mathbf{X}_0 = \bar{\mathbf{b}}'\bar{\mathbf{U}}_0$. Hence $\mathbf{X}_0, \bar{\mathbf{U}}_0$ yield optimal solutions to the primal and dual problems respectively. Q.E.D.

By way of interpretation, (6.14.a) informs us that if the primal slack variable $\bar{x}_{n+i}^0$, $i=1,\ldots,m$, corresponding to the ith primal structural constraint is positive, then the associated dual variable $\bar{u}_i^0$, $i=1,\ldots,m$, must equal zero. Conversely, if $\bar{u}_i^0$, $i=1,\ldots,m$, is positive, then the ith primal structural constraint must hold as an equality with $\bar{x}_{n+i}^0 = 0$, $i=1,\ldots,m$. Before turning to (6.14.b) let us examine the form of the jth dual structural constraint. Since $\mathbf{A} = [\boldsymbol{\alpha}_1, \ldots, \boldsymbol{\alpha}_n]$,

$$\bar{\mathbf{A}}'\bar{\mathbf{U}} = \left[\mathbf{A}' \vdots -\mathbf{I}_n\right]\begin{bmatrix}\mathbf{U}\\ \bar{u}_{m+1}\\ \vdots\\ \bar{u}_{m+n}\end{bmatrix} = \begin{bmatrix}\alpha_1' & \vdots & \\ \vdots & \vdots & -\mathbf{I}_n\\ \alpha_n' & \vdots & \end{bmatrix}\begin{bmatrix}\mathbf{U}\\ \bar{u}_{m+1}\\ \vdots\\ \bar{u}_{m+n}\end{bmatrix} = \mathbf{C}$$

and thus the jth dual structural constraint appears as $\alpha_j'\mathbf{U} - \bar{u}_{m+j} = c_j$, j=1,...,n. In this light, from (6.14.b), if the jth primal variable x_j^0, j=1,...,n, is different from zero, then the dual surplus variable $\bar{u}_{m+j}^0$, j=1,...,n, associated with the jth dual structural constraint $\alpha_j'\mathbf{U} - \bar{u}_{m+j} = c_j$, j=1,...,n, equals zero. Conversely, if $\bar{u}_{m+j}^0 > 0$, j=1,...,n, then the jth structural constraint is not binding and thus the jth primal variable x_j^0, j=1,...,n, must be zero.

It is important to note that the weak complementary slackness conditions are a consequence of the Karush-Kuhn-Tucker theorem for linear programs. That is to say,

6.15. COROLLARY. If (6.4) holds so that the primal maximum and dual minimum problems have optimal solutions $(\mathbf{X}_0, (\mathbf{X}_s)_0)$, $(\mathbf{U}_0, (\mathbf{U}_s)_0)$ respectively, then the (weak) complementary slackness conditions

$$(\mathbf{U}_s)_0'\mathbf{X}_0 + \mathbf{U}_0'(\mathbf{X}_s)_0 = 0$$

$$\text{or} \quad \begin{cases}\text{(a) } u_i^0\, x_{n+i}^0 = 0,\ i=1,\ldots,m;\\ \text{(b) } x_j^0\, \bar{u}_{m+j}^0 = 0,\ j=1,\ldots,n.\end{cases} \tag{6.15}$$

also hold, *i.e.*, for optimal solutions $(\mathbf{X}_0, (\mathbf{X}_s)_0)$, $(\mathbf{U}_0, (\mathbf{U}_s)_0)$ to the primal maximum and dual minimum problems respectively:

(a) if a primal slack variable x_{n+i}^0, i=1,...,m, is positive, then the ith dual variable u_i^0, i=1,...,m, must equal zero. Conversely, if u_i^0, i=1,...,m, is positive, then x_{n+i}^0, i=1,...,m, equals zero;

(b) if the jth primal variable x_j^0, j=1,...,n, is positive, then the dual surplus variable u_{m+j}^0, j=1,...,n, equals zero. Conversely, if u_{m+j}^0, j=1,...,n, is positive, then x_j^0, j=1,...,n, must equal zero.

PROOF. If $\mathbf{X}_s$ represents an $(m \times 1)$ vector of nonnegative primal slack variables and $\mathbf{U}_s$ depicts an $(n \times 1)$ vector of nonnegative dual surplus variables, then (6.4.b,c) become $\mathbf{X}_0'(-\mathbf{U}_s)_0 = \mathbf{U}_0'\mathbf{X}_s$ or $(\mathbf{U}_s)_0'\mathbf{X}_0 + \mathbf{U}_0'(\mathbf{X}_s)_0 = 0$. Alternatively, (6.5.a) reveals that $\mathbf{U}_s = -\mathbf{V}_0$ so that the (weak) complementary slackness condition may be rewritten in terms of the primal variables and Lagrange multipliers as $\mathbf{U}_0'(\mathbf{X}_s)_0 - \mathbf{V}_0'\mathbf{X}_0 = 0$. Q.E.D.

7 COMPUTATIONAL ASPECTS OF LINEAR PROGRAMMING

7.1. The Primal Simplex Method Reexamined

In this chapter we shall be concerned with the development of a computationally efficient version of the simplex procedure for solving a primal linear program. Let us assume that we are faced with the now familiar problem of determining the maximum of

$$f(\mathbf{X}) = \mathbf{C}'\mathbf{X} \quad \text{s.t.}$$
$$\mathbf{A}\mathbf{X} = \mathbf{b}, \ \mathbf{X} \geqq \mathbf{0}.$$

Given that $\rho(\mathbf{A}) = m$, our previous approach was to first extract m linearly independent columns from $\mathbf{A}$ so as to form an $(m \times m)$ non-singular basis matrix $\mathbf{B}$. We next determined $\mathbf{B}^{-1}$ and thus

$$\begin{aligned} \mathbf{X}_B &= \mathbf{B}^{-1}\mathbf{b} - \mathbf{B}^{-1}\mathbf{R}\mathbf{X}_R = \bar{\mathbf{b}} - \mathbf{B}^{-1}\mathbf{R}\mathbf{X}_R, \\ f(\mathbf{X}) &= \mathbf{C}'_B\mathbf{B}^{-1}\mathbf{b} + (\mathbf{C}'_R - \mathbf{C}'_B\mathbf{B}^{-1}\mathbf{R})\mathbf{X}_R \\ &= \mathbf{C}'_B\bar{\mathbf{b}} + \bar{\mathbf{C}}'_R\mathbf{X}_R. \end{aligned}$$

Upon setting $\mathbf{X}_R = \mathbf{0}$, we then obtained our first basic feasible solution of the form

$$\begin{aligned} \mathbf{X}_B &= \bar{\mathbf{b}}, \\ f(\mathbf{X}) &= \mathbf{C}'_B\bar{\mathbf{b}}. \end{aligned}$$

(If $\mathbf{B} = \mathbf{I}_m$, then our initial basic feasible solution is just $\mathbf{X}_B = \mathbf{b}$, $f(\mathbf{X}) = \mathbf{C}'_B\mathbf{b}$.) As an alternative, let us form the following simultaneous system involving $m+1$ equations and the $n+1$variables $\mathbf{X}$, f, namely

$$\left.\begin{array}{ll}\mathbf{AX} & =\mathbf{b}\\ -\mathbf{C}'\mathbf{X}+f & =0\end{array}\right\} \text{ or } \begin{bmatrix}\mathbf{A} & \mathbf{0}\\ -\mathbf{C}' & 1\end{bmatrix}\begin{bmatrix}\mathbf{X}\\ f\end{bmatrix} = \begin{bmatrix}\mathbf{b}\\ 0\end{bmatrix}. \tag{7.1}$$

Here the objective function is treated *as if* it were an additional structural constraint. In this formulation of the primal linear programming problem, we desire to obtain a solution to (7.1) wherein f is as large as possible (and unrestricted in sign) with $\mathbf{X} \geqq \mathbf{0}$. To consolidate our notation a bit, let

$$\bar{\mathbf{A}} = \begin{bmatrix}\mathbf{A} & \mathbf{0}\\ -\mathbf{C}' & 1\end{bmatrix}, \ \bar{\mathbf{X}} = \begin{bmatrix}\mathbf{X}\\ f\end{bmatrix}, \text{ and } \mathbf{b}_0 = \begin{bmatrix}\mathbf{b}\\ 0\end{bmatrix},$$

where $\bar{\mathbf{A}}$ is of order $(m+1 \times n+1)$, $\bar{\mathbf{X}}$ is of order $(n+1 \times 1)$, and $\mathbf{b}_0$ is of order $(m+1 \times 1)$. Then (7.1) can be treated more succinctly as

$$\bar{\mathbf{A}}\bar{\mathbf{X}} = \mathbf{b}_0. \tag{7.1.1}$$

Since the columns within $\bar{\mathbf{A}}$ contain $m+1$ components, a basis for $\mathbf{R}^{m+1}$ may be formed by selecting $m+1$ linearly independent vectors from $\bar{\mathbf{A}}$. Let the said collection of vectors constitute the columns of the $m+1^{st}$ order nonsingular basis matrix $\bar{\mathbf{B}} = \left[\bar{\mathbf{b}}_1, \ldots, \bar{\mathbf{b}}_{m+1}\right]$. Given that the primal objective function is to be maximized, the variable f must be a member of every basic solution. In this regard, since the last column of $\bar{\mathbf{A}}$ corresponds to the variable f, the unit column vector $\mathbf{e}_{m+1}$ will be a member of every basis matrix. If we adopt the convention of letting $\mathbf{e}_{m+1}$ always appear in the last column of the basis matrix, then $\bar{\mathbf{B}} = \left[\bar{\mathbf{b}}_1, \ldots, \bar{\mathbf{b}}_m, \bar{\mathbf{e}}_{m+1}\right]$.

How does the task of obtaining a basic solution to $\bar{\mathbf{A}}\bar{\mathbf{X}} = \mathbf{b}_0$ compare with our previous problem involving the determination of a basic solution to $\mathbf{AX} = \mathbf{b}$? To answer this question let us see exactly how the vectors $\bar{\mathbf{b}}_i$, $i=1,\ldots,m$, are to be selected. Let the first n columns of $\bar{\mathbf{A}}$ be denoted as

$$\bar{\mathbf{a}}_j = \begin{bmatrix}\mathbf{a}_j\\ -c_j\end{bmatrix}, \ j=1,\ldots,n, \tag{7.2}$$

where $\mathbf{a}_j$ is the jth column of $\mathbf{A}$. If we can find m linearly independent vectors $\mathbf{b}_i$ within $\mathbf{A}$, then upon substituting them into (7.2), m of the $\bar{\mathbf{a}}_j$ can correspondingly be represented as

$$\bar{\mathbf{b}}_i = \begin{bmatrix} \mathbf{b}_i \\ -c_{Bi} \end{bmatrix}, \ i=1,\ldots,m,$$

where, as before, $\mathbf{B} = [\mathbf{b}_1,\ldots,\mathbf{b}_m]$. Hence

$$\bar{\mathbf{B}} = \begin{bmatrix} \mathbf{B} & \mathbf{0} \\ -\mathbf{C}'_B & 1 \end{bmatrix},$$

i.e., since m linearly independent $\mathbf{a}_j$ uniquely determine m linearly independent $\bar{\mathbf{a}}_j$, there exists a one-to-one correspondence between the basis matrix $\mathbf{B}$ for $\mathbf{AX} = \mathbf{b}$ and the basis matrix $\bar{\mathbf{B}}$ for $\bar{\mathbf{A}}\bar{\mathbf{X}} = \mathbf{b}_0$. If the variables associated with the $m+1$ columns of $\bar{\mathbf{B}}$ are denoted as $x_{\bar{B}i}$, $i=1,\ldots,m$, and f, then a basic feasible solution to $\bar{\mathbf{A}}\bar{\mathbf{X}} = \mathbf{b}_0$ will be defined as one for which f is as large as possible and $x_{\bar{B}i} \geq 0$, $i=1,\ldots,m$. In order to guarantee that the variant of the primal simplex process to be presented below converges to an optimal basic feasible solution in a finite number of iterations, let us again impose the nondegeneracy assumption set out at the beginning of the previous chapter, namely, that every set of m columns from $\mathbf{A}$ is linearly independent.

To determine an initial basic feasible solution to $\bar{\mathbf{A}}\bar{\mathbf{X}} = \mathbf{b}_0$, let us form the $(m+1 \times n+2)$ augmented matrix associated with this system as

$$[\bar{\mathbf{A}} \,\vdots\, \mathbf{b}_0] = \begin{bmatrix} \mathbf{A} & \mathbf{0} & \mathbf{b} \\ -\mathbf{C}' & 1 & 0 \end{bmatrix} \tag{7.3}$$

$$= \begin{array}{c} \begin{matrix} x_1 & x_2 & & x_p & x_{p+1} & x_{p+2} & & x_{p+m} & f & \end{matrix} \\ \begin{bmatrix} a_{11} & a_{12} & \cdots & a_{1p} & 1 & 0 & \cdots & 0 & 0 & b_1 \\ a_{21} & a_{22} & \cdots & a_{2p} & 0 & 1 & \cdots & 0 & 0 & b_2 \\ \cdots & \cdots & \cdots & \cdots & \cdots & \cdots & \cdots & \cdots & \cdots & \cdots \\ a_{m1} & a_{m2} & \cdots & a_{mp} & 0 & 0 & \cdots & 1 & 0 & b_m \\ -c_1 & -c_2 & \cdots & -c_p & 0 & 0 & \cdots & 0 & 1 & 0 \end{bmatrix} \end{array}.$$

(For purposes of exposition, the variables corresponding to the first $n+1$ columns of (7.3) appear directly above this matrix.) Henceforth this augmented matrix will be referred to as the **primal simplex matrix**.[1] For example, if we consider the problem presented at the outset of the previous chapter, *i.e.*, if we desire to

$$\begin{aligned} &max\ f = 2x_1 + 3x_2 + 4x_3 \quad \text{s.t.} \\ &2x_1 + \ \ x_2 + \ 4x_3 \leq 100 \\ &x_1 + \ 3x_2 + \ \ x_3 \leq \ 40 \\ &x_i \geq 0,\ i{=}1,2,3, \end{aligned}$$

then the simplex matrix becomes

$$\begin{bmatrix} 2 & 1 & 4 & 1 & 0 & 0 & 100 \\ 1 & 3 & 1 & 0 & 1 & 0 & 40 \\ -2 & -3 & -4 & 0 & 0 & 1 & 0 \end{bmatrix}. \tag{7.4}$$

If the basic variables are taken to be those corresponding to the $m{+}1$ unit column vectors in (7.3), then we may express these $m{+}1$ variables $x_{p+1},\ldots,x_{p+m}$, f in terms of the p remaining or nonbasic variables $x_1,\ldots,x_p$ as

$$\begin{aligned} x_{p+1} &= b_1 - \sum_{j=1}^{p} a_{1j}x_j \\ x_{p+2} &= b_2 - \sum_{j=1}^{p} a_{2j}x_j \\ &\cdots\cdots\cdots\cdots\cdots \\ x_{p+m} &= b_m - \sum_{j=1}^{p} a_{mj}x_j \\ f &= \sum_{j=1}^{p} c_j x_j\ . \end{aligned} \tag{7.5}$$

[1]In the calculations which follow, our first step in the determination of an optimal basic feasible solution will be to form the simplex matrix. To establish the connection between the computational algorithm presented in this chapter and our previous results, we shall demonstrate that the simplex matrix can also serve as a vehicle with which to derive the representation of $\mathbf{X}_B$ and $f(\mathbf{X})$ in terms of the nonbasic variables $\mathbf{X}_R$, *i.e.*, if we premultiply

$$\begin{bmatrix} \mathbf{A} & \mathbf{0} & \mathbf{b} \\ -\mathbf{C}' & 1 & 0 \end{bmatrix} = \begin{bmatrix} \mathbf{B} & \mathbf{R} & \mathbf{0} & \mathbf{b} \\ -\mathbf{C}'_B & -\mathbf{C}'_R & 1 & 0 \end{bmatrix}$$ by the inverse of

$$\bar{\mathbf{B}} = \begin{bmatrix} \mathbf{B} & \mathbf{0} \\ -\mathbf{C}'_B & 1 \end{bmatrix},\ \bar{\mathbf{B}}^{-1} = \begin{bmatrix} \mathbf{B}^{-1} & \mathbf{0} \\ \mathbf{C}'_B\mathbf{B}^{-1} & 1 \end{bmatrix},$$

the resulting product $\begin{bmatrix} \mathbf{I}_m & \mathbf{B}^{-1}\mathbf{R} & \mathbf{0} & \mathbf{B}^{-1}\mathbf{b} \\ \mathbf{0} & \mathbf{C}'_B\mathbf{B}^{-1}\mathbf{R}-\mathbf{C}'_R & 1 & \mathbf{C}'_B\mathbf{B}^{-1}\mathbf{b} \end{bmatrix}$ has, as its first row,

$\mathbf{X}_B + \mathbf{B}^{-1}\mathbf{R}\mathbf{X}_R = \mathbf{B}^{-1}\mathbf{b}$ or $\mathbf{X}_B = \mathbf{B}^{-1}\mathbf{b} - \mathbf{B}^{-1}\mathbf{R}\mathbf{X}_R$ and, as its second row, $(\mathbf{C}'_B\mathbf{B}^{-1}\mathbf{R} - \mathbf{C}'_R)\mathbf{X}_R + f = \mathbf{C}'_B\mathbf{B}^{-1}\mathbf{b}$ or $f = \mathbf{C}'_B\mathbf{B}^{-1}\mathbf{b} + (\mathbf{C}'_R - \mathbf{C}'_B\mathbf{B}^{-1}\mathbf{R})\mathbf{X}_R$.

Our initial basic feasible solution is then determined by setting the nonbasic variables in (7.5) equal to zero, to wit $x_{p+1} = b_1, x_{p+2} = b_2, \ldots, x_{p+m} = b_m$, and $f = 0$. Have we obtained an optimal (maximal) basic feasible solution? To answer this question let us examine the last equation in (7.5). It is obvious that f will increase in value if any nonbasic variable with a strictly positive coefficient changes from its current zero level by some positive amount. But if any nonbasic variable has a positive coefficient in the $m+1^{st}$ equation in system (7.5), its associated coefficient in the $m+1^{st}$ row of (7.3) must be negative. Thus if any negative values appear in the last or objective function row of the simplex matrix, the current basic feasible solution is not optimal. In terms of (7.4), if x_4 and x_5 are taken to be basic variables with x_1, x_2, and x_3 deemed nonbasic, then

$$\begin{aligned} x_4 &= 100 - 2x_1 - x_2 - 4x_3 \\ x_5 &= 40 - x_1 - 3x_2 - x_3 \\ f &= 2x_1 + 3x_2 + 4x_3 \end{aligned}$$

and thus, for $x_1 = x_2 = x_3 = 0$, $x_4 = 100$, $x_5 = 40$, and $f = 0$. Clearly this initial basic feasible solution is not optimal since there are three negative entries in the last row of (7.4). In general, a particular basic feasible solution to the primal problem is optimal when all of the coefficients on the nonbasic variables within the last row of the primal simplex matrix are nonnegative or $-\bar{c}_j \geq 0$, $j=1,\ldots,n-m$. Moreover, an optimal primal basic feasible solution is unique if $-\bar{c}_j > 0$, $j=1,\ldots,n-m$. In this regard, if the coefficient on some particular nonbasic variable within the last row of the simplex matrix is zero, then if that variable increases in value to some positive level, no change in f occurs. Hence the objective function assumes the same value for more than one basic feasible solution so that the case of multiple optimal solutions obtains. (As the reader has probably already recognized, the optimality criterion just stated is simply a variation of the one provided by optimality theorem 5.1.)

Given that our initial basic feasible solution is not optimal, the next section sets out an iterative scheme which enables us to generate an optimal basic feasible solution using (7.3) as the starting point.

7.2. Improving a Basic Feasible Solution

Before proceeding directly to the primal simplex matrix, let us first indicate some of the specific steps that will be executed in our search for an optimal basic feasible solution. Given that the simplex matrix has within it a set of $m+1$ unit column vectors, let us consider a set of operations on this equation system which transforms it into an equivalent system whose associated simplex matrix also contains $m+1$ unit column vectors. In particular, the following two types of elementary row operations will transform (7.3) into a new simplex matrix which is row-equivalent to the latter:

TYPE 1. Multiply any equation E_h by a nonzero scaler k and replace E_h by the new equation $kE_h, h=1,\ldots,m$. (Note that since $h \neq m+1$, this type of operation may not be performed on the objective function row of the simplex matrix.)

TYPE 2. Replace any equation E_i by the new equation E_i+kE_h, $i=1,\ldots,m+1$; $h=1,\ldots,m$.

In actual practice (7.3) will be transformed into an equivalent system of equations yielding a new basic feasible solution, with an objective function value which is at least as great as that obtained from the previous one, by performing a particular sequence of elementary row operations called a **pivot operation**. Such an operation consists of an appropriate sequence of at most $m+1$ elementary row operations which transform (7.3) into an equivalent system in which a specified variable has a coefficient of unity in one equation and zeros elsewhere. The implementation of the pivot process then proceeds as follows:

(1) select a **pivot term** $a_{rs} \neq 0$, $r \leq m$, $s \leq p+m = n$, within the simplex matrix (since $r \neq m+1, s \neq n+1, n+2$, the elements within the last row and last two columns of this matrix may not serve as pivot elements). Next,

(2) replace E_r by $\frac{1}{a_{rs}} E_r$. Finally,

(3) replace E_i by $E_i - \frac{a_{is}}{a_{rs}} E_r$, $i=1,\ldots,m+1$; $i \neq r$.

In matrix terms, a pivot operation is equivalent to premultiplying the simplex matrix by the $m+1^{st}$ order **pivot matrix**

$$\mathbf{P} = \begin{bmatrix} 1 & 0 & \cdots & \frac{-a_{1s}}{a_{rs}} & \cdots & 0 \\ 0 & 1 & \cdots & \frac{-a_{2s}}{a_{rs}} & \cdots & 0 \\ \vdots & \vdots & & \vdots & & \vdots \\ 0 & 0 & \cdots & \frac{1}{a_{rs}} & \cdots & 0 \\ \vdots & \vdots & & & & \vdots \\ 0 & 0 & \cdots & \frac{-a_{m+1,s}}{a_{rs}} & \cdots & 1 \end{bmatrix}.$$

Given that at least one $-\bar{c}_j < 0$ in (7.3), our first step will be to determine which nonbasic variable is to become basic. We shall adopt the convention of choosing the nonbasic variable $x_j, j=1,\ldots,p$, whose associated coefficient $-\bar{c}_j$ within the objective function row of the simplex matrix is *most negative*, *i.e.*, if

$$\text{ENTRY CRITERION:} \qquad -c_k = \min_j\{-\bar{c}_j, -\bar{c}_j < 0\} \tag{7.6}$$

then x_k, the nonbasic variable corresponding to the kth column of (7.3), becomes basic. In terms of (7.5), such a choice produces the largest increment in f per unit change in $x_j, j=1,\ldots,p$. (Note that this new entry criterion is consistent with our previous one involving (5.11).) Since only one nonbasic variable will be introduced into the set of basic variables at a time, $x_1,\ldots,x_{k-1}, x_{k+1},\ldots,x_p$ are all held fixed at their current zero level. Hence, from (7.5),

$$\begin{aligned} x_{p+1} &= b_1 - a_{1k}x_k \\ x_{p+2} &= b_2 - a_{2k}x_k \\ &\cdots\cdots\cdots\cdots \\ x_{p+m} &= b_m - a_{mk}x_k \\ f &= c_k x_k. \end{aligned} \tag{7.5.1}$$

As x_k increases in value from zero, the new values of $x_{p+i}, i=1,\ldots,m$, must still satisfy the nonnegativity conditions, *i.e.*, we require that

$$x_{p+i} = b_i - a_{ik}x_k \geq 0, i{=}1,\ldots,m. \tag{7.7}$$

Under what conditions will (7.7) be satisfied? If $a_{ik} \leq 0$, then $x_{p+i} > 0$ so that, for the moment, we need only concern ourselves with those $a_{ik} > 0$. In this regard, when at least one of the coefficients a_{ik} within the kth column of (7.3) is positive, there exists an upper bound to an increase in x_k which will not violate (7.7). Rearranging (7.7) yields $\frac{b_i}{a_{ik}} \geq x_k$ so that any increase in x_k which equals

$$\text{EXIT CRITERION: } \hat{\theta} = \min_i\left\{\frac{b_i}{a_{ik}}, a_{ik} > 0\right\} = \frac{b_r}{a_{rk}} \tag{7.8}$$

preserves feasibility. So when $i{=}r$, in terms of (7.3), column k replaces column $p{+}r$ in the basis and thus x_k becomes a basic variable while x_{p+r} turns nonbasic. From (7.7), our new basic feasible solution becomes

$$\begin{aligned} x_k &= \frac{b_r}{a_{rk}}, \\ x_{p+i} &= b_i - a_{ik}\frac{b_r}{a_{rk}}, \; i{=}1,\ldots,m, \; i \neq r, \text{ and} \\ f &= c_k\frac{b_r}{a_{rk}}. \end{aligned}$$

Turning to the simplex matrix, if x_k is to replace x_{p+r} in the set of basic variables, a pivot operation with a_{rk} as the pivot element may be performed so as to express the new set of basic variables $x_{p+i}, i{=}1,\ldots,m, \; i \neq r, \; x_k$, and f in terms of the nonbasic variables $x_1,\ldots,x_{k-1}, \; x_{k+1},\ldots,x_p$, and x_{p+r}. To this end let us first divide E_r in (7.3) by a_{rk} to obtain

$$\begin{array}{cccccccccccccccccc} x_1 & x_2 & & x_{k-1} & x_k & x_{k+1} & & x_p & x_{p+1} & x_{p+2} & & x_{p+r} & & x_{p+m} & f & \\ a_{11} & a_{12} & \cdots & a_{1,k-1} & a_{1k} & a_{1,k+1} & \cdots & a_{1p} & 1 & 0 & \cdots & 0 & \cdots & 0 & 0 & b_1 \\ \multicolumn{16}{c}{\cdots\cdots\cdots\cdots\cdots\cdots\cdots\cdots\cdots\cdots\cdots\cdots\cdots\cdots\cdots\cdots} \\ \frac{a_{r1}}{a_{rk}} & \frac{a_{r2}}{a_{rk}} & \cdots & \frac{a_{r,k-1}}{a_{rk}} & 1 & \frac{a_{r,k+1}}{a_{rk}} & \cdots & \frac{a_{rp}}{a_{rk}} & 0 & 0 & \cdots & \frac{1}{a_{rk}} & \cdots & 0 & 0 & \frac{b_r}{a_{rk}} \\ \multicolumn{16}{c}{\cdots\cdots\cdots\cdots\cdots\cdots\cdots\cdots\cdots\cdots\cdots\cdots\cdots\cdots\cdots\cdots} \\ a_{m1} & a_{m2} & \cdots & a_{m,k-1} & a_{mk} & a_{m,k+1} & \cdots & a_{mp} & 0 & 0 & \cdots & 0 & \cdots & 1 & 0 & b_m \\ -c_1 & -c_2 & \cdots & -c_{k-1} & -c_k & -c_{k+1} & \cdots & -c_p & 0 & 0 & \cdots & 0 & \cdots & 0 & 1 & 0 \end{array} \tag{7.3.1}$$

Next, to eliminate x_k from the remaining m equations $E_i, i=1,\ldots,m+1,\ i\neq r$, let us: (1) multiply E_r by $-a_{ik}$ and replace E_i by the new equation $E_i - a_{ik}E_r$, $i=1,\ldots,m, i\neq r$; and (2) multiply E_r by c_k and replace E_{m+1} by the new equation $E_{m+1}+c_kE_r$. Hence (7.3.1) is ultimately transformed to

$$\begin{bmatrix} \bar{a}_{11} & \bar{a}_{12} & \cdots & \bar{a}_{1,k-1} & 0 & \bar{a}_{1,k+1} & \cdots & \bar{a}_{1p} & 1 & 0 \cdots & \bar{a}_{1,p+r} & \cdots & 0 & 0 & \bar{b}_1 \\ \cdots & \cdots & \cdots & \cdots & \cdots & \cdots & \cdots & \cdots & \cdots & \cdots & \cdots & \cdots & \cdots & \cdots & \cdots \\ \frac{a_{r1}}{a_{rk}} & \frac{a_{r2}}{a_{rk}} & \cdots & \frac{a_{r,k-1}}{a_{rk}} & 1 & \frac{a_{r,k+1}}{a_{rk}} & \cdots & \frac{a_{rp}}{a_{rk}} & 0 & 0 \cdots & \frac{1}{a_{rk}} & \cdots & 0 & 0 & \frac{b_r}{a_{rk}} \\ \cdots & \cdots & \cdots & \cdots & \cdots & \cdots & \cdots & \cdots & \cdots & \cdots & \cdots & \cdots & \cdots & \cdots & \cdots \\ \bar{a}_{m1} & \bar{a}_{m2} & \cdots & \bar{a}_{m,k-1} & 0 & \bar{a}_{m,k+1} & \cdots & \bar{a}_{mp} & 0 & 0 \cdots & \bar{a}_{m,p+r} & \cdots & 1 & 0 & \bar{b}_m \\ c'_1 & c'_2 & \cdots & c'_{k-1} & 0 & c'_{k+1} & \cdots & c'_p & 0 & 0 \cdots & c'_{p+r} & \cdots & 0 & 1 & c_k\frac{b_r}{a_{rk}} \end{bmatrix} \tag{7.3.2}$$

where

$$\begin{aligned} \bar{a}_{ij} &= a_{ij} - a_{ik}\frac{a_{rj}}{a_{rk}},\ i=1,\ldots,m,\ i\neq r;\ j=1,\ldots,p+m \\ \bar{b}_i &= b_i - a_{ik}\frac{b_r}{a_{rk}},\ i=1,\ldots,m,\ i\neq r \\ c'_j &= -c_j + c_k\frac{a_{rj}}{a_{rk}},\ j=1,\ldots,p+m. \end{aligned}$$

(The reader should ponder the connection between the c'_j's and the $\bar{c}_j$'s defined earlier.) If the basic variables are again taken to be those corresponding to the $m+1$ unit column vectors in the simplex matrix, then, from (7.3.2), these variables may be expressed in terms of the remaining or nonbasic variables as

$$\begin{aligned} x_k &= \frac{b_r}{a_{rk}} - \sum_{\substack{j=1 \\ j\neq k}}^{p} \frac{a_{rj}}{a_{rk}} x_j - \frac{1}{a_{rk}} x_{p+r} \\ x_{p+i} &= \bar{b}_i - \sum_{\substack{j=1 \\ j\neq k}}^{p} \bar{a}_{ij}x_j - \bar{a}_{i,p+r}\, x_{p+r},\ i=1,\ldots,m,\ i\neq r \\ f &= c_k\frac{b_r}{x_{rk}} - \sum_{\substack{j=1 \\ j\neq k}}^{p} c'_j x_j - c'_{p+r}x_{p+r}. \end{aligned} \tag{7.9}$$

Upon setting the nonbasic variables equal to zero, our new primal basic feasible solution appears as

$$x_k = \frac{b_r}{a_{rk}}$$

$$x_{p+i} = \bar{b}_i,\ i=1,\ldots,m,\ i\neq r, \text{ and}$$

$$f = c_k \frac{b_r}{a_{rk}}.$$

(Note that this basic feasible solution may be obtained directly from the simplex matrix by ignoring all columns within (7.3.2) save those corresponding to the $m+1$ unit column vectors and the (last) column of constants.) In addition, this solution is optimal if each $c'_j, j=1,\ldots,k-1,\ k+1,\ldots,p,p+r,$ within the objective function row of (7.2.3) is nonnegative. If negative coefficients still appear within the last row of (7.3.2), then the above pivot process is repeated until an optimal primal basic feasible solution is attained.

We now return to the maximization problem started earlier. Let us duplicate the associated simplex matrix (7.4) as

$$\left[\begin{array}{rrrrrrr} 2 & 1 & \boxed{4} & 1 & 0 & 0 & 100 \\ 1 & 3 & 1 & 0 & 1 & 0 & 40 \\ -2 & -3 & -4 & 0 & 0 & 1 & 0 \end{array}\right] \begin{array}{l} 100/4 = 25 \\ 40/1 = 40 \\ . \end{array} \tag{(7.4)}$$

Remembering that the basic variables are those corresponding to the $m+1(=3)$ unit column vectors in this matrix, our initial basic feasible solution consists of $x_4 = 100$, $x_5 = 40$, and $f = 0$. Given that there are some negative coefficients within the last row of (7.4), the basic feasible solution just obtained is not optimal. To generate an improved solution let us undertake

ROUND 1. Since -4 is the most negative entry in this row, the nonbasic variable x_3 now becomes basic. What is the largest increment in x_3 which preserves feasibility? To determine this let us take the positive coefficients on x_3 in the first two rows of (7.4) (values 4 and 1) and divide them, respectively, into the first two constants in the last column of this matrix (namely 100 and 40). Carrying out the indicated division to the right of (7.4) yields $25 = min\{25, 40\}$ as the maximum allowable increase in x_3. Hence the (blocked) pivot element within simplex matrix (7.4) is 4. In the calculations to follow, column three of (7.4) will be replaced by a unit column vector with unity in the pivot position and zeros elsewhere. Such a unit column vector lies in the fourth column of (7.4). Hence x_4 turns nonbasic in the sense that it ultimately loses its unit column $\mathbf{e}_1$ to x_3. In order to achieve this transformation, the pivot operation on (7.4) proceeds as follows. If E_1 is replaced by $\frac{1}{4}E_1$, we obtain

$$\begin{bmatrix} \frac{1}{2} & \frac{1}{4} & \boxed{1} & \frac{1}{4} & 0 & 0 & 25 \\ 1 & 3 & 1 & 0 & 1 & 0 & 40 \\ -2 & -3 & -4 & 0 & 0 & 1 & 0 \end{bmatrix}. \tag{7.4.1}$$

Next, replacing E_2, E_3 in (7.4.1) by $E_2 - E_1$, $E_3 + 4E_1$ respectively produces

$$\left[\begin{array}{ccccccc} \frac{1}{2} & \frac{1}{4} & 1 & \frac{1}{4} & 0 & 0 & 25 \\ \frac{1}{2} & \boxed{\frac{11}{4}} & 0 & -\frac{1}{4} & 1 & 0 & 15 \\ 0 & -2 & 0 & 1 & 0 & 1 & 100 \end{array}\right] \begin{array}{l} 25/(1/4) = 100 \\ 15/(11/14) = 5.46 \\ . \end{array} \tag{7.4.2}$$

Our second basic feasible solution is thus $x_3 = 25$, $x_5 = 15$, and $f = 100$. Given that a negative value appears in the last row of (7.4.2), this basic feasible solution is still not optimal.

ROUND 2. To obtain a maximal solution, let us change the status of x_2 from nonbasic to basic. Since the largest increment in x_2 which does not violate the nonnegativity conditions is $5.46 = min\{100, 5.46\}$, the pivot element within (7.4.2) is $\frac{11}{4}$. This time column two of (7.4.2) is to be replaced by the unit column vector appearing in column five of the same so that x_5 will be removed from the basis. To this end we now perform a pivot operation on (7.4.2). First, let us replace E_2 by $\frac{4}{11}E_2$ to obtain

$$\begin{bmatrix} \frac{1}{2} & \frac{1}{4} & 1 & \frac{1}{4} & 0 & 0 & 25 \\ \frac{2}{11} & \boxed{1} & 0 & -\frac{1}{11} & \frac{4}{11} & 0 & 5.46 \\ 0 & -2 & 0 & 1 & 0 & 1 & 100 \end{bmatrix}. \tag{7.4.3}$$

Secondly, upon replacing E_1, E_3 in (7.4.3) by $E_1 - \frac{1}{4}E_2, E_3 + 2E_2$ respectively, we ultimately have

$$\begin{bmatrix} \frac{5}{11} & 0 & 1 & \frac{3}{11} & -\frac{1}{11} & 0 & 23.63 \\ \frac{2}{11} & 1 & 0 & -\frac{1}{11} & \frac{4}{11} & 0 & 5.46 \\ \frac{4}{11} & 0 & 0 & \frac{9}{11} & \frac{8}{11} & 1 & 110.91 \end{bmatrix}. \tag{7.4.4}$$

This particular basic feasible solution consisting of $x_2 = 5.46$, $x_3 = 23.63$, and $f = 110.91$ is optimal since only nonnegative entries appear in the objective function row of (7.4.4).

7.3. The Cases of Multiple Optimal, Unbounded, and Degenerate Solutions

Let us begin with

EXAMPLE 7.1. Our objective is to

$$\begin{aligned} max\ f &= 2x_1 + 3x_2 \qquad \text{s.t.} \\ x_1 \qquad &\leq 10 \\ x_2 &\leq 8 \\ \tfrac{2}{3}x_1 + x_2 &\leq 12 \\ x_1, x_2 &\geq 0. \end{aligned}$$

For this problem the starting simplex matrix has the form

$$\begin{bmatrix} 1 & 0 & 1 & 0 & 0 & 0 & 10 \\ 0 & 1 & 0 & 1 & 0 & 0 & 8 \\ \frac{2}{3} & 1 & 0 & 0 & 1 & 0 & 12 \\ -2 & -3 & 0 & 0 & 0 & 1 & 0 \end{bmatrix}$$

and our initial basic feasible solution is $x_3=10$, $x_4=8$, $x_5=12$, and $f=0$. Since this solution is not optimal, we undertake two rounds of pivoting to obtain

$$\begin{bmatrix} 0 & 0 & 1 & \boxed{\frac{3}{2}} & -\frac{3}{2} & 0 & 4 \\ 0 & 1 & 0 & 1 & 0 & 0 & 8 \\ 1 & 0 & 0 & -\frac{3}{2} & \frac{3}{2} & 0 & 6 \\ 0 & 0 & 0 & 0 & 3 & 1 & 36 \end{bmatrix}.$$

Given that all negative entries have been purged from the last row of our simplex matrix, we have obtained an optimal primal basic feasible solution wherein $x_1=6$, $x_2=8$, $x_3=4$, and $f=36$.

Further examination of the objective function row of this last simplex matrix indicates that since the coefficient on the nonbasic variable x_4 in this row is zero, we may increase x_4 by some positive amount without affecting the optimal value of the objective function, *i.e.*, the status of x_4 can change from nonbasic to basic as we move from one extreme point to another and yet f will still equal 36. To see this let us introduce x_4 into the set of basic variables. Pivoting on the element $\frac{3}{2}$ yields

$$\begin{bmatrix} 0 & 0 & \frac{2}{3} & 1 & -1 & 0 & \frac{8}{3} \\ 0 & 1 & -\frac{2}{3} & 0 & 1 & 0 & \frac{16}{3} \\ 1 & 0 & 1 & 0 & 0 & 0 & 10 \\ 0 & 0 & 0 & 0 & 3 & 1 & 36 \end{bmatrix}.$$

The second optimal primal basic feasible solution is $x_1=10$, $x_2=\frac{16}{3}$, $x_4=\frac{8}{3}$, and $f=36$. Graphically, our first basic feasible solution corresponds to point A in Figure 7.1 while the second is located at point B. In addition, the set of all convex combinations of points A and B gives rise to an infinity of optimal nonbasic feasible solutions.

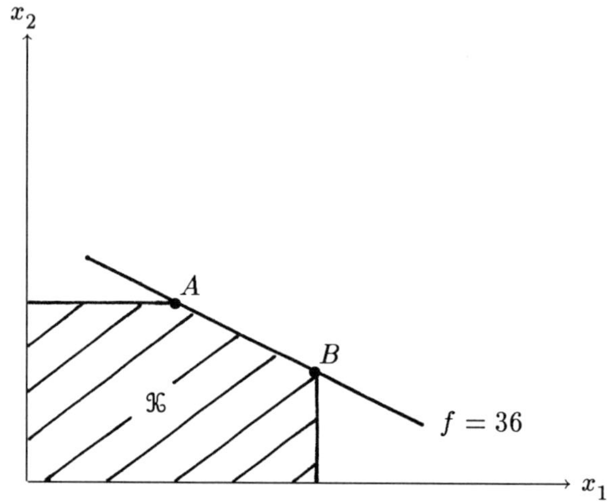

Figure 7.1

To summarize: the case of multiple optimal primal basic feasible solutions emerges if, in the objective function row of the optimal simplex matrix, a nonbasic variable has a zero coefficient. Is there another way of recognizing that the objective function will be tangent to the feasible region $\mathcal{K}$ at more than one extreme point? A glance at the objective function and the structural constraint inequalities indicates that the optimal objective function equation is three times the third structural constraint when the latter is treated as a strict equality. In this regard, when the objective function can be written as a multiple of some particular structural constraint, the case of multiple optimal solutions obtains.

EXAMPLE 7.2. We desire to

$$max\ f = -x_1 + 2x_2 \quad \text{s.t.}$$
$$-2x_1 + x_2 \leq 3$$
$$-\tfrac{1}{2}x_1 + x_2 \leq 6$$
$$x_1, x_2 \geq 0.$$

Our initial simplex matrix appears as

$$\begin{bmatrix} -2 & 1 & 1 & 0 & 0 & 3 \\ -\frac{1}{2} & 1 & 0 & 1 & 0 & 6 \\ 1 & -2 & 0 & 0 & 1 & 0 \end{bmatrix}$$

and our initial basic feasible solution is $x_3=3$, and $x_4=6$, and $f=0$. Since it is not optimal, we ultimately pivot to

$$\begin{bmatrix} 0 & 1 & -\frac{1}{3} & \frac{4}{3} & 0 & 7 \\ 1 & 0 & -\frac{2}{3} & \frac{2}{3} & 0 & 2 \\ 0 & 0 & 0 & 2 & 1 & 12 \end{bmatrix}.$$

The absence of any negative entries in the last row of this final simplex matrix indicates that this particular basic feasible solution consisting of $x_1=2$, $x_2=7$ and $f=12$ is optimal.

We noted earlier that if the objective function coefficient on a nonbasic variable in the optimal simplex matrix is zero, that variable (in this case x_3) can be increased from its current nonbasic level of zero to yield an alternative solution which has the same value of f. While this alternative solution is also optimal, it is, in this particular example, nonbasic, the reason being that the x_3 coefficients in the first two rows of the above matrix are negative. To see this let us write the basic variables x_1, x_2 in terms of x_3 as

$$\begin{aligned} x_1 &= 2+\tfrac{2}{3}x_3 \\ x_2 &= 7+\tfrac{1}{3}x_3. \end{aligned}$$

In this instance x_3, as well as x_1 and x_2, can be increased without bound so that an infinity of alternative optimal nonbasic feasible solutions obtains. For each of these alternative solutions we have three positive variables (excluding f) instead of only two, thus violating our requirement for an optimal basic feasible solution. As Figure 7.2 demonstrates, there exists but a single optimal basic feasible solution (point A) and an infinity of optimal nonbasic feasible solutions.

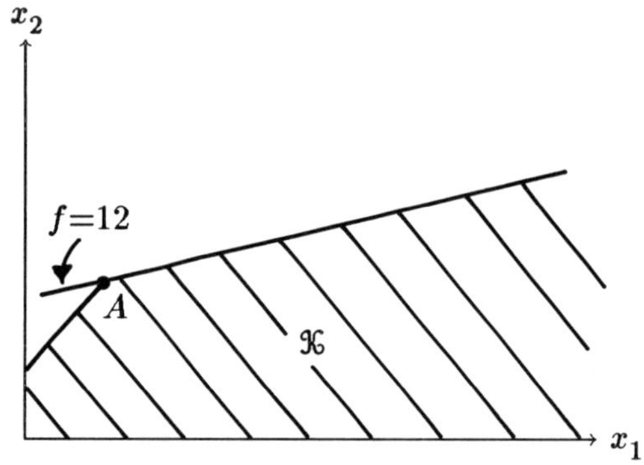

Figure 7.2

EXAMPLE 7.3. Let us

$$
\begin{aligned}
max\ f &= 2x_1 + 3x_2 \qquad \text{s.t.}\\
-2x_1 + x_2 &\le 3\\
-\tfrac{1}{2}x_1 + x_2 &\le 6\\
x_1, x_2 &\ge 0.
\end{aligned}
$$

The associated simplex matrix

$$
\begin{bmatrix}
-2 & 1 & 1 & 0 & 0 & 3\\
-\frac{1}{2} & 1 & 0 & 1 & 0 & 6\\
-2 & -3 & 0 & 0 & 1 & 0
\end{bmatrix}
$$

yields the nonoptimal initial basic feasible solution composed of x_3=3, x_4=6, and f=0. Two successive pivot steps give us the matrix

$$
\begin{bmatrix}
0 & 1 & -\frac{1}{3} & \frac{4}{3} & 0 & 7\\
1 & 0 & -\frac{2}{3} & \frac{2}{3} & 0 & 2\\
0 & 0 & -\frac{7}{3} & \frac{16}{3} & 1 & 25
\end{bmatrix}.
$$

Our third basic feasible solution is thus x_1=2, x_2=7, and f=25. Even though this particular basic feasible solution is not optimal, we shall terminate the simplex process, the reason being that an examination of column three of the last simplex matrix indicates that the value of the objective function is unbounded. To see this let us express the basic variables x_1, x_2 along with f in terms of x_3, the variable to be introduced into the basis. That is, since

$$\begin{aligned} x_2 &= 7+\tfrac{1}{3}x_3 \\ x_1 &= 2+\tfrac{2}{3}x_3 \\ f &= 25+\tfrac{7}{3}x_3, \end{aligned}$$

x_3, as well as x_1, x_2 and f, may be increased without limit (Figure 7.3). Hence we obtain a feasible though nonbasic solution to this problem since the three variables x_1, x_2, and x_3 each assume positive values. In sum, if the only possible pivotal column within the simplex matrix corresponding to a nonbasic variable x_{Rk} has each of its first m components nonpositive and the objective function coefficient (the $m+1^{st}$ component) $-\bar{c}_k$ is negative, the value of the objective function is unbounded.

We may relate this observation to our preceding theoretical results (corollary 5.2) by noting that, from (nonbasic) column three of the preceding simplex matrix, we may form $\mathbf{Y}_1' = (-\frac{1}{3}, -\frac{2}{3}) \leqq \mathbf{0}'$. Then

$$\mathbf{d}_1 = \begin{bmatrix} -\mathbf{Y}_1 \\ \mathbf{e}_1 \end{bmatrix} \geqq \mathbf{0}$$

serves as an extreme direction along which f increases without bound since

$$\mathbf{C}'\mathbf{d}_1 = -\mathbf{C}_B'\mathbf{Y}_1 + c_{R1} = -(3,2)\begin{bmatrix} -\frac{1}{3} \\ -\frac{2}{3} \end{bmatrix} + c_3 = \frac{7}{3}+0 = \frac{7}{3} = \bar{c}_1 > 0.$$

However, within the simplex matrix, the objective function coefficient on $x_{R1} = x_3$ is $-\bar{c}_1 = -\frac{7}{3} < 0$.

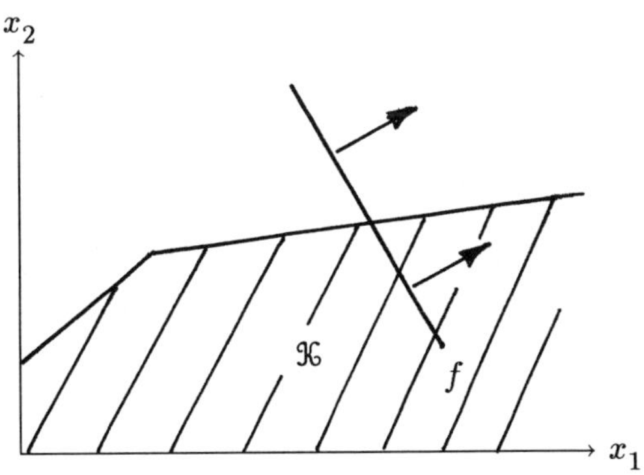

Figure 7.3

EXAMPLE 7.4.

$$\begin{aligned} \max f &= 2x_1 + 3x_2 \qquad \text{s.t.} \\ x_1 + x_2 &\le 60 \\ x_1 + 2x_2 &\le 120 \\ 2x_1 + x_2 &\le 80 \\ x_1, x_2 &\ge 0. \end{aligned}$$

The initial simplex matrix

$$\begin{bmatrix} 1 & 1 & 1 & 0 & 0 & 0 & 60 \\ 1 & 2 & 0 & 1 & 0 & 0 & 120 \\ 2 & 1 & 0 & 0 & 1 & 0 & 80 \\ -2 & -3 & 0 & 0 & 0 & 1 & 0 \end{bmatrix}$$

yields the nonoptimal basic feasible solution x_3=60, x_4=120, x_5=80, and f=0. Execution of the primal simplex entry and exit criteria is illustrated in the following matrix:

$$\begin{bmatrix} 1 & 1 & 1 & 0 & 0 & 0 & 60 \\ 1 & \boxed{2} & 0 & 1 & 0 & 0 & 120 \\ 2 & 1 & 0 & 0 & 1 & 0 & 80 \\ -2 & -3 & 0 & 0 & 0 & 1 & 0 \end{bmatrix} \begin{matrix} 60 \\ 60 \\ 80 \\ \\ \end{matrix} .$$

Since there exists a tie for the minimum of the ratios to the right of this last matrix, we have the choice of replacing either x_3 or x_4 by x_2 in the set of basic variables. Let us choose x_4. Hence the pivot is the second element in the second column of the above array. Pivoting produces

$$\begin{bmatrix} \boxed{\frac{1}{2}} & 0 & 1 & -\frac{1}{2} & 0 & 0 & 0 \\ \frac{1}{2} & 1 & 0 & \frac{1}{2} & 0 & 0 & 60 \\ \frac{3}{2} & 0 & 0 & -\frac{1}{2} & 1 & 0 & 20 \\ -\frac{1}{2} & 0 & 0 & \frac{3}{2} & 0 & 1 & 180 \end{bmatrix} .$$

This new basic feasible solution involving x_2=60, x_3=0, x_5=20, and f=180, while being nonoptimal, is also degenerate since one of the basic variables, namely x_3, vanishes. As anticipated by our previous theoretical discussion, the next round of the simplex process should also yield a degenerate basic feasible solution with the same value of f as at the previous degenerate basic feasible solution. Pivoting on $\frac{1}{2}$ above generates

$$\begin{bmatrix} 1 & 0 & 2 & -1 & 0 & 0 & 0 \\ 0 & 1 & -1 & 1 & 0 & 0 & 60 \\ 0 & 0 & -3 & 1 & 1 & 0 & 20 \\ 0 & 0 & 1 & 1 & 0 & 1 & 180 \end{bmatrix}.$$

As readily seen from this last simplex matrix, the indicated degenerate basic feasible solution is optimal at x_1=0, x_2=60, x_5=20, and f=180.

7.4. Summary of the Primal Simplex Method

Let us now summarize the basic rudiments of the above iterative process. If we abstract from degeneracy, the principal steps leading to an optimal basic feasible solution may be stated as follows:

STEP 0. Initialize.

Form the simplex matrix $[\bar{\mathbf{A}}, \mathbf{b}_0]$ ((7.3)) from system (7.1). To obtain an initial basic feasible solution to $\bar{\mathbf{A}}\bar{\mathbf{X}} = \mathbf{b}_0$, let the m basic variables be those corresponding to the $m+1$ unit column vectors in the simplex matrix. The values of these variables along with the objective function value may be determined directly by examining the last column of (7.3).

STEP 1. Examine the entries in the last or objective function row of the simplex matrix:

(a) If they are all positive, the current basic feasible solution is optimal. Next, determine whether any coefficient corresponding to a nonbasic variable is zero. If all are positive, the optimal basic feasible solution is unique. If one happens to be zero, the case of multiple optimal basic feasible solutions obtains. Determine the alternative basic feasible solution.

(b) If the coefficient on some nonbasic variable is negative and all other components in the column associated with this variable are nonpositive, there exists a feasible solution in positive variables which is nonbasic and for which the value of the objective function is unbounded. In this instance the simplex process should be terminated.

(c) If one or more of the coefficients on the nonbasic variables are negative and for each such variable there is at least one positive component in its associated column, proceed to the next step.

STEP 2. Locate the most negative coefficient in the last row of the simplex matrix in order to determine which nonbasic variable is to become basic.

STEP 3. If the kth nonbasic variable is to turn basic, take the positive components of the column associated with this variable and divide them into their corresponding components in the last column of the simplex matrix.

STEP 4. Compute $\underset{i}{min}\left\{\frac{b_i}{a_{ik}}, a_{ik} > 0\right\}$ so as to determine which basic variable is to become nonbasic. Let this minimum correspond to $\widehat{\theta} = \frac{b_r}{a_{rk}}$.

STEP 5. Given that the rth basic variable is to become nonbasic, a_{rk}, the divisor that yields the smallest quotient in step (6), serves as the pivot element and a pivot operation is performed.

STEP 6. Return to step (1).

By repeating the pivot process outlined above in steps $(1)-(6)$, each iteration takes us to a new primal basic feasible solution with the new value of the objective function being an improvement over the old one. The process is terminated when there are no negative coefficients remaining in the last row of the simplex matrix.

7.5. Obtaining the Optimal Dual Solution from the Optimal Primal Matrix

From a computational viewpoint, how are the optimal values of the dual variables obtained from the optimal primal solution? The simultaneous augmented form of the above primal maximum problem is

$$\begin{aligned} \mathbf{AX} + \mathbf{I}_m\mathbf{X}_s \qquad &= \mathbf{b} \\ -\mathbf{C}'\mathbf{X} + \mathbf{0}'\mathbf{X}_s + f \qquad &= 0 \end{aligned}$$

with simplex matrix

$$\left[\begin{array}{cccc} \mathbf{A} & \mathbf{I}_m & \mathbf{0} & \mathbf{b} \\ -\mathbf{C}' & \mathbf{0}' & 1 & 0 \end{array}\right].$$

For this system the $(m+1 \times m+1)$ basis matrix and its inverse appear respectively as

$$\bar{\mathbf{B}} = \left[\begin{array}{cc} \mathbf{B} & \mathbf{0} \\ -\mathbf{C}'_B & 1 \end{array}\right] \qquad \bar{\mathbf{B}}^{-1} = \left[\begin{array}{cc} \mathbf{B}^{-1} & \mathbf{0} \\ \mathbf{C}'_B\mathbf{B}^{-1} & 1 \end{array}\right],$$

where $\mathbf{B}$ represents the $(m \times m)$ basis matrix associated with the primal structural constraint system. Then

$$\left[\begin{array}{cc} \mathbf{B}^{-1} & \mathbf{0} \\ \mathbf{C}'_B\mathbf{B}^{-1} & 1 \end{array}\right]\left[\begin{array}{cccc} \mathbf{A} & \mathbf{I}_m & \mathbf{0} & \mathbf{b} \\ -\mathbf{C}' & \mathbf{0}' & 1 & 0 \end{array}\right] =$$

$$\left[\begin{array}{cccc} \mathbf{B}^{-1}\mathbf{A} & \mathbf{B}^{-1} & \mathbf{0} & \mathbf{B}^{-1}\mathbf{b} \\ \boxed{\boxed{\mathbf{C}'_B\mathbf{B}^{-1}\mathbf{A}-\mathbf{C}'}\ \boxed{\mathbf{C}'_B\mathbf{B}^{-1}}} & & 1 & \mathbf{C}'_B\mathbf{B}^{-1}\mathbf{b} \end{array}\right] \tag{7.10}$$

yields a basic feasible solution to the primal maximum problem. Assuming that the optimality criterion is satisfied, the optimal values of the dual variables appear as the blocked array of elements in the last row of (7.10), *i.e.*, $\mathbf{U}' = \mathbf{C}'_B\mathbf{B}^{-1} \geqq \mathbf{0}'$. Moreover, from the augmented dual structural constraint system $\mathbf{A}'\mathbf{U} - \mathbf{I}_n\mathbf{U}_s = \mathbf{C}$, we may solve for the dual surplus variables as $\mathbf{U}'_s = \mathbf{U}'\mathbf{A} - \mathbf{C}'$. But this expression is just the double-blocked array of elements $\mathbf{U}'_s = \mathbf{C}'_B\mathbf{B}^{-1}\mathbf{A} - \mathbf{C}' \geqq \mathbf{0}'$ in the last row of (7.10). In sum, our rules for determining the optimal values of the dual variables (both original and surplus) may be stated as follows:

(a) to find the values of the dual variables $\mathbf{U}$ which correspond to an optimal basic feasible solution of the primal maximum problem, look in the last row of the final primal simplex matrix under those columns which were, in the original simplex matrix, the columns of the identity matrix. The ith dual variable $u_i, i=1,\ldots,m$, will appear under the column which was initially the ith column of the identity matrix;

(b) to find the values of the dual surplus variables $\mathbf{U}_s$ which correspond to an optimal basic feasible solution of the primal maximum problem, look in the last row of the final primal simplex matrix under those columns which were, in the original simplex matrix, the columns of the matrix $\mathbf{A}$. The jth dual surplus variable $u_{m+j}, j=1,\ldots,n$, will appear under the column which was initially the jth column of $\mathbf{A}$.

EXAMPLE 7.5. Let us

$$\begin{aligned} max\ f &= x_1 + x_2 \quad \text{s.t.} \\ 2x_1 + \ x_2 &\leq 20 \\ x_1 \qquad &\leq 8 \\ x_2 &\leq 6 \\ x_1, x_2 &\geq 0\,. \end{aligned}$$

Given that x_3, x_4, and x_5 depict nonnegative slack variables, it is easily demonstrated that the optimal simplex matrix is of the form

$$\begin{bmatrix} 0 & 1 & 0 & 0 & 1 & 0 & 6 \\ 1 & 0 & \frac{1}{2} & 0 & -\frac{1}{2} & 0 & 7 \\ 0 & 0 & -\frac{1}{2} & 1 & \frac{1}{2} & 0 & 1 \\ 0 & 0 & \frac{1}{2} & 0 & \frac{1}{2} & 1 & 13 \end{bmatrix}$$

with $\mathbf{X}_0' = (7,6)$ and $f^0{=}13$ (see point A of Figure 7.4). In addition, $(\mathbf{X}_s)_0' = (0,1,0)$ depicts the optimal vector of primal slack variables. Moreover, according to our previous discussion concerning the construction of the dual solution, $\mathbf{U}_0' = \left(\frac{1}{2}, 0, \frac{1}{2}\right)$ and $(\mathbf{U}_s)_0' = (0,0)$, where the latter vector represents the set of dual surplus variables.

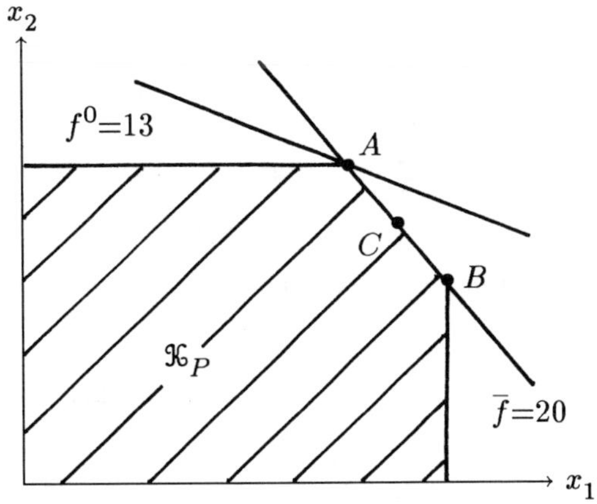

Figure 7.4

Turning now to the complementary slackness conditions we have, for the weak and strong cases respectively,

Weak	Strong
$u_1^0 x_{s,1}^0 = \frac{1}{2} \cdot 0 = 0$	$u_1^0 + x_{s,1}^0 = \frac{1}{2} + 0 = \frac{1}{2} > 0$
$u_2^0 x_{s,2}^0 = 0 \cdot 1 = 0$	$u_2^0 + x_{s,2}^0 = 0 + 1 = 1 > 0$
$u_3^0 x_{s,3}^0 = \frac{1}{2} \cdot 0 = 0$	$u_3^0 + x_{s,3}^0 = \frac{1}{2} + 0 = \frac{1}{2} > 0$
and	and
$x_1^0 u_{s,1}^0 = 7 \cdot 0 = 0$	$x_1^0 + u_{s,1}^0 = 7 + 0 = 7 > 0$
$x_2^0 u_{s,2}^0 = 6 \cdot 0 = 0$	$x_2^0 + u_{s,2}^0 = 6 + 0 = 6 > 0$

Hence at point A both strong and weak complementary slackness prevails.

Let us now recast this problem by changing the objective function to $\bar{f} = 2x_1 + x_2$ with the structural constraints and nonnega-tivity conditions unchanged. For this modified problem the optimal simplex matrix is

$$\begin{bmatrix} 0 & 1 & 1 & -2 & 0 & 0 & 4 \\ 1 & 0 & 0 & 1 & 0 & 0 & 8 \\ 0 & 0 & -1 & 2 & 1 & 0 & 2 \\ 0 & 0 & 1 & 0 & 0 & 1 & 20 \end{bmatrix}.$$

with $\mathbf{X}_1' = (8,4)$ and $\bar{f}^1$=20 (point B of Figure 7.4). (Actually, the case of multiple optimal solutions emerges, as an examination of the last row of the preceding simplex matrix reveals. Here $\bar{f}(A) = \bar{f}(B) = \bar{f}(C) = 20$, where $C = \lambda B + (1-\lambda)A, 0 \leq \lambda \leq 1$.) Moreover, $(\mathbf{X}_s)_1' = (0,0,2)$, $\mathbf{U}_1' = (1,0,0)$, and $(\mathbf{U}_s)_1' = (0,0)$. For this problem the complementary slackness conditions appear as

Weak	Strong
$u_1^1 x_{s,1}^1 = 1 \cdot 0 = 0$	$u_1^1 + x_{s,1}^1 = 1 + 0 = 1 > 0$
$u_2^1 x_{s,2}^1 = 0 \cdot 0 = 0$	$u_2^1 + x_{s,2}^1 = 0 + 0 = 0$
$u_3^1 x_{s,3}^1 = 0 \cdot 2 = 0$	$u_3^1 + x_{s,3}^1 = 0 + 2 = 2 > 0$
and	and
$x_1^1 u_{s,1}^1 = 8 \cdot 0 = 0$	$x_1^1 + u_{s,1}^1 = 8 + 0 = 8 > 0$
$x_2^1 u_{s,2}^1 = 4 \cdot 0 = 0$	$x_2^1 + u_{s,2}^1 = 4 + 0 = 4 > 0$

In this instance only the weak complementary slackness conditions hold since an additional zero-valued dual variable appears in the optimal set of the same, *i.e.* now $u_3^1 = 0$. To offset the effect of this extra zero-valued dual variable on the strong complementary slackness conditions, let us set $\lambda = \frac{1}{2}$ and thus choose point $C = \lambda B + (1-\lambda)A, \;\; 0 \leq \lambda \leq 1$ (Figure 7.4), as our optimal primal solution for which $\bar{f}^c = 2\left(\frac{15}{2}\right) + 5 = 20$. Then the complementary slackness conditions become

Weak	Strong
$u_1^c x_{s,1}^c = 1 \cdot 0 = 0$	$u_1^c + x_{s,1}^c = 1 + 0 = 1 > 0$
$u_2^c x_{s,2}^c = 0 \cdot (0.5) = 0$	$u_2^c + x_{s,2}^c = 0 + 0.5 = 0.5 > 0$
$u_3^c x_{s,3}^c = 0 \cdot 1 = 0$	$u_3^c + x_{s,3}^c = 0 + 1 = 1 > 0$
and	and
$x_1^c u_{s,1}^c = (7.5) \cdot 0 = 0$	$x_1^c + u_{s,1}^c = 7.5 + 0 = 7.5 > 0$
$x_2^c u_{s,2}^c = 5 \cdot 0 = 0$	$x_2^c + u_{s,2}^c = 5 + 0 = 5 > 0$

Notice that now both weak and strong complementary slackness holds.

8 ONE-PHASE, TWO-PHASE, AND COMPOSITE METHODS OF LINEAR PROGRAMMING

8.1. Artificial Variables

In this section we shall return to the general primal linear programming problem in standard form set out in Chapter 3, namely

$$\begin{aligned} max f &= \mathbf{C}'\mathbf{X} \quad \text{s. t.} \\ \mathbf{AX} &= \mathbf{b},\ \mathbf{X} \geqq \mathbf{0} \end{aligned}$$

with $\mathbf{X} \epsilon \mathbf{R}^n$. If the objective function in this formulation is treated as an additional structural constraint, then as indicated in the preceding chapter, we seek to determine a solution of

$$\left.\begin{aligned} \mathbf{AX} &= \mathbf{b} \\ -C'\mathbf{X} + f &= 0 \end{aligned}\right\} \quad \text{or} \quad \bar{\mathbf{A}}\bar{\mathbf{X}} = \mathbf{b}_0,$$

where

$$\bar{\mathbf{A}} = \begin{bmatrix} \mathbf{A} & \mathbf{0} \\ -\mathbf{C}' & 1 \end{bmatrix},\ \bar{\mathbf{X}} = \begin{bmatrix} \mathbf{X} \\ f \end{bmatrix},\ \text{and } \mathbf{b}_0 = \begin{bmatrix} \mathbf{b} \\ 0 \end{bmatrix}.$$

Here f is to be made as large as possible (and unrestricted in sign) with $\mathbf{X} \geqq \mathbf{0}$. Then the $(m+1 \times n+2)$ simplex matrix associated with this simultaneous system is

$$\left[\bar{\mathbf{A}},\ \mathbf{b}_0\right].$$

For instance, given that we desire to

$$\begin{aligned} max f &= x_1 + 4x_2 \quad \text{s.t.} \\ x_1 - \ x_2 &\leq 6 \\ 2x_1 + 3x_2 &\geq 2 \\ x_1 + \ x_2 &= 4 \\ x_1, x_2 &\geq 0, \end{aligned}$$

the corresponding simplex matrix has the form

$$\left[\bar{\mathrm{A}}, \mathbf{b}_0\right] = \begin{bmatrix} 1 & -1 & 1 & 0 & 0 & 6 \\ 2 & 3 & 0 & -1 & 0 & 2 \\ 1 & 1 & 0 & 0 & 0 & 4 \\ -1 & -4 & 0 & 0 & 1 & 0 \end{bmatrix}.$$

Since this matrix does not contain within it an $m+1 = 4^{th}$ order identity matrix, an initial basic feasible solution does not exist, *i.e.*, we lack a starting point from which to initiate the various rounds of the simplex method, the reason being that not all of the original structural constraints are of the " $\leq$ " variety. So in situations where some or all of the original structural constraints are of the " $\geq$ " or " $=$ " form, specific measures must be taken to obtain an identity matrix and thus a starting basic feasible solution. To this end our procedure will be to first transform the augmented structural constrain system $\mathbf{AX} = \mathbf{b}$ or

$$\begin{aligned} x_1 - x_2 + x_3 \qquad &= 6 \\ 2x_1 + 3x_2 + \quad - x_4 &= 2 \\ x_1 + x_2 \qquad &= 4 \end{aligned}$$

into what is termed the **artificial augmented structural constraint system**

$$\begin{aligned} x_1 - x_2 + x_3 \qquad\qquad &= 6 \\ 2x_1 + 3x_2 \quad - x_4 + x_5 \quad &= 2 \\ x_1 + x_2 \qquad\qquad + x_6 &= 4 \end{aligned}$$

by introducing the nonnegative **artificial variables** x_5 and x_6, so named because these variables are meaningless in terms of the formulation of the original structural constraint system which contains only real or **legitimate variables.** Artificial variables are introduced solely to initialize the simplex method, with the number of such variables corresponding to the number of " $\geq$ " and " $=$ " structural constraints present. In standard form, the preceding system of equalities appears as

$$\mathbf{A}_*\mathbf{X}_* = \mathbf{b}, \tag{8.1}$$

where

$$\mathbf{A}_* = \begin{bmatrix} 1 & -1 & 1 & 0 & 0 & 0 \\ 2 & 3 & 0 & -1 & 1 & 0 \\ 1 & 1 & 0 & 0 & 0 & 1 \end{bmatrix}, \mathbf{X}_* = \begin{bmatrix} \mathbf{X} \\ \mathbf{X}_a \end{bmatrix} = \begin{bmatrix} x_1 \\ x_2 \\ x_3 \\ x_4 \\ x_5 \\ x_6 \end{bmatrix},$$

and $\mathbf{X}_a$ represents a vector of artificial variables with elements x_5, x_6. Here the individual vectors within $\mathbf{A}_*$ corresponding to artificial variables are termed **artificial vectors** and appear as a set of unit column vectors. Note that columns three, five, and six of $\mathbf{A}_*$ yield an $m = 3^{rd}$ order identity matrix and thus an initial basic feasible solution to (8.1), namely $x_3 = 6$, $x_5 = 2$, and $x_6 = 4$ when $x_1 = x_2 = x_4 = 0$.

While we have obtained an initial basic feasible solution to (8.1), it is not a feasible solution to the original structural constraint system $\mathbf{AX} = \mathbf{b}$, the reason being that $\mathbf{X}_a \neq \mathbf{0}$. Hence any basic feasible solution to the latter system must not admit any artificial variable at a positive level. In this regard, if we are to obtain a basic feasible solution to the original linear programming problem, we must utilize the simplex process to remove from the initial basis (*i.e.*, from the identity matrix) the artificial vectors and substitute in their stead an alternative set of vectors chosen from those remaining in $\mathbf{A}_*$. In doing so we simply obtain the original structural constraint system $\mathbf{AX} = \mathbf{b}$ itself. Hence all artificial variables are to become nonbasic or zero with the result that a basic feasible solution to $\mathbf{AX} = \mathbf{b}$ emerges wherein the basis vectors are chosen from the coefficient matrix of this system. To summarize: any basic feasible solution to $\mathbf{A}_*\mathbf{X}_* = \mathbf{b}$ which is also a basic feasible solution to $\mathbf{AX} = \mathbf{b}$ has the artificial vector $\mathbf{X}_a = \mathbf{0}$. In the remaining sections of this chapter we shall develop two important techniques which serve to systematically purge the basis of any artificial vectors (the one-phase and two-phase methods) and a routine which economizes on the use of artificial variables (the composite method).

EXAMPLE 8.1. Determine an initial basic feasible solution to $\mathbf{A}_*\mathbf{X}_* = \mathbf{b}$ when the initial set of structural constraints appears as

$$\begin{aligned} x_1 + 2x_2 &\geq 1 \\ x_1 + \tfrac{1}{2}x_2 &\geq 4 \\ \tfrac{1}{2}x_1 + 6x_2 &= 2 \\ x_1, x_2 &\geq 0. \end{aligned}$$

If $x_3, x_4 \geq 0$ denote surplus variables and $x_5, x_6, x_7 \geq 0$ depict artificial variables, the preceding system becomes

$$\begin{aligned} x_1 + 2x_2 - x_3 \qquad + x_5 \qquad\qquad &= 1 \\ x_1 + \tfrac{1}{2}x_2 \qquad - x_4 \qquad + x_6 \qquad &= 4 \\ \tfrac{1}{2}x_1 + 6x_2 \qquad\qquad\qquad\qquad + x_7 &= 2 \\ x_1, \ldots, x_7 \geq 0. \end{aligned}$$

Since

$$\mathbf{A}_* = \begin{bmatrix} 1 & 2 & -1 & 0 & 1 & 0 & 0 \\ 1 & \frac{1}{2} & 0 & -1 & 0 & 1 & 0 \\ \frac{1}{2} & 6 & 0 & 0 & 0 & 0 & 1 \end{bmatrix}$$

contains within it the columns of a 3rd order identity matrix, the variables associated with these columns yield an initial basic feasible solution to $\mathbf{A}_*\mathbf{X}_* = \mathbf{b}$ once the remaining variables are set equal to zero, *i.e.*, $x_5 = 1$, $x_6 = 4$, and $x_7 = 2$.

8.2. The One-Phase Method

The first technique under consideration which serves to remove the artificial vectors from the basis in but a single phase is the M-penalty method (A. Charnes, *et al.* (1953)). Here extremely large unspecified negative coefficients are assigned to the artificial variables in the objective function in order to preclude the objective function from attaining a maximum so long as any artificial variable remains in the set of basic variables at a positive level. In this regard, for a sufficiently large scalar $M > 0$, the **artificial linear programming problem** may be framed as: maximize the **artificial objective function**

$$f(\mathbf{X}_*) = \mathbf{C}'\mathbf{X} - M\mathbf{1}'\mathbf{X}_a \quad \text{s.t.}$$
$$\mathbf{A}_*\mathbf{X}_* = \mathbf{b}, \mathbf{X}_* \geqq \mathbf{0}.$$

For example, the problem appearing at the outset of this chapter has the artificial form

$$\begin{aligned} & max f(x_1, \ldots, x_6) = x_1 + 4x_2 - Mx_5 - Mx_6 \quad \text{s.t.} \\ & x_1 + x_2 - x_3 = 6 \\ & 2x_1 + 3x_2 - x_4 + x_5 = 2 \\ & x_1 + x_2 + x_6 = 4 \\ & x_1, \ldots, x_6 \geq 0. \end{aligned}$$

The simplex matrix associated with this problem may now be directly expressed as

$$\left[\bar{\mathbf{A}}_*, \mathbf{b}_0\right] = \begin{bmatrix} 1 & -1 & 1 & 0 & 0 & 0 & 0 & 6 \\ 2 & 3 & 0 & -1 & 1 & 0 & 0 & 2 \\ 1 & 1 & 0 & 0 & 0 & 1 & 0 & 4 \\ -1 & -4 & 0 & 0 & M & M & 1 & 0 \end{bmatrix}. \tag{8.2}$$

Before we examine some example problems dealing with the implementation of this technique, a few summary comments pertaining to the salient features of the M-penalty method will be advanced. First, once an artificial vector leaves the basis, it never reenters the basis. Furthermore, once an artificial variable turns nonbasic, it remains so throughout the remaining rounds of the simplex process since, if an artificial variable increases in value from zero by a positive amount, a penalty of $-M$ is incurred and f concomitantly decreases. Since the simplex process is one for which f is nondecreasing between succeeding rounds, this possibility is nonexistent. So as the various rounds of the simplex process are executed, we may delete from the simplex matrix those columns corresponding to artificial vectors that have turned nonbasic. Next, if the maximal solution to the artificial problem has all artificial variables equal to zero, then its $\mathbf{X}$ component is a maximal solution to the original problem with the values of the artificial and original objective functions being the same. Finally, if the maximal solution to the artificial problem contains at least one positive artificial variable, then the original problem possesses no feasible solution (we shall return to this last point later on when a detailed analysis pertaining to the questions of inconsistency and redundancy of the structural constraints is advanced).

EXAMPLE 8.2. Using the M-penalty method, determine the simplex matrix and an initial basic feasible solution to the problem:

$$\begin{aligned} max f &= 4x_1 + 5x_2 \quad \text{s.t.} \\ x_1 + 3x_2 &\leq 4 \\ 3x_1 - x_2 &\geq 5 \\ x_1 + x_2 &= 8 \\ x_1, x_2 &\geq 0. \end{aligned}$$

If x_3 depicts a slack variable; x_4 represents a surplus variable; and x_5, x_6 denote artificial variables, we obtain

$$\begin{aligned} x_1 + 3x_2 + x_3 \qquad\qquad\qquad\qquad &= 4 \\ 3x_1 - x_2 \qquad - x_4 + x_5 \qquad\qquad &= 5 \\ x_1 + x_2 \qquad\qquad\qquad + x_6 \qquad &= 8 \\ -4x_1 - 5x_2 \qquad\qquad + Mx_5 + Mx_6 + f &= 0 \end{aligned}$$

with associated simplex matrix

$$[\bar{\mathbf{A}}_*, \mathbf{b}_0] = \begin{bmatrix} 1 & 3 & 1 & 0 & 0 & 0 & 0 & 4 \\ 3 & -1 & 0 & -1 & 1 & 0 & 0 & 5 \\ 1 & 1 & 0 & 0 & 0 & 1 & 0 & 8 \\ -4 & -5 & 0 & 0 & M & M & 1 & 0 \end{bmatrix}.$$

While this matrix does not yield an *obvious* initial basic feasible solution, we may readily transform it into an equivalent representation that does by multiplying every row containing an artificial variable by $-M$ and adding their sum to the last row, e.g., when the sum

$$\begin{array}{rrrrrrrr}
-3M & M & 0 & M & -M & 0 & 0 & -5M \\
-M & -M & 0 & 0 & 0 & -M & 0 & -8M \\
\hline
-4M & 0 & 0 & M & -M & -M & 0 & -13M
\end{array}$$

is added to the objective function row of the simplex matrix, we obtain

$$\begin{bmatrix}
1 & 3 & 1 & 0 & 0 & 0 & 0 & 4 \\
3 & -1 & 0 & -1 & 1 & 0 & 0 & 5 \\
1 & 1 & 0 & 0 & 0 & 1 & 0 & 8 \\
-4 & -5 & 0 & 0 & 0 & 0 & 1 & 0 \\
-4M & 0 & 0 & M & 0 & 0 & 0 & -13M
\end{bmatrix}$$

To facilitate the execution of subsequent rounds of the simplex process, each entry in the last row has been separated into two parts, one which does not involve M, and one appearing directly beneath it which does. Hence this *double-row* representation actually depicts the *single* objective function row of the simplex matrix *i.e.*, the objective row is the sum of the last two rows in the preceding matrix. It is at this point in our discussion of the one-phase method that an initial basic feasible solution to the artificial linear programming problem emerges since the existence of seven unit column vectors within the simplex matrix is now readily apparent. In this regard, our initial basic feasible solution is $x_3 = 4$, $x_4 = 3$, $x_7 = 6$, $x_8 = 5$, $x_9 = 8$, $x_{10} = 3$, and $f = -13M$. Note that since f is expressed in terms of M, the objective function cannot possibly attain a maximum with artificial vectors in the basis.

EXAMPLE 8.3. We seek to

$$\begin{aligned}
max\ f &= x_1 + x_2 \qquad \text{s.t.} \\
x_1 + 2x_2 &\leq 12 \\
x_1 + x_2 &\geq 4 \\
-x_1 + x_2 &\geq 1 \\
x_1, x_2 &\geq 0.
\end{aligned}$$

If we let x_3 denote a slack variable; x_4, x_5 depict surplus variables; and x_6, x_7 represent artificial variables, we obtain the simplex matrix

$$\left[\bar{\mathbf{A}}_*, \mathbf{b}_0\right] = \begin{bmatrix} 1 & 2 & 1 & 0 & 0 & 0 & 0 & 0 & 12 \\ 1 & 1 & 0 & -1 & 0 & 1 & 0 & 0 & 4 \\ -1 & 1 & 0 & 0 & -1 & 0 & 1 & 0 & 1 \\ -1 & -1 & 0 & 0 & 0 & M & M & 1 & 0 \end{bmatrix}.$$

To generate an initial basic feasible solution to this artificial linear programming problem, let us add

$$\begin{array}{ccccccccc} -M & -M & 0 & M & 0 & -M & 0 & 0 & -4M \\ M & -M & 0 & 0 & M & 0 & -M & 0 & -M \\ \hline 0 & -2M & 0 & M & M & -M & -M & 0 & -5M \end{array}$$

to the last row of the simplex matrix to obtain

$$\begin{bmatrix} 1 & 2 & 1 & 0 & 0 & 0 & 0 & 0 & 12 \\ 1 & 1 & 0 & -1 & 0 & 1 & 0 & 0 & 4 \\ -1 & 1 & 0 & 0 & -1 & 0 & 1 & 0 & 1 \\ -1 & -1 & 0 & 0 & 0 & 0 & 0 & 1 & 0 \\ 0 & -2M & 0 & M & M & 0 & 0 & 0 & -5M \end{bmatrix}.$$

The first basic feasible solution to the artificial problem thus becomes $x_3 = 12$, $x_6 = 4$, $x_1 = 1$, and $f = -5M$. Since this solution is clearly not optimal, let us undertake

ROUND 1.

$$\left[\begin{array}{ccccccccc} 1 & 2 & 1 & 0 & 0 & 0 & 0 & 0 & 12 \\ 1 & 1 & 0 & -1 & 0 & 1 & 0 & 0 & 4 \\ -1 & \boxed{1} & 0 & 0 & -1 & 0 & 1 & 0 & 1 \\ -1 & -1 & 0 & 0 & 0 & 0 & 0 & 1 & 0 \\ 0 & -2M & 0 & M & M & 0 & 0 & 0 & -5M \end{array}\right] \begin{array}{c} 6 \\ 4 \\ 1 \\ \\ \\ \end{array} \longrightarrow$$

$$\begin{bmatrix} 3 & 0 & 1 & 0 & 2 & 0 & -2 & 0 & 10 \\ 2 & 0 & 0 & -1 & 1 & 1 & -1 & 0 & 3 \\ -1 & 1 & 0 & 0 & -1 & 0 & 1 & 0 & 1 \\ -2 & 0 & 0 & 0 & -1 & 0 & 1 & 1 & 1 \\ -2M & 0 & 0 & M & -M & 0 & 2M & 0 & -3M \end{bmatrix}.$$

Here the second basic feasible solution to the artificial problem is $x_2 = 1$, $x_3 = 10$, $x_6 = 3$, and $f = 1 - 3M$. Since this solution is also not optimal, we proceed to

ROUND 2 (column seven is deleted).

$$\left[\begin{array}{rrrrrrrr|c} 3 & 0 & 1 & 0 & 2 & 0 & 0 & 10 & 10/3 \\ \boxed{2} & 0 & 0 & -1 & 1 & 1 & 0 & 3 & 3/2 \\ -1 & 1 & 0 & 0 & -1 & 0 & 0 & 1 & — \\ -2 & 0 & 0 & 0 & -1 & 0 & 1 & 1 & \\ -2M & 0 & 0 & M & -M & 0 & 0 & -3M & \end{array}\right] \longrightarrow \left[\begin{array}{cccccccc} 0 & 0 & 1 & \frac{3}{2} & \frac{1}{2} & -\frac{3}{2} & 0 & \frac{11}{2} \\ 1 & 0 & 0 & -\frac{1}{2} & \frac{1}{2} & \frac{1}{2} & 0 & \frac{3}{2} \\ 0 & 1 & 0 & -\frac{1}{2} & -\frac{1}{2} & \frac{1}{2} & 0 & \frac{5}{2} \\ 0 & 0 & 0 & -1 & 0 & 1 & 1 & 4 \\ 0 & 0 & 0 & 0 & 0 & M & 0 & 0 \end{array}\right].$$

The third basic feasible solution to the artificial problem thus appears as $x_1 = \frac{2}{3}$, $x_2 = \frac{5}{2}$, $x_3 = \frac{11}{2}$, and $f = 4$. Since all artificial vectors have been removed from the basis, this particular basic feasible solution to the artificial problem represents an initial basic feasible solution to the original problem. In this regard, let us now apply the simplex method to this initial basic feasible solution.

ROUND 3 (column six and the M-row are deleted).

$$\left[\begin{array}{ccccccc|c} 0 & 0 & 1 & \boxed{\frac{3}{2}} & \frac{1}{2} & 0 & \frac{11}{2} & 11/3 \\ 1 & 0 & 0 & -\frac{1}{2} & \frac{1}{2} & 0 & \frac{3}{2} & — \\ 0 & 1 & 0 & -\frac{1}{2} & -\frac{1}{2} & 0 & \frac{5}{2} & — \\ 0 & 0 & 0 & -1 & 0 & 1 & 4 & \end{array}\right] \longrightarrow \left[\begin{array}{ccccccc} 0 & 0 & \frac{2}{3} & 1 & \frac{1}{3} & 0 & \frac{11}{3} \\ 1 & 0 & \frac{1}{3} & 0 & \frac{2}{3} & 0 & \frac{10}{3} \\ 0 & 1 & \frac{1}{3} & 0 & -\frac{2}{3} & 0 & \frac{13}{3} \\ 0 & 0 & \frac{2}{3} & 0 & \frac{1}{3} & 1 & \frac{23}{3} \end{array}\right].$$

Since the optimality criterion is satisfied, the optimal basic feasible solution to the original linear programming problem is $x_1 = \frac{10}{3}$, $x_2 = \frac{13}{3}$, $x_4 = \frac{11}{3}$, and $f = \frac{23}{3}$.

8.3. Inconsistency and Redundancy

We now turn to a discussion of the circumstances under which all artificial vectors appearing in the initial basis of the artificial linear programming problem may be removed from the basis in order to proceed to an optimal primal basic feasible solution to the original problem. In addition, we shall also determine whether the original augmented structural constraint equalities $\mathbf{AX} = \mathbf{b}$ are consistent, and whether any subset of them is redundant. Let us assume that from an initial basic feasible solution to the artificial problem we obtain a basic solution for which the optimality criterion is satisfied. Here one of three mutually exclusive and collectively exhaustive results will hold:

(1) the basis contains no artificial vectors;

(2) the basis contains one or more artificial vectors at the zero level;

(3) the basis contains one or more artificial vectors at a positive level.

Let us discuss the implications of each case in turn.

First, if the basis contains no artificial vectors, then, given that the optimality criterion is satisfied, it is evident that we have actually obtained an optimal basic feasible solution to the original primal problem. In this instance the original structural constraints are consistent and none are redundant, *i.e.*, if there emerges at least one basic feasible solution to the original structural constraint system, the simplex process will remove all artificial vectors from the basis and ultimately reach an optimal solution provided that one exists (a case in point was example 8.3, above).

We next turn to the case where the basis contains one or more artificial vectors at the zero level. With all artificial variables equal to zero, we have a feasible solution to $\mathbf{AX} = \mathbf{b}$, *i.e.*, the original structural constraint equations are consistent. However, there still exists the possibility of redundancy in the original structural constraint system. Upon addressing ourselves to this latter point we find that two alternative situations merit our consideration. To set the stage for this discussion, let $\hat{\mathbf{Y}}_j$ denote the jth legitimate nonbasic column of the optimal simplex matrix. In this regard we first assume that y_{ij}, the ith component of $\hat{\mathbf{Y}}_j$, is different from zero for one or more j and for one or more $i \epsilon \mathcal{A}$, where $\mathcal{A}$ denotes an index set consisting of all i's corresponding to artificial basis vectors $\mathbf{e}_i$. Now, if for some $j = j'$ we find that $y_{ij'} \neq 0$, it follows that the associated artificial basis vector $\mathbf{e}_i$ may be removed from the basis and replaced by $\hat{\mathbf{Y}}_{j'}$. Since the artificial vector currently appears in the basis at the zero level, $\hat{\mathbf{Y}}_{j'}$ also enters the basis at the zero level, with the result that the optimal value of the objective function is unchanged in the new basic feasible solution. If this process is repeated until all artificial vectors have been removed from the basis, we obtain a degenerate optimal basic feasible solution to the original problem involving only real or legitimate variables. In this instance none of the original structural constraints within $\mathbf{AX} = \mathbf{b}$ is redundant. Next, if this procedure does not remove all artificial vectors from the basis, we ultimately reach a state where $y_{ij} = 0$ for all $\hat{\mathbf{Y}}_j$ and all remaining $i \in \mathcal{A}$. Under this circumstance we cannot replace any of the remaining artificial vectors by some $\hat{\mathbf{Y}}_j$ and still maintain a basis. If we assume that there are k artificial vectors in the basis at the zero level, every column of the $(m \times n)$ matrix $\mathbf{A}$ may be written as a linear combination of the $m-k$ linearly independent columns of $\mathbf{A}$ appearing in the basis, *i.e.*, the k artificial vectors are not needed to express any column of $\mathbf{A}$ in terms of the basis vectors. Hence $\rho(\mathbf{A}) = m-k$ so that only $m-k$ rows of $\mathbf{A}$ are linearly independent, thus implying that k of the original structural constraints in $\mathbf{AX} = \mathbf{b}$ are redundant. [As a practical matter, since inequality structural constraints can never be redundant (each is converted into an equality by the introduction of its *own* slack or surplus variable), our search for redundant structural constraints must be limited to only some subset of equations appearing in $\mathbf{AX} = \mathbf{b}$, namely those expressed as $\sum_{j=1}^{p} a_{tj}x_j = b_t$, $t \in \mathcal{T} = \{1, \ldots, m\}$.] To identify the specific structural constraints within $\mathbf{AX} = \mathbf{b}$ that are redundant, we note briefly that if the artificial basis vector $\mathbf{e}_h$ remains in the basis at the zero level and $y_{hj} = 0$ for all legitimate nonbasis vectors $\hat{\mathbf{Y}}_j$, then the hth structural constraint of $\mathbf{AX} = \mathbf{b}$ is redundant. In this regard, if at some stage of the simplex process we find that the artificial vector $\mathbf{e}_h$ appears in the basis at the zero level with $y_{hj} = 0$ for all $\hat{\mathbf{Y}}_j$, then before executing the next round of the simplex routine, we may delete the hth row of the simplex matrix containing the zero-valued artificial variable along with the associated artificial basis vector $\mathbf{e}_h$. The implication here is that any redundant constraint may be omitted from the original structural constraint system without losing any information contained within the latter. (See example 8.4 below.)

We finally encounter the case where one or more artificial vectors appear in the basis at a positive level. In this instance the basic solution generated is not meaningful, *i.e.*, the original problem possesses no feasible solution since otherwise the artificial vectors would all enter the basis at the zero level. Here the original system either has no solution (the structural constraints are inconsistent) or has solutions that are not feasible. To distinguish between these two alternatives we first note that since the optimality criterion is satisfied, $y_{ij} \leq 0$ for legitimate nonbasis vectors $\widehat{\mathbf{Y}}_j$ and for those $i \in \mathcal{A}$. If for some $j = j'$ we find that $y_{ij'} < 0$, we can insert $\widehat{\mathbf{Y}}_{j'}$ into the basis and remove the associated artificial basis vector $\mathbf{e}_i$ and still maintain a basis. However, with $y_{ij'} < 0$, the new basic solution will not be feasible since $\widehat{\mathbf{Y}}_{j'}$ enters the basis at a negative level. If this process is repeated until all artificial vectors have been driven from the basis, we obtain a basic though nonfeasible solution to the original problem involving only legitimate variables (see example 8.5, below). Next, if this procedure does not remove all artificial vectors from the basis, we ultimately obtain a state where $y_{ij} = 0$ for all $\widehat{\mathbf{Y}}_j$ and for all i remaining in $\mathcal{A}$. Assuming that there are k artificial vectors in the basis, every column of $\mathbf{A}$ may be written as a linear combination of the $m-k$ legitimate basis vectors so that $\rho(\mathbf{A}) = m-k$. But the fact that k artificial columns from $\mathbf{A}$ appear in the basis at positive levels implies that $\mathbf{b}$ is expressible as a linear combination of more than $m-k$ basis vectors so that $\rho(\mathbf{A}) \neq \rho[\mathbf{A}, \mathbf{b}]$. In this regard, the original structural constraint system $\mathbf{AX} = \mathbf{b}$ is inconsistent and thus does not posses a solution (see example 8.6 below).

A consolidation of the results obtained in this section now follows. Given that the optimality criterion is satisfied: (1) if no artificial vectors appear in the basis, then the solution is an optimal basic feasible solution to the original primal linear programming problem. In this instance the original structural constraints $\mathbf{AX} = \mathbf{b}$ are consistent and none are redundant; (2) if at least one artificial vector appears in the basis at a zero level, then $\mathbf{AX} = \mathbf{b}$ is consistent and we either obtain a degenerate optimal basic feasible solution to the original primal problem or at least one of the original structural constraints is redundant; and (3) if at least one artificial vector appears in the basis at a positive level, the original primal program has no feasible solution since either $\mathbf{AX} = \mathbf{b}$ represents an inconsistent system or there are solutions but none is feasible.

EXAMPLE 8.4. Let us

$$\begin{aligned} max\, f &= x_1 + x_2 + x_3 \quad \text{s.t.} \\ x_1 + 4x_2 + 3x_3 &= 6 \\ 3x_1 + 12x_2 + 9x_3 &= 18 \\ x_1, x_2, x_3 &\geq 0. \end{aligned}$$

Letting x_4, x_5 represented artificial variables, we obtain the simplex matrix

$$[\bar{\mathbf{A}}_*, \mathbf{b}_0] = \begin{bmatrix} 1 & 4 & 3 & 1 & 0 & 0 & 6 \\ 3 & 12 & 9 & 0 & 1 & 0 & 18 \\ -1 & -1 & -1 & M & M & 1 & 0 \end{bmatrix} \longrightarrow$$

$$\begin{bmatrix} 1 & 4 & 3 & 1 & 0 & 0 & 6 \\ 3 & 12 & 9 & 0 & 1 & 0 & 18 \\ -1 & -1 & -1 & 0 & 0 & 1 & 0 \\ -4M & -16M & -12M & 0 & 0 & 0 & -24M \end{bmatrix}.$$

Our initial basic feasible solution to this artificial problem is $x_4 = 6$, $x_5 = 18$, and $f = -24M$. Since the optimality criterion is not satisfied, we proceed to

ROUND 1.

$$\begin{bmatrix} 1 & \boxed{4} & 3 & 1 & 0 & 0 & 6 \\ 3 & 12 & 9 & 0 & 1 & 0 & 18 \\ -1 & -1 & -1 & 0 & 0 & 1 & 0 \\ -4M & -16M & -12M & 0 & 0 & 0 & -24M \end{bmatrix} \begin{matrix} 3/2 \\ 3/2 \\ \\ \\ \end{matrix} \longrightarrow$$

$$\begin{matrix} \hat{\mathbf{Y}}_1 & & \hat{\mathbf{Y}}_2 & & & & \end{matrix}$$
$$\begin{bmatrix} \frac{1}{4} & 1 & \frac{1}{4} & \frac{1}{4} & 0 & 0 & \frac{3}{2} \\ 0 & 0 & 0 & -3 & 1 & 0 & 0 \\ -\frac{3}{4} & 0 & -\frac{1}{4} & \frac{1}{4} & 0 & 1 & \frac{3}{2} \\ 0 & 0 & 0 & 4M & 0 & 0 & 0 \end{bmatrix}.$$

Then our second basic feasible solution to the artificial problem is $x_2 = \frac{3}{2}$, $x_5 = 0$, and $f = \frac{3}{2}$. Since the optimality criterion is still not satisfied, an additional round of the simplex routine is warranted. Before moving on to round two, let us pause for a moment to inspect the preceding simplex matrix. Given that the artificial vector $\mathbf{e}_2$ enters the basis at the zero level ($\mathcal{A} = \{2\}$) and $y_{21} = y_{22} = 0$ for legitimate nonbasis vectors $\widehat{\mathbf{Y}}_1$, $\widehat{\mathbf{Y}}_2$, it is obvious that the second structural constraint is redundant so that we may at this point delete row two and column five. Moreover, since all artificial vectors have been eliminated from the basis, round two starts with an initial basic feasible solution to the original problem. We now look to

ROUND 2 (rows two and four along with columns four and five are deleted).

$$\left[\begin{array}{ccccc|c} \boxed{\frac{1}{4}} & 1 & \frac{3}{4} & 0 & \frac{3}{2} & 6 \\ -\frac{3}{4} & 0 & -\frac{1}{4} & 1 & \frac{3}{2} & \end{array}\right] \longrightarrow \left[\begin{array}{ccccc} 1 & 4 & 3 & 0 & 6 \\ 0 & 3 & 2 & 1 & 6 \end{array}\right].$$

A glance at this last matrix indicates that we have obtained an optimal basic feasible solution to the original problem consisting of $x_1 = 6$ and $f = 6$.

EXAMPLE 8.5. Next, we seek to

$$\begin{aligned} max f &= x_1 + x_2 \quad \text{s.t.} \\ 4x_1 - x_2 &\leq 1 \\ x_1 - x_2 &\geq 3 \\ x_1, x_2 &\geq 0. \end{aligned}$$

If x_3 denotes a slack variable, x_4 represents a surplus variable, and x_5 depicts an artificial variable, then we may form the simplex matrix

$$[\bar{\mathbf{A}}_*, \mathbf{b}_0] = \left[\begin{array}{ccccccc} 4 & -1 & 1 & 0 & 0 & 0 & 1 \\ 1 & -1 & 0 & -1 & 1 & 0 & 3 \\ -1 & -1 & 0 & 0 & M & 1 & 0 \end{array}\right] \longrightarrow \left[\begin{array}{ccccccc} 4 & -1 & 1 & 0 & 0 & 0 & 1 \\ 1 & -1 & 0 & -1 & 1 & 0 & 3 \\ -1 & -1 & 0 & 0 & 0 & 1 & 0 \\ -M & M & 0 & M & 0 & 0 & -3M \end{array}\right].$$

Since the initial basic feasible solution consisting of $x_3 = 1$, $x_5 = 3$, and $f = -3M$ is obviously not optimal, we proceed to

ROUND 1.

$$\left[\begin{array}{ccccccc|c} \boxed{4} & -1 & 1 & 0 & 0 & 0 & 1 & \frac{1}{4} \\ 1 & -1 & 0 & -1 & 1 & 0 & 3 & 3 \\ -1 & -1 & 0 & 0 & 0 & 1 & 0 & \\ -M & M & 0 & M & 0 & 0 & -3M & \end{array}\right] \longrightarrow \begin{array}{c} \quad\quad \widehat{\mathbf{Y}}_1 \; \widehat{\mathbf{Y}}_2 \, \widehat{\mathbf{Y}}_3 \\ \left[\begin{array}{ccccccc} 1 & -\frac{1}{4} & \frac{1}{4} & 0 & 0 & 0 & \frac{1}{4} \\ 0 & -\frac{3}{4} & -\frac{1}{4} & -1 & 1 & 0 & \frac{11}{4} \\ 0 & -\frac{5}{4} & \frac{1}{4} & 0 & 0 & 1 & \frac{1}{4} \\ 0 & \frac{3}{4}M & \frac{1}{4}M & M & 0 & 0 & -\frac{11}{4}M \end{array}\right] \end{array}.$$

The second basic feasible solution to this artificial problem thus appears as $x_1 = \frac{1}{4}$, $x_5 = \frac{11}{4}$, and $f = \frac{1}{4} - \frac{11M}{4}$. Can the artificial basis vector $\mathbf{e}_2(\mathcal{A}=\{2\})$ be eliminated from the basis? As we shall now see, the answer is yes. From the second row of the preceding simplex matrix we find that

$$x_5 = \frac{11}{4} + \frac{3}{4}x_2 + \frac{1}{4}x_3 + x_4.$$

Hence, x_5 may be driven to zero only if x_2, x_3, or x_4 assumes a negative value since $y_{21}, y_{22}, y_{23} < 0$. More formally, since the artificial basis vector $\mathbf{e}_k$ remains in the basis at a positive level and $y_{kj} < 0$ for all legitimate nonbasis vectors $\widehat{\mathbf{Y}}_j$, the original problem has no feasible solution in the sense that at least one basic variable is negative (as Figure 8.1 indicates).

EXAMPLE 8.6. We now desire to

$$\begin{aligned} max f &= x_1 - x_2 \quad \text{s.t.} \\ x_1 + x_2 &= 6 \\ x_1 + x_2 &= 4 \\ x_1, x_2 &\geq 0. \end{aligned}$$

If x_3, x_4 represent artificial variables, obtain

$$\left[\bar{\mathbf{A}}_*, \mathbf{b}_0\right] = \left[\begin{array}{cccccc} 1 & 1 & 1 & 0 & 0 & 6 \\ 1 & 1 & 0 & 1 & 0 & 4 \\ -1 & 1 & M & M & 1 & 0 \end{array}\right] \longrightarrow \left[\begin{array}{cccccc} 1 & 1 & 1 & 0 & 0 & 6 \\ 1 & 1 & 0 & 1 & 0 & 4 \\ -1 & 1 & 0 & 0 & 1 & 0 \\ -2M & -2M & 0 & 0 & 0 & -10M \end{array}\right].$$

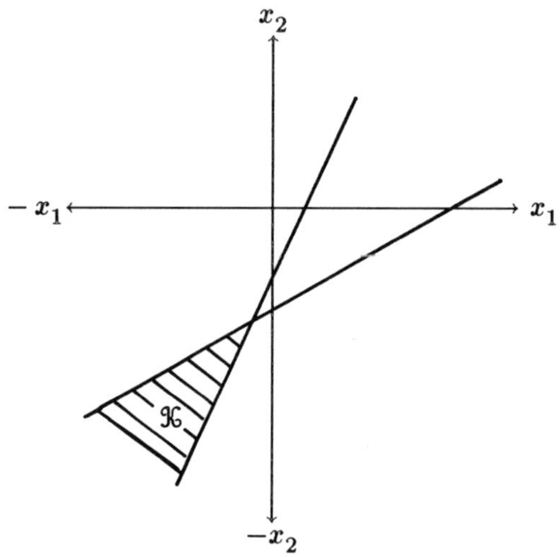

Figure 8.1

and thus an initial basic feasible solution consisting of $x_3 = 6$, $x_4 = 4$, and $f = -10M$. Since the optimality criterion is not satisfied, we look to

ROUND 1.

$$\begin{bmatrix} 0 & 1 & 1 & 0 & 0 & 6 \\ \boxed{1} & 1 & 0 & 1 & 0 & 4 \\ -1 & 1 & 0 & 0 & 1 & 0 \\ -2M & -2M & 0 & 0 & 0 & -10M \end{bmatrix} \begin{matrix} 6 \\ 4 \\ \\ \\ \end{matrix} \longrightarrow \begin{bmatrix} 0 & \overset{\hat{\mathbf{Y}}_1}{0} & 1 & -1 & 0 & 2 \\ 1 & 1 & 0 & 1 & 0 & 4 \\ 0 & 2 & 0 & 1 & 1 & 4 \\ 0 & 0 & 0 & 2M & 0 & -2M \end{bmatrix}.$$

Hence the optimal basic feasible solution to the artificial problem is $x_1 = 4$, $x_3 = 2$, and $f = 4 - 2M$. Since $y_{11} = 0$, $\hat{\mathbf{Y}}_1$ cannot replace $\mathbf{e}_1$ (here $\mathcal{A} = \{1\}$) in the basis. In sum, if the artificial basis vector $\mathbf{e}_k$ remains in the basis at a positive level and $y_{kj} = 0$ for all legitimate nonbasis vectors $\hat{\mathbf{Y}}_j$, then the original structural constraints $\mathbf{AX} = \mathbf{b}$ are inconsistent so that no feasible solution to the original problem exists (Figure 8.2).

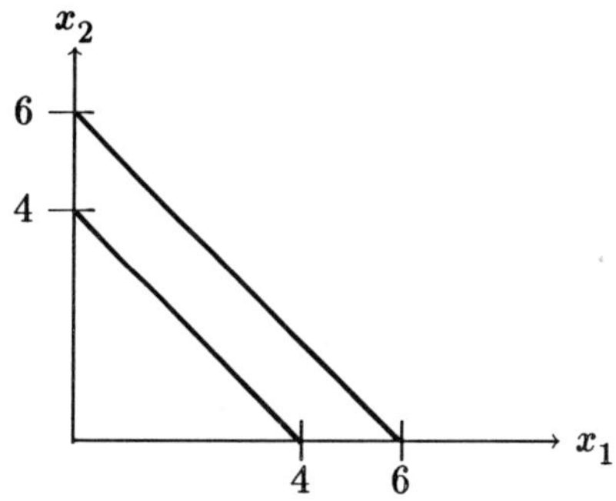

Figure 8.2

8.4. Unbounded Solutions to the Artificial Problem

It may be the case that at some point during the solution of the artificial problem we encounter a vector $\hat{\mathbf{Y}}_k$ for which $y_{ik} \leq 0$, $i=1,\ldots,m$, with $-\bar{c}_k = \min_j\{-\bar{c}_j, -\bar{c}_j < 0\}$. Then clearly the artificial objective function value is unbounded. In this circumstance, what can be said about the behavior of the original problem? Here two subcases present themselves:

1. if $-\bar{c}_k = \min_j\{-\bar{c}_j, -\bar{c}_j < 0\}$, $y_{ik} \leq 0, i=1,\ldots,m,$ for some vector $\hat{\mathbf{Y}}_k$, and all artificial variables are zero, then the original linear program also has an unbounded objective value.
2. if $-\bar{c}_k = \min_j\{-\bar{c}_j, -\bar{c}_j < 0\}$, $y_{ik} \leq 0$, $i=1,\ldots,m$, for some vector $\hat{\mathbf{Y}}_k$, and at least one artificial variable is positive, then the original problem is infeasible.

For instance, consider

EXAMPLE 8.7. Let us

$$\begin{aligned} max\ f &= x_1 + x_2 \quad \text{s.t.} \\ x_1 - x_2 + x_3 &= 1 \\ -x_1 + x_2 &\geq 1 \\ x_1, x_2, x_3 &\geq 0. \end{aligned}$$

With x_4, x_5 nonnegative surplus and artificial variables respectively, we form

$$\left[\bar{\mathbf{A}}_*, \mathbf{b}_0\right] = \begin{bmatrix} 1 & -1 & 1 & 0 & 0 & 0 & 1 \\ -1 & 1 & 0 & -1 & 1 & 0 & 1 \\ -1 & -1 & 0 & 0 & \mathrm{M} & 1 & 0 \end{bmatrix} \rightarrow \begin{bmatrix} 1 & -1 & 1 & 0 & 0 & 0 & 1 \\ -1 & 1 & 0 & -1 & 1 & 0 & 1 \\ -1 & -1 & 0 & 0 & 0 & 1 & 0 \\ \mathrm{M} & -\mathrm{M} & 0 & \mathrm{M} & 0 & 0 & -\mathrm{M} \end{bmatrix}$$

and thus an initial basic feasible solution consisting of $x_3 = 1$, $x_5 = 1$, and $f = -M$. Since this current solution is not optimal, we look to

ROUND 1.

$$\begin{bmatrix} 1 & -1 & 1 & 0 & 0 & 0 & 1 \\ -1 & \boxed{1} & 0 & -1 & 1 & 0 & 1 \\ -1 & -1 & 0 & 0 & 0 & 1 & 0 \\ \mathrm{M} & -\mathrm{M} & 0 & \mathrm{M} & 0 & 0 & -\mathrm{M} \end{bmatrix} 1 \longrightarrow \begin{bmatrix} \overset{\widehat{\mathbf{Y}}_1}{0} & 0 & 1 & -1 & 1 & 0 & 2 \\ -1 & 1 & 0 & -1 & 1 & 0 & 1 \\ -2 & 0 & 0 & -1 & 1 & 1 & 1 \\ 0 & 0 & 0 & 0 & \mathrm{M} & 0 & 0 \end{bmatrix}.$$

Since $-\bar{c}_1 = -2 = \underset{j}{min}\{-\bar{c}_j, -\bar{c}_j < 0\}$ and each component of $\widehat{\mathbf{Y}}_1$ is nonpositive, it follows that the artificial problem is unbounded.

ROUND 2 (row four and column five are deleted).

$$\begin{bmatrix} 0 & 0 & 1 & -1 & 0 & 2 \\ -1 & 1 & 0 & -1 & 0 & 1 \\ -2 & 0 & 0 & -1 & 1 & 1 \end{bmatrix}.$$

Given that the artificial variable x_5 is zero, a glance at column one of this matrix (which houses an initial basic feasible solution to the original problem) indicates that the original program is also unbounded.

What this example problem has revealed is that when the artificial problem is unbounded and all artificial variables are zero, then there is an admissible direction $\mathbf{d}' = (\mathbf{d}_1', \mathbf{d}_2') \geqq \mathbf{0}'$ of the feasible set for the artificial problem $\{\mathbf{X}_* \mid \mathbf{A}_*\mathbf{X}_* = \mathbf{b}, \mathbf{X}_* \geqq \mathbf{0}\}$ such that $\mathbf{C}'\mathbf{d}_1 - M\mathbf{1}'\mathbf{d}_2 > 0$. Since M is large and positive and $\mathbf{d}_2 \geqq \mathbf{0}$, the preceding inequality implies that $\mathbf{d}_2 = \mathbf{0}$ and thus

$\mathbf{C}'\mathbf{d}_1 > 0$. Thus $\mathbf{d}_1$ is an extreme direction for the original feasible region $\{\mathbf{X} \mid \mathbf{AX} = \mathbf{b},\ \mathbf{X} \geqq \mathbf{0}\}$ along which f increases without bound. In fact, from column one of the preceding simplex matrix, $\mathbf{Y}_1' = (0, -1)$ and thus

$$\mathbf{d}_1 = \begin{bmatrix} -\mathbf{Y}_1 \\ \mathbf{e}_1 \end{bmatrix} \geqq \mathbf{0}$$

as required.

8.5. The Two-Phase Method

In this section we shall develop an alternative procedure for solving a primal linear programming problem involving artificial variables. Our approach will be to frame the simplex method in terms of two successive phases, phase I and phase II (Dantzig, *et al.* (1955)). In phase I we seek to drive all artificial variables to zero by maximizing a surrogate objective function involving only artificial variables. If the optimal basis for the phase I problem contains no artificial vectors or contains one or more artificial vectors at the zero level, then phase II is initiated. However, if the optimal phase I basis contains at least one artificial vector at a positive level, the process is terminated. If phase II is warranted, we proceed by maximizing the original objective function using the optimal phase I basis as a starting point, *i.e.*, the optimal phase I basis provides an initial though possibly nonbasic feasible solution to the original linear programming problem.

PHASE I. In constructing the phase I objective function, the legitimate variables will be assigned zero coefficients while each artificial variable is given a coefficient of minus one. Then the phase I **surrogate linear programming problem** may be formulated as: maximize the **infeasibility form**

$$\begin{aligned} g(\mathbf{X}_*) &= \mathbf{0}'\mathbf{X} - \mathbf{1}'\mathbf{X}_a \quad \text{s.t.} \\ \mathbf{A}_*\mathbf{X}_* &= \mathbf{b},\ \mathbf{X}_* \geqq \mathbf{0}. \end{aligned}$$

For instance, the phase I surrogate version of the problem on page is

$$\begin{aligned} &max\ g = -x_5 - x_6 \quad \text{s.t.} \\ &x_1 - x_2 + x_3 = 6 \\ &2x_1 + 3x_2 - x_4 + x_5 = 2 \\ &x_1 + x_2 + x_6 = 4 \\ &x_1, \ldots, x_6 \geq 0 \end{aligned}$$

with simplex matrix

$$\left[\hat{\mathbf{A}}_{*}, \mathbf{b}_0\right] = \begin{bmatrix} 1 & -1 & 1 & 0 & 0 & 0 & 0 & 6 \\ 2 & 3 & 0 & -1 & 1 & 0 & 0 & 2 \\ 1 & 1 & 0 & 0 & 0 & 1 & 0 & 4 \\ 0 & 0 & 0 & 0 & 1 & 1 & 0 & 0 \end{bmatrix}. \tag{8.3}$$

Note that with $\mathbf{X}_a \geqq \mathbf{0}$, it must be the case that $max\{g\} \leq 0$ since g, which initially assumes a negative value, can never turn positive. In addition, the largest possible value that g can attain is zero; and this occurs only if the optimal phase I basis contains no artificial vectors or if those artificial vectors remaining in the said basis do so at the zero level. So when $max\{g\} < 0$, at least one artificial vector remains in the optimal phase I basis at a positive level.

Before discussing the variety of possible solutions that may emerge during phase I, let us look to some prominent aspects of the phase I routine. First, as in the M-penalty method, an artificial vector never reenters the basis once it has been removed. Here too the columns of the simplex matrix corresponding to artificial vectors that have turned nonbasic may be deleted as pivoting is carried out. Moreover, during phase I, the sequence of vectors which enters and leaves the basis is the same as in the M-penalty method. Finally, if g becomes zero before the optimality criterion is satisfied, we may terminate phase I and proceed directly to phase II.

As indicated above, phase I ends when the optimality criterion is satisfied. In this instance, the optimal basis will exhibit one of three mutually exclusive and collectively exhaustive characteristics:

(1) $max\{g\} = 0$ and no artificial vectors remain in the basis;

(2) $max\{g\} = 0$ and one or more artificial vectors remain in the basis at the zero level;

(3) $max\{g\} < 0$ and one or more artificial vectors remain in the basis at a positive level.

If case one obtains, it is evident that the optimal phase I basis provides a basic feasible solution to the original primal problem. Here the original structural constraints are consistent and none are redundant. Next, if case two prevails, we obtain a feasible solution to the original problem, *i.e.*, the original structural constraints are consistent although possible redundant. Finally, for case three, the surrogate problem has no feasible solution with $\mathbf{X}_a = \mathbf{0}$ and thus the original problem possesses no feasible solution either. In sum, if either case one or two emerges at the end of phase I, there exists at least one feasible solution to the original problem. In this regard, we proceed to

PHASE II. Let us first specify the appropriate phase II objective function. Since we ultimately desire to maximize $f(\mathbf{X})$, the optimal phase I simplex matrix is transformed into the initial phase II simplex matrix by respecifying the objective function coefficients. That is to say, during phase II, the coefficients on the legitimate variables are the same as those appearing in the original objective function while the coefficients on the artificial variables are zero, *i.e.*, the phase II objective function is the original objective function $f(\mathbf{X})$ itself. In this light, the sequence of operations underlying the two-phase method may be expressed as

$$\begin{aligned} max\ g(\mathbf{X}_*) &= \mathbf{0}'\mathbf{X} - \mathbf{1}'\mathbf{X}_a \quad \text{s.t.} \quad [\text{Phase I}] \\ max\ f(\mathbf{X}_*) &= \mathbf{C}'\mathbf{X} + \mathbf{0}'\mathbf{X}_a \quad \text{s.t.} \quad [\text{Phase II}] \\ \mathbf{A}_*\mathbf{X}_* &= \mathbf{b},\ \mathbf{X}_* \geqq \mathbf{0}. \end{aligned}$$

We now turn to an analysis concerning the implications for phase II of the aforementioned characteristics of the optimal phase I simplex matrix.

If case one above occurs, the optimal phase I basis provides an initial basic feasible solution to the original linear programming problem. Once the modified objective function is introduced, pivoting proceeds until an optimal basic feasible solution to the original problem is attained.

Turning now to case two, while we have obtained a feasible though nonbasic solution to the original problem, there still remains the question of redundancy in the original structural constraints $\mathbf{AX} = \mathbf{b}$ and the development of a safeguard against the possibility that one or more artificial vectors appearing in the initial phase II basis at the zero level will abide in some future basis at a positive level as the phase II pivot operations are carried out.

Our discussion of redundancy here is similar to that advanced for the M-penalty method. To this end, let us assume that within the optimal phase I simplex matrix $y_{ij} = 0$ for all legitimate nonbasis vectors $\hat{\mathbf{Y}}_j$ and for all $i \in \mathcal{A}$. In this instance, none of the artificial basis vectors $\mathbf{e}_i$ can be replaced by the $\hat{\mathbf{Y}}_j$ so that there is redundancy in the original structural constraint equalities. In this regard, if there are k artificial vectors in the basis at the zero level, then, at the end of phase I, we may simply delete the k rows of the simplex matrix containing the zero-valued artificial variables along with the associated artificial basis vectors $\mathbf{e}_i$. Then phase II begins with a basis of reduced size and pivoting is carried out until an optimal basic feasible solution to the original problem is attained.

We next examine a procedure which serves to guard against the possibility that an artificial vector appearing in the initial phase II basis at the zero level manifests itself in some future basis at a positive level. Let us assume that for some $j = j'$ the vector $\hat{\mathbf{Y}}_{j'}$ is to enter the basis and $y_{ij'} \leq 0$ with $y_{ij'} < 0$ for at least one $i \in \mathcal{A}$. In this circumstance the artificial vectors for which $y_{ij'} < 0$ will not be considered as candidates for removal from the basis. Rather, some legitimate vector will be removed with the result that the artificial basis vectors for which $y_{ij'} < 0$ appear in the new basis at a positive level. To avoid this difficulty, instead of removing a legitimate vector from the basis, let us adopt the policy of arbitrarily removing any one of the artificial basis vectors $\mathbf{e}_i$ with $y_{ij'} < 0$. The result will be a feasible solution to the original problem wherein $\hat{\mathbf{Y}}_{j'}$ enters the new basis at the zero level (since the artificial vector that it replaced was formerly at the zero level.) With the value of the incoming variable equal to zero, the values of the other basic variables for the new solution are the same as in the preceding solution. Hence, the value of the objective function for the new solution must be the same as at the previous one (see example 8.9) below).

EXAMPLE 8.8. Solve the problem appearing in example 8.3 using the two-phase method. Forming the phase I infeasibility form $g = -x_6 - x_7$ we easily obtain the simplex matrix

$$\left[\widehat{\mathbf{A}}_{*}, \mathbf{b}_0\right] = \begin{bmatrix} 1 & 2 & 1 & 0 & 0 & 0 & 0 & 0 & 12 \\ 1 & 1 & 0 & -1 & 0 & 1 & 0 & 0 & 4 \\ -1 & 1 & 0 & 0 & -1 & 0 & 1 & 0 & 1 \\ 0 & 0 & 0 & 0 & 0 & 1 & 1 & 1 & 0 \end{bmatrix}.$$

To generate an initial basic feasible solution to this phase I surrogate problem, let us add

$$\begin{array}{rrrrrrrrr} -1 & -1 & 0 & 1 & 0 & -1 & 0 & 0 & -4 \\ 1 & -1 & 0 & 0 & 1 & 0 & -1 & 0 & -1 \\ \hline 0 & -2 & 0 & 1 & 1 & -1 & -1 & 0 & -5 \end{array}$$

to the last row of the above matrix to obtain

$$\begin{bmatrix} 1 & 2 & 1 & 0 & 0 & 0 & 0 & 0 & 12 \\ 1 & 1 & 0 & -1 & 0 & 1 & 0 & 0 & 4 \\ -1 & 1 & 0 & 0 & -1 & 0 & 1 & 0 & 1 \\ 0 & -2 & 0 & 1 & 1 & 0 & 0 & 1 & -5 \end{bmatrix}$$

The first basic feasible solution to the surrogate problem thus becomes $x_3 = 12$, $x_6 = 4$, $x_7 = 1$, and $g = -5$. Since this solution is obviously not optimal, we undertake

PHASE I, ROUND 1.

$$\left[\begin{array}{rrrrrrrrr} 1 & 2 & 1 & 0 & 0 & 0 & 0 & 0 & 12 \\ 1 & 1 & 0 & -1 & 0 & 1 & 0 & 0 & 4 \\ -1 & \boxed{1} & 0 & 0 & -1 & 0 & 1 & 0 & 1 \\ 0 & -2 & 0 & 1 & 1 & 0 & 0 & 1 & -5 \end{array}\right] \begin{array}{r} 6 \\ 4 \\ 1 \\ \\ \end{array} \longrightarrow$$

$$\begin{bmatrix} 3 & 0 & 1 & 0 & 2 & 0 & -2 & 0 & 10 \\ 2 & 0 & 0 & -1 & 1 & 1 & -1 & 0 & 3 \\ -1 & 1 & 0 & 0 & -1 & 0 & 1 & 0 & 1 \\ -2 & 0 & 0 & 1 & -1 & 0 & 2 & 1 & -3 \end{bmatrix}.$$

Here the second basic feasible solution to the surrogate problem is $x_1 = 1$, $x_3 = 10$, $x_6 = 3$, and $g = -3$. Since this solution is also not optimal, we turn to

PHASE I, ROUND 2 (column seven is deleted).

$$\begin{bmatrix} 3 & 0 & 1 & 0 & 2 & 0 & 0 & 10 \\ \boxed{2} & 0 & 0 & -1 & 1 & 1 & 0 & 3 \\ -1 & 1 & 0 & 0 & -1 & 0 & 0 & 1 \\ -2 & 0 & 0 & 1 & -1 & 0 & 1 & -3 \end{bmatrix} \begin{matrix} 10/3 \\ 3/2 \\ \\ \\ \end{matrix} \longrightarrow$$

$$\begin{bmatrix} 0 & 0 & 1 & \frac{3}{2} & \frac{1}{2} & -\frac{3}{2} & 0 & \frac{11}{2} \\ 1 & 0 & 0 & -\frac{1}{2} & \frac{1}{2} & \frac{1}{2} & 0 & \frac{3}{2} \\ 0 & 1 & 0 & -\frac{1}{2} & -\frac{1}{2} & \frac{1}{2} & 0 & \frac{5}{2} \\ 0 & 0 & 0 & 0 & 0 & 1 & 1 & 0 \end{bmatrix}.$$

The third basic feasible solution to the surrogate problem thus appears as $x_1 = \frac{3}{2}$, $x_2 = \frac{5}{2}$, $x_3 = \frac{11}{2}$, and $g = 0$. Since it is clearly optimal, phase I ends. Moreover, since all artificial vectors have been removed from the basis, this particular basic feasible solution to the surrogate problem represents an initial basic feasible solution to the original problem. In this regard, phase II begins by deleting column six of the final phase I simplex matrix and replacing the last row of the same by the original objective function $f = x_1 + x_2$. That is,

$$\begin{bmatrix} 0 & 0 & 1 & \frac{3}{2} & \frac{1}{2} & 0 & \frac{11}{2} \\ 1 & 0 & 0 & -\frac{1}{2} & -\frac{1}{2} & 0 & \frac{3}{2} \\ 0 & 1 & 0 & -\frac{1}{2} & -\frac{1}{2} & 0 & \frac{5}{2} \\ -1 & -1 & 0 & 0 & 0 & 1 & 0 \end{bmatrix}$$

or, upon adding

$$\begin{array}{ccccccc} 1 & 0 & 0 & -\frac{1}{2} & \frac{1}{2} & 0 & \frac{3}{2} \\ 0 & 1 & 0 & -\frac{1}{2} & -\frac{1}{2} & 0 & \frac{5}{2} \\ \hline 1 & 1 & 0 & -1 & 0 & 1 & 4 \end{array}$$

to the last row of the preceding simplex matrix,

$$\begin{bmatrix} 0 & 0 & 1 & \frac{3}{2} & \frac{1}{2} & 0 & \frac{11}{2} \\ 1 & 0 & 0 & -\frac{1}{2} & \frac{1}{2} & 0 & \frac{3}{2} \\ 0 & 1 & 0 & -\frac{1}{2} & -\frac{1}{2} & 0 & \frac{5}{2} \\ 0 & 0 & 0 & -1 & 0 & 1 & 4 \end{bmatrix}.$$

Thus the initial basic feasible solution to the phase II original problem appears as $x_1 = \frac{3}{2}$, $x_2 = \frac{5}{2}$, $x_3 = \frac{11}{2}$, and $f = 4$. Since the optimality criterion is not satisfied, we proceed to

PHASE II, ROUND 1.

$$\begin{bmatrix} 0 & 0 & 1 & \boxed{\frac{3}{2}} & \frac{1}{2} & 0 & \frac{11}{2} \\ 1 & 0 & 0 & -\frac{1}{2} & \frac{1}{2} & 0 & \frac{3}{2} \\ 0 & 1 & 0 & -\frac{1}{2} & -\frac{1}{2} & 0 & \frac{5}{2} \\ 0 & 0 & 0 & -1 & 0 & 1 & 4 \end{bmatrix} \begin{matrix} 11/3 \\ — \\ — \\ \\ \end{matrix} \longrightarrow \begin{bmatrix} 0 & 0 & \frac{2}{3} & 1 & \frac{1}{3} & 0 & \frac{11}{3} \\ 1 & 0 & \frac{1}{3} & 0 & \frac{2}{3} & 0 & \frac{10}{3} \\ 0 & 1 & \frac{1}{3} & 0 & -\frac{1}{3} & 0 & \frac{13}{3} \\ 0 & 0 & \frac{2}{3} & 0 & \frac{1}{3} & 1 & \frac{23}{3} \end{bmatrix}.$$

An inspection of the last row of this simplex matrix indicates that we have obtained an optimal basic feasible solution to the original problem consisting of $x_1 = \frac{10}{3}$, $x_2 = \frac{13}{3}$, $x_4 = \frac{11}{3}$, and $f = \frac{23}{3}$.

As an exercise the reader should resolve example problems 8.4-8.6 above using the two-phase method so as to verify that the same conclusions regarding redundancy, infeasibility, and inconsistency respectively hold.

EXAMPLE 8.9. Let us

$$\begin{aligned} max f &= 5x_1 + 4x_2 + x_3 + x_4 \quad \text{s.t.} \\ x_1 + x_2 + x_3 \quad &= 2 \\ x_2 + x_3 \quad &= 2 \\ x_1 + 2x_2 + 3x_3 + 6x_4 &= 4 \\ x_1, x_2, x_3, x_4 &\geq 0. \end{aligned}$$

Constructing the infeasibility form $g = -x_5 - x_6 - x_7$ we obtain

$$\left[\hat{\mathbf{A}}_*, \mathbf{b}_0\right] = \begin{bmatrix} 1 & 1 & 1 & 0 & 1 & 0 & 0 & 0 & 2 \\ 0 & 1 & 1 & 0 & 0 & 1 & 0 & 0 & 2 \\ 1 & 2 & 3 & 6 & 0 & 0 & 1 & 0 & 4 \\ 0 & 0 & 0 & 0 & 1 & 1 & 1 & 1 & 0 \end{bmatrix} \longrightarrow$$

$$\begin{bmatrix} 1 & 1 & 1 & 0 & 1 & 0 & 0 & 0 & 2 \\ 0 & 1 & 1 & 0 & 0 & 1 & 0 & 0 & 2 \\ 1 & 2 & 3 & 6 & 0 & 0 & 1 & 0 & 4 \\ -2 & -4 & -5 & -6 & 0 & 0 & 0 & 1 & -8 \end{bmatrix},$$

where x_5, x_6 and x_7 are artificial variables. Here the initial basic feasible solution to this surrogate problem is $x_5 = 2$, $x_6 = 2$, $x_7 = 4$, and $g = -8$. To maximize g

we turn to

PHASE I, ROUND 1.

$$\left[\begin{array}{rrrrrrrrr|l} 1 & 1 & 1 & 0 & 1 & 0 & 0 & 0 & 2 & — \\ 0 & 1 & 1 & 0 & 0 & 1 & 0 & 0 & 2 & — \\ 1 & 2 & 3 & \boxed{6} & 0 & 0 & 1 & 0 & 4 & 2/3 \\ -2 & -4 & -5 & -6 & 0 & 0 & 0 & 1 & -8 & \end{array}\right] \longrightarrow$$

$$\left[\begin{array}{rrrrrrrrr} 1 & 1 & 1 & 0 & 1 & 0 & 0 & 0 & 2 \\ 0 & 1 & 1 & 0 & 0 & 1 & 0 & 0 & 2 \\ \frac{1}{6} & \frac{1}{3} & \frac{1}{2} & 1 & 0 & 0 & \frac{1}{6} & 0 & \frac{2}{3} \\ -1 & -2 & -2 & 0 & 0 & 0 & 1 & 1 & -4 \end{array}\right].$$

The second basic feasible solution to the surrogate problem consists of $x_4 = \frac{2}{3}$, $x_5 = 2$, $x_6 = 2$, and $g = -4$. Given that the optimality criterion is still not satisfied, we proceed to

PHASE I, ROUND 2 (column seven is deleted)

$$\left[\begin{array}{rrrrrrrr|l} 1 & \boxed{1} & 1 & 0 & 1 & 0 & 0 & 2 & 2 \\ 0 & 1 & 1 & 0 & 0 & 1 & 0 & 2 & 2 \\ \frac{1}{6} & \frac{1}{3} & \frac{1}{2} & 1 & 0 & 0 & 0 & \frac{2}{3} & 2 \\ -1 & -2 & -2 & 0 & 0 & 0 & 1 & -4 & \end{array}\right] \longrightarrow$$

$$\left[\begin{array}{rrrrrrrr} 1 & 1 & 1 & 0 & 1 & 0 & 0 & 2 \\ -1 & 0 & 0 & 0 & -1 & 1 & 0 & 0 \\ -\frac{1}{6} & 0 & \frac{1}{6} & 1 & -\frac{1}{3} & 0 & 0 & 0 \\ 1 & 0 & 0 & 0 & 2 & 0 & 1 & 0 \end{array}\right].$$

With the optimality criterion satisfied, the optimal basic feasible solution to the phase I surrogate problem is $x_2 = 2$, $x_4 = 0$, $x_6 = 0$, and $g = 0$. Turning now to phase II, let us delete column five of the final phase I simplex matrix and replace the last row of the same by the original objective function $f = 5x_1 + 4x_2 + x_3 + x_4$. To this end we obtain

$$\begin{bmatrix} 1 & 1 & 1 & 0 & 0 & 0 & 2 \\ -1 & 0 & 0 & 0 & 1 & 0 & 0 \\ -\frac{1}{6} & 0 & \frac{1}{6} & 1 & 0 & 0 & 0 \\ -5 & -4 & -1 & -1 & 0 & 1 & 0 \end{bmatrix} \longrightarrow \overset{\hat{\mathbf{Y}}_1}{\begin{bmatrix} 1 & 1 & 1 & 0 & 0 & 0 & 2 \\ -1 & 0 & 0 & 0 & 1 & 0 & 0 \\ -\frac{1}{6} & 0 & \frac{1}{6} & 1 & 0 & 0 & 0 \\ -\frac{7}{6} & 0 & \frac{19}{6} & 0 & 0 & 1 & 8 \end{bmatrix}}.$$

Since the initial phase II basic feasible solution to the original problem consisting of $x_2 = 2$, $x_4 = 0$, $x_6 = 0$, and $f = 8$ is not optimal, round one of phase II is warranted. However, before attempting to generate an improved solution to the phase II objective function, we note that since $y_{21} < 0$, the artificial vector $\mathbf{e}_2$ will not be considered for removal from the basis. Rather, simplex pivoting dictates that the vector $\mathbf{e}_1$ will be deleted from the basis with the result that the current artificial basis vector $\mathbf{e}_2$ will appear in the new basis at a positive level. To preclude the latter from occurring, instead of removing $\mathbf{e}_1$, let us remove the artificial basis vector $\mathbf{e}_2$. To this end we proceed to

PHASE II, ROUND 1.

$$\begin{bmatrix} 1 & 1 & 1 & 0 & 0 & 0 & 2 \\ \boxed{-1} & 0 & 0 & 0 & 1 & 0 & 0 \\ -\frac{1}{6} & 0 & \frac{1}{6} & 1 & 0 & 0 & 0 \\ -\frac{7}{6} & 0 & \frac{19}{6} & 0 & 0 & 1 & 8 \end{bmatrix} \longrightarrow \begin{bmatrix} 0 & 1 & 1 & 0 & 1 & 0 & 2 \\ 1 & 0 & 0 & 0 & -1 & 0 & 0 \\ 0 & 0 & \frac{1}{6} & 1 & -\frac{1}{6} & 0 & 0 \\ 0 & 0 & \frac{19}{6} & 0 & -\frac{7}{6} & 1 & 8 \end{bmatrix}.$$

Thus the optimal (though degenerate) basic feasible solution to the phase II original problem consists of $x_1 = 0$, $x_2 = 2$, $x_4 = 0$, and $f = 8$.

8.6. Obtaining the Optimal Primal Solution from the Optimal Dual Matrix

We noted earlier that the simplex method can be applied to either the primal or dual formulation of a linear programming problem, with the choice of the version actually solved determined by considerations of computational ease. In this regard, if the dual problem is selected, how may we determine the optimal values of the primal variables from the optimal dual solution? Given that the dual problem appears as

$$\begin{aligned} \min g(\mathbf{U}) &= \mathbf{b}'\mathbf{U} \quad \text{s.t.} \\ \mathbf{A}'\mathbf{U} &\geqq \mathbf{C},\ \mathbf{U} \geqq \mathbf{0} \end{aligned}$$

or

$$\begin{aligned} \max h(\mathbf{U}) &= -g(\mathbf{U}) = -\mathbf{b}'\mathbf{U} \quad \text{s.t.} \\ \mathbf{A}'\mathbf{U} &\geqq \mathbf{C},\ \mathbf{U} \geqq \mathbf{0}, \end{aligned}$$

we may utilize the M-penalty method to derive the artificial simultaneous augmented form

$$\begin{aligned} \mathbf{A}'\mathbf{U} - \mathbf{I}_n\mathbf{U}_s + \mathbf{I}_n\mathbf{U}_a \qquad\qquad &= \mathbf{C} \\ \mathbf{b}'\mathbf{U} \qquad\qquad + M\mathbf{1}'\mathbf{U}_a + h &= 0, \end{aligned}$$

where $\mathbf{U}_a$ denotes an $(m \times 1)$ vector of nonnegative artificial variables and the associated dual simplex matrix is

$$\begin{bmatrix} \mathbf{A}' & -\mathbf{I}_n & \mathbf{I}_n & \mathbf{0} & \mathbf{C} \\ \mathbf{b}' & \mathbf{0}' & \mathrm{M}\mathbf{1}' & 1 & 0 \end{bmatrix} \longrightarrow \begin{bmatrix} \mathbf{A}' & -\mathbf{I}_n & \mathbf{I}_n & \mathbf{0} & \mathbf{C} \\ \mathbf{b}'-\mathrm{M}\mathbf{1}'\mathbf{A}' & \mathrm{M}\mathbf{1}' & \mathbf{0}' & 1 & -\mathrm{M}\mathbf{1}'\mathbf{C} \end{bmatrix}.$$

Relative to this system the $(p+1 \times p+1)$ basis matrix and its inverse may be written respectively as

$$\bar{\mathbf{D}} = \begin{bmatrix} \mathbf{D} & \mathbf{0} \\ \mathbf{b}_D'-\mathrm{M}\mathbf{1}'\mathbf{D} & 1 \end{bmatrix}, \quad \bar{\mathbf{D}}^{-1} = \begin{bmatrix} \bar{\mathbf{D}}^{-1} & \mathbf{0} \\ -\mathbf{b}_D'\mathbf{D}^{-1}+\mathrm{M}\mathbf{1}' & 1 \end{bmatrix},$$

where $\mathbf{D}$ depicts the $(p \times p)$ basis matrix associated with the dual artificial structural constraint system. Then

$$\begin{bmatrix} \bar{\mathbf{D}}^{-1} & \mathbf{0} \\ -\mathbf{b}_D'\mathbf{D}^{-1}+\mathrm{M}\mathbf{1}' & 1 \end{bmatrix} \begin{bmatrix} \mathbf{A}' & -\mathbf{I}_n & \mathbf{I}_n & \mathbf{0} & \mathbf{C} \\ \mathbf{b}'-\mathrm{M}\mathbf{1}'\mathbf{A}' & \mathrm{M}\mathbf{1}' & \mathbf{0}' & 1 & -\mathrm{M}\mathbf{1}'\mathbf{C} \end{bmatrix} =$$

$$\begin{bmatrix} \mathbf{D}^{-1}\mathbf{A}' & -\mathbf{D}^{-1} & \mathbf{0} & \mathbf{D}^{-1}\mathbf{C} \\ \boxed{\boxed{-\mathbf{b}_D'\mathbf{D}^{-1}\mathbf{A}' + \mathbf{b}'}} & \boxed{\mathbf{b}_D'\mathbf{D}^{-1}} & 1 & -\mathbf{b}_{\bar{D}}^{-1}\mathbf{D}^{-1}\mathbf{C} \end{bmatrix} \tag{8.4}$$

yields a basic feasible solution to the dual minimum problem, where the vectors associated with the artificial variables, *i.e.*, those corresponding to the third partitioned column of (8.4), have been deleted since the artificial vectors have been driven from the basis. If we assume that the optimality criterion is satisfied, then the optimal values of the primal variables appear as the blocked array of elements in the objective function row of (8.4), *i.e.*, $\mathbf{X}' = \mathbf{b}'_D\mathbf{D}^{-1} \geqq \mathbf{0}'$. In addition, from the augmented primal structural constraint system $\mathbf{AX} + \mathbf{I}_m\mathbf{X}_s = \mathbf{b}$ we may isolate the primal slack variables as $\mathbf{X}'_s = \mathbf{b}' - \mathbf{X}'\mathbf{A}'$. But this expression corresponds to the double-blocked array of elements $\mathbf{X}'_s = -\mathbf{b}'_D\mathbf{D}^{-1}\mathbf{A}' + \mathbf{b}' \geqq \mathbf{0}'$ in the objective function row of (8.4). To summarize, our rules for isolating the optimal values of the primal variables (original and slack) may be determined as follows:

(a) to find the values of the primal variables $\mathbf{X}$ which correspond to an optimal basic feasible solution of the dual minimum problem, look in the last row of the final dual simplex matrix under those columns which were, in the original simplex matrix, the columns of the negative of the identity matrix. The jth primal variable $x_j, j=1,\ldots,p$, will appear under the column which was initially the jth column of the negative of the identity matrix;

(b) to find the values of the primal slack variables $\mathbf{X}_s$ which correspond to an optimal basic feasible solution of the dual minimum problem, look in the last row of the final dual simplex matrix under those columns which were, in the original simplex matrix, the columns of the matrix $\mathbf{A}'$. The ith primal slack variable x_{p+i}, $i=1,\ldots,m$, will appear under the column which was initially the ith column of $\mathbf{A}'$.

In this section we have demonstrated how the optimal primal variables may be obtained from the optimal dual simplex matrix. Earlier in Chapter 6 we extracted the optimal dual variables from the optimal primal simplex matrix. Hence no matter which of the primal-dual pair of problems is solved, *all* of the pertinent information generated by an optimal basic feasible solution is readily available, *i.e.*, from either the final primal or dual simplex matrix we can isolate both the primal and dual original variables as well as the primal slack and dual surplus variables.

EXAMPLE 8.10. The simultaneous artificial augmented form of the dual minimum problem presented in example 4.3 is

$$\begin{aligned} u_1 + 4u_2 - u_3 \qquad\qquad + u_5 \qquad\qquad &= 24 \\ 5u_1 + u_2 \qquad - u_4 \qquad\qquad + u_6 &= 25 \\ u_1 + 3u_2 \qquad\qquad + Mu_5 + Mu_6 &= 0, \end{aligned}$$

where $h = -g$ (here we are employing the transformation $min\, g = -max\{-g\}$ given in section 9.1); $u_3, u_4 \geq 0$ represent surplus variables; and $u_5, u_6 \geq 0$ depict artificial variables. In addition, the associated simplex matrix is

$$\begin{bmatrix} 1 & 4 & -1 & 0 & 1 & 0 & 0 & 24 \\ 5 & 1 & 0 & -1 & 0 & 1 & 0 & 25 \\ 1 & 3 & 0 & 0 & M & M & 1 & 0 \end{bmatrix} \longrightarrow$$

$$\begin{bmatrix} 1 & 4 & -1 & 0 & 1 & 0 & 0 & 24 \\ 5 & 1 & 0 & -1 & 0 & 1 & 0 & 25 \\ 1 & 3 & 0 & 0 & 0 & 0 & 1 & 0 \\ -6M & -5M & M & M & 0 & 0 & 0 & -49M \end{bmatrix}.$$

It is easily shown that the optimal simplex matrix associated with this problem is

$$\begin{bmatrix} 0 & 1 & -\frac{5}{19} & \frac{1}{19} & 0 & 5 \\ 1 & 0 & \frac{1}{19} & -\frac{4}{19} & 1 & 4 \\ 0 & 0 & \frac{14}{19} & \frac{1}{19} & 0 & -19 \end{bmatrix},$$

where the columns associated with the artificial variables and the M-row have been deleted. Upon examining the objective function row of this matrix we see that $\mathbf{X}_0' = (14/19,\ 1/19)$ and $\mathbf{X}_s' = (0,0)$ as expected.

To isolate the values of the original primal variables $\mathbf{X}$ along with the primal slack variables $\mathbf{X}_s$ when the two-phase method is used to solve the dual problem, it will simply be noted that if the optimal phase I basis contains no artificial vectors, then the optimal phase II simplex matrix is of the form exhibited by (8.4). Hence the same rules (a), (b) above for locating $\mathbf{X}$, $\mathbf{X}_s$ apply for the two-phase method as well.

8.7. The Composite Simplex Method

As was the case for both the M-penalty and two-phase methods, the composite simplex algorithm is an iterative technique for obtaining an initial basic feasible solution to a linear programming problem wherein a number of " $\geq$ " and/or " $=$ " structural constraints are present. Our previous approach in this circumstance was to transform $\mathbf{AX} = \mathbf{b}$ into $\mathbf{A}_*\mathbf{X}_* = \mathbf{b}$ by inserting into the former system a sufficient number of artificial variables so that an initial basic feasible solution to the latter system could be obtained. The M-penalty and two-phase methods then operated to systematically drive the artificial vectors out of the basis so that an optimal solution to the original problem could eventually be determined (if, of course, it existed). And as the results of the preceding sections indicate, a considerable amount of computational effort is often expended in implementing the M-penalty and two-phase routines even if only a few artificial variables are present. To circumvent this difficulty the ***composite simplex method*** (Wolfe (1965)) was designed to curtail the use of artificial variables by allowing some of the basic variables to initially assume negative values. These variables are ultimately driven to feasible levels (and any artificial vectors within the initial basis are routinely forced from the same) so that an optimal solution to the original problem is finally attained. Let us see exactly how the composite simplex method operates.

Since a basic variable is not precluded from initially assuming a negative value, let us convert any " $\geq$ " structural constraints to the form $-\sum_{j=1}^{p} a_{sj}x_j \leq -b_s$, where $-b_s$ is now unrestricted in sign. (Note that if the original structural constraint system is composed entirely of inequalities, we have an initial basic — though not necessarily feasible — solution in slack variables.) Moreover, if equality structural constraints are also present, a non-negative artificial variable is added to each so that again we have no difficulty in obtaining an initial basic feasible solution. In view of this approach, it is obvious that unless the original problem exhibits a substantial number of equality structural constraints, the number of artificial vectors within the starting basis will be relatively small.

Given that our starting point is an infeasible basic solution, it is clear that there are two possible sources of infeasibility, namely: (1) non-negative artificial variables (negative artificial variables will not be admitted to our discussion since, if an equality constraint exhibits a negative right-hand side, it may be multiplied by minus one so that the associated artificial variable enters the set of basic variables at a positive level); and (2) negative legitimate variables. In this regard, the ith unit column vector $\mathbf{e}_i$ of the current basic solution is said to belong to the index set $\mathcal{E}$ if the associated basic variable x_{Bi} is non-negative and artificial while $\mathbf{e}_i$ belongs to index set $\mathcal{F}$ if x_{Bi} is negative and legitimate, *i.e.*,

$$\begin{aligned} \mathcal{E} &= \{i \mid x_{Bi} \geq 0,\ x_{Bi} \text{ artificial}\}, \\ \mathcal{F} &= \{i \mid x_{Bi} < 0,\ x_{Bi} \text{ legitimate}\}. \end{aligned}$$

To take account of all infeasibilities simultaneously for this particular basic solution, let us define the ***total infeasibility*** of the same as

$$I = \sum_{i \in \mathcal{E} \cup \mathcal{F}} |x_{Bi}| = \sum_{i \in \mathcal{E}} x_{Bi} - \sum_{i \in \mathcal{F}} x_{Bi}. \tag{8.5}$$

It is evident that I is always non-negative while I equals zero if and only if the current solution is a basic feasible solution to the original linear programming problem. Moreover, I is non-increasing at each iteration *i.e.*, the criteria by which vectors are chosen to enter and leave the basis guarantee that I decreases at each round of the composite simplex process until an initial basic feasible solution to the original problem with $I = 0$ is found.

To ascertain the effect on I of introducing a non-basic vector $\mathbf{r}_j$ into the basis, let us express (8.5) in terms of the non-basic variables $x_{Rj}, j=1,\ldots,n-m$, as

$$\begin{aligned} I &= \sum_{i\epsilon\mathcal{E}}\left(b_i - \sum_{j=1}^{n-m} y_{ij}x_{Rj}\right) - \sum_{i\epsilon\mathcal{F}}\left(b_i - \sum_{j=1}^{n-m} y_{ij}x_{Rj}\right) \\ &= \sum_{i\epsilon\mathcal{E}} b_i - \sum_{i\epsilon\mathcal{F}} b_i + \sum_{j=1}^{n-m}\left(\sum_{i\epsilon\mathcal{F}} y_{ij} - \sum_{i\epsilon\mathcal{E}} y_{ij}\right) x_{Rj} \qquad (8.5.1) \\ &= \text{constant} + \sum_{j=1}^{n-m} d_j x_{Rj}, \end{aligned}$$

where $d_j = \sum_{i\epsilon\mathcal{F}} y_{ij} - \sum_{i\epsilon\mathcal{E}} y_{ij}$. Since the constant term in (8.5.1) is non-negative, I will decrease if any non-basic variable x_{Rj} for which $d_j < 0$ enters the set of basic variables. (If $d_j \geq 0$ for all j but the constant term in (8.5.1) is positive, the original problem has no feasible solution, *i.e.*, I cannot be made smaller for any nonzero values of the non-basic variables and yet must be positive for any feasible solution. So if the problem has a feasible solution, until it finally is found, it must be the case that $d_j < 0$ for at least one j.) In this regard, d_j represents the reduction in feasibility achieved by changing the status of x_{Rj} from non-basic to basic and will consequently be termed the jth ***infeasibility reduction term.***

To determine which non-basic vector $\mathbf{r}_j$ enters the basis let us adopt the convention that

$$\text{ENTRY CRITERION: } d_k = \min_j\{d_j, d_j < 0\}, \qquad (8.6)$$

i.e., x_{Rk} enters the set of basic variables. Clearly this choice criterion yields the greatest decrease in I per unit increase in any of the non-basic variables. Having specified the selection of the pivot column, let us now look to the determination of the pivot row. Here we seek to select the vector to be deleted from the current basis. From our previous discussion we know that

$$\begin{aligned} x_{\widehat{B}i} &= x_{Bi} - \widehat{\theta} y_{ik}, \; i \neq r, \\ x_{\widehat{B}r} &= \frac{x_{Bi}}{y_{rk}} = \widehat{\theta}. \end{aligned} \tag{8.7}$$

Now, if the total infeasibility I of the problem is to decrease at each round of the composite simplex process, then, in order to preclude the creation of any new infeasibilities, it must be the case that:

(a) no non-negative legitimate or artificial variable changes sign; and

(b) as many negative legitimate variables as possible are to change sign,

where rule (a) has priority over rule (b). To translate these rules into potential pivot-row-selection criteria, let us relate our discussion to the previous two guidelines. The following cases then emerge:

(a.1) Given that not all x_{Bi} may be non-negative, no non-negative variable changes sign so long as the vector to be deleted from the basis is chosen according to

$$\widehat{\theta}_1 = \frac{x_{Br_1}}{y_{r_1k}} = \min_i \left\{ \frac{x_{Bi}}{y_{ik}} \middle| x_{Bi} \geq 0 \text{ and legitimate or artificial, } y_{ik} > 0 \right\}, \tag{8.8}$$

i.e., the r_1^{th} column of the basis matrix $\mathbf{B}$ is replaced by $\mathbf{r}_k$ (here (8.8) is the standard simplex exit criterion). Moreover, if there is a tie for the minimum, choose r_1 so that x_{Br_1} is an artificial variable, if at all possible. Next,

(b.1) To eliminate the infeasibility caused by a negative legitimate basic variable x_{Bi}, let us: (1) remove x_{Bi} from the set of basic variables and, at the same time; (2) increase the values of all other negative legitimate basic variables as much as possible by selecting the vector to be deleted from the current basis according to

$$\widehat{\theta}_2 = \frac{x_{Br_2}}{y_{r_2k}} = \max_i \left\{ \frac{x_{Bi}}{y_{ik}} \middle| x_{Bi} < 0 \text{ and legitimate, } y_{ik} < 0 \right\}. \tag{8.9}$$

Here x_{Br_2} is to be replaced by x_{Rk} in the set of basic variables.

What about the relative effect of the choice of a particular $\widehat{\theta}_i$, i=1,2, value on the total infeasibility I of the problem? A moment's reflection indicates that if $\widehat{\theta}_1 \leq \widehat{\theta}_2$, x_{Br_1} may be deleted from the set of basic variables without creating any new infeasibilities, while if $\widehat{\theta}_1 > \widehat{\theta}_2$, x_{Br_1} cannot turn non-basic without at least one of the remaining non-negative basic variables becoming negative and this increasing the value of the total infeasibility I. In the light of these remarks we shall now consolidate the above two rules into what will be called the composite

EXIT CRITERION: to determine which current basic variable is to turn non-basic, choose x_{Br_2} if $\hat{\theta}_2 < \hat{\theta}_1 < +\infty$; otherwise select x_{Br_1} if $\hat{\theta}_1 \leq \hat{\theta}_2$. (8.10)

Clearly this choice criterion guarantees that the total infeasibility I decreases at each iteration and that, if possible, either a negative legitimate variable will be transformed to zero or a non-negative artificial variable will be removed from the set of basic variables. We note briefly that this criterion fails when $\hat{\theta}_1 = \hat{\theta}_2 = +\infty$, a result which emerges only if all positive artificial variables have $y_{ik} \leq 0$, and all negative legitimate variables have $y_{ik} \geq 0$. In this circumstance $d_k \geq 0$, in which case column k would not have been selected as the pivot column in the first place.

EXAMPLE 8.11. To see how the composite method works, let us

$$
\begin{aligned}
max\, f &= x_1 + x_2 \quad \text{s.t.} \\
x_1 + 2x_2 &\leq 12 \\
x_1 + x_2 &\geq 4 \\
-x_1 + x_2 &\geq 1 \\
-13x_1 + 4x_2 &= -26 \\
x_1, x_2 &\geq 0.
\end{aligned}
$$

Upon multiplying the second, third, and fourth structural constraints by -1 with x_3, x_4 and x_5 depicting non-negative slack variables and x_6 representing a non-negative artificial variable, the above system becomes

$$
\begin{aligned}
x_1 + 2x_2 + x_3 \qquad\qquad\qquad &= 12 \\
-x_1 - x_2 \qquad + x_4 \qquad\qquad &= -4 \\
x_1 - x_2 \qquad\qquad + x_5 \qquad &= -1 \\
13x_1 - 4x_2 \qquad\qquad\qquad + x_6 &= 26.
\end{aligned}
$$

The simplex matrix associated with this problem is then

$$\begin{bmatrix} 1 & 2 & 1 & 0 & 0 & 0 & 0 & 12 \\ -1 & -1 & 0 & 1 & 0 & 0 & 0 & -4 \\ 1 & -1 & 0 & 0 & 1 & 0 & 0 & -1 \\ \boxed{13} & -4 & 0 & 0 & 0 & 1 & 0 & 26 \\ -1 & -1 & 0 & 0 & 0 & 0 & 1 & 0 \end{bmatrix}$$

and our initial basic, though infeasible, solution is $x_{B1} = x_3 = 12$, $x_{B2} = x_4 = -4$, $x_{B3} = x_5 = -1$, $x_{B4} = x_6 = 26$, and $f = 0$ (here x_{Bi} is the basic variable associated with the unit column vector $\mathbf{e}_i$). Since $\mathcal{E} = \{4\}$ and $\mathcal{F} = \{2,3\}$, the total infeasibility associated with the current solution is, from (8.5),

$$I = x_{B4} - x_{B2} - x_{B3} = x_6 - x_4 - x_5 = 31.$$

In addition, from (8.5.1),

$$d_1 = \sum_{i\epsilon\mathcal{F}} y_{i1} - \sum_{i\epsilon\mathcal{E}} y_{i1} = y_{21} + y_{31} - y_{41} = -1+1-13 = -13,$$

$$d_2 = \sum_{i\epsilon\mathcal{F}} y_{i2} - \sum_{i\epsilon\mathcal{E}} y_{i2} = y_{22} + y_{32} - y_{42} = -1-1+4 = 2.$$

Since $d_1 = \min_j\{d_j, d_j < 0\} = -13$, the non-basic variable $x_{R1} = x_1$ enters the set of basic variables. To determine the current basic variable which is to turn non-basic, let us compare

$$\hat{\theta}_1 = \min_i\left\{\frac{x_{Bi}}{y_{i1}}\middle| x_{Bi} \geq 0 \text{ and legitimate or artificial, } y_{i1} > 0\right\}$$

$$= \min_i\left\{\frac{x_{B1}}{y_{11}} = 12, \frac{x_{B4}}{y_{41}} = 2\right\} = 2,$$

$$\hat{\theta}_2 = \max_i\left\{\frac{x_{Bi}}{y_{i1}}\middle| x_{Bi} < 0 \text{ and legitimate, } y_{i1} < 0\right\} = \frac{x_{B2}}{y_{21}} = 4.$$

Looking to our composite exit criterion (8.10), we see that since $\hat{\theta}_1 = 2 < \hat{\theta}_2 = 4$, $x_{B4} = x_6$ will be removed from the set of basic variables. Upon pivoting on the (boxed) element $y_{41} = 13$ we obtain

$$\begin{bmatrix} 0 & \boxed{\frac{30}{13}} & 1 & 0 & 0 & -\frac{1}{13} & 0 & 10 \\ 0 & -\frac{17}{13} & 0 & 1 & 0 & \frac{1}{13} & 0 & -2 \\ 0 & -\frac{9}{13} & 0 & 0 & 1 & -\frac{1}{13} & 0 & -3 \\ 1 & -\frac{4}{13} & 0 & 0 & 0 & \frac{1}{13} & 0 & 2 \\ 0 & -\frac{17}{13} & 0 & 0 & 0 & \frac{1}{13} & 1 & 2 \end{bmatrix}.$$

Hence our second basic (infeasible) solution consists of $x_{B1} = x_3 = 10$, $x_{B2} = x_4 = -2$, $x_{B3} = x_5 = -3$, $x_{B4} = x_1 = 2$, and $f = 2$. Since the artificial variable x_6 has been driven from the set of basic variables, its associated column within the simplex matrix will be deleted in subsequent calculations. Since now $\mathcal{E} = \{\phi\}$, $\mathcal{F} = \{2,3\}$, $I = -x_{B2} - x_{B3} = -x_4 - x_5 = 5$, and $d_2 = \sum_{i \in \mathcal{F}} y_{i2} = y_{22} + y_{32} = -2$, $x_{R1} = x_2$ is the only candidate for entry into the set of basic variables. We next compare

$$\hat{\theta}_1 = \min_i \left\{ \frac{x_{B1}}{y_{12}} = \frac{13}{3} \right\} = \frac{13}{3},$$

$$\hat{\theta}_2 = \max_i \left\{ \frac{x_{B2}}{y_{22}} = \frac{26}{17}, \frac{x_{B3}}{y_{32}} = \frac{13}{3} \right\} = \frac{13}{3}.$$

Since $\hat{\theta}_1 = \hat{\theta}_2 = \frac{13}{3}$, we select $x_{B1} = x_3$ to leave the set of basic variables. Pivoting on the element $y_{12} = \frac{30}{13}$ yields

$$\begin{bmatrix} 0 & 1 & \frac{13}{30} & 0 & 0 & 0 & \frac{13}{3} \\ 0 & 0 & \frac{17}{30} & 1 & 0 & 0 & \frac{11}{3} \\ 0 & 0 & \frac{9}{30} & 0 & 1 & 0 & 0 \\ 1 & 0 & \frac{4}{30} & 0 & 0 & 0 & \frac{10}{3} \\ 0 & 0 & \frac{17}{30} & 0 & 0 & 1 & \frac{23}{3} \end{bmatrix}$$

and thus an optimal (though degenerate) basic feasible solution to the original problem wherein $x_{B1} = x_2 = \frac{13}{3}$, $x_{B2} = x_4 = \frac{11}{3}$, $x_{B4} = x_5 = 0$, $x_{B4} = x_1 = \frac{10}{3}$, and $f = \frac{23}{3}$.

9 COMPUTATIONAL ASPECTS OF LINEAR PROGRAMMING: SELECTED TRANSFORMATIONS

In this chapter our efforts will be directed towards solving a variety of linear programming problems which do not exactly fit the basic structure of the general problem considered in earlier chapters, *i.e.*, a problem of the form

$$max\, f(\mathbf{X}) \;=\; \mathbf{C}'\mathbf{X} \quad \text{s.t.}$$
$$\mathbf{AX} \leqq \mathbf{b},\ \mathbf{X} \geqq \mathbf{0}.$$

For instance, we may want to find the minimum rather than the maximum of a hyperplane subject to $\mathbf{AX} \geqq \mathbf{b}$, $\mathbf{X} \geqq \mathbf{0}$, or quite possibly the variables may be unrestricted in sign rather than nonnegative, or the variables may be bounded from below or above, and so on.

9.1. Minimizing the Objective Function

Let us assume that we desire to

$$min\, f(\mathbf{X}) \;=\; \mathbf{C}'\mathbf{X} \quad \text{s.t.}$$
$$\mathbf{AX} \geqq \mathbf{b},\ \mathbf{X} \geqq \mathbf{0}.$$

As we shall now see, to solve this minimization problem, transform it to a maximization problem, *i.e.*, we shall employ the relationship $min\, f = -max\{-f\}$. To this end we state

THEOREM 9.1. The minimum of $f(\mathbf{X}) = \mathbf{C}'\mathbf{X}$ occurs at the same point $\mathbf{X}_0$ as the maximum of $-f(\mathbf{X}) = -\mathbf{C}'\mathbf{X}$ or

$$min\, f(\mathbf{X}) = -max\{-f(\mathbf{X})\}. \tag{9.1}$$

PROOF. If $\mathbf{X}_0$ maximizes $-f(\mathbf{X}) = -\mathbf{C}'\mathbf{X}$, then $-\mathbf{C}'\mathbf{X}_0 \geq -\mathbf{C}'\mathbf{X}$ for all $\mathbf{X} \epsilon \mathcal{K} = \{\mathbf{X} \mid \mathbf{A}\mathbf{X} \geqq \mathbf{b}, \mathbf{X} \geqq \mathbf{0}\}$. Hence $\mathbf{C}'\mathbf{X}_0 \leq \mathbf{C}'\mathbf{X}$ for all $\mathbf{X} \epsilon \mathcal{K}$ so that $\mathbf{X}_0$ minimizes $f(\mathbf{X})$. Thus the optimal values of f and $-f$ are the negative of each other so that (9.1) holds. Q.E.D.

EXAMPLE 9.1. To apply this transformation we shall

$$\begin{aligned} min\, f = x_1 + x_2 \quad \text{s.t.} \\ 2x_1 + 4x_2 \geq 28 \\ 5x_1 + x_2 \geq 25 \\ x_1, x_2 \geq 0. \end{aligned}$$

From (9.1), let us maximize $g = -f = -x_1 - x_2$ subject to the above structural constraints and non-negativity conditions. For x_3, x_4 surplus variables and x_5, x_6 artificial variables (each non-negative), the initial and optimal simplex matrices are respectively

$$\begin{bmatrix} 2 & 4 & -1 & 0 & 1 & 0 & 0 & 28 \\ 5 & 1 & 0 & -1 & 0 & 1 & 0 & 25 \\ 1 & 1 & 0 & 0 & 0 & 0 & 1 & 0 \\ -7M & -5M & M & M & 0 & 0 & 0 & -53M \end{bmatrix} \longrightarrow$$

$$\begin{bmatrix} 0 & 1 & -\frac{5}{18} & \frac{1}{9} & 0 & 5 \\ 1 & 0 & \frac{1}{18} & -\frac{2}{9} & 0 & 4 \\ 0 & 0 & \frac{2}{9} & \frac{1}{9} & 1 & -9 \end{bmatrix}.$$

Thus the optimal basic feasible solution consists of $x_1 = 4$, $x_2 = 5$, and $max\, g = max\{-f\} = -9$ or $min f = -max\{-f\} = -(-9) = 9$.

9.2. Unrestricted Variables

Another variation of the basic linear programming problem is that (one or more of) the components of $\mathbf{X}$ may be unrestricted in sign, *i.e.*, may be positive, negative, or zero. In this instance we seek to

$$\begin{aligned} max\, f(\mathbf{X}) = \mathbf{C}'\mathbf{X} \quad \text{s.t.} \\ \mathbf{A}\mathbf{X} \leqq \mathbf{b}, \ \mathbf{X} \text{ unrestricted.} \end{aligned}$$

To handle situations of this sort, the variable x_k may be expressed as the difference between two non-negative variables or

$$x_k = x_k^1 - x_k^2, \text{ with } x_k^1, x_k^2 \geq 0. \tag{9.2}$$

So by virtue of this transformation, the above problem may be rewritten, for $\mathbf{X} = \mathbf{X}_1 - \mathbf{X}_2$, as

$$\begin{aligned} max\, f(\mathbf{X}_1, \mathbf{X}_2) &= \mathbf{C}'\mathbf{X}_1 - \mathbf{C}'\mathbf{X}_2 \quad \text{s.t.} \\ &\mathbf{A}\mathbf{X}_1 - \mathbf{A}\mathbf{X}_2 \leqq \mathbf{b} \\ &\mathbf{X}_1, \mathbf{X}_2 \geqq \mathbf{0}, \end{aligned}$$

where $\mathbf{X}_1' = (x_1^1, \ldots, x_p^1)$, $\mathbf{X}_2' = (x_1^2, \ldots, x_p^2)$. Then the standard simplex technique is applicable since all variables are now non-negative. Upon examining the various subcases which emerge, namely:

(a) if $x_k^1 > x_k^2$, then $x_k > 0$;

(b) if $x_k^1 < x_k^2$, then $x_k < 0$; and

(c) if $x_k^1 = x_k^2$, then $x_k = 0$,

we see that x_k, depending upon the relative magnitude of x_k^1 and x_k^2, is truly unrestricted in sign.

One important observation relative to using (9.2) to handle unrestricted variables is that any basic feasible solution in non-negative variables cannot have both x_k^1, x_k^2 in the set of basic variables since, for the kth column of $\mathbf{A}$, $\mathbf{a}_k$, $(x_k^1 - x_k^2)\mathbf{a}_k = x_k^1\mathbf{a}_k + x_k^2(-\mathbf{a}_k)$ so that $\mathbf{a}_k, -\mathbf{a}_k$ are not linearly independent and thus cannot both constitute columns of the basis matrix.

EXAMPLE 9.2. Let us

$$\begin{aligned} max\, f &= 3x_1 + x_2 + x_3 \quad \text{s.t.} \\ 2x_1 + 5x_2 - x_3 &\leq 10 \\ x_1 + 4x_2 - x_3 &\leq 12 \\ x_1, x_3 &\text{ unrestricted, } x_2 > 0. \end{aligned}$$

Setting

$$\begin{aligned} x_1 &= x_1^1 - x_1^2, \\ x_3 &= x_3^1 - x_3^2, \end{aligned}$$

we desire to

$$\begin{aligned} max\, f &= 3x_1^1 - 3x_1^2 + x_2 + x_3^1 - x_3^2 \quad \text{s.t.} \\ 2x_1^1 - 2x_1^2 + 5x_2 - x_3^1 + x_3^2 &\leq 10 \\ x_1^1 - x_1^2 + 4x_2 + x_3^1 - x_3^2 &\leq 12 \\ x_1^1, x_1^2, x_2, x_3^1, x_3^2 &\geq 0. \end{aligned}$$

The standard simplex routine in all non-negative variables now applies, with $x_j^0 = (x_j^1)^0 - (x_j^2)^0, j=1,3.$

9.3. Bounded Variables

Oftentimes it may be the case that in addition to the usual structural constraints and non-negativity conditions, a supplemental set of inequality structural constraints of the form $x_j \geq \ell_j$ or $x_j \leq u_j$, $j=1,\ldots,n$, are introduced into a linear programming problem. Here ℓ_j is a finite lower bound on the variable x_j while u_j is termed a finite upper bound on x_j. (Previously, we held $\ell_j = 0$ and $u_j = +\infty$.) In this regard, we may either

$$\left.\begin{array}{l}\text{[Variables bounded below]} \\ max\, f(\mathbf{X}) = \mathbf{C}'\mathbf{X} \quad \text{s.t.} \\ \mathbf{AX} = \mathbf{b},\ \mathbf{X} \geqq \mathbf{L} \geqq \mathbf{0}\end{array}\right\} \text{ or } \left\{\begin{array}{l}\text{[Variables bounded above]} \\ max\, f(\mathbf{X}) = \mathbf{C}'\mathbf{X} \quad \text{s.t.} \\ \mathbf{AX} = \mathbf{b},\ \mathbf{0} \leqq \mathbf{X} \leqq \mathbf{U},\end{array}\right.$$

where both

$$\mathbf{L} = \begin{bmatrix} \ell_1 \\ \ell_2 \\ \vdots \\ \ell_n \end{bmatrix}, \quad \mathbf{U} = \begin{bmatrix} u_1 \\ u_2 \\ \vdots \\ u_n \end{bmatrix}$$

are of order $(n \times 1)$ (and a set of m non-negative slack variables have already been incorporated into the original structural constraint system). Here both $\mathbf{AX} = \mathbf{b}$, $\mathbf{X} \geqq \mathbf{L} \geqq \mathbf{0}$ and $\mathbf{AX} = \mathbf{b}$, $\mathbf{0} \leqq \mathbf{X} \leqq \mathbf{U}$ will be termed **capacitated equation systems.** Moreover, it is evident from the presence of these lower or upper bounds that now a *feasible solution* to the linear programming problem is one for which each variable x_j assumes a non-negative value not less than its lower bound ℓ_j or not in excess of its upper bound $u_j, j=1,\ldots,n$.

Let us assume for the moment that we desire to

$$max\, f(\mathbf{X}) \;=\; \mathbf{C}'\mathbf{X} \quad \text{s.t.}$$
$$\mathbf{AX} = \mathbf{b},\ \mathbf{0} \leqq \mathbf{X} \leqq \mathbf{U}.$$

If this problem is handled in the usual fashion, we would, according to the standard simplex technique, introduce n additional non-negative slack variables into the problem — one for each upper bound constraint $x_j \leq u_j$, j=1,...,n. (Here each upper bound constraint is treated as a separate structural constraint.) The resulting problem would then have m+n structural constraints in $2n$ variables. In this regard, we would need a basis matrix of order m+n. And if we are to

$$max\, f(\mathbf{X}) \;=\; \mathbf{C}'\mathbf{X} \quad \text{s.t.}$$
$$\mathbf{AX} = \mathbf{b},\ \ \mathbf{X} \geqq \mathbf{L} \geqq \mathbf{0},$$

we would need a non-negative surplus variable and a non-negative artificial variable for each lower bound constraint $x_j \geq \ell_j$, j=1,...,n. In this instance the final problem would involve m+n structural constraints in $3n$ variables, again with a basis matrix of order m+n. Fortunately, since these lower and upper bound constraints possess a very simplified structure, we may exploit this special structure to our advantage by modifying the standard simplex method so that only the coefficients on the first m structural constraints $\mathbf{AX} = \mathbf{b}$ appear explicitly in the simplex matrix while the lower or upper bound constraints are handled indirectly. This being the case, we need only work with a basis matrix of order m instead of m+n.

Let us first consider the case of a finite set of lower bounds on the variables x_j, j=1,...,n. That is, let us

$$max\, f(\mathbf{X}) \;=\; \mathbf{C}'\mathbf{X} \quad \text{s.t.}$$
$$\mathbf{AX} = \mathbf{b},\ \mathbf{X} \geqq \mathbf{L} \geqq \mathbf{0}.$$

To solve this problem, we shall utilize a transformation which converts the original variables x_j with finite lower bounds ℓ_j, j=1,...,n, to a new set of variables with lower bounds of zero. In particular, we may measure each variable x_j from its lower bound ℓ_j, j=1,...,n, *i.e.*, let us transform $x_j \geq \ell_j$ to $\bar{x}_j = x_j - \ell_j \geq 0$, j=1,...,n. In matrix terms, $\bar{\mathbf{X}} = \mathbf{X} - \mathbf{L}$ or $\mathbf{X} = \mathbf{L} + \bar{\mathbf{X}}$. Then the original problem becomes

$$\max f(\mathbf{X}) = \mathbf{C}'(\mathbf{L}+\bar{\mathbf{X}}) \quad \text{s.t.}$$
$$\mathbf{A}(\mathbf{L}+\bar{\mathbf{X}}) = \mathbf{b},\ \bar{\mathbf{X}} \geqq \mathbf{0},$$

or

$$\max f(\bar{\mathbf{X}}) = \mathbf{C}'\bar{\mathbf{X}} \quad \text{s.t.}$$
$$\mathbf{A}\bar{\mathbf{X}} = \bar{\mathbf{b}},\ \bar{\mathbf{X}} \geqq \mathbf{0}, \tag{9.3}$$

where $f(\bar{\mathbf{X}}) = f(\mathbf{X}) - \mathbf{C}'\mathbf{L}$ and $\bar{\mathbf{b}} = \mathbf{b} - \mathbf{AL}$. Here $\mathbf{AL}$ represents the portion of $\mathbf{b}$ which is necessary to support the minimum allowable level of $\mathbf{X}$, namely $\mathbf{L}$. The final solution is then obtained as $\mathbf{X}_0 = \mathbf{L} + \bar{\mathbf{X}}_0$, $f(\mathbf{X}_0) = \mathbf{C}'\mathbf{L} + \mathbf{C}'\bar{\mathbf{X}}_0$.

EXAMPLE 9.3. To employ this technique, let us

$$\begin{aligned}
\max f &= x_1 + x_2 + x_3 \quad \text{s.t.}\\
2x_1 + x_2 + x_4 &= 10\\
2x_1 + 3x_2 + x_3 + x_5 &= 16\\
x_1 &\geq 2 > 0\\
x_2 &\geq 1 > 0\\
x_3 &\geq 0.5 > 0\\
x_4 &\geq 1.5 > 0\\
x_5 &\geq 2 > 0,
\end{aligned}$$

where x_4, x_5 are slack variables. By virtue of (9.3), we desire to

$$\begin{aligned}
\max f &= \bar{x}_1 + \bar{x}_2 + \bar{x}_3 \quad \text{s.t.}\\
2\bar{x}_1 + \bar{x}_2 + \bar{x}_4 &= 3.5\\
2\bar{x}_1 + 3\bar{x}_2 + \bar{x}_3 + \bar{x}_5 &= 6.5\\
\bar{x}_j \geq 0,\ j&=1,\ldots,5.
\end{aligned}$$

Setting up the simplex matrix we have

$$\begin{bmatrix} 2 & 1 & 0 & 1 & 0 & 0 & 3.5\\ 2 & 3 & 1 & 0 & 1 & 0 & 6.5\\ -1 & -1 & -1 & 0 & 0 & 1 & 0 \end{bmatrix} \tag{9.4}$$

and an initial basic feasible solution consisting of $\bar{x}_4 = 3.5$, $\bar{x}_5 = 6.5$, and f=0. If we introduce $\bar{x}_3$ into the set of basic variables, (9.4) becomes

$$\begin{bmatrix} 2 & 1 & 0 & 1 & 0 & 0 & 3.5 \\ 2 & 3 & 1 & 0 & 1 & 0 & 6.5 \\ 1 & 2 & 0 & 0 & 1 & 1 & 6.5 \end{bmatrix} \tag{9.5}$$

and thus the optimal basic feasible solution to the modified or transformed problem is $\bar{x}_3 = 6.5$, $\bar{x}_4 = 3.5$ and $f(\bar{\mathbf{X}}) = 6.5$. Moreover, the optimal basic feasible solution to the original problem is

$$\mathbf{X}_0 = \mathbf{L} + \bar{\mathbf{X}}_0 = \begin{bmatrix} 2 \\ 1 \\ 0.5 \\ 1.5 \\ 2 \end{bmatrix} + \begin{bmatrix} 0 \\ 0 \\ 6.5 \\ 3.5 \\ 0 \end{bmatrix} = \begin{bmatrix} 2 \\ 1 \\ 7 \\ 5 \\ 2 \end{bmatrix},$$

$$f(\mathbf{X}_0) = \mathbf{C}'\mathbf{L} + \mathbf{C}'\bar{\mathbf{X}}_0 = 10.$$

We next examine the case where each variable x_j possesses a finite upper bound u_j, $j=1,\ldots,n$. That is, let us

$$max\, f(\mathbf{X}) = \mathbf{C}'\mathbf{X} \quad \text{s.t.}$$
$$\mathbf{AX} = \mathbf{b},\ \mathbf{0} \leqq \mathbf{X} \leqq \mathbf{U}.$$

As we shall now see, this situation is a bit more complicated to handle computationally than the previous one which admitted lower bound restrictions on the variables. To develop the (upper) bounded variable algorithm, let us start with the simplex matrix

$$\begin{array}{ccccccccccc} x_1 & x_2 & & x_p & x_{p+1} & x_{p+2} & & x_{p+m} & f & \\ \end{array}$$
$$\begin{bmatrix} a_{11} & a_{12} & \cdots & a_{1p} & 1 & 0 & \cdots & 0 & 0 & b_1 \\ a_{21} & a_{22} & \cdots & a_{2p} & 0 & 1 & \cdots & 0 & 0 & b_2 \\ \cdots & \cdots & \cdots & \cdots & \cdots & \cdots & \cdots & \cdots & \cdots & \cdots \\ a_{m1} & a_{m2} & \cdots & a_{mp} & 0 & 0 & \cdots & 1 & 0 & b_m \\ -c_1 & -c_2 & \cdots & -c_p & 0 & 0 & \cdots & 0 & 1 & 0 \end{bmatrix} \tag{9.6}$$

from which we obtain

$$
\begin{aligned}
x_{p+1} &= b_1 - a_{11}x_1 - a_{12}x_2 - \cdots - a_{1p}x_p \\
x_{p+2} &= b_2 - a_{21}x_1 - a_{22}x_2 - \cdots - a_{2p}x_p \\
&\cdots\cdots\cdots\cdots\cdots\cdots\cdots\cdots \\
x_{p+m} &= b_m - a_{m1}x_1 - a_{m2}x_2 - \cdots - a_{mp}x_p \\
f &= \qquad c_1x_1 + c_2x_2 + \cdots + c_px_p
\end{aligned}
\tag{9.7}
$$

and thus an initial basic feasible solution consisting of $x_{p+i} = b_i$, $i{=}1,\ldots,m$; $f = 0$. Since the variables $x_1,\ldots,x_{p+m}$ have upper bounds $u_1,\ldots,u_{p+m}$ respectively, it is assumed that none of the basic variables exceed their upper bounds. Hence the current solution is assumed feasible for both the non-negativity conditions and the upper bound constraints.

Assume that x_1 is selected by the simplex entry criterion to enter the set of basic variables. Hence (9.7) becomes

$$
\begin{aligned}
x_{p+1} &= b_1 - a_{11}x_1 \\
x_{p+2} &= b_2 - a_{21}x_1 \\
&\cdots\cdots\cdots\cdots \\
x_{p+m} &= b_m - a_{m1}x_1 \\
f &= \qquad c_1x_1 \, .
\end{aligned}
\tag{9.7.1}
$$

We noted earlier that, according to the standard simplex exit criterion, the largest allowable increment in x_1 which does not violate the non-negativity conditions and the upper bound restriction is $\min_i\{b_i/a_{i1}, a_{i1} > 0\}$. However, if we explicitly incorporate such a restriction on x_1 into our discussion, then the largest allowable increment in x_1 which does not violate the non-negativity conditions and the upper bound restriction is

$$
\min\Big\{\min_i\{b_i/a_{i1}, a_{i1} > 0\}, u_1\Big\}. \tag{9.8}
$$

It is evident that if the minimum in (9.8) is not attained for u_1, then the standard simplex pivot process is warranted, with any of the elements $a_{i1} > 0$, $i{=}1,\ldots,m$, eligible for selection as the pivot. If the minimum in (9.8) is u_1, then

x_1 must equal u_1 after an iteration is performed. To this end let us transform the upper bound constraint $x_1 \leq u_1$ to

$$x_1 + x_1^* = u_1, \tag{9.9}$$

where x_1^* is a non-negative slack variable. (Note that $x_1 = u_1$ for $x_1^* = 0$, while $x_1^* = u_1$ when $x_1 = 0$.) Upon substituting $x_1 = u_1 - x_1^*$ into (9.7) we obtain

$$\begin{aligned} x_{p+1} &= (b_1 - a_{11}u_1) + a_{11}x_1^* - a_{12}x_2 - \cdots - a_{1p}x_p \\ x_{p+2} &= (b_2 - a_{21}u_1) + a_{21}x_1^* - a_{22}x_2 - \cdots - a_{2p}x_p \\ &\cdots\cdots\cdots\cdots\cdots\cdots\cdots\cdots \\ x_{p+m} &= (b_m - a_{m1}u_1) + a_{m1}x_1^* - a_{m2}x_2 - \cdots - a_{mp}x_p \\ f &= c_1u_1 - c_1x_1^* + c_2x_2 + \cdots + c_px_p \,. \end{aligned} \tag{9.10}$$

And since x_1^* is non-basic, $x_1^* = 0$ so that $x_1 = u_1$. (Here all of the values of the basic variables $b_i - a_{i1}u_1$ are non-negative under this substitution since each of these differences may be expressed as $a_{i1}\left(\frac{b_i}{a_{i1}} - u_i\right) \geq 0$ for $a_{i1} > 0$ and $\frac{b_i}{a_{i1}} > u_1$ by virtue of (9.8). Actually, for $a_{i1} < 0$, these values are still non-negative.) In terms of the simplex matrix, the preceding transformation yields

$$\begin{array}{cccccccccc} x_1^* & x_2 & & x_p & x_{p+1} & x_{p+2} & & x_{p+m} & f & \\ \end{array}$$
$$\left[\begin{array}{cccccccccc} -a_{11} & a_{12} & \cdots & a_{1p} & 1 & 0 & \cdots & 0 & 0 & (b_1 - a_{11}u_1) \\ -a_{21} & a_{22} & \cdots & a_{2p} & 0 & 1 & \cdots & 0 & 0 & (b_2 - a_{21}u_1) \\ \cdots & \cdots & \cdots & \cdots & \cdots & \cdots & \cdots & \cdots & \cdots & \cdots \\ -a_{m1} & a_{m2} & \cdots & a_{mp} & 0 & 0 & \cdots & 1 & 0 & (b_m - a_{m1}u_1) \\ c_1 & -c_2 & \cdots & -c_p & 0 & 0 & \cdots & 0 & 1 & c_1u_1 \end{array}\right]. \tag{9.11}$$

So to obtain (9.11), in (9.6):

(1) update the first m components of the last column to $b_i - a_{i1}u_1$ and the objective function value to c_1u_1;

(2) multiply the first column by -1; and

(3) change the heading of the first column from x_1 to x_1^*.

It is conceivable that as the simplex iterations progress x_1^* may be selected for entry into the set of basic variables at the level of its upper bound. In this instance, the preceding transformation (9.9) with $x_1^* = u_1 - x_1$ is again utilized, with x_1 now assuming non-basic status. That is, steps (1) - (3) above are carried out to obtain the new simplex matrix. In view of these remarks, it is evident that, in general, either x_i or x_i^* may appear in the final or optimal simplex matrix. If x_i appears, then its optimal value x_i^0 is located in the usual fashion in the last column of the optimal simplex matrix. And if x_i^* appears, then $(x_i^*)^0 = u_i - x_i^0$ so that $x_i^0 = u_i - (x_i^*)^0$.

One additional contingency remains. It may be the case that, when some variable changes its status from non-basic to basic and thus increases from zero to some positive level, one or more of the other basic variables may exceed their upper bounds. How may we insure that this situation will never occur? For instance, in (9.7) with $x_2, \ldots, x_p$ all non-basic we again have

$$\begin{aligned} x_{p+1} &= b_1 - a_{11}x_1 \\ x_{p+2} &= b_2 - a_{21}x_1 \\ &\ldots\ldots\ldots\ldots \\ x_{p+m} &= b_m - a_{m1}x_1 \\ f &= c_1x_1, \end{aligned}$$

where the non-basic variable x_1 is to enter the set of basic variables. Note first that the values of the basic variables x_{p+i} increase for negative values of a_{i1}, $i=1,\ldots,m$. Moreover, the upper bounds u_{p+i} can be attained if $b_i - a_{i1}x_1 = u_{p+i}$, $i=1,\ldots,m$. Thus x_1 cannot be increased above the smallest $(b_i - u_{p+i})/a_{i1}$ value, $a_{i1} < 0$. In this regard, the largest allowable increment in the entering variable x_1 which preserves the upper bound restrictions on the basic variables x_{p+i}, $i=1,\ldots,m$, is

$$\min_i\left\{(b_i - u_{p+i})/a_{i1}, a_{i1} < 0\right\}. \tag{9.12}$$

If, for instance, the minimum in (9.12) occurs for $i=1$, *i.e.*, x_{p+1} reaches its upper bound u_{p+1} after an iteration is performed, then x_{p+1} is set at its upper bound by substituting $x_{p+1} = u_{p+1} - x^*_{p+1}$ into (9.6) and then pivoting on the element a_{11}. Here (9.9) has again been applied, where x^*_{p+1} is a non-negative slack variable. Relative to (9.12), (9.6) is transformed to

$$\begin{array}{cccccccccc} x_1 & x_2 & & x_p & x^*_{p+1} & x_{p+2} & & x_{p+m} & f & \\ \end{array}$$
$$\left[\begin{array}{cccccccccc} 1 & \frac{a_{12}}{a_{11}} & \cdots & \frac{a_{1p}}{a_{11}} & \frac{1}{a_{11}} & 0 & \cdots & 0 & 0 & \frac{b_1-u_{p+1}}{a_{11}} \\ 0 & a_{22}-a_{21}\frac{a_{12}}{a_{11}} & \cdots & a_{2p}-a_{21}\frac{a_{1p}}{a_{11}} & -\frac{a_{21}}{a_{11}} & 1 & \cdots & 0 & 0 & b_2-a_{21}\left(\frac{b_1-u_{p+1}}{a_{11}}\right) \\ \vdots & \vdots & & \vdots & \vdots & \vdots & & \vdots & \vdots & \vdots \\ 0 & a_{m2}-a_{m1}\frac{a_{12}}{a_{11}} & \cdots & a_{mp}-a_{m1}\frac{a_{1p}}{a_{11}} & -\frac{a_{m1}}{a_{11}} & 0 & \cdots & 1 & 0 & b_m-a_{m1}\left(\frac{b_1-u_{p+1}}{a_{11}}\right) \\ 0 & -c_2+c_1\frac{a_{12}}{a_{11}} & \cdots & -c_p+c_1\frac{a_{1p}}{a_{11}} & \frac{c_1}{a_{11}} & 0 & \cdots & 0 & 1 & c_1\left(\frac{b_1-u_{p+1}}{a_{11}}\right) \end{array}\right]. \tag{9.13}$$

And since x^*_{p+1} is now nonbasic, $x^*_{p+1} = 0$ so that $x_{p+1} = u_{p+1}$. So to obtain (9.13), in (9.6):

(1) update the first component of the last column to $b_1 - u_{p+1}$;

(2) multiply the $(p+1)^{st}$ column by -1;

(3) pivot on the element a_{11}; and

(4) change the heading on the $(p+1)^{st}$ column from x_{p+1} to x^*_{p+1}.

Upon combining all of the aforementioned exit criteria, we see that the largest feasible increase in x_1 which preserves both the non-negativity conditions and all of the upper bound constraints is

$$min\Big\{\underset{i}{min}\{b_i/a_{i1}, a_{i1} > 0\},\ \underset{i}{min}\{(b_i - u_{p+i})/a_{i1}, a_{i1} < 0\}, u_1\Big\}. \tag{9.14}$$

To summarize, the complete bounded variable exit criterion may be expressed as follows: given that the simplex entry criterion selects column k as the pivot column, the pivot row is determined according to

$$\theta = min\{\theta_1, \theta_2, \theta_3\},$$

where

$$\theta_1 = \min_i \{b_i/a_{ik}, a_{ik} > 0\},$$

$\theta_2 = \min_i \{(b_i - u_i)/a_{ik}, a_{ik} < 0\}$, where u_i is the upper bound of the basic variable associated with row i; and

$$\theta_3 = u_k.$$

If row r is chosen as the pivot row, then:

(1) if $\theta = \theta_1$, the standard simplex exit criterion emerges with the element a_{rk} serving as the pivot.

(2) for $\theta = \theta_2$:

(a) if x_t is the outgoing basic variable, update the rth com-ponent of the last column to $b_r - u_t$;

(b) multiply the outgoing basis vector by -1;

(c) pivot on the element a_{rk};

(d) change the heading of the column associated with the out-going basis vector from x_t to x_t^*.

(3) if $\theta = \theta_3$:

(a) update the first m components of the last column to $b_i - a_{ik}u_k$ and add $c_k u_k$ to the current objective function value.

(b) multiply the kth column by -1; and

(c) change the heading of the kth column from x_k to x_k^*.

EXAMPLE 9.4. To implement this routine we shall

$$\begin{aligned} \max f &= x_1 + \tfrac{3}{2}x_2 + 2x_3 \quad \text{s.t.} \\ x_1 - 2x_2 + 2x_3 + x_4 &= 5 \\ -\tfrac{1}{3}x_1 + x_2 - x_3 + x_5 &= 3 \end{aligned}$$

$$\begin{gathered} 0 \le x_1 \le 1 \\ 0 \le x_2 \le 2 \\ 0 \le x_3 \le 3 \\ 0 \le x_4 \le 4 \\ 0 \le x_5 \le 5, \end{gathered}$$

where x_4, x_5 are slack variables. The associated simplex matrix thus appears as

$$\begin{bmatrix} 1 & -2 & 2 & 1 & 0 & 0 & 5 \\ -\frac{1}{3} & 1 & -1 & 0 & 1 & 0 & 3 \\ -1 & -\frac{3}{2} & -2 & 0 & 0 & 1 & 0 \end{bmatrix} \tag{9.15}$$

with an initial basic but infeasible (in terms of the upper bound restrictions) solution consisting of $x_4 = 5$, $x_5 = 3$, and $f = 0$.

Let x_3 enter the set of basic variables. With $\theta_1 = \frac{5}{2}$, $\theta_2 = (3-5)/(-1) = 2$, and $\theta_3 = 2$, let us select $\theta = min\{\theta_1, \theta_2, \theta_3\} = \theta_3 = 2$ (since this choice of θ involves less computational work than if $\theta = \theta_2$ were selected). Then (9.15) becomes

$$\begin{bmatrix} 1 & \boxed{-2} & \overset{x_3^*}{-2} & 1 & 0 & 0 & \boxed{1} \\ -\frac{1}{3} & 1 & 1 & 0 & 1 & 0 & \boxed{5} \\ -1 & -\frac{3}{2} & 2 & 0 & 0 & 1 & \boxed{4} \end{bmatrix} \begin{array}{l} \longleftarrow 5-2(2) \\ \longleftarrow 3-(-1)2 \\ \longleftarrow 2(2) \end{array} \tag{9.16}$$

This time a basic feasible, although non-optimal, solution emerges where-in $x_4 = 1$, $x_5 = 5$, and $f = 4$.

Next, given that $\theta_1 = 5$, $\theta_2 = (1-4)/(-2) = \frac{3}{2}$, and $\theta_3 = 3$, it follows that $\theta = \theta_2 = \frac{3}{2}$. Hence the (boxed) pivot element in (9.16) is -2. In this regard, (9.16) transforms to

$$\begin{bmatrix} 1 & \boxed{-2} & \overset{x_3^*}{-2} & -1 & 0 & 0 & \boxed{-3} \\ -\frac{1}{3} & 1 & 1 & 0 & 1 & 0 & 5 \\ -1 & -\frac{3}{2} & 2 & 0 & 0 & 1 & 4 \end{bmatrix} \begin{array}{l} \longleftarrow 1-4 \\ \longrightarrow \\ \end{array} \begin{bmatrix} -\frac{1}{2} & 1 & \overset{x_3^*}{1} & \overset{x_4^*}{\frac{1}{2}} & 0 & 0 & \frac{3}{2} \\ \frac{1}{6} & 0 & 0 & -\frac{1}{2} & 1 & 0 & \frac{7}{2} \\ -\frac{7}{4} & 0 & \frac{7}{2} & \frac{3}{4} & 0 & 1 & \frac{25}{4} \end{bmatrix}. \tag{9.17}$$

Here our third basic feasible solution is $x_2 = \frac{3}{2}$, $x_5 = \frac{7}{2}$, and $f = \frac{25}{4}$.

Finally, for $\theta_1 = 21$, $\theta_2 = (\frac{3}{2}-3)/(-\frac{1}{2}) = 3$, and $\theta_3 = 1$, we have $\theta = \theta_3 = 1$. Then (9.17) becomes

$$\begin{array}{ccccccc} x_1^* & & x_3^* & x_4^* & & & \\ \end{array}$$

$$\left[\begin{array}{ccccccc} \frac{1}{2} & 1 & 1 & \frac{1}{2} & 0 & 0 & \boxed{2} \\ -\frac{1}{6} & 0 & 0 & -\frac{1}{2} & 1 & 0 & \boxed{\frac{10}{3}} \\ \frac{7}{4} & 0 & \frac{7}{2} & \frac{3}{4} & 0 & 1 & \boxed{\frac{29}{4}} \end{array}\right] \begin{array}{l} \leftarrow \frac{3}{2}-(-\frac{1}{2})(1) \\ \leftarrow \frac{7}{2}-(\frac{1}{6})(1) \\ \leftarrow \frac{25}{4}+1 \end{array}$$

with the optimal basic feasible solution involving $x_2 = 2$, $x_5 = \frac{10}{3}$, and $f = \frac{29}{4}$. Moreover, since $x_i = u_i - x_i^*$ and $x_1^* = x_3^* = x_4^* = 0$, it follows that $x_1 = 1$, $x_3 = 2$, and $x_4 = 4$.

Finally, it may be the case that the variables x_j are bounded both from below and above, *i.e.*, x_j is constrained to lie within a range of values or $0 \leq \ell_j \leq x_j \leq u_j$, $j=1,\ldots,n$. Specifically, the problem at hand is to

$$\begin{aligned} \max f(\mathbf{X}) &= \mathbf{C}'\mathbf{X} \quad \text{s.t.} \\ \mathbf{AX} = \mathbf{b},\ &\mathbf{0} \leqq \mathbf{L} \leqq \mathbf{X} \leqq \mathbf{U}. \end{aligned}$$

To handle this situation, let us measure x_j from its lower bound. Hence $\ell_j \leq x_j \leq u_j$ is transformed to $0 \leq \bar{x}_j = x_j - \ell_j \leq u_j - \ell_j = \bar{u}_j$, $j=1,\ldots,n$. In this instance we have converted the original finite interval constraint on x_j to one involving only an upper bound restriction on the transformed variables $\bar{x}_j$, $j=1,\ldots,n$. In matrix notation, the preceding problem becomes, for $\bar{\mathbf{X}} = \mathbf{X} - \mathbf{L}$ or $\mathbf{X} = \mathbf{L} + \bar{\mathbf{X}}$,

$$\begin{aligned} \max f(\bar{\mathbf{X}}) &= \mathbf{C}'\bar{\mathbf{X}} \quad \text{s.t.} \\ \mathbf{A}\bar{\mathbf{X}} = \bar{\mathbf{b}},\ &\mathbf{0} \leqq \bar{\mathbf{X}} \leqq \bar{\mathbf{U}}, \end{aligned}$$

where $\bar{\mathbf{U}} = \mathbf{U} - \mathbf{L}$, $\bar{\mathbf{b}} = \mathbf{b} - \mathbf{AL}$, $\mathbf{X}_0 = \mathbf{L} + \bar{\mathbf{X}}_0$, and $f(\mathbf{X}_0) = \mathbf{C}'\mathbf{L} + \mathbf{C}'\bar{\mathbf{X}}_0$. To generate a solution to this latter problem, the upper bound algorithm is then utilized.

EXAMPLE 9.5. To illustrate this transformation, let us

$$\begin{aligned}
max\ f &= x_1 + 3x_2 + x_3 \quad \text{s.t.} \\
x_1 - x_2 + x_3 + x_4 & = 10 \\
2x_1 + x_2 - 4x_3 + x_5 & = 8 \\
x_1 + 3x_2 - 2x_3 + x_6 & = 12
\end{aligned}$$

$$\begin{gathered}
1 \leq x_1 \leq 3 \\
3 \leq x_2 \leq 4 \\
2 \leq x_3 \leq 5 \\
0.5 \leq x_4 \leq 2.5 \\
1 \leq x_5 \leq 3 \\
4 \leq x_6 \leq 6,
\end{gathered}$$

where x_4, x_5, and x_6 are non-negative slack variables. By virtue of the preceding discussion, we seek to

$$\begin{aligned}
max\ f &= \bar{x}_1 + 3\bar{x}_2 + \bar{x}_3 \quad \text{s.t.} \\
\bar{x}_1 - \bar{x}_2 + \bar{x}_3 + \bar{x}_4 & = 9.5 \\
2\bar{x}_1 + \bar{x}_2 - 4\bar{x}_3 + \bar{x}_5 & = 10 \\
\bar{x}_1 + 3\bar{x}_2 - 2\bar{x}_3 + \bar{x}_6 & = 2
\end{aligned}$$

$$\begin{gathered}
0 \leq \bar{x}_1 \leq 2 \\
0 \leq \bar{x}_2 \leq 1 \\
0 \leq \bar{x}_3 \leq 3 \\
0 \leq \bar{x}_4 \leq 2 \\
0 \leq \bar{x}_5 \leq 2 \\
0 \leq \bar{x}_6 \leq 2.
\end{gathered}$$

The upper bound technique may now be used to generate a solution to this transformed problem.

9.4. Interval Linear Programming

An interval linear programming problem is one for which we seek to

$$\begin{aligned}
max\ f(\mathbf{X}) &= \mathbf{C}'\mathbf{X} \quad \text{s.t.} \\
&\mathbf{b}_1 \leqq \mathbf{AX} \leqq \mathbf{b}_2,\ \mathbf{b}_1 \leqq \mathbf{b}_2 \\
&\mathbf{X} \geqq \mathbf{0},
\end{aligned}$$

where $\mathbf{b}_1' = (b_1^1, \ldots, b_m^1)$, $\mathbf{b}_2' = (b_1^2, \ldots, b_m^2)$. To generate a solution to the same, let us write the above interval structural constraint system as

$$\left.\begin{array}{l}\mathbf{AX} \leqq \mathbf{b}_2 \\ \mathbf{AX} \geqq \mathbf{b}_1\end{array}\right\} \text{ or } \left\{\begin{array}{lll}\mathbf{AX} + \mathbf{X}_{s1} & & = \mathbf{b}_2 \\ \mathbf{AX} & - \mathbf{X}_{s2} + \mathbf{X}_a & = \mathbf{b}_1,\end{array}\right.$$

where $\mathbf{X}_{s1}$, $\mathbf{X}_{s2}$, and $\mathbf{X}_a$ are $(m \times 1)$ vectors of non-negative slack, surplus, and artificial variables respectively. Then either the M-penalty or two-phase method may be used to solve the final problem.

Interestingly enough, if $\mathbf{A}$ is a pth order non-singular matrix, then we may obtain an explicit solution to the interval linear program-ming problem in unrestricted variables according to

THEOREM 9.2.[1] The optimal solution to the problem

$$\begin{array}{ll} max\ f(\mathbf{X}) = \mathbf{C}'\mathbf{X} & \text{s.t.} \\ \quad \mathbf{b}_1 \leqq \mathbf{AX} \leqq \mathbf{b}_2,\ \mathbf{b}_1 \leqq \mathbf{b}_2 & \\ \quad \mathbf{X} \text{ unrestricted} & \end{array} \tag{9.18}$$

with $\mathbf{A}$ of order $(p \times p)$ and non-singular, is given by

$$\mathbf{X}_0 = \mathbf{A}^{-1}\bar{\mathbf{X}}_0, \tag{9.19}$$

where the jth component of $\bar{\mathbf{X}}_0$ is

$$\bar{x}_j^0 = \left\{\begin{array}{c} b_j^2 \\ \theta_j b_j^1 + (1-\theta_j)b_j^2,\ 0 \leq \theta_j \leq 1, \\ b_j^1 \end{array}\right\} \text{if } \mathbf{C}'\mathbf{a}_j \left\{\begin{array}{c} > 0 \\ = 0 \\ < 0 \end{array}\right\},\ j=1,\ldots,p, \tag{9.20}$$

and $\mathbf{a}_j$ is the jth column of $\mathbf{A}^{-1}$.

PROOF. Substituting $\bar{\mathbf{X}} = \mathbf{AX}$ into (9.18) yields the equivalent problem

$$max\ f(\bar{\mathbf{X}}) = \mathbf{C}'\mathbf{A}^{-1}\bar{\mathbf{X}} = \sum_{j=1}^{p} (\mathbf{C}'\mathbf{a}_j)\bar{x}_j$$

$$\text{s.t.} \left\{\begin{array}{l} \mathbf{b}_1 \leqq \bar{\mathbf{X}} \leqq \mathbf{b}_2,\ \mathbf{b}_1 \leqq \mathbf{b}_2,\ \bar{\mathbf{X}} \text{ unrestricted, or} \\ b_j^1 \leq \bar{x}_j \leq b_j^2,\ b_j^1 \leq b_j^2,\ \bar{x}_j \text{ unrestricted},\ j=1,\ldots,p, \end{array}\right.$$

[1] Ben-Israel, *et al.* (1969). A more general formulation of this theorem which allows $\mathbf{A}$ to be of order $(m \times p)$ and thus employs the generalized inverse appears in Ben-Israel, Charnes (1968).

whose optimal solution is provided by (9.20). This latter expression may then be used to determine (9.19). Q.E.D.

EXAMPLE 9.6. Let us

$$\begin{aligned} max\ f &= x_1 + 8x_2 \quad \text{s.t.} \\ 3 \leq\ & x_1 + x_2 \leq 12 \\ 4 \leq\ & 2x_1 + 4x_2 \leq 20 \\ & x_1, x_2 \text{ unrestricted.} \end{aligned}$$

Since

$$\mathbf{A} = \begin{bmatrix} 1 & 1 \\ 2 & 4 \end{bmatrix},\ \mathbf{A}^{-1} = \begin{bmatrix} 2 & -\frac{1}{2} \\ -1 & \frac{1}{2} \end{bmatrix},\ \text{and } \mathbf{C}'\mathbf{A}^{-1} = \left(-6, \tfrac{7}{2}\right),$$

this problem is equivalent to

$$\begin{aligned} max\ f &= -6\bar{x}_1 + \tfrac{7}{2}\bar{x}_2 \quad \text{s.t.} \\ & 3 \leq \bar{x}_1 \leq 12,\ 4 \leq \bar{x}_2 \leq 20 \\ & \bar{x}_1,\ \bar{x}_2 \text{ unrestricted.} \end{aligned}$$

With $\mathbf{C}'\mathbf{a}_1 = -6$ and $\mathbf{C}'\mathbf{a}_2 = \frac{7}{2}$, (9.20) yields $\bar{x}_1^0 = 3$, $\bar{x}_2^0 = 20$. Then from (9.19), $x_1^0 = -4$, $x_2^0 = 7$.

9.5. Absolute Value Functionals

The linear programming problem to which we shall now address ourselves is of the form

$$\begin{aligned} max\ f(\mathbf{X}) &= \sum_{j=1}^{n} |x_j| \quad \text{s.t.} \\ \mathbf{AX} &= \mathbf{b},\ \mathbf{X} \text{ unrestricted.} \end{aligned}$$

Since $|x_j|$ is essentially non-linear, the objective function may be linearized by utilizing the transformation $x_j = x_j^1 - x_j^2$, with x_j^1, $x_j^2 \geq 0$. Then for $\mathbf{X} = \mathbf{X}_1 - \mathbf{X}_2$, where again $\mathbf{X}_1' = (x_1^1, \ldots, x_n^1)$ and $\mathbf{X}_2' = (x_1^2, \ldots, x_n^2)$, the preceding problem may be rewritten as

$$\begin{aligned} max\ f(\mathbf{X}_1, \mathbf{X}_2) &= \mathbf{1}'(\mathbf{X}_1 - \mathbf{X}_2) = \sum_{j=1}^{p} (x_j^1 - x_j^2) \quad \text{s.t.} \\ \mathbf{AX}_1 - \mathbf{AX}_2 &= \mathbf{b} \\ \mathbf{X}_1, \mathbf{X}_2 &\geqq \mathbf{0}. \end{aligned}$$

As pointed out earlier, x_j^1 and x_j^2 cannot both appear in any basic solution since their associated columns $\mathbf{a}_j, -\mathbf{a}_j$ are not linearly indenpendent.

And if the initial problem appears as

$$\begin{aligned} max\ f(\mathbf{X}) &= \sum_{j=1}^{n} c_j \mid x_j \mid \quad \text{s.t.} \\ \mathbf{AX} &= \mathbf{b},\ \mathbf{X} \text{ unrestricted,} \end{aligned}$$

then the above transformation $\mathbf{X} = \mathbf{X}_1 - \mathbf{X}_2$, with $\mathbf{X}_1, \mathbf{X}_2 \geqq \mathbf{0}$, allows us to rewrite this problem as

$$\begin{aligned} max\ f(\mathbf{X}_1, \mathbf{X}_2) &= \mathbf{C}'\mathbf{X}_1 - \mathbf{C}'\mathbf{X}_2 \quad \text{s.t.} \\ \mathbf{AX}_1 - \mathbf{AX}_2 &= \mathbf{b} \\ \mathbf{X}_1, \mathbf{X}_2 &\geqq \mathbf{0}. \end{aligned}$$

Interestingly enough, this latter problem possesses a standard simplex solution only if each c_j term is non-positive (for a minimization problem, the c_j's must all be non-negative) (Shanno, Weil (1971); Rao (1972)).

10 THE DUAL SIMPLEX, PRIMAL-DUAL, AND COMPLEMENTARY PIVOT METHODS

In this chapter we shall first consider two important computational variants of the standard simplex method, namely the **dual simplex** and **primal-dual** routines. These techniques are essentially modifications of the simplex method because their implementation involves a sequence of pivot operations applied to the primal problem, but based upon alternative pivot selection or entry and exit criteria. These methods were developed to: (1) reduce the number of iterations needed to achieve optimality when the number of variables (n) is large; and (2) exploit situations involving the availability of an initial basic feasible solution to the dual problem

Both of these algorithms start with an infeasible basic solution to the primal problem and, through a sequence of pivot operations, strive for feasibility in a fashion such that, once attained, the resulting basic solution is also optimal. In this regard, there is no need for the *usual*-Phase I calculation, *i.e.*, since the initial basic feasible solution to the original primal problem found at the end of Phase I may be far from optimal (Phase I ignores primal optimality and seeks only primal feasibility by driving the artificial variables to zero), these alternative methods are constructed so that they simultaneously work towards primal feasibility and optimality. The starting point for each is a dual feasible but primal infeasible solution and, at each iteration, dual feasibility as well as complementary slackness is maintained. When dual as well as primal feasibility is achieved, the algorithms are terminated.

A third simplex variant, called the **complementary pivot method**, actually utilizes a formal complementarity requirement (constraint) to facilitate pivoting. A linear program is first converted to an equivalent linear complementarity problem. Then under a complementary pivot rule, we move from what is called an *almost complementary basic solution* to another until we attain a *complementary basic solution* (containing exactly one basic variable from each complementary pair of variables). At this point we have an optimal solution to the original linear program and the algorithm is terminated.

10.1. Dual Simplex Method

In attempting to generate a optimal basic feasible solution to a linear programming problem, our procedure, as outlined in chapters five and seven, was to employ the primal simplex method. To briefly review this process, we start from an initial basic feasible solution for which the optimality criterion is not satisfied, *i.e.*, not all $\bar{c}_j \leq 0$, $j=1,\ldots,n-m$ (the dual problem is infeasible). We then make changes in the basis, one vector at a time, maintaining nonnegativity of the basic variables at each iteration, until we obtain a basic feasible solution for which all $\bar{c}_j \leq 0$, $j=1,\ldots,n-m$ (so that dual feasibility now holds). Hence the primal simplex method is a systematic technique that preserves primal feasibility while striving for primal optimality, thereby reducing, at each round, dual infeasibility, until primal optimality, and thus dual feasibility, is ultimately achieved. This last observation precipitates a very interesting question. Specifically, can we solve a linear programming problem by a procedure which works in a fashion opposite to that of the primal simplex algorithm? That is to say, by a method which relaxes primal feasibility while maintaining primal optimality. As we shall now see, the answer is yes. In this regard, we now look to an alternative simplex routine called the **dual simplex method** (Lemke (1954)), so named because it involves an application of the simplex method to the dual problem, yet is constructed so as to have its pivot operations carried out within the standard primal simplex matrix. Here the starting point is an initial basic but nonfeasible solution to the primal problem for which optimality (and thus dual feasibility) holds. As is the usual case, we move from this solution to an optimal basic feasible solution through a sequence of pivot operations which involve changing the status of a single basis vector at a time. Hence the dual simplex method preserves primal optimality (dual feasibility) while reducing, at each iteration, primal infeasibility, so that an optimal basic feasible solution to both the primal and dual problems is simultaneously achieved.

Let us now determine exactly how the dual simplex method works. Given the primal maximum problem in standard form

$$\begin{aligned} max\ f(\mathbf{X}) &= \mathbf{C}'\mathbf{X} \quad \text{s.t.} \\ \mathbf{A}\mathbf{X} &= \mathbf{b},\ \mathbf{X} \geqq \mathbf{0}, \end{aligned}$$

the dual minimum problem is of the form

$$\begin{aligned} min\ g(\mathbf{U}) &= \mathbf{b}'\mathbf{U} \quad \text{s.t.} \\ \mathbf{A}'\mathbf{U} &\geqq \mathbf{C},\ \mathbf{U} \text{ unrestricted.} \end{aligned}$$

Moreover, let $\mathbf{X}_B = \mathbf{B}^{-1}\mathbf{b} = \bar{\mathbf{b}}$ represent a basic solution to the primal maximum problem for which not all $x_{Bi} \geq 0$, $i=1,\ldots,m$. Since $\mathbf{X}_B$ is not feasible, the solution vector $\mathbf{U}$ to the dual minimum problem is not optimal. In this regard, we should be able to find a new vector $\widehat{\mathbf{U}}$ such that $\mathbf{A}'\widehat{\mathbf{U}} \geqq \mathbf{C}$ and $g(\widehat{\mathbf{U}}) < g(\mathbf{U})$. Now,

$$\mathbf{X}_B = \mathbf{B}^{-1}\mathbf{b} = \begin{bmatrix} \boldsymbol{\beta}_1 \\ \vdots \\ \boldsymbol{\beta}_m \end{bmatrix} \mathbf{b},$$

where $\boldsymbol{\beta}_i$, $i=1,\ldots,m$, denotes the ith row of $\mathbf{B}^{-1}$. Hence $x_{Bi} = \boldsymbol{\beta}_i\mathbf{b}$, $i=1,\ldots,m$. Moreover, since

$$\mathbf{B}^{-1}\mathbf{B} = \begin{bmatrix} \boldsymbol{\beta}_1 \\ \vdots \\ \boldsymbol{\beta}_m \end{bmatrix} [\mathbf{b}_1,\ldots,\mathbf{b}_m] = \mathbf{I}_m,$$

it follows that

$$\boldsymbol{\beta}_i\mathbf{b}_i = \delta_{ij} = \begin{cases} 1, & i=j; \\ 0, & i\neq j, \end{cases}$$

where i, j=1,...,m. For some $i=r$ let $x_{Br} = \beta_r\mathbf{b} < 0$. Since we desire to ultimately obtain an optimal basic feasible solution to both the primal and dual problems simultaneously, let us: (1) eliminate x_{Br} from the current set of primal basic variables (and thus $\mathbf{b}_r$ from the current primal basis); and (2) use $\mathbf{U}$ to obtain a new solution vector $\hat{\mathbf{U}}$ for which $g(\hat{\mathbf{U}}) < g(\mathbf{U})$ and $\mathbf{A}'\hat{\mathbf{U}} \geqq \mathbf{C}$. If we form

$$\hat{\mathbf{U}}' = \mathbf{U} - \theta\beta_r,$$

then

$$g(\hat{\mathbf{U}}) = \hat{\mathbf{U}}'\mathbf{b} = \mathbf{U}'\mathbf{b} - \theta\beta_r\mathbf{b} = g(\mathbf{U}) - \theta x_{Br} \tag{10.1}$$

and thus, for $\theta < 0$, $g(\hat{\mathbf{U}}) < g(\mathbf{U})$ since $x_{Br} < 0$.

Let us pause momentarily to consider two important aspects of the dual simplex algorithm. First, since $g(\hat{\mathbf{U}}) < g(\mathbf{U})$, it is evident that g starts out at a superoptimal level and monotonically decreases to its optimal feasible value. Second, to obtain the largest possible decrease in g at each iteration, it follows that the difference $g(\mathbf{U}) - g(\hat{\mathbf{U}}) = \theta x_{Bi}$, $x_{Bi} < 0$, should be as large as possible. In this regard, to select the variable to leave the set of basic variables, let us adopt the convention

$$\text{EXIT CRITERION: } x_{Bk} = \min_i\{x_{Bi}, x_{Bi} < 0\}, \tag{10.2}$$

in which case the kth column of the primal basis matrix $\mathbf{B}$, $\mathbf{b}_k$, is removed from the basis. So from this point on, we shall assume that the variable x_{Br} in (10.1) satisfies (10.2).

Under what circumstances will $\hat{\mathbf{U}}$ satisfy the dual structural constraint system $\mathbf{A}'\mathbf{U} \geqq \mathbf{C}$? Since $\mathbf{U}' = \mathbf{C}'_B\mathbf{B}^{-1}$, $\mathbf{U}'\mathbf{B} = \mathbf{C}'_B$ or $\mathbf{U}'\mathbf{b}_i = c_{Bi}$, i=1,...,m. Then for those vectors $\mathbf{b}_i$ within the primal basis,

$$\hat{\mathbf{U}}'\mathbf{b}_i = \mathbf{U}'\mathbf{b}_i - \theta\beta_r\mathbf{b}_i = c_{Bi} - \theta\delta_{ij} = \begin{cases} c_{Br} - \theta \geq c_{Br}, \ \theta \leq 0, \ i=r\ ; \\ c_{Bi}, \ i \neq r, \end{cases}$$

where i=1,...,m. And for all vectors $\mathbf{r}_j$ not in the primal basis,

$$\hat{\mathbf{U}}'\mathbf{r}_j = \mathbf{U}'\mathbf{r}_j - \theta\beta_r\mathbf{r}_j.$$

Since $\mathbf{r}_j = \mathbf{B}\mathbf{Y}_j$, $\mathbf{Y}_j = \mathbf{B}^{-1}\mathbf{r}_j$ has as its rth component $y_{rj} = \beta_r\mathbf{r}_j$. Then

$$\hat{\mathbf{U}}'\mathbf{r}_j = \mathbf{U}'\mathbf{r}_j - \theta y_{rj}.$$

Here two cases emerge. First, if $y_{rj} \geq 0$ for all j, then for any $\theta \leq 0$,

$$\hat{\mathbf{U}}'\mathbf{r}_j = c_{Rj} - \theta y_{rj} \geq c_{Rj},\ j=1,\ldots,n-m.$$

In this instance θ can be made an arbitrarily large negative number while still maintaining the feasibility of $\hat{\mathbf{U}}$ so that $g(\hat{\mathbf{U}}) = g(\mathbf{U}) - \theta x_{Br}$ becomes arbitrarily small, *i.e.*, the dual minimum problem has an unbounded feasible solution. But if this is the case, our previous results indicate that the primal maximum problem has no feasible solution. Next, let us assume that there exists at least one nonbasis vector $\mathbf{r}_j$ for which $y_{rj} < 0$. For all such $y_{rj} < 0$ to satisfy

$$\hat{\mathbf{U}}'\mathbf{r}_j = \mathbf{U}'\mathbf{r}_j - \theta y_{rj} \geq c_{Rj}$$

it must be true that

$$\mathbf{U}'\mathbf{r}_j - c_{Rj} \geq \theta y_{rj}.$$

In this instance θ cannot be made an arbitrarily large negative number without violating the feasibility of $\hat{\mathbf{U}}$. From the previous expression

$$\theta \geq \frac{\mathbf{U}'\mathbf{r}_j - c_{Rj}}{y_{rj}}$$

and thus, to preserve dual feasibility, let us choose a θ, call it $\bar{\theta}$, according to

$$\bar{\theta} = \max_j \left\{ \frac{\mathbf{U}'\mathbf{r}_j - c_{Rj}}{y_{rj}},\ y_{rj} < 0 \right\}.$$

Since $\mathbf{U}'\mathbf{r}_j - c_{Rj} = \mathbf{C}_B'\mathbf{B}^{-1}\mathbf{r}_j - c_{Rj} = -\bar{c}_j \geq 0$, the preceding expression may be alternatively written as

$$\text{ENTRY CRITERION: } \bar{\theta} = \frac{-\bar{c}_k}{y_{rk}} = \max_j \left\{ \frac{-\bar{c}_j}{y_{rj}},\ y_{rj} < 0 \right\}. \qquad (10.3)$$

We now have a solution to the dual problem for which

$$\hat{\mathbf{U}}'\mathbf{b}_i = \begin{cases} c_{Br} - \bar{\theta} \geq c_{Br},\ \bar{\theta} \leq 0,\ i=r\ ; \\ c_{Bi},\ i \neq r,\ i=1,\ldots,m\ ; \end{cases}$$

$$\hat{\mathbf{U}}'\mathbf{r}_i \geq c_{Ri},\ j \neq k,\ j=1,\ldots,n-m,$$

$$\hat{\mathbf{U}}'\mathbf{r}_k = \mathbf{U}'\mathbf{r}_k - \bar{\theta} y_{rk} = c_{Rk},\ \bar{\theta} \leq 0;$$

$$g(\hat{\mathbf{U}}) = g(\mathbf{U}) - \bar{\theta} x_{Br}.$$

With $\bar{\theta} \leq 0$, $g(\hat{\mathbf{U}}) < g(\mathbf{U})$ if $x_{Br} < 0$. And with $y_{rk} \neq 0$, $\mathbf{r}_k$ replaces $\mathbf{b}_r$ in the basis to yield a new basis matrix

$$\hat{\mathbf{B}} = [\mathbf{b}_1,\ldots,\mathbf{b}_{r-1},\mathbf{r}_k,\ \mathbf{b}_{r+1},\ldots,\mathbf{b}_m].$$

10.2. Computational Aspects of the Dual Simplex Method

Having discussed the theoretical underpinnings of the dual simplex method, let us now examine the type of problems to which it is applicable along with the basic rudiments of their solution. Specifically, some of the more common instances in which a dual feasible solution is readily obtainable are:

(1) the primal problem is such that a surplus variable is added to each " $\geq$ " structural constraint (as we shall see below, the dual simplex method enables us to circumvent the procedure of adding artificial variables to the primal structural constraints); and

(2) an additional structural constraint is added to the optimal simplex matrix.

For instance, the dual simplex method is readily applicable if the primal problem appears as:

$$\begin{gathered} \min f(\mathbf{X}) = \mathbf{C}'\mathbf{X} \quad \text{s.t.} \\ \mathbf{AX} \geqq \mathbf{b} \ \text{ with } \mathbf{C}, \mathbf{X} \geqq \mathbf{0} \\ \text{or} \\ \max h(\mathbf{X}) = -f(\mathbf{X}) = -\mathbf{C}'\mathbf{X} \quad \text{s.t.} \\ -\mathbf{AX} \leqq -\mathbf{b} \ \text{ with } \mathbf{C}, \mathbf{X} \geqq \mathbf{0}. \end{gathered}$$

To see this let us consider

EXAMPLE 10.1. Using the dual simplex method,

$$\begin{aligned} \min f &= 2x_1 + 4x_2 \quad \text{s.t.} \\ x_1 + 3x_2 &\geq 20 \\ x_1 + x_2 &\geq 15 \\ 2x_1 + x_2 &\geq 25 \\ x_1, x_2 &\geq 0. \end{aligned}$$

Upon converting this problem to a maximization problem and subtracting non-negative surplus variables x_3, x_4, and x_5 from the left-hand sides of the structural constraints, we obtain the simplex matrix

$$\begin{bmatrix} 1 & 3 & -1 & 0 & 0 & 0 & 20 \\ 1 & 1 & 0 & -1 & 0 & 0 & 15 \\ 2 & 1 & 0 & 0 & -1 & 0 & 25 \\ 2 & 4 & 0 & 0 & 0 & 1 & 0 \end{bmatrix} \longrightarrow \begin{bmatrix} -1 & -3 & 1 & 0 & 0 & 0 & -20 \\ -1 & -1 & 0 & 1 & 0 & 0 & -15 \\ \boxed{-2} & -1 & 0 & 0 & 1 & 0 & -25 \\ 2 & 4 & 0 & 0 & 0 & 1 & 0 \end{bmatrix}$$

and an initial primal optimal (dual feasible) basic solution $x_3 = -20$, $x_4 = -15$, $x_5 = -25$, and $h = 0$, where $h = -f$. Since the current solution is not primal feasible, let us undertake the first round of the dual simplex method by determining

$$-25 = \min_i\{x_{Bi},\ x_{Bi} < 0\} \ = \ \min\{-20,\ -15,\ -25\},$$

$$\bar{\theta} = -1 = \max_j\left\{\frac{-\bar{c}_j}{-y_{rj}},\ -y_{rj} < 0\right\} \ = \ \max\{2/(-2) = -1,\ 4/(-1) = -4\}.$$

Here x_1 is to replace x_5 in the set of primal basic variable (alternatively, $\mathbf{r}_1$ replaces $\mathbf{b}_3$ in the primal basis) so that the (blocked) pivot element is -2. An appropriate pivot operation yields

$$\begin{bmatrix} 0 & \boxed{-\frac{5}{2}} & 1 & 0 & -\frac{1}{2} & 0 & -\frac{15}{2} \\ 0 & -\frac{1}{2} & 0 & 1 & -\frac{1}{2} & 0 & -\frac{5}{2} \\ 1 & \frac{1}{2} & 0 & 0 & -\frac{1}{2} & 0 & \frac{25}{2} \\ 0 & 3 & 0 & 0 & 1 & 1 & -25 \end{bmatrix}$$

and a second primal basic solution consisting of $x_1 = \frac{25}{2}$, $x_3 = -\frac{15}{2}$, $x_4 = -\frac{5}{2}$, and $h = -25$. Again primal feasibility does not hold so that we next compute:

$$-\frac{15}{2} = \min_i\{x_{Bi},\ x_{Bi} < 0\} = \min\left\{-\frac{15}{2}, -\frac{5}{2}\right\},$$

$$\begin{aligned} \bar{\theta} \ = \ -\frac{6}{5} \ &= \ \max_j\left\{\frac{-\bar{c}_j}{-y_{rj}}, -y_{rj} < 0\right\} \\ &= \ \max\left\{3\Big/\left(-\frac{5}{2}\right) = -\frac{6}{5},\ 1\Big/\left(-\frac{1}{2}\right) = -2\right\}. \end{aligned}$$

If we now undertake an additional pivot operation with $-\frac{5}{2}$ as the pivot we obtain

$$\begin{bmatrix} 0 & 1 & -\frac{2}{5} & 0 & \frac{1}{5} & 0 & 3 \\ 0 & 0 & -\frac{1}{5} & 1 & \boxed{-\frac{2}{5}} & 0 & -1 \\ 1 & 0 & \frac{1}{5} & 0 & -\frac{3}{5} & 0 & 11 \\ 0 & 0 & \frac{6}{5} & 0 & \frac{2}{5} & 1 & -34 \end{bmatrix}$$

and thus a third primal basic solution wherein $x_1 = 11$, $x_2 = 3$, $x_4 = -1$, and $h = -34$. Since primal infeasibility is still evident, we again look to our respective exit and entry criteria

$$-1 = \underset{i}{min}\{x_{Bi}, x_{Bi} < 0\} = min\{-1\},$$

$$\bar{\theta} = -1 = \underset{j}{max}\left\{\frac{-\bar{c}_j}{-y_{rj}}, -y_{rj} < 0\right\} = max\left\{\frac{(6/5)}{(-1/5)} = -6, \frac{(2/5)}{(-2/5)} = -1\right\}$$

to obtain a pivot of $-\frac{2}{5}$. A final iteration yields

$$\begin{bmatrix} 0 & 1 & -\frac{1}{2} & \frac{1}{2} & 0 & 0 & \frac{5}{2} \\ 0 & 0 & \frac{1}{2} & -\frac{5}{2} & 1 & 0 & \frac{5}{2} \\ 1 & 0 & \frac{1}{2} & -\frac{3}{2} & 0 & 0 & \frac{25}{2} \\ 0 & 0 & 1 & 1 & 0 & 1 & -35 \end{bmatrix}$$

and an optimal primal basic feasible solution with $x_1 = \frac{25}{2}$, $x_2 = \frac{5}{2}$, $x_5 = \frac{5}{2}$, and $f = 35$.

It was pointed out at the beginning of this section that the dual simplex method is particularly useful in instances where " $\geq$ " structural constraints are present. In this regard, the dual simplex method can be conveniently used to solve the dual problem

$$\left.\begin{array}{ll} min\ g(\mathbf{U}) = \mathbf{b}'\mathbf{U} & \text{s.t.} \\ \mathbf{A}'\mathbf{U} \geqq \mathbf{C}, \text{ with } \mathbf{b}, \mathbf{U} \geqq \mathbf{0} & \end{array}\right\} \text{ or } \left\{\begin{array}{ll} max\ h(\mathbf{U}) = -g(\mathbf{U}) = -\mathbf{b}'\mathbf{U} & \text{s.t.} \\ -\mathbf{A}'\mathbf{U} \leqq -\mathbf{C}, \text{ with } \mathbf{b}, \mathbf{U} \geqq \mathbf{0} & \end{array}\right.$$

for both the primal as well as the dual variables. The simultaneous augmented form of this problem appears as

$$\begin{aligned} -\mathbf{A}'\mathbf{U} + \ \mathbf{I}_n \qquad\quad &= -\mathbf{C} \\ \mathbf{b}'\mathbf{U} \qquad\quad + h &= 0 \end{aligned}$$

so that the associated simplex matrix is

$$\begin{bmatrix} -\mathbf{A}' & \mathbf{I}_n & \mathbf{0} & -\mathbf{C} \\ \mathbf{b}' & \mathbf{0}' & 1 & 0 \end{bmatrix}.$$

Relative to this system the $(n+1 \times n+1)$ basis matrix and its inverse may be expressed respectively as

$$\bar{\mathbf{E}} = \begin{bmatrix} -\mathbf{E} & \mathbf{0} \\ \mathbf{b}'_E & 1 \end{bmatrix}, \quad \bar{\mathbf{E}}^{-1} = \begin{bmatrix} -\mathbf{E}^{-1} & \mathbf{0} \\ \mathbf{b}'_E\mathbf{E}^{-1} & 1 \end{bmatrix},$$

where $\mathbf{E}$ is an $(n \times n)$ basis matrix for the dual structural constraint system. Then

$$\begin{bmatrix} -\mathbf{E}^{-1} & \mathbf{0} \\ \mathbf{b}'_E\mathbf{E}^{-1} & 1 \end{bmatrix}\begin{bmatrix} -\mathbf{A}' & \mathbf{I}_n & \mathbf{0} & -\mathbf{C} \\ \mathbf{b}' & \mathbf{0}' & 1 & 0 \end{bmatrix} = \begin{bmatrix} \mathbf{E}^{-1}\mathbf{A}' & -\mathbf{E}^{-1} & \mathbf{0} & \mathbf{E}^{-1}\mathbf{C} \\ -\mathbf{b}'_E\mathbf{E}^{-1}\mathbf{A}'+\mathbf{b}' & \mathbf{b}'_E\mathbf{E}^{-1} & 1 & -\mathbf{b}'_E\mathbf{E}^{-1}\mathbf{C} \end{bmatrix}$$

Notice that this latter matrix is of the same basic form as (8.4) save for the fact that $\mathbf{E}^{-1}$ appears instead of $\mathbf{D}^{-1}$. Hence the primal vectors $\mathbf{X}$, $\mathbf{X}_s$ are readily obtainable as before.

10.3. Dual Degeneracy

We now turn to some additional remarks concerning the value of θ which constitutes the maximum of (10.3). Let us assume that at some point in the dual simplex routine it turns out that $\bar{\theta}$ is not unique. Hence we may introduce into the primal basis any one of the columns of $\mathbf{R}$ for which the same maximum value of θ is assumed and obtain a new (call it the kth) primal basic solution. In this instance $\bar{c}_j = 0$ for all j corresponding to those columns that have the same $\theta = \bar{\theta}$ as the incoming basis vector and thus the $k+1^{st}$ iteration is executed with $\bar{\theta} = 0$, *i.e.*, the case of **dual degeneracy** emerges. Moreover, for the $k+1^{st}$ primal basic solution, the primal objective value is unchanged. (Note that the dual objective value must also remain invariant since, from (10.1), $g(\hat{\mathbf{U}}) = g(\mathbf{U})$). In fact the primal objective value may remain unchanged for several successive iterations. However, in the absence of cycling, the dual simplex algorithm will eventually converge to an optimal basic feasible solution to both the primal and dual problems simultaneously.

10.4. Summary of the Dual Simplex Method

Let us now consolidate our discussion of the theory and major computaional aspects of the dual simplex method. If we abstract from the possibility of degeneracy, then a summary of the principal steps leading to an optimal basic feasible solution to both the primal and dual problems simultaneously is provided by the:

10.1. DUAL SIMPLEX METHOD.

STEP 0. Initialize.

Obtain an initial dual feasible (primal optimal) solution with $\bar{c}_j \leq 0$, j=1,...,$n-m$.

STEP 1. If primal feasibility also holds, then the current solution is an optimal basic feasible solution to both the primal and dual problems. If not all $x_{Bi} \geq 0$, i=1,...,m, (primal feasibility is relaxed), then proceed with the next step.

STEP 2. Find $x_{Br} = \min_i\{x_{Bi}, x_{Bi} < 0\}$. The vector $\mathbf{b}_r$ corresponding to the minimum is to be removed from the basis.

STEP 3. Determine whether $-y_{rj} \geq 0$, j=1,...,$n-m$. If the answer is yes, the process is terminated since, in this instance, the primal problem is infeasible and thus the dual objective function is unbounded. If the answer is no, step 4 is warranted.

STEP 4. Compute $\bar{\theta} = \max_j\left\{\frac{-\bar{c}_j}{-y_{rj}}, -y_{rj} < 0\right\}$. If $\frac{-\bar{c}_k}{-y_{rk}} = \bar{\theta}$, then $\mathbf{r}_k$ replaces $\mathbf{b}_r$ in the basis.

At this point let us contrast the pivot operation involved in the dual simplex method with the one encountered previously in the primal simplex method. First, only negative elements can serve as pivot elements in the dual simplex method whereas only positive pivot elements are admitted in the primal simplex routine. Second, the dual simplex method first decides which vector is to leave the basis and next determines the vector to enter the basis. However, just the opposite order of operations for determining those vectors entering and leaving the basis holds for the primal simplex method. Next:

STEP 5. Determine the new dual feasible (primal optimal) solution by pivoting on $-y_{rk}$. Return to step 1.

By continuing this process of replacing a single vector in the basis at a time, we move to a new dual feasible solution with the value of the dual objective function diminished and primal infeasibility reduced. The process is terminated when we attain our first primal feasible solution, in which case we attain an optimal basic feasible solution to both the primal and dual problems.[1]

10.5. Generating an Initial Primal-Optimal Basic Solution: The Artificial Constraint Method

One of the major difficulties associated with the implementation of the dual simplex method is that an initial primal-optimal (dual-feasible) basic solution may not be readily available. In this regard, steps must be taken to construct such a solution and to this end we now proceed. For instance, let us assume that the primal problem appears as

$$max\ f(x_1,\ldots,x_p) = -\sum_{j=1}^{k} c_j x_j + \sum_{j=k+1}^{p} c_j x_j \quad \text{s.t.}$$

$$\sum_{j=1}^{p} a_{ij}x_j \geq b_i,\ i=1,\ldots,m, \text{ with } c_j,\ x_j \geq 0,\ j=1,\ldots,p.$$

Upon introducing non-negative surplus variables x_{p+i}, $i=1,\ldots,m$, we can readily obtain an initial primal-basic but not primal-optimal (and thus dual infeasible) solution $x_{p+i} = -b_i, i=1,\ldots,m$, with $f = 0$. At this point let us add to the original problem the **artificial structural constraint**

$$\sum_{j\in\mathcal{A}} x_j \leq M$$

(artificial in the sense that it is meaningless in terms of the formulation of the original structural constraint system which contains only legitimate constraints), where the index set $\mathcal{A} = \{j \mid \bar{c}_j < 0\}$ and M is an arbitrarily large positive number. For x_M a **non-negative artificial slack variable**, the previous inequality becomes

[1]For the details on the steepest edge modification of the dual simplex method see appendix B.

$$\sum_{j \in \mathcal{A}} x_j + x_M = M.$$

Since our immediate goal is primal optimality, let us employ the criterion

$$|\bar{c}_l| = \max_{j \in \mathcal{A}} \{ |\bar{c}_j|, \bar{c}_j < 0\} \tag{10.4}$$

to express the ℓth legitimate variable as

$$x_l = M - x_M - \sum_{\substack{j \in \mathcal{A} \\ j \neq \ell}} x_j.$$

This expression for x_l is then substituted into the original objective function and structural constraint system and also included in the latter as a supplementary structural constraint. Once this is accomplished, primal optimality is attained. Moreover, if one or more of the primal basic variables $x_{p+i} = -b_i + a_{il}M$, $i=1,\ldots,m$, is negative, then the dual simplex method is directly applicable.

As far as the salient features of the artificial constraint method are concerned, we briefly note that: if at some point in the dual simplex routine the vector corresponding to x_M enters the primal basis, then a basic solution to the original problem is obtained by ignoring both the row and column to which it corresponds; moreover, the row and column associated with x_M may be deleted from the simplex matrix and the dual simplex iterations continued until primal feasibility is attained.

EXAMPLE 10.2. To view the artificial-constraint method in action,

$$\begin{aligned} \max f &= x_1 + 3x_2 \quad \text{s.t.} \\ x_1 + x_2 &\leq 15 \\ x_1 + 4x_2 &\geq 20 \\ 5x_1 + x_2 &\geq 25 \\ x_1, x_2 &\geq 0. \end{aligned}$$

With x_3 a slack variable and x_4, x_5 surplus variables (each non-negative) we form

$$
\begin{aligned}
x_1 + x_2 + x_3 \quad &= 15 \\
-x_1 - 4x_2 + x_4 \quad &= -20 \\
-5x_1 - x_2 + x_5 \quad &= -25 \\
-x_1 - 3x_2 + f \quad &= 0
\end{aligned}
$$

Clearly an initial basic solution derived from the system is both primal and dual infeasible. To achieve dual feasibility, let us introduce the artificial structural constraint

$$x_1 + x_2 + x_M = M.$$

Since $|\bar{c}_2| = \max_{j \in \mathcal{A}}\{|\bar{c}_j|, \bar{c}_j < 0\}$, we substitute $x_2 = M - x_M - x_1$ into the above simultaneous form to obtain

$$
\begin{aligned}
x_3 - x_M \quad &= 15 - M \\
3x_1 + x_4 + 4x_M \quad &= -20 + 4M \\
-4x_1 + x_5 + x_M \quad &= -25 + M \\
x_1 + x_2 + x_M \quad &= M \\
2x_1 + 3x_M + f \quad &= 3M
\end{aligned}
$$

with simplex matrix

$$
\begin{bmatrix}
0 & 0 & 1 & 0 & 0 & \boxed{-1} & 0 & 15-M \\
3 & 0 & 0 & 1 & 0 & 4 & 0 & -20+4M \\
-4 & 0 & 0 & 0 & 1 & 1 & 0 & -25+M \\
1 & 1 & 0 & 0 & 0 & 1 & 0 & M \\
2 & 0 & 0 & 0 & 0 & 3 & 1 & 3M
\end{bmatrix}
$$

and initial primal-optimal (dual-feasible) basic solution $x_2 = M$, $x_3 = 15-M$, $x_4 = -20+4M$, $x_5 = -25+M$, and $f = 3M$. Since the current solution is not primal feasible, we look to

$$
\begin{aligned}
15 - M &= \min_i\{x_{Bi}, x_{Bi} < 0\}, \\
\bar{\theta} = -3 &= \max_j\left\{\frac{-\bar{c}_j}{-y_{rj}}, -y_{rj} < 0\right\},
\end{aligned}
$$

which yields a pivot of -1, to wit x_M replaces x_3 in the set of primal basic variables. Pivoting on -1 results in

$$\begin{bmatrix} 0 & 0 & -1 & 0 & 0 & 1 & 0 & -15+M \\ 3 & 0 & 4 & 1 & 0 & 0 & 0 & 40 \\ \boxed{-4} & 0 & 1 & 0 & 1 & 0 & 0 & -10 \\ 1 & 1 & 1 & 0 & 0 & 0 & 0 & 15 \\ 2 & 0 & 3 & 0 & 0 & 0 & 1 & 45 \end{bmatrix}.$$

Thus our second primal basic solution becomes $x_2 = 15$, $x_4 = 40$, $x_5 = -10$, $x_M = -15+M$, and $f = 45$. Since x_M is feasible, row one and column six may be deleted from this latter simplex matrix. And with

$$\begin{aligned} -10 &= \min_i\{x_{Bi}, \ x_{Bi} < 0\}, \\ \bar{\theta} &= -\frac{1}{2} = \max_j\left\{\frac{-\bar{c}_j}{-y_{rj}}, \ -y_{rj} < 0\right\}, \end{aligned}$$

we pivot on the element -4 to obtain

$$\begin{bmatrix} 0 & 0 & \frac{19}{4} & 1 & \frac{3}{4} & 0 & \frac{65}{2} \\ 1 & 0 & -\frac{1}{4} & 0 & -\frac{1}{4} & 0 & \frac{5}{2} \\ 0 & 1 & \frac{5}{4} & 0 & \frac{1}{4} & 0 & \frac{25}{2} \\ 0 & 0 & \frac{7}{2} & 0 & \frac{1}{2} & 1 & 40 \end{bmatrix}$$

and an optimal primal basic feasible solution consisting of $x_1 = \frac{5}{2}$, $x_2 = \frac{25}{2}$, $x_4 = \frac{65}{2}$ and $f = 40$.

10.6. Primal-Dual Method

The primal-dual algorithm (Dantzig, Ford, Fulkerson (1956)) requires that we solve an adjusted Phase I problem. The legitimate vectors selected to replace the artificial vectors in the basis are chosen in a way that is quite different from the *usual* Phase I entry criterion. In this regard, we begin with a feasible solution to the dual problem from which we attempt to reduce primal infeasibility by inserting only certain selected legitimate primal vectors into the primal basis, *i.e.*, we utilize what is called *restricted basis entry* when solving the primal problem. We next employ the standard simplex method as a subalgorithm until optimality is achieved for those vectors marked for entry into the primal basis. An improved solution to the dual (in the sense that the dual objective value is made to decrease) is thus obtained which in turn targets a new set of legitimate primal vectors as candidates for (restricted) entry into the primal basis. This process is repeated with complementary slackness maintained at each step so that the first primal feasible solution obtained is also optimal. Hence the primal-dual routine, so named because we attempt to increase the value of the Phase I infeasibility form and, at the same time, decrease the dual objective function value, works towards primal feasibility and optimality at the same time.

Let the primal-dual pair of problems under consideration appear as

PRIMAL	DUAL
$max\ f(\mathbf{X}) = \mathbf{C}'\mathbf{X}$ s.t.	$min\ g(\mathbf{U}) = \mathbf{b}'\mathbf{U}$ s.t.
$\mathbf{AX} = \mathbf{b},\ \mathbf{X} \geqq \mathbf{0}$	$\mathbf{A}'\mathbf{U} \geqq \mathbf{C}$, $\mathbf{U}$ unrestricted. (10.5)

If $\mathbf{C} \leq \mathbf{0}$, then it is easily seen that $\mathbf{U} = \mathbf{0}$ is a feasible solution to the above dual problem. If $c_j \nleq 0$ for all j, then a feasible solution to the same is not readily available. However, to generate an initial dual feasible solution, let us modify (10.5) slightly by introducing the **artificial constraint** $x_0 + \mathbf{1}'\mathbf{X} = b_0$ into the primal problem (Beale (1954)). Here $x_0 \geq 0$ and b_0 is taken to be an unspecified large positive constant. In addition, the primal objective function coefficient associated with x_0 (call it c_0) is zero. Clearly the introduction of this constraint does not affect the values of the primal variables. In this regard, the modification of (10.5) appears as

MODIFIED PRIMAL

$$max\ f(\mathbf{X}) = \mathbf{C}'\mathbf{X} \quad \text{s.t.}$$

$$\begin{bmatrix} \mathbf{0} & \mathbf{A} \\ 1 & \mathbf{1}' \end{bmatrix}\begin{bmatrix} x_0 \\ \mathbf{X} \end{bmatrix} = [\bar{\mathbf{a}}_0, \bar{\mathbf{a}}_1, \ldots, \bar{\mathbf{a}}_n]\begin{bmatrix} x_0 \\ \mathbf{X} \end{bmatrix} = \begin{bmatrix} \mathbf{b} \\ b_0 \end{bmatrix}$$

$$x_0 \geq 0,\ \mathbf{X} \geqq \mathbf{0},\ c_0 = 0$$

MODIFIED DUAL

$$min\ \bar{g} = \mathbf{b}'\mathbf{U} + b_0 u_0 \quad \text{s.t.}$$

$$\begin{bmatrix} \mathbf{0}' & 1 \\ \mathbf{A}' & \mathbf{1} \end{bmatrix}\begin{bmatrix} \mathbf{U} \\ u_0 \end{bmatrix} \geqq \begin{bmatrix} 0 \\ \mathbf{C} \end{bmatrix} \tag{10.6}$$

$$u_0 \geq 0,\ \mathbf{U} \text{ unrestricted},$$

where $\bar{\mathbf{a}}_j$, j=0,1,...,n, denotes the jth column of the modified primal coefficient matrix, *i.e.*, $\bar{\mathbf{a}}_0 = \mathbf{e}_{m+1}$, $\bar{\mathbf{a}}_j' = (\mathbf{a}_j', 1)$, and $\mathbf{a}_j$ is the jth column of $\mathbf{A}$, j=1,...,n. An obvious feasible solution to the modified dual is thus $u_0 \geq 0$, $\mathbf{U} = \mathbf{0}$, and $\mathbf{1}u_0 \geqq \mathbf{C}$ or $u_0 = \underset{j}{max}\{c_j\}$, j=1,...,n.

If we now introduce nonnegative dual surplus variables u_{s0}, $\mathbf{U}_s$ into the modified dual program, the latter becomes

$$\begin{aligned} & min\ \bar{g} \ = \ \mathbf{b}'\mathbf{U} + b_0 u_0 \quad \text{s.t.} \\ & \begin{bmatrix} \mathbf{0}' & 1 \\ \mathbf{A}' & \mathbf{1} \end{bmatrix} \begin{bmatrix} \mathbf{U} \\ u_0 \end{bmatrix} - \begin{bmatrix} u_{s0} \\ \mathbf{U}_s \end{bmatrix} = \begin{bmatrix} 0 \\ \mathbf{C} \end{bmatrix} \\ & u_0 \geq 0,\ \mathbf{U} \text{ unrestricted},\ u_{s0} \geq 0,\ \mathbf{U}_s \geqq \mathbf{0}. \end{aligned} \tag{10.7}$$

Note that if $u_0 = 0$, any feasible solution to (10.7) is also a feasible solution to the original dual problem appearing in (10.5).

Let us now return to the modified primal problem and introduce the artificial variables $x_{a0} \geq 0$, $\mathbf{X}_a \geqq \mathbf{0}$ as we would for the usual Phase I calculation. This yields what is called the **extended primal problem**

$$\begin{aligned} & max\ f^* \ = \ -\mathbf{1}'\mathbf{X}_a - x_{a0} \quad \text{s.t.} \\ & \begin{bmatrix} \mathbf{0} & \mathbf{A} \\ 1 & \mathbf{1}' \end{bmatrix} \begin{bmatrix} x_0 \\ \mathbf{X} \end{bmatrix} + \begin{bmatrix} \mathbf{X}_a \\ x_{a0} \end{bmatrix} = \begin{bmatrix} \mathbf{b} \\ b_0 \end{bmatrix} \\ & x_0 \geq 0,\ \mathbf{X} \geqq \mathbf{0},\ x_{a0} \geq 0,\ \mathbf{X}_a \geqq \mathbf{0}. \end{aligned} \tag{10.8}$$

In what follows we shall use a variation of this problem to carry out the primal-dual pivot operations.

To see exactly how the primal-dual method works, we note first that if the vectors $\mathbf{X}, \mathbf{U}$ represent feasible solutions to the primal-dual pair of problems in (10.5), then they are also optimal solutions to the same if and only if the following weak complementary slackness conditions hold:

$$\begin{aligned} &\text{(a)}\ \ x_0 u_{s0} + \mathbf{X}'(\mathbf{C}-\mathbf{A}'\mathbf{U}) \ = \ x_0 u_{s0} + \mathbf{X}'\mathbf{U}_s \ = \ 0; \\ &\text{(b)}\ \ u_0(b_0 - x_0 - \mathbf{1}'\mathbf{X}) + \mathbf{U}'(\mathbf{b}-\mathbf{A}\mathbf{X}) \ = \ 0. \end{aligned} \tag{10.9}$$

(Since the structural constraints of the primal are strict equalities, (10.9.b) is automatically satisfied when a feasible primal solution x_0, $\mathbf{X}$ is obtained.) Relative to the current feasible solution to the original dual in (10.5), some of the structural constraints of this problem are binding and some are not. For those which are satisfied as equalities (u_{sj}=0), the corresponding x_j can be any positive value and (10.9.a) holds. If one or more of these constraints hold as a strict inequality, then the corresponding $x_j = 0$ so that again (10.9.a) obtains. How can we find a vector $\mathbf{X}$ that satisfies (10.9.a) (complementary slackness is preserved) and the constraints of the primal problem in (10.5)? We noted above that an initial feasible solution to the modified dual (and thus to (10.7)) is $\mathbf{U} = \mathbf{0}$, $u_0 = \underset{j}{max}\{c_j\}, j=1,\ldots,n$. Then from the partitioned rows of (10.7) we have

$$u_{s0} = u_0 = \underset{j}{max}\{c_j\}, \mathbf{U}_s = \mathbf{1}u_0 - \mathbf{C} = \mathbf{1}\ \underset{j}{max}\{c_j\} - \mathbf{C},\ j=1,\ldots,n. \tag{10.10}$$

Clearly at least one component of $(u_{s0}, \mathbf{U}_s')$ must vanish by virtue of the definition of u_0. Moreover, for this solution, $\bar{g} = \mathbf{b}'\mathbf{U} + b_0 u_0 = b_0 u_0$. For the current feasible solution to the dual, let $\mathfrak{J} = \{j \mid u_{sj} = 0\}$, *i.e.*, $\mathfrak{J}$ represents the set of subscripts j corresponding to the dual binding structural constraints. The importance of index set $\mathfrak{J}$ is that it provides us with a device for using information about the dual with which to select legitimate vectors $\bar{\mathbf{a}}_j$ to enter the primal basis. (Initially the primal basis for (10.8) contains only artificial vectors.) That is, given that the complementary slackness conditions must be maintained throughout all of our computations, if $j \in \mathfrak{J}$, then any $x_j \geq 0$ will satisfy (10.9.a) and thus $\bar{\mathbf{a}}_j$ can be admitted to the primal basis of (10.8); and if $j \notin \mathfrak{J}$ so that $u_{sj} > 0$, then (10.9.a) is satisfied only if $x_j = 0$ so that $\bar{\mathbf{a}}_j$ remains nonbasic. (Since $u_0 = \underset{j}{max}\{c_j\}$, $j=1,\ldots,n$, $\mathfrak{J} \neq \phi$ initially so that at least one $\bar{\mathbf{a}}_j$ is eligible for restricted basis entry in (10.8).) Hence we allow only those $\bar{\mathbf{a}}_j$, $j \in \mathfrak{J}$, to enter the primal basis at each iteration. These observations on **restricted basis entry** thus lead us to construct what is called the **restricted primal problem** (a modification of the standard Phase I calculation): we must find a nonnegative linear combination of the vectors $\bar{\mathbf{a}}_j$, $j \in \mathfrak{J}$, which equals $(\mathbf{b}', b_0)$ by solving

$$\begin{aligned} max\ f_r^* &= -\mathbf{1}'\mathbf{X}_a - x_{a0} \quad \text{s.t.} \\ \sum_{j \in \mathfrak{J}} \bar{\mathbf{a}}_j x_j + \begin{bmatrix} \mathbf{X}_a \\ x_{a0} \end{bmatrix} &= \begin{bmatrix} \mathbf{b} \\ b_0 \end{bmatrix} = \bar{\mathbf{b}} \\ x_j \geq 0,\ j=0,1,\ldots,n,\ x_{a0} &\geq 0,\ \mathbf{X}_a \geqq \mathbf{0}. \end{aligned} \tag{10.11}$$

To generate a solution to (10.11) we can apply the primal simplex method in the usual fashion; terminating the calculations when $\bar{c}_j \leq 0$ for all $\bar{\mathbf{a}}_j$, $j \epsilon \mathfrak{J}$, entering the primal basis.

If $f_r^* = 0$, then $x_{ai} = 0$, $i=1,\ldots,m$, and the resulting set of x_j's for which $j \epsilon \mathfrak{J}$ along with those x_j's for $j \notin \mathfrak{J}$ yield a feasible solution to the primal in (10.6) (since the constraints in (10.11) with $x_{ai} = 0$ are the same as those in (10.6) with the columns $\bar{\mathbf{a}}_j$, $j \epsilon \mathfrak{J}$, deleted). Moreover, this solution is also optimal since the x_j's satisfy the complementary slackness conditions. Hence $f = \bar{g}$ for the pair of problems in (10.6). How does this discussion relate to the solution of the original pair of problems in (10.5)? Specifically, if x_0, $\mathbf{X}$ is a feasible solution to the modified primal and (10.9.a) holds, then x_0, $\mathbf{X}$ is also an optimal solution to this problem and thus u_0, $\mathbf{U}$ is an optimal solution to the modified dual. And if $u_0 = 0$, then $f = \bar{g} = g$ with $\mathbf{X}'\mathbf{U}_s = 0$. Thus $\mathbf{X}$ is an optimal solution to the original primal in (10.5) and $\mathbf{U}$ is an optimal solution to its dual. (However, if there exists a solution to the modified dual for which $u_0 \neq 0$ and (10.9.a) is in effect, then with $f = \bar{g} = \mathbf{b}'\mathbf{U} + b_0 u_0$, b_0 and thus f can be made arbitrarily large so that the modified and thus the original primal objective is unbounded.) These comments may be summarized by the

10.2. PRIMAL-DUAL OPTIMALITY THEOREM. Solutions $(\mathbf{U}', u_0)$, $(x_0, \mathbf{X}')$ to the modified dual and restricted primal problems respectively with $f^*=0$, $u_0=0$, and $x_j=0$, $j \notin \mathfrak{J}$, are optimal for the original primal-dual pair of problems.

PROOF[2]. With $f^* = 0$ we have $\mathbf{X}_a = \mathbf{0}$, $x_a = 0$. And with $u_0 = 0$, the vectors $\mathbf{X}, \mathbf{U}$ satisfy the original primal and dual structural constraints, *i.e.*, if we premultiply both sides of the structural constraint system in (10.11) by $(\mathbf{U}', u_0) = (\mathbf{U}', 0)$ we obtain, for $\mathbf{X}_a = \mathbf{0}$, $x_{ao} = 0$,

$$\sum_{j \epsilon \mathfrak{J}} \mathbf{U}'\mathbf{a}_j x_j \ = \ \mathbf{U}'\mathbf{b}.$$

Since for $j \epsilon \mathfrak{J}$ we have $\mathbf{U}'\mathbf{a}_j = c_j$, the preceding expression becomes

[2]This proof is a slight variation of the one provided by Dantzig, et al. (1956).

$$\sum_{j \in \mathfrak{J}} c_j x_j = \mathbf{b}'\mathbf{U}.$$

Thus $\mathbf{X}, \mathbf{U}$ are optimal solutions to the original primal and dual problems respectively. Q.E.D.

If $f_r^* < 0$ and $\bar{\mathbf{V}}$ is an optimal solution to the dual of (10.11), *i.e.*, to the **restricted dual problem**

$$\begin{aligned} min\ g_r^* &= \bar{\mathbf{b}}'\mathbf{V} \quad \text{s.t.} \\ \bar{\mathbf{a}}_j'\mathbf{V} &\geq 0,\ j \in \mathfrak{J} \\ \mathbf{V} &\geqq \mathbf{1}, \end{aligned} \tag{10.12}$$

where $\mathbf{V}' = (\mathbf{U}', u_0)$, then we can demonstrate that there exists some $\theta > 0$ such that $\mathbf{V}^* = \mathbf{V} + \theta\bar{\mathbf{V}}$ is a new feasible solution to the dual in (10.6) and that $\mathbf{V}^*$ produces a decrease in the dual objective, *i.e.*, $\bar{\mathbf{b}}'\mathbf{V}^* > \bar{\mathbf{b}}'\mathbf{V}$, where $\mathbf{V}$ is a feasible solution to the said dual. To this end let us consider the following two possible ways in which this can occur.

First, it may be the case that $\bar{\mathbf{a}}_j'\bar{\mathbf{V}} \geq 0$, for $j \notin \mathfrak{J}$. In this case the primal in (10.6) has no feasible solution. To see this we may note that if $\bar{\mathbf{V}}$ is an optimal solution to (10.12), then $\bar{\mathbf{a}}_j'\bar{\mathbf{V}} \geq 0$, $j \in \mathfrak{J}$; and since by assumption $\bar{\mathbf{a}}_j'\bar{\mathbf{V}} \geq 0$, $j \notin \mathfrak{J}$, it follows that $\bar{\mathbf{a}}_j'\bar{\mathbf{V}} \geq 0$ for all $j = 0, 1, \ldots, n$. If $\mathbf{V}$ is a feasible solution to the dual problem in (10.6), then so is $\mathbf{V}^* = \mathbf{V} + \theta\bar{\mathbf{V}}$, $\theta > 0$, *i.e.*,

$$\bar{\mathbf{a}}_j'\mathbf{V}^* = \bar{\mathbf{a}}_j'\mathbf{V} + \theta\bar{\mathbf{a}}_j'\bar{\mathbf{V}} \geq c_j$$

since $\bar{\mathbf{a}}_j'\mathbf{V} \geq c_j$ and $\bar{\mathbf{a}}_j'\bar{\mathbf{V}} \geq 0$, j=$0, 1, \ldots, n$. Furthermore, the dual objective in (10.6) can be made arbitrarily small, *i.e.*, since $\bar{\mathbf{b}}'\mathbf{V}^* = \bar{\mathbf{b}}'\mathbf{V} + \theta\bar{\mathbf{b}}'\bar{\mathbf{V}}$ and $\bar{\mathbf{b}}'\bar{\mathbf{V}} = g_r^* = f_r^* < 0$, it follows that as θ becomes arbitrarily large, $\bar{\mathbf{b}}'\mathbf{V}^*$ decreases without bound. Thus from theorem 4.9, the modified primal in (10.6) has no feasible solution.

Second, let $\bar{\mathbf{a}}_j'\bar{\mathbf{V}} < 0$ for at least one $j \notin \mathfrak{J}$. Since $\bar{\mathbf{V}}$ is an optimal solution to (10.12), we know that $\bar{\mathbf{a}}_j'\bar{\mathbf{V}} \geq 0$, $j \in \mathfrak{J}$. And for a suitably restricted value of $\theta > 0$, $\mathbf{V}^* = \mathbf{V} + \theta\bar{\mathbf{V}}$ is feasible for the dual in (10.6). To see this let us increase the value of θ to the point where one of the inequalities $\bar{\mathbf{a}}_j'\mathbf{V}^* > c_j$, $j \notin \mathfrak{J}$, j=$0, 1, \ldots, n$, becomes an equality, *i.e.*, the largest allowable increase in θ which does not violate dual feasibility in (10.6) is

$$0 \le \bar{\theta} = \frac{c_k - \bar{\mathbf{a}}_k'\mathbf{V}}{\bar{\mathbf{a}}_k'\bar{\mathbf{V}}} = \min_j\left\{\frac{c_j - \bar{\mathbf{a}}_j'\mathbf{V}}{\mathbf{a}_j'\bar{\mathbf{V}}},\ j \notin \mathfrak{J},\ \bar{\mathbf{a}}_j'\bar{\mathbf{V}} < 0\right\}. \tag{10.13}$$

Then for $\theta \le \bar{\theta}$,

$$\begin{aligned}\bar{\mathbf{a}}_j'\mathbf{V}^* = \bar{\mathbf{a}}_j'\mathbf{V} + \theta\bar{\mathbf{a}}_j'\bar{\mathbf{V}} &\ge \bar{\mathbf{a}}_j'\mathbf{V} + \bar{\theta}\bar{\mathbf{a}}_j'\bar{\mathbf{V}} \\ &\ge \bar{\mathbf{a}}_j'\mathbf{V} + (c_j - \bar{\mathbf{a}}_j'\mathbf{V}) = c_j \quad \text{for} \quad \bar{\mathbf{a}}_j'\bar{\mathbf{V}} \le 0;\end{aligned}$$

$$\bar{\mathbf{a}}_j'\mathbf{V}^* = \bar{\mathbf{a}}_j'\mathbf{V} + \theta\bar{\mathbf{a}}_j'\bar{\mathbf{V}} \ge \bar{\mathbf{a}}_j'\mathbf{V} \ge c_j \quad \text{for} \quad \bar{\mathbf{a}}_j'\bar{\mathbf{V}} > 0.$$

In addition, $\mathbf{V}^*$ yields an improved solution to the dual in (10.6) relative to the original feasible vector $\mathbf{V}$, *i.e.*, $\bar{g}^* = \bar{\mathbf{b}}'\mathbf{V}^* = \bar{\mathbf{b}}'\mathbf{V} + \theta\bar{\mathbf{b}}'\bar{\mathbf{V}} = \bar{g} + \theta g_r^* = \bar{g} + \theta f_r^* \le \bar{g}$ for $\theta \ge 0$, $f_r^* = g_r^* < 0$. Clearly $\theta = \bar{\theta}$ yields the largest possible decrease in g_r^*.

If $\mathbf{B}_r$ depicts the current basis matrix for the restricted primal in (10.12) when the optimality criterion is satisfied for all $\bar{\mathbf{a}}_j, j \in \mathfrak{J}$, and $\mathbf{C}_{B_r}$ has as its components the objective function coefficients (being either 0 or -1) associated with the basic variables, then $\bar{\mathbf{V}}' = \mathbf{C}_{B_r}'\mathbf{B}_r^{-1}$. Thus the optimality evaluators may be expressed as $\bar{c}_j^* = c_{Rj} - \mathbf{C}_{B_r}'\mathbf{B}_r^{-1}\bar{\mathbf{a}}_j = -\bar{\mathbf{a}}_j'\bar{\mathbf{V}}$ and, with $u_{sj} = \bar{\mathbf{a}}_j'\mathbf{V} - c_j$, it follows that (10.13) can be rewritten as

$$0 \le \bar{\theta} = \frac{u_{sk}}{\bar{c}_k^*} = \min_j\left\{\frac{u_{sj}}{\bar{c}_j^*},\ j \notin \mathfrak{J},\ -\bar{c}_j^* < 0\right\}. \tag{10.13.1}$$

Based upon the new solution vector $\bar{\mathbf{V}}' = (\bar{\mathbf{U}}', \bar{u}_0)$ we can obtain a new set of dual surplus variables $u_{sj}^* = \bar{\mathbf{a}}_j'\mathbf{V}^* - c_j = u_{sj} + \bar{\theta}(-\bar{c}_j^*)$, $j=0,1,\ldots,n$, which enable us to redefine the index set $\mathfrak{J}$ and subsequently carry out an additional set of iterations until the optimal solutions to the problems in (10.6) are obtained or it is revealed that the original primal in (10.5) has no feasible solution. It is important to note that the new index set contains at least one subscript j which was not included in the old one. In fact, the revised $\mathfrak{J}$ contains the subscript k appearing in (10.13) (or (10.13.1)) along with any other index corresponding to a positive value of x_j emerging from the solution to (10.11) because of the complementary slackness requirement. For any such j we have $\bar{\mathbf{a}}_j'\bar{\mathbf{V}} = 0$ and thus $\bar{\mathbf{a}}_j'\mathbf{V}^* = \bar{\mathbf{a}}_j'\mathbf{V} + \theta\bar{\mathbf{a}}_j'\bar{\mathbf{V}} = \bar{\mathbf{a}}_j'\mathbf{V} = c_j$ so that $u_{sj} = 0$. Hence at least one new vector $\bar{\mathbf{a}}_j$

which was previously not a candidate for restricted basis entry will be released for entry into the primal basis. Once the new index set $\mathfrak{J}$ has been determined, we solve the new restricted primal problem, the outcome of which allows us to obtain another solution to the dual problem in (10.6) with a smaller objective value than before. This new dual solution again updates $\mathfrak{J}$, *i.e.*, chooses one or more vectors for restricted basis entry, and the process is repeated. Assuming the absence of degeneracy, no basis of the primal in (10.6) is ever repeated and thus the primal-dual method converges to an optimal basic feasible solution in a finite number of steps. When $f^* = 0$ we have an optimal basic feasible solution to the modified primal and, for $u_0 = 0$, $f = \overline{g} = g$. If $u_0 \neq 0$, the modified primal has an unbounded solution. And if at some iteration we have $f^* < 0$ and $\overline{c}_j^* \leq 0$ for all j, then the primal in (10.6) has no feasible solution.

EXAMPLE 10.3. Using the primal-dual technique,

$$\begin{aligned} max\ f &= 2x_1 + 10x_2 \quad \text{s.t.} \\ 2x_1 + 4x_2 &\leq 20 \\ 4x_1 + x_2 &\geq 20 \\ x_1, x_2 &\geq 0. \end{aligned}$$

Since $\mathbf{C} \nleq \mathbf{0}$, we must solve the extend primal problem (10.8) or

$$\begin{aligned} max\ f^* &= -x_5 - x_{a0} \quad \text{s.t.} \\ 2x_1 + 4x_2 + x_3 &= 20 \\ 4x_1 + x_2 - x_4 + x_5 &= 20 \\ x_0 + x_1 + x_2 + x_3 + x_4 + x_{a0} &= b_0 \\ x_0, x_1, \ldots, x_5, x_{a0} &\geq 0, \end{aligned} \tag{10.14}$$

where x_3 is a slack variable, x_4 is a surplus variable, and x_5, x_{a0} are artificial variables. As indicated earlier, a starting feasible solution to the modified dual problem (10.6) is

$$\mathbf{U} = \mathbf{0},\ u_0 = \underset{j}{max}\{c_j\} = 10 = u_{s0},$$

$$\mathbf{U}_s = \mathbf{1}u_0 - \mathbf{C} = \begin{bmatrix} u_{s1} \\ u_{s2} \\ u_{s3} \\ u_{s4} \end{bmatrix} = \begin{bmatrix} 10-2 \\ 10-10 \\ 10-0 \\ 10-0 \end{bmatrix} = \begin{bmatrix} 8 \\ 0 \\ 10 \\ 10 \end{bmatrix}, \text{ and}$$

$$\bar{g} = b_0 u_0 + \mathbf{b}'\mathbf{U} = 10b_0.$$

Given that complementary slackness must be preserved at each iteration, it is evident that with $u_{s2} = 0$, x_2 can become positive and thus $\mathfrak{J} = \{2\}$. From (10.14), the initial simplex matrix appears as

$$\begin{bmatrix} 0 & 2 & 4 & 1 & 0 & 0 & 0 & 0 & 20 \\ 0 & 4 & 1 & 0 & -1 & 1 & 0 & 0 & 20 \\ 1 & 1 & 1 & 1 & 1 & 0 & 1 & 0 & b_0 \\ 0 & 0 & 0 & 0 & 0 & 1 & 1 & 1 & 0 \end{bmatrix} \longrightarrow$$

$$\begin{matrix} & & x_2 & & & & & & \end{matrix}$$
$$\begin{bmatrix} 0 & 2 & \boxed{4} & 1 & 0 & 0 & 0 & 0 & 20 \\ 0 & 4 & 1 & 0 & -1 & 1 & 0 & 0 & 20 \\ 1 & 1 & 1 & 1 & 1 & 0 & 1 & 0 & b_0 \\ -1 & -5 & -2 & -1 & 0 & 0 & 0 & 1 & -20-b_0 \end{bmatrix}, \tag{10.15}$$

wherein only the artificial variables are basic, *i.e.*, $x_5 = 20$, $x_{a0} = b_0$. Under restricted basis entry, the (blocked) pivot in (10.15) must be 4 (since b_0 is arbitrarily large). Then (10.15) is transformed to

$$\begin{matrix} & x_1 & x_2 & & & & & & \end{matrix}$$
$$\begin{bmatrix} 0 & \frac{1}{2} & 1 & \frac{1}{4} & 0 & 0 & 0 & 0 & 5 \\ 0 & \boxed{\frac{7}{2}} & 0 & -\frac{1}{4} & -1 & 1 & 0 & 0 & 15 \\ 1 & \frac{1}{2} & 0 & \frac{3}{4} & 1 & 0 & 1 & 0 & b_0-5 \\ -1 & -4 & 0 & -\frac{1}{2} & 0 & 0 & 0 & 1 & -10-b_0 \end{bmatrix} \tag{10.16}$$

with $x_2 = 5$, $x_5 = 15$, $x_{a0} = b_0 - 5$, and $f_r^* = -10 - b_0$ depicting the optimal basic feasible solution to the restricted primal.

To find the values of the new dual surplus variables we can use the relationship $u_{sj}^* = u_{sj} + \bar{\theta}(-\bar{c}_j^*)$, where

$$\bar{\theta} = \min_j \left\{ \frac{u_{sj}}{\bar{c}_j^*}, \ j \notin \mathcal{J}, \ -\bar{c}_j^* < 0 \right\}.$$

With

$$\frac{u_{s0}}{\bar{c}_0^*} = \frac{10}{1} = 10, \ \frac{u_{s1}}{\bar{c}_1^*} = \frac{8}{4} = 2, \text{ and } \frac{u_{s4}}{\bar{c}_4^*} = \frac{10}{1/2} = 20,$$

it follows that $\bar{\theta} = 2$ and thus

$$\begin{aligned}
u_{s0}^* &= u_{s0} + \bar{\theta}(-\bar{c}_0^*) = 10 + 2(-1) = 8 \\
u_{s1}^* &= u_{s1} + \bar{\theta}(-\bar{c}_1^*) = 8 + 2(-4) = 0 \\
u_{s2}^* &= u_{s2} + \bar{\theta}(-\bar{c}_2^*) = 0 + 2(0) = 0 \\
u_{s3}^* &= u_{s3} + \bar{\theta}(-\bar{c}_3^*) = 10 + 2(-\tfrac{1}{2}) = 9 \\
u_{s4}^* &= u_{s4} + \bar{\theta}(-\bar{c}_4^*) = 10 + 2(0) = 10.
\end{aligned}$$

Since $u_{s1}^* = u_{s2}^* = 0$, it is evident, by virtue of the complementary slackness requirement, that both x_1 and x_2 can assume positive values and thus $\mathcal{J} = \{1, 2\}$. Moreover, the new modified dual objective value is $\bar{g}^* = \bar{g} + \bar{\theta} f_r^* = 10b_0 + 2(-10 - b_0) = -20 + 8b_0$.

Again employing restricted basis entry we can easily determine that the pivot in (10.16) is $\frac{7}{2}$ and thus the new optimal basic feasible solution to the restricted primal appears as

$$\begin{bmatrix} x_0 & x_1 & x_2 & & & & & \\ 0 & 0 & 1 & \frac{2}{7} & \frac{1}{7} & 0 & 0 & \frac{20}{7} \\ 0 & 1 & 0 & -\frac{1}{14} & -\frac{2}{7} & 0 & 0 & \frac{30}{7} \\ \boxed{1} & 0 & 0 & \frac{11}{14} & \frac{8}{7} & 1 & 0 & b_0 - \frac{50}{7} \\ -1 & 0 & 0 & -\frac{11}{14} & -\frac{8}{7} & 0 & 1 & \frac{50}{7} - b_0 \end{bmatrix}, \tag{10.17}$$

where $x_1 = \frac{30}{7}$, $x_2 = \frac{20}{7}$, $x_{a0} = b_0 - \frac{50}{7}$, and $f_r^* = \frac{50}{7} - b_0$. (Note that the column corresponding to the outgoing artificial basis vector associated with x_5 has been deleted from the current simplex matrix.) Proceeding as above, since

$$\frac{u_{s0}}{\overline{c}_0^*} = 8, \quad \frac{u_{s3}}{\overline{c}_3^*} = \frac{126}{11}, \text{ and } \frac{u_{s4}}{\overline{c}_4^*} = \frac{35}{4},$$

it follows that $\overline{\theta} = 8$ and thus the new set of dual surplus variables appears as

$$\begin{aligned} u_{s0}^* &= 8 + 8(-1) = 0 \\ u_{s1}^* &= 0 + 8(0) = 0 \\ u_{s2}^* &= 0 + 8(0) = 0 \\ u_{s3}^* &= 9 + 8(-11/14) = 19/7 \\ u_{s4}^* &= 10 + 8(-8/7) = 6/7 \end{aligned}$$

with $\overline{g}^* = \overline{g} + \overline{\theta} f_r^* = \frac{260}{7}$. Given that $u_{s0}^* = u_{s1}^* = u_{s2}^* = 0$, $\mathfrak{J} = \{0, 1, 2\}$. Upon pivoting in (10.17) we ultimately obtain

$$\begin{bmatrix} 0 & 0 & 1 & \frac{2}{7} & \frac{1}{7} & 0 & \frac{20}{7} \\ 0 & 1 & 0 & -\frac{1}{14} & -\frac{2}{7} & 0 & \frac{30}{7} \\ 1 & 0 & 0 & \frac{11}{14} & \frac{8}{7} & 0 & b_0 - \frac{50}{7} \\ 0 & 0 & 0 & 0 & 0 & 1 & 0 \end{bmatrix},$$

where the column associated with the outgoing artificial basic variable x_{a0} has been deleted. Here $x_1 = \frac{30}{7}$, $x_2 = \frac{20}{7}$, $f_r^* = 0$, and $\overline{g}^* = \overline{g} + \overline{\theta}(0) = \frac{260}{7} = f$.

10.7. Summary of the Primal-Dual Method

At this point we can review the salient computational features of the primal-dual technique. A summary of the main steps which underlie a process which strives for primal feasibility and optimality simultaneously is incorporated in the:

10.3. PRIMAL-DUAL METHOD.

STEP 0. Initialize.

Obtain an initial dual feasible (primal optimal) solution $\mathbf{U}$. This determines a set of dual surplus variables $\mathbf{U}_s$ and thus the elements of the index set $\mathfrak{J} = \{j \mid u_{sj} = 0\}$. The associated restricted primal problem can then be constructed.

STEP 1. Solve the restricted primal problem. If $f_r^* = 0$ (Phase I terminates), the corresponding solution is also optimal for the original primal problem.

STEP 2. $f_r^* < 0$ and $\overline{c}_j^* \leq 0$ for all j, then the primal problem has no feasible solution and thus the process is terminated. If $f_r^* < 0$ and $-\overline{c}_j^* < 0$ for at least one $j \notin \mathcal{J}$, then step 3 is warranted.

STEP 3. Determine a new set of dual surplus variables according to $u_{sj}^* = u_{sj} + \overline{\theta}(-\overline{c}_j^*)$, $j = (0), 1, \ldots, n$, where

$$\overline{\theta} = \min_j \left\{ \frac{u_{sj}}{\overline{c}_j^*},\ j \notin \mathcal{J},\ -\overline{c}_j^* < 0 \right\},$$

along with an updated objective value $\overline{g}^* = \overline{g} + \overline{\theta} f_r^*$ (or $g^* = g + \theta f_r^*$ if an artificial constraint is not included.)

STEP 4. Use the u_{sj}^*'s to specify a new index set $\mathcal{J}$. Return to step 1.

So to find an optimal **X** given a starting feasible **U**, we solve, under weak complementarity, the restricted primal problem, obtain information from its dual, and then use this information to update the initial dual solution. The entire cycle is repeated as required or terminated. (The following schematic illustrates the main steps of this iterative process.)

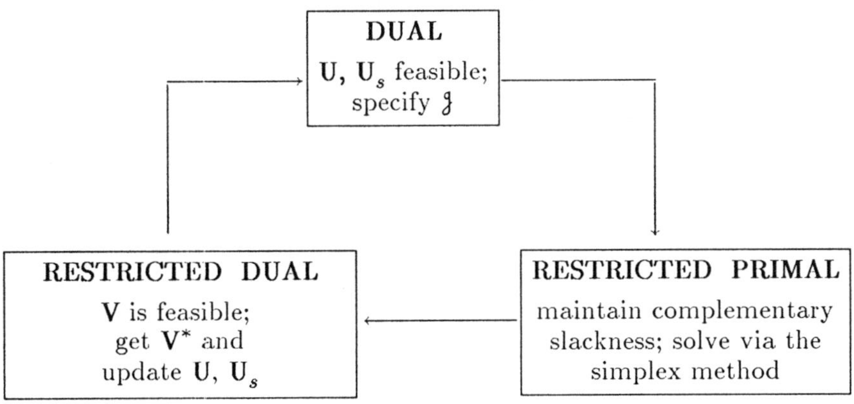

An appealing aspect of the primal-dual routine is that it can be initiated using *any* dual feasible solution (not necessarily basic). This then allows us to immediately specify $\mathcal{J}$ and in turn the corresponding restricted primal problem.

By solving the latter we obtain an improved solution to the dual. From this we can form a new restricted primal program, the solution of which leads to a further improvement in the dual objective. After a finite number of such improvements, an optimal basic feasible solution is generated for the original primal problem. Thus the primal-dual process may be viewed as a way of starting with an infeasible basic solution to the primal problem and using an available dual feasible solution to decrease the infeasibility form of the extended primal problem in a prescribed manner such that, when a primal basic feasible solution is obtained, it will also be optimal.

10.8. A Robust Primal-Dual Algorithm

In the preceding sections of this chapter we have utilized either the dual simplex or primal-dual algorithms exclusively to solve linear programming problems. We shall now examine a solution method which exploits certain features of both the primal simplex and dual simplex techniques (thus accounting for its *robustness*), with the actual routine selected to carry out a particular iteration dependent upon whether a primal simplex or dual simplex pivot operation is feasible (or desirable). In this regard, the robust or mixed method relaxes both primal and dual feasibility simultaneously (here not all $-\bar{c}_j \geq 0$ and not all $x_{Bi} \geq 0$ at some round in the simplex process) and works towards primal optimality and feasibility.

Ignoring the possibility of degeneracy, we may now summarize the salient features of the

10.4. ROBUST PRIMAL-DUAL METHOD.

STEP 0. Initialize.

Transform all structural constraints to the "$\leq$" variety. Thus no artificial variables are needed since: (a) any "$\geq$" structural constraint can be converted to "$\leq$" form by admitting primal infeasibility; and (b) if any equality constraints

$$\sum_{j=1}^{p} a_{ij}x_j = b_i, i=1,\ldots,r, \tag{10.18}$$

are present, the latter may be replaced by the equivalent system of inequalities

$$\begin{aligned} &\text{(a)} \quad \sum_{j=1}^{p} a_{ij}x_j \leq b_i, i=1,\ldots,r, \\ &\text{(b)} \quad -\sum_{i=1}^{r}\left(\sum_{j=1}^{p} a_{ij}x_j\right) \leq -\sum_{i=1}^{r} b_i. \end{aligned} \tag{10.19}$$

The substitution of (10.19) for (10.18) in the linear program again allows us to economize on the use of artificial variables at the modest cost of including (10.19.b) as an additional structural constraint.

STEP 1. Evaluate the effect of both primal simplex *and* dual simplex pivots on the primal objective function, *i.e.*,

(a) if a primal simplex pivot is possible, denote its effect on f as

$$\Delta f_P = \bar{c}_j \hat{\theta}, \; \hat{\theta} = \min_i \left\{ \frac{x_{Bi}}{y_{ij}}, \; y_{ij} > 0 \right\};$$

(b) if a dual simplex pivot is possible, denote its effect on f as

$$\Delta f_D = x_{Bi}\bar{\theta}, \; \bar{\theta} = \max_j \left\{ \frac{-\bar{c}_j}{-y_{ij}}, \; y_{ij} < 0 \right\}.$$

If a pivot is not available for the smallest $-\bar{c}_j$ value (in the case of primal simplex pivoting) or the smallest $-x_{Bi}$ term (in the case of dual simplex pivoting), choose the next smallest $-\bar{c}_j$ or $-x_{Bi}$ value for pivot selection.

STEP 2. Perform either a primal or dual simplex pivot according to which has the greatest impact of f, *i.e.*, the largest increase in f is determined as $max\{\Delta f_P, \Delta f_D\}$. If neither a primal nor dual simplex pivot is possible, even though the current solution is primal infeasible and/or nonoptimal, terminate the process — the original problem has no feasible solution. Otherwise return to step 1.

10.9. The Complementary Pivot Method

One of the salient features of the primal-dual method is that at each round of calculations the (weak) complementary slackness conditions are satisfied. In a similar vein, a computational routine which explicitly requires that a complementarity condition holds at each iteration, namely the complementary pivot method (Lemke (1965)), has been used to generate a solution to the so-called **linear complementarity problem** (hereafter abbreviated as *LCP*): given a vector $\mathbf{q} \epsilon \mathbf{R}^n$ and an nth order matrix $\mathbf{M}$, find a vector $\mathbf{Z} \epsilon \mathbf{R}^n$ such that

$$\begin{aligned} &\text{(a)}\quad \mathbf{q}+\mathbf{MZ} \geqq \mathbf{0} \\ &\text{(b)}\quad \mathbf{Z} \geqq \mathbf{0} \\ &\text{(c)}\quad \mathbf{Z}'(\mathbf{q}+\mathbf{MZ}) = 0 \end{aligned} \tag{10.20}$$

Following convention, the *LCP* will be denoted by the pair $(\mathbf{q}, \mathbf{M})$.

If we let $\mathbf{W} = \mathbf{q}+\mathbf{MZ}$, then the $LCP(\mathbf{q}, \mathbf{M})$ may be written in the alternative fashion: find vectors $\mathbf{W}$, $\mathbf{Z} \epsilon \mathbf{R}^n$ such that

$$\begin{aligned} &\text{(a)}\quad \mathbf{W}-\mathbf{MZ} = \mathbf{q} \\ &\text{(b)}\quad \mathbf{W} \geqq \mathbf{0},\ \mathbf{Z} \geqq \mathbf{0} \\ &\text{(c)}\quad \mathbf{W}'\mathbf{Z} = 0. \end{aligned} \tag{10.21}$$

Here $\mathbf{W}'\mathbf{Z} = \sum_{i=1}^{n} w_i z_i = 0$ or $w_i z_i = 0$, $i=1,\ldots,n$, is called the **complementarity condition**, where the variables w_i and z_i, called a **complementary pair**, are said to be **complements** of each other, *i.e.*, at least one variable in each pair w_i, z_i must equal zero.

A nonnegative solution $(\mathbf{W}, \mathbf{Z})$ to the linear system (10.21.a) is called a **feasible solution** to the *LCP*. Furthermore, a feasible solution $(\mathbf{W}, \mathbf{Z})$ to the *LCP* which also satisfies the complementarity condition $\mathbf{W}'\mathbf{Z} = 0$ is called a **complementary solution.** Interestingly enough, problems (10.20), (10.21) contain no objective function; and there is but a single structural constraint in the variables vectors $\mathbf{W}$ and $\mathbf{Z}$, namely the complementarity restriction $\mathbf{W}'\mathbf{Z} = 0$.

If $\mathbf{q} \geqq \mathbf{0}$, then there exists an obvious (trivial) **complementary basic solution** given by $\mathbf{W} = \mathbf{q}$, $\mathbf{Z} = \mathbf{0}$. Hence the *LCP* has a nontrivial solution only when

$\mathbf{q} \ngeq \mathbf{0}$. In this circumstance, the initial basic solution given by $\mathbf{W} = \mathbf{q}$, $\mathbf{Z} = \mathbf{0}$ is infeasible even though the complementarity condition $\mathbf{W}'\mathbf{Z} = 0$ holds.

We now demonstrate that any symmetric primal-dual pair of problems

$$\begin{array}{ll} \underline{\text{PRIMAL}} & \underline{\text{DUAL}} \\ max\ f(\mathbf{X}) = \mathbf{C}'\mathbf{X} \quad \text{s.t.} & min\ g(\mathbf{U}) = \mathbf{b}'\mathbf{U} \quad \text{s.t.} \\ \mathbf{A}\mathbf{X} \leqq \mathbf{b},\ \mathbf{X} \geqq \mathbf{0} & \mathbf{A}'\mathbf{U} \geqq \mathbf{C},\ \mathbf{U} \geqq \mathbf{0} \end{array} \tag{(10.5)}$$

where $\mathbf{X} \in R^p$, $\mathbf{U} \in R^m$, and $\mathbf{A}$ is of order $(m \times p)$, can always be written as a $LCP(\mathbf{q}, \mathbf{M})$. To this end let $\mathbf{X}_s \in R^m$ and $\mathbf{U}_s \in R^p$ be vectors of primal slack and dual surplus variables respectively. Then our aim is to find vectors $\mathbf{X}$, $\mathbf{U}$, $\mathbf{X}_s$, and $\mathbf{U}_s$ such that

$$\begin{aligned} &\mathbf{X}_s = -\mathbf{A}\mathbf{X} + \mathbf{b} \\ &\mathbf{U}_s = \mathbf{A}'\mathbf{U} - \mathbf{C} \\ &\mathbf{X},\ \mathbf{U},\ \mathbf{X}_s,\ \mathbf{U}_s \geqq \mathbf{0} \\ &\mathbf{X}'\mathbf{U}_s + \mathbf{U}'\mathbf{X}_s = 0. \end{aligned} \tag{10.22}$$

If we set

$$\mathbf{W} = \begin{bmatrix} \mathbf{X}_s \\ \mathbf{U}_s \end{bmatrix}, \quad \mathbf{q} = \begin{bmatrix} \mathbf{b} \\ -\mathbf{C} \end{bmatrix}, \quad \mathbf{Z} = \begin{bmatrix} \mathbf{U} \\ \mathbf{X} \end{bmatrix}, \quad \text{and} \quad \mathbf{M} = \begin{bmatrix} \mathbf{0} & -\mathbf{A} \\ \mathbf{A}' & \mathbf{0} \end{bmatrix},$$

then (10.22) can be easily rewritten as

$$\begin{aligned} &\begin{bmatrix} \mathbf{X}_s \\ \mathbf{U}_s \end{bmatrix} - \begin{bmatrix} \mathbf{0} & -\mathbf{A} \\ \mathbf{A}' & \mathbf{0} \end{bmatrix} \begin{bmatrix} \mathbf{U} \\ \mathbf{X} \end{bmatrix} = \begin{bmatrix} \mathbf{b} \\ -\mathbf{C} \end{bmatrix} \\ &\begin{bmatrix} \mathbf{X}_s \\ \mathbf{U}_s \end{bmatrix} \geqq \mathbf{0}, \begin{bmatrix} \mathbf{U} \\ \mathbf{X} \end{bmatrix} \geqq \mathbf{0} \\ &(\mathbf{X}_s', \mathbf{U}_s') \begin{bmatrix} \mathbf{U} \\ \mathbf{X} \end{bmatrix} = 0 \end{aligned} \tag{10.22.1}$$

or (10.21). Hence we may solve a linear program by converting it into an equivalent LCP, *i.e.*, it follows from the weak complementary slackness theorem that solving the symmetric primal-dual pair of problems above is equivalent to finding vectors $\mathbf{W}$, $\mathbf{Z}$ such that (10.22.1) holds.

Looking to the geometry of the $LCP(\mathbf{q}, \mathbf{M})$ let us consider the $(n \times 2n)$ matrix

$$\left[\ \mathbf{I}_n, -\mathbf{M}\ \right] = \left[\ \mathbf{e}_1,\ldots,\mathbf{e}_n, -\mathbf{m}_1,\ldots,-\mathbf{m}_n\ \right]$$

associated with the equation system $\mathbf{I}_n\mathbf{W} - \mathbf{MZ} = \mathbf{q}$. Since $\mathbf{q}$ is expressable as a nonnegative linear combination of the columns of $\left[\ \mathbf{I}_n, -\mathbf{M}\ \right]$,

$$\sum_{i=1}^{n} w_i\mathbf{e}_i + \sum_{i=1}^{n} z_i(-\mathbf{m}_i) = \mathbf{q}, \tag{10.23}$$

and $w_i z_i = 0$, i=1,...,n, it is evident that if either w_i or z_i is positive, its complement must be zero. Hence $\mathbf{e}_i$, $-\mathbf{m}_i$, i=1,...,n, cannot both be used in (10.23). Here $\mathbf{e}_i$, $-\mathbf{m}_i$, $1 \leq i \leq n$, will be called the **i^{th} complementary pair of vectors.** In general, to form (10.23), for each i=1,...,n, select either the column $\mathbf{e}_i$ of $\mathbf{I}_n$ or the ith column $-\mathbf{m}_i$ of $\mathbf{M}$ and denote the column actually chosen by $\mathbf{b}^{(i)}$. Then the convex cone formed from the **complementary set of vectors $\mathbf{b}^{(i)}$**, i=1,...,n, is called the **complementary cone of M** and denoted as

$$coni\left\{\mathbf{b}^{(1)},\ldots,\mathbf{b}^{(n)}\right\} = \left\{\mathbf{V} \mid \mathbf{V} = \sum_{i=1}^{n} \theta_i\mathbf{b}^{(i)},\ 0 \leq \theta_i \in \mathbf{R}\right\}.$$

Since $\mathbf{b}^{(i)}$ equals $\mathbf{e}_i$ or $-\mathbf{m}_i$ for each $1 \leq i \leq n$, it is evident that there exists 2^n possible complementary cones of an $(n \times n)$ matrix $\mathbf{M}$. Let $\mathcal{C}(\mathbf{M})$ denote the class of complementary cones of $\mathbf{M}$. Then solving the $LCP(\mathbf{q},\mathbf{M})$ (10.22) (or (10.22.1)) amounts to finding a complementary cone in $\mathcal{C}(\mathbf{M})$ that contains the vector $\mathbf{q}$, *i.e.*, we desire to find a complementary set of vectors $\left\{\mathbf{b}^{(1)},\ldots,\mathbf{b}^{(n)}\right\}$ such that:

(1) $\mathbf{b}^{(i)} = \mathbf{e}_i$ or $-\mathbf{m}_i$, $1 \leq i \leq n$; and

(2) $\mathbf{q} \in coni\left\{\mathbf{b}^{(1)},\ldots,\mathbf{b}^{(n)}\right\} \subset \mathcal{C}(\mathbf{M})$.

EXAMPLE 10.4. For n=2, let

$$\mathbf{M} = \begin{bmatrix} 2 & -1 \\ -1 & 4 \end{bmatrix}.$$

To find $\mathcal{C}(\mathbf{M})$, let us form

$$\left[\mathbf{e}_1, \mathbf{e}_2, -\mathbf{m}_1, -\mathbf{m}_2\right] = \begin{bmatrix} 1 & 0 & -2 & 1 \\ 0 & 1 & 1 & -4 \end{bmatrix}.$$

From this matrix we may obtain $2^2 = 4$ complementary sets of vectors and thus four complementary cones (Figure 10.1):

$$coni\{\mathbf{e}_1, \mathbf{e}_2\},\ coni\{-\mathbf{m}_1, -\mathbf{m}_2\},$$
$$coni\{\mathbf{e}_1, -\mathbf{m}_2\}, \text{ and } coni\{\mathbf{e}_2, -\mathbf{m}_1\}.$$

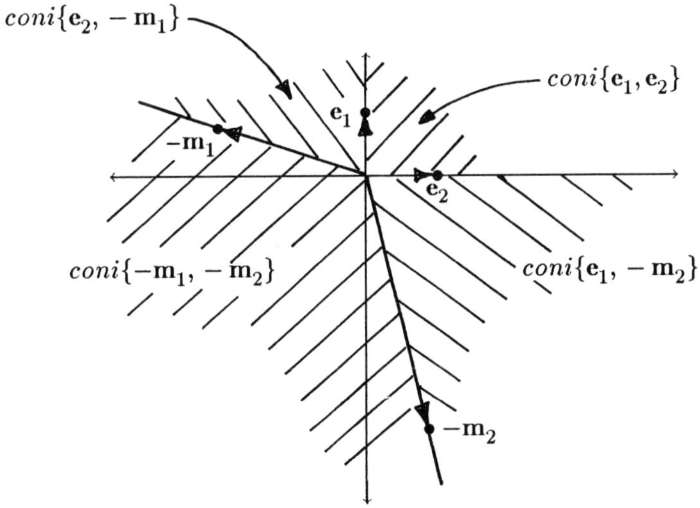

Figure 10.1

EXAMPLE 10.5. For n=2, let

$$\mathbf{M} = \begin{bmatrix} 3 & -1 \\ -1 & 2 \end{bmatrix},\quad \mathbf{q} = \begin{bmatrix} -1 \\ -2 \end{bmatrix}.$$

Then $LCP(\mathbf{q}, \mathbf{M})$ appears as

$$\begin{aligned} w_1 \quad\quad - 3z_1 + z_2 &= -1 \\ w_2 + z_1 - 2z_2 &= -2 \\ w_1, w_2, z_1, z_2 &\geq 0 \\ w_1 z_1 = 0,\ w_2 z_2 &= 0. \end{aligned}$$

If we set $w_1 = w_2 = 0$, then the preceding system becomes

$$\begin{aligned} -3z_1 + z_2 &= -1 \\ z_1 - 2z_2 &= -2 \\ z_1, z_2 &\geq 0 \end{aligned}$$

with solution $z_1 = \frac{4}{5}$, $z_2 = \frac{7}{5}$. Hence $\mathbf{q}$ lies within the complementary cone $coni\{-\mathbf{m}_1, -\mathbf{m}_2\}$ (Figure 10.2) and the complete solution to the $LCP(\mathbf{q}, \mathbf{M})$ is $\mathbf{W}' = \mathbf{0}'$, $\mathbf{Z}' = \left(\frac{4}{5}, \frac{7}{5}\right)$.

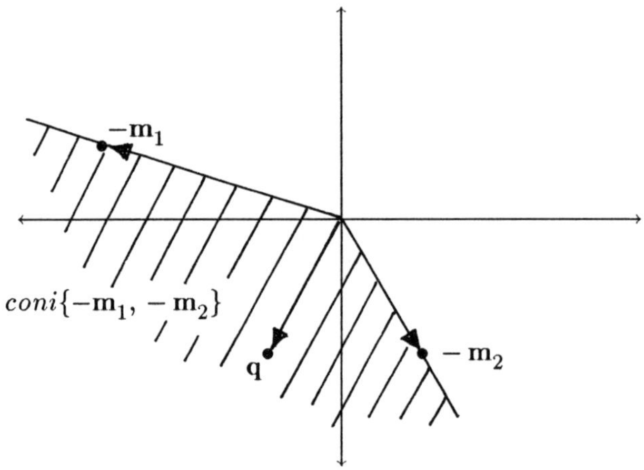

Figure 10.2

In order to solve the $LCP(\mathbf{q}, \mathbf{M})$ associated with a linear program, let us examine the workings of the complementary pivot method. This algorithm starts with an initial basic solution to (10.21) of the form $\mathbf{W} = \mathbf{q}$, $\mathbf{Z} = \mathbf{0}$, where $\mathbf{q} \ngeq \mathbf{0}$. To avoid having at least one $w_i < 0$, a nonnegative artificial variable z_0 is introduced at a sufficiently positive level (take $z_0 = -\underset{i}{min}\{q_i, i{=}1,\ldots,n\}$) into the left-hand side of each equation in $\mathbf{W} - \mathbf{MZ} = \mathbf{q}$. In this regard, we now desire to find vectors $\mathbf{W}$, $\mathbf{Z}$ and a variable z_0 such that

$$\begin{aligned} &\text{(a)}\quad \mathbf{W} - \mathbf{MZ} - z_0\mathbf{1} = \mathbf{q} \\ &\text{(b)}\quad \mathbf{W},\, \mathbf{Z} \geqq \mathbf{0},\; z_0 \geq 0 \\ &\text{(c)}\quad \mathbf{W}'\mathbf{Z} = 0 \text{ or } w_i z_i = 0 \end{aligned} \tag{10.24}$$

for at least $n{-}1$ of the i values given $z_0 \geq 0$.

With each new right-hand side value $q_i + z_0$ nonnegative in (10.24), a basic solution to the same amounts to $\mathbf{W} = \mathbf{q} + z_0\mathbf{1} \geqq \mathbf{0}$, $\mathbf{Z} = \mathbf{0}$. But while this basic solution is nonnegative and satisfies all of the relationships in (10.24) (including complementarity), it is nevertheless infeasible for the original LCP (10.21) since $z_0 > 0$. A solution such as this will be termed an **almost complementary basic**

solution. More specifically, while a complementary basic solution to (10.21) is one which contains exactly one basic variable from each complementary pair of variables (w_i, z_i), $i=1,\ldots,n$, an almost complementary basic solution of (10.21) is a basic solution for (10.24) in which z_0 is a basic variable and there is exactly one basic variable from each of only $n-1$ complementary pairs of variables.

In this light, the complementary pivot method has us move from one *almost complementary basic solution* of (10.21) to another until we reach a *complementary basic solution* to (10.21), in which case we have $w_i z_i = 0$ for all $i=1,\ldots,n$. At this point the algorithm is terminated. To see this let us rewrite (10.24) as

$$\begin{array}{ccccccccccccccc}
w_1 & & & - m_{11}z_1 & - & \cdots & - & m_{1r}z_r & - & \cdots & - & m_{1n}z_n & - & z_0 & = & q_1 \\
& w_2 & & - m_{21}z_1 & - & \cdots & - & m_{2r}z_r & - & \cdots & - & m_{2n}z_n & - & z_0 & = & q_2 \\
\cdots & \cdots & \cdots & \cdots & \cdots & \cdots & \cdots & \cdots & \cdots & \cdots & \cdots & \cdots & \cdots & \cdots & \cdots & \cdots \\
& & w_n & - m_{n1}z_1 & - & \cdots & - & m_{nr}z_r & - & \cdots & - & m_{nn}z_n & - & z_0 & = & q_n
\end{array}$$

with associated *simplex matrix*

$$\begin{array}{c} \begin{array}{cccccccccccccc} w_1 & w_2 & & w_r & & w_n & z_1 & & z_r & & z_n & z_0 & \end{array} \\ \left[\begin{array}{cccccccccccccc}
1 & 0 & \cdots & 0 & \cdots & 0 & -m_{11} & \cdots & -m_{1r} & \cdots & -m_{1n} & -1 & q_1 \\
0 & 1 & \cdots & 0 & \cdots & 0 & -m_{21} & \cdots & -m_{2r} & \cdots & -m_{2n} & -1 & q_2 \\
\cdots & \cdots & \cdots & \cdots & \cdots & \cdots & \cdots & \cdots & \cdots & \cdots & \cdots & \cdots & \cdots \\
0 & 0 & \cdots & 1 & \cdots & 0 & -m_{r1} & \cdots & -m_{rr} & \cdots & -m_{rn} & \boxed{-1} & q_r \\
\cdots & \cdots & \cdots & \cdots & \cdots & \cdots & \cdots & \cdots & \cdots & \cdots & \cdots & \cdots & \cdots \\
0 & 0 & \cdots & 0 & \cdots & 1 & -m_{n1} & \cdots & -m_{nr} & \cdots & -m_{nn} & -1 & q_n
\end{array}\right] \end{array} \tag{10.25}$$

To find the initial almost complementary basic solution the artificial variable z_0 is made basic by having it replace the current basic variable with the most negative q_i value. Assume $q_r = \min_i\{q_i, i=1,\ldots,n\}$. Hence z_0 replaces w_r in the set of basic variables. To accomplish this we pivot on the (boxed) element -1 in (10.25). This renders

$$\begin{bmatrix} \overset{w_1}{1} & \overset{w_2}{0} & \cdots & \overset{w_r}{-1} & \cdots & \overset{w_n}{0} & \overset{z_1}{\overline{m}_{11}} & \cdots & \overset{z_r}{\overline{m}_{1r}} & \cdots & \overset{z_n}{\overline{m}_{1n}} & \overset{z_0}{0} & \overline{q}_1 \\ 0 & 1 & \cdots & -1 & \cdots & 0 & \overline{m}_{21} & \cdots & \overline{m}_{2r} & \cdots & \overline{m}_{2n} & 0 & \overline{q}_2 \\ \cdots & & & & & & & & & & & & \cdots \\ 0 & 0 & \cdots & -1 & \cdots & 0 & \overline{m}_{r1} & \cdots & \overline{m}_{rr} & \cdots & \overline{m}_{rn} & 1 & \overline{q}_r \\ \cdots & & & & & & & & & & & & \cdots \\ 0 & 0 & \cdots & -1 & \cdots & 0 & \overline{m}_{n1} & \cdots & \overline{m}_{nr} & \cdots & \overline{m}_{nn} & 0 & \overline{q}_n \end{bmatrix} \tag{10.26}$$

where

$$\begin{aligned} \overline{q}_r &= -q_r; \quad \overline{q}_i = q_i - q_r, \quad i{\neq}r, \;\; i{=}1,\ldots,n, \\ \overline{m}_{rj} &= m_{rj}, \quad j{=}1,\ldots,n, \\ \overline{m}_{ij} &= m_{rj} - m_{ij}, \; i{\neq}r, \;\; j{=}1,\ldots,n. \end{aligned}$$

As (10.26) reveals, an almost complementary basic solution to (10.21) is $w_i = \overline{q}_i$, $i \neq r$, $z_0 = \overline{q}_r$, and $z_i = 0$, $i{=}1,\ldots,n$.

The complementary pivot algorithm generates a sequence of almost complementary basic solutions until z_0 becomes nonbasic or zero. Moreover, pivoting must be done in a fashion such that: (1) complementarity between the variables w_i, z_i is maintained at each iteration; and (2) each successive basic solution is nonnegative.

At the moment both w_r, z_r are nonbasic ($w_r z_r = 0$ is satisfied). Since w_r turned nonbasic in (10.26), the appropriate variable to choose for entry into the set of basic variables is its complement z_r. In fact, this selection criterion for determining the nonbasic variable to become basic is referred to as the

COMPLEMENTARY PIVOT RULE: choose as the incoming basic variable the one complementary to the basic variable which just turned nonbasic.

Once the entering variable is selected, the outgoing variable can readily be determined from the standard simplex exit criterion, *i.e.*, w_k is replaced by z_k in the set of basic variables if

$$\hat{\theta} = \frac{\overline{q}_k}{\overline{m}_{kr}} = \min_i \left\{ \frac{\overline{q}_i}{\overline{m}_{ir}}, \; \overline{m}_{ir} > 0, \; i{=}1,\ldots,n \right\}. \tag{10.27}$$

A pivot operation is then performed using $\overline{m}_{kr}$ as the pivot element. Once w_k

turns nonbasic, the complementary pivot rule next selects z_k for entry into the set of basic variables. Complementary pivoting continues until one of two possible outcomes obtains, at which point the complementary pivot algorithm terminates:

(1) the simplex exit criterion selects row r as the pivot row and z_0 turns nonbasic. The resulting solution is a complementary basic solution to (10.21); or

(2) no $\bar{m}_{ir} > 0$ in (10.27). In this instance the complementary pivot algorithm fails to yield a complementary basic solution and **ray termination** occurs, *i.e.*, an extreme ray $\begin{bmatrix} \mathbf{W} \\ \mathbf{Z} \\ z_0 \end{bmatrix} + \lambda \mathbf{d}$, $0 \leq \lambda \in \mathbf{R}$, $\mathbf{d}$ an extreme direction, of almost complementary basic solutions is determined.

A few additional comments pertaining to the complementary pivot algorithm are in order. First, in a linear programming setting, $\mathbf{M}$ is a positive semidefinite matrix.[3] In this instance the said algorithm always terminates with a complementary basic solution if the original linear programming problem has a solution. Next, if ray termination obtains, the linear program either has no feasible solution or, if a primal (dual) feasible solution exists, the primal (dual) objective function is unbounded.

EXAMPLE 10.6. For the primal-dual pair of problems

PRIMAL	DUAL
$max\ f = x_1 + 3x_2 + 4x_3$ s.t.	$min\ g = 100u_1 + 40u_2$ s.t.
$2x_1 + x_2 + 4x_3 \leq 100$	$2u_1 + u_2 \geq 1$
$x_1 + 3x_2 + x_3 \leq 40$	$u_1 + 3u_2 \geq 3$
$x_1, x_2, x_3 \geq 0$	$4u_1 + u_2 \geq 4$
	$u_1, u_2 \geq 0$

we may form

[3] A sufficient condition for an $(n \times n)$ matrix $\mathbf{A}$ to be **positive semidefinite** is that: (a) all principal diagonal elements are nonnegative; and (b) all principal minors are nonnegative. (A principal minor of order k is a determinant of a principal submatrix of order k, where the latter is a $(k \times k)$ submatrix obtained from $\mathbf{A}$ by deleting any $n-k$ variables and their corresponding columns.)

$$\mathbf{W} = \begin{bmatrix} \mathbf{X}_s \\ \mathbf{U}_s \end{bmatrix} = \begin{bmatrix} x_4 \\ x_5 \\ u_3 \\ u_4 \\ u_5 \end{bmatrix}, \ \mathbf{q} = \begin{bmatrix} \mathbf{b} \\ -\mathbf{C} \end{bmatrix} = \begin{bmatrix} 100 \\ 40 \\ -1 \\ -3 \\ -4 \end{bmatrix}, \ \mathbf{Z} = \begin{bmatrix} \mathbf{U} \\ \mathbf{X} \end{bmatrix} = \begin{bmatrix} u_1 \\ u_2 \\ x_1 \\ x_2 \\ x_3 \end{bmatrix}, \text{ and}$$

$$\mathbf{M} = \begin{bmatrix} 0 & 0 & -2 & -1 & -4 \\ 0 & 0 & -1 & -3 & -1 \\ 2 & 1 & 0 & 0 & 0 \\ 1 & 3 & 0 & 0 & 0 \\ 4 & 1 & 0 & 0 & 0 \end{bmatrix}.$$

Then (10.24) becomes

$$\begin{aligned} x_4 \qquad\qquad\qquad\qquad\qquad\qquad &+ 2x_1 + x_2 + 4x_3 - z_0 = 100 \\ x_5 \qquad\qquad\qquad\qquad\qquad &+ x_1 + 3x_2 + x_3 - z_0 = 40 \\ u_3 \qquad - 2u_1 - u_2 \qquad\qquad &- z_0 = -1 \\ u_4 \quad - u_1 - 3u_2 \qquad\qquad &- z_0 = -3 \\ u_5 - 4u_1 - u_2 \qquad\qquad &- z_0 = -4 \end{aligned}$$

$$x_1, \ldots, x_5 \geq 0; \quad u_1, \ldots, u_5 \geq 0$$

$$x_4u_1 = x_5u_2 = u_3x_1 = u_4x_2 = u_5x_3 = 0$$

The simplex matrix associated with this problem is

$$\begin{array}{c} \begin{matrix} x_4 & x_5 & u_3 & u_4 & u_5 & u_1 & u_2 & x_1 & x_2 & x_3 & z_0 & \end{matrix} \\ \begin{bmatrix} 1 & 0 & 0 & 0 & 0 & 0 & 0 & 2 & 1 & 4 & -1 & 100 \\ 0 & 1 & 0 & 0 & 0 & 0 & 0 & 1 & 3 & 1 & -1 & 40 \\ 0 & 0 & 1 & 0 & 0 & -2 & -1 & 0 & 0 & 0 & -1 & -1 \\ 0 & 0 & 0 & 1 & 0 & -1 & -3 & 0 & 0 & 0 & -1 & -3 \\ 0 & 0 & 0 & 0 & 1 & -4 & -1 & 0 & 0 & 0 & \boxed{-1} & -4 \end{bmatrix} \longrightarrow \end{array}$$

$$\begin{array}{c} \begin{matrix} & & & & u_5 & & & & & x_3 & & \end{matrix} \\ \begin{bmatrix} 1 & 0 & 0 & 0 & -1 & 4 & 1 & 2 & 1 & \boxed{4} & 0 & 104 \\ 0 & 1 & 0 & 0 & -1 & 4 & 1 & 1 & 3 & 1 & 0 & 44 \\ 0 & 0 & 1 & 0 & -1 & 2 & 0 & 0 & 0 & 0 & 0 & 3 \\ 0 & 0 & 0 & 1 & -1 & 3 & -2 & 0 & 0 & 0 & 0 & 1 \\ 0 & 0 & 0 & 0 & -1 & 4 & 1 & 0 & 0 & 0 & 1 & 4 \end{bmatrix}. \end{array}$$

Here z_0 has been made basic by pivoting on the (boxed) element -1. Thus the resulting almost complementary basic solution is $x_4 = 104$, $x_5 = 44$, $u_3 = 3$, $u_4 = 1$, $z_0 = 4$, and $u_5 = u_1 = u_2 = x_1 = x_2 = x_3 = 0$. Since u_5 turned nonbasic, its complementary variable x_3 will be chosen for entry into the set of basic variables.

In this regard, pivoting on the (boxed) element 4 yields

$$\begin{array}{c} \begin{matrix} x_4 & & & & & u_1 & & & & & & \end{matrix} \\ \begin{bmatrix} \frac{1}{4} & 0 & 0 & 0 & -\frac{1}{4} & 1 & \frac{1}{4} & \frac{1}{2} & \frac{1}{4} & 1 & 0 & 26 \\ -\frac{1}{4} & 1 & 0 & 0 & -\frac{3}{4} & 3 & \frac{3}{4} & \frac{1}{2} & \frac{11}{4} & 0 & 0 & 18 \\ 0 & 0 & 1 & 0 & -1 & 2 & 0 & 0 & 0 & 0 & 0 & 3 \\ 0 & 0 & 0 & 1 & -1 & \boxed{3} & -2 & 0 & 0 & 0 & 0 & 1 \\ 0 & 0 & 0 & 0 & -1 & 4 & 1 & 0 & 0 & 0 & 1 & 4 \end{bmatrix} \end{array}$$

Our second almost complementary solution is $x_5 = 18$, $u_3 = 3$, $u_1 = 4$, $x_3 = 26$, $z_0 = 4$, and $x_4 = u_5 = u_1 = u_2 = x_1 = x_2 = 0$. Given that x_4 turned nonbasic, its complementary variable u_1 will now become basic. To this end we now pivot on the (boxed) element 3. The resulting matrix is

$$\begin{array}{c} \begin{matrix} & & & u_4 & & & & & x_2 & & & \end{matrix} \\ \begin{bmatrix} \frac{1}{4} & 0 & 0 & -\frac{1}{3} & \frac{1}{12} & 0 & \frac{11}{12} & \frac{1}{2} & \frac{1}{4} & 1 & 0 & \frac{77}{3} \\ -\frac{1}{4} & 1 & 0 & -1 & \frac{1}{4} & 0 & \frac{11}{4} & \frac{1}{2} & \boxed{\frac{11}{4}} & 0 & 0 & 17 \\ 0 & 0 & 1 & -\frac{2}{3} & -\frac{1}{3} & 0 & \frac{4}{3} & 0 & 0 & 0 & 0 & \frac{7}{3} \\ 0 & 0 & 0 & \frac{1}{3} & -\frac{1}{3} & 1 & -\frac{2}{3} & 0 & 0 & 0 & 0 & \frac{1}{3} \\ 0 & 0 & 0 & -\frac{4}{3} & \frac{1}{3} & 0 & \frac{11}{3} & 0 & 0 & 0 & 1 & \frac{8}{3} \end{bmatrix} \end{array}.$$

As this matrix reveals, the third almost complementary basic solution amounts to $x_5 = 17$, $u_3 = \frac{7}{3}$, $u_1 = \frac{1}{3}$, $x_3 = \frac{77}{3}$, $z_0 = \frac{8}{3}$, and $x_4 = u_4 = u_5 = u_2 = x_1 = x_2 = 0$. Since u_4 turned nonbasic, its complementary variable x_2 becomes basic, *i.e.*, pivoting on $\frac{11}{4}$ above gives

$$\begin{array}{c} \begin{array}{cccccccccccc} & x_5 & & & & & u_2 & & & & & \end{array} \\ \left[\begin{array}{cccccccccccc} \frac{3}{11} & -\frac{1}{11} & 0 & -\frac{8}{33} & \frac{2}{33} & 0 & \frac{1}{6} & \frac{5}{11} & 0 & 1 & 0 & \frac{796}{33} \\ -\frac{1}{11} & \frac{4}{11} & 0 & -\frac{4}{11} & \frac{1}{11} & 0 & 1 & \frac{2}{11} & 1 & 0 & 0 & \frac{68}{11} \\ 0 & 0 & 1 & -\frac{2}{3} & -\frac{1}{3} & 0 & \frac{4}{3} & 0 & 0 & 0 & 0 & \frac{7}{3} \\ 0 & 0 & 0 & \frac{1}{3} & -\frac{1}{3} & 1 & -\frac{2}{3} & 0 & 0 & 0 & 0 & \frac{1}{3} \\ 0 & 0 & 0 & -\frac{4}{3} & \frac{1}{3} & 0 & \boxed{\frac{11}{3}} & 0 & 0 & 0 & 1 & \frac{8}{3} \end{array}\right] \end{array}.$$

Hence the current almost complementary basic solution amounts to $u_3 = \frac{7}{3}$, $u_1 = \frac{1}{3}$, $x_2 = \frac{68}{11}$, $x_3 = \frac{796}{33}$, $z_0 = \frac{8}{3}$, and $x_4 = x_5 = u_4 = u_5 = u_2 = x_1 = 0$. With x_5 now nonbasic, its complementary variable u_2 is selected for entry into the set of basic variables. Pivoting on $\frac{11}{3}$ above yields

$$\begin{array}{c} \begin{array}{cccccccccccc} & & & & & & & & & & z_0 & \end{array} \\ \left[\begin{array}{cccccccccccc} \frac{3}{11} & -\frac{1}{11} & 0 & -\frac{2}{11} & \frac{1}{22} & 0 & 0 & \frac{5}{11} & 0 & 1 & -\frac{1}{33} & 23.630 \\ -\frac{1}{11} & \frac{4}{11} & 0 & 0 & 0 & 0 & 0 & \frac{2}{11} & 1 & 0 & -\frac{3}{11} & 5.464 \\ 0 & 0 & 1 & -\frac{2}{11} & -\frac{5}{11} & 0 & 0 & 0 & 0 & 0 & -\frac{4}{11} & 1.360 \\ 0 & 0 & 0 & \frac{1}{11} & -\frac{3}{11} & 1 & 0 & 0 & 0 & 0 & \frac{2}{11} & 0.818 \\ 0 & 0 & 0 & -\frac{4}{11} & \frac{1}{11} & 0 & 1 & 0 & 0 & 0 & \frac{3}{11} & 0.727 \end{array}\right] \end{array}.$$

Since z_0 has turned nonbasic, we now have a complementary basic and thus optimal solution consisting of $u_3 = 1.36$, $u_1 = 0.818$, $u_2 = 0.727$, $x_2 = 5.464$, $x_3 = 23.63$, and $x_4 = x_5 = u_4 = u_5 = x_1 = 0$.

11 POSTOPTIMALITY ANALYSIS I

Our point of departure in this chapter involves the assumption that we have already obtained an optimal basic feasible solution to a linear programming problem in standard form, *i.e.*, to a problem of the form

$$\begin{aligned} max f(\mathbf{X}) &= \mathbf{C}'\mathbf{X} \quad \text{s.t.} \\ \mathbf{AX} &= \mathbf{b},\ \mathbf{X} \geqq \mathbf{0}. \end{aligned}$$

We then seek to employ the information contained within the optimal simplex matrix to gain some insight into the solution of a variety of linear programming problems which are essentially slight modifications of the original one. Specifically, two broad classes of variations in the original problem may be identified. First, there may exist instances in which a ***sensitivity analysis*** is appropriate, *e.g.*, there may occur discrete changes in any of the components of the matrices **C**, **b**, or **A**, in which case the values of c_j, b_i, or a_{ij}, $i=1,\ldots,m$; $j=1,\ldots,n-m$, respectively are altered (increased or decreased) in order to determine the extent to which the original problem may be modified without violating the feasibility or optimality of the original solution. Second, we may address ourselves to the problem of adding or deleting certain variables or structural constraints, in which case this latter class of problems will simply be 3termed the analysis of ***structural changes***.

11.1. Sensitivity Analysis

Turning again to our linear programming problem in standard form, let us

$$\begin{aligned} max\, f(\mathbf{X}) &= \mathbf{C}'_B\mathbf{X}_B + \mathbf{C}'_R\mathbf{X}_R \quad \text{s.t.} \\ \mathbf{BX}_B + \mathbf{RX}_R &= \mathbf{b},\ \text{with } \mathbf{X}_B,\ \mathbf{X}_R \geqq \mathbf{0}. \end{aligned}$$

Then as previously determined, with

$$\mathbf{X}_B = \mathbf{B}^{-1}\mathbf{b} - \mathbf{B}^{-1}\mathbf{R}\mathbf{X}_R,$$
$$f(\mathbf{X}) = \mathbf{C}'_B\mathbf{B}^{-1}\mathbf{b} + (\mathbf{C}'_R - \mathbf{C}'_B\mathbf{B}^{-1}\mathbf{R})\mathbf{X}_R,$$

it follows that if $\mathbf{X}_R = \mathbf{0}$, the resulting solution is basic (primal) feasible if $\mathbf{X}_B = \mathbf{B}^{-1}\mathbf{b} \geqq \mathbf{0}$ and optimal (dual feasible) if $\mathbf{C}'_R - \mathbf{C}'_B\mathbf{B}^{-1}\mathbf{R} \leqq \mathbf{0}'$.

(1) CHANGING AN OBJECTIVE FUNCTION COEFFICIENT. If we now change $\mathbf{C}' = (\mathbf{C}'_B, \mathbf{C}'_R)$ to $\mathbf{C}' + \Delta\mathbf{C}' = (\mathbf{C}'_B + \Delta\mathbf{C}'_B, \mathbf{C}'_R + \Delta\mathbf{C}'_R)$, then

$$\begin{aligned} f(\mathbf{X}) &= (\mathbf{C}_B + \Delta\mathbf{C}_B)'\mathbf{X}_B + (\mathbf{C}_R + \Delta\mathbf{C}_R)'\mathbf{X}_R \\ &= (\mathbf{C}_B + \Delta\mathbf{C}_B)'(\mathbf{B}^{-1}\mathbf{b} - \mathbf{B}^{-1}\mathbf{R}\mathbf{X}_R) + (\mathbf{C}_R + \Delta\mathbf{C}_R)'\mathbf{X}_R \\ &= \mathbf{C}'_B\mathbf{B}^{-1}\mathbf{b} + \Delta\mathbf{C}'_B\mathbf{B}^{-1}\mathbf{b} + (\mathbf{C}'_R - \mathbf{C}'_B\mathbf{B}^{-1}\mathbf{R} + \Delta\mathbf{C}'_R - \Delta\mathbf{C}'_B\mathbf{B}^{-1}\mathbf{R})\mathbf{X}_R. \end{aligned}$$

Since $\mathbf{X}_B = \mathbf{B}^{-1}\mathbf{b}$ is independent of $\mathbf{C}$, this new basic solution remains primal feasible while if $(\mathbf{C}'_R - \mathbf{C}'_B\mathbf{B}^{-1}\mathbf{b}) + (\Delta\mathbf{C}'_R - \Delta\mathbf{C}'_B\mathbf{B}^{-1}\mathbf{R}) \leqq \mathbf{0}'$, dual feasibility or primal optimality holds. Since our discussion from this point on will be framed in terms of examining selected components within the optimal simplex matrix, it is evident from the last row of the same (footnote 1, Chapter 7) that primal optimality or dual feasibility is preserved if $\mathbf{C}'_B\mathbf{B}^{-1}\mathbf{R} - \mathbf{C}'_R + (\Delta\mathbf{C}'_B\mathbf{B}^{-1}\mathbf{R} - \Delta\mathbf{C}'_R) \geqq \mathbf{0}'$. Let us now consider the various subcases which may obtain.

If $\Delta\mathbf{C}_B = \mathbf{0}$ (so that an objective function coefficient associated with a non-basic variable changes), it follows that optimality is preserved if $\mathbf{C}'_B\mathbf{B}^{-1}\mathbf{R} - \mathbf{C}'_R - \Delta\mathbf{C}'_R \geqq \mathbf{0}'$, *i.e.*, if

$$-\bar{c}_j - \Delta c_{Rj} \geq 0, \; j=1,\ldots,n-m. \tag{11.1}$$

How may changes in c_{Rj} affect (11.1)? First, if c_{Rj} decreases in value so that $\Delta c_{Rj} < 0$, then (11.1) is still satisfied. In fact, c_{Rj} may decrease without bound and still not violate (11.1), *i.e.*, this lower limit may be expressed as

$$\underline{lim}\Delta c_{Rj} = -\infty. \tag{11.2.a}$$

Next, if c_{Rj} increases so that $\Delta c_{Rj} > 0$, then (11.1) still holds so long as c_{Rj} increases by no more than $-\bar{c}_j$ since (11.1) requires that $-\bar{c}_j \geq \Delta c_{Rj}$, $j=1,\ldots,n-m$. In this instance the upper limit to Δc_{Rj} is

$$\overline{lim}\Delta c_{Rj} = -\bar{c}. \tag{11.2.b}$$

Hence the only circumstance in which x_{Rj} becomes basic is if its objective function coefficient c_{Rj} increases by more than $-\bar{c}_j$. We may represent the upper limit depicted by (11.2.b) in terms of the new level of c_{Rj}, denoted as $\hat{c}_{Rj}$, since for $\Delta c_{Rj} = \hat{c}_{Rj} - c_{Rj}$,

$$\overline{lim}\hat{c}_{Rj} = c_{Rj} - \bar{c}_j. \tag{11.2.c}$$

Hence this equation provides the upper limit to which c_{Rj} can increase without violating primal optimality.

If $\Delta c_R = \mathbf{0}$ (so that an objective function coefficient associated with a basic variable changes), it follows that the optimality criterion is not violated if $\mathbf{C}'_B\mathbf{B}^{-1}\mathbf{R} - \mathbf{C}'_R + \Delta\mathbf{C}'_B\mathbf{B}^{-1}\mathbf{R} \geqq \mathbf{0}'$, *i.e.*, if

$$-\bar{c}_j + \sum_{i=1}^{m} y_{ij}\Delta c_{Bi} \geq 0,\ j=1,\ldots,n-m. \tag{11.3}$$

Let us assume that c_{Bk} changes. Hence (11.3) becomes

$$-\bar{c}_j + y_{kj}\Delta c_{Bk} \geq 0,\ j=1,\ldots,n-m. \tag{11.3.1}$$

We now consider the circumstances under which changes in c_{Bk} preserve the sense of this inequality. First, if $y_{kj} = 0$, then (11.3.1) holds regardless of the direction or magnitude of the change in c_{Bk}. Second, if $y_{kj} > 0$, then (11.3.1) is satisfied if $\Delta c_{Bk} \geq 0$, *i.e.*, Δc_{Bk} may increase without bound so that its upper limit becomes

$$\overline{lim}\Delta c_{Bk} = +\infty,\ y_{kj} > 0. \tag{11.4.a}$$

Hence (11.3.1) is affected only when c_{Bk} decreases of $\Delta c_{Bk} < 0$. Since (11.3.1) implies that $\bar{c}_j/y_{kj} \leq \Delta c_{Bk}$, $j=1,\ldots,n-m$, it is evident that the lower limit to which Δc_{Bk} can decrease without violating optimality is

$$\underline{lim}\Delta c_{Bk} = \bar{c}_j/y_{kj},\ y_{kj} > 0. \tag{11.4.b}$$

Hence x_{Rj} remains non-basic so long as its objective function coefficient does not decrease by more than $\bar{c}_j/y_{kj}$. Now, it may be the case that $y_{kj} > 0$ for two or more non-basic variables. To insure that (11.3.1) is satisfied for all such j, the lower limit to which Δc_{Bk} can decrease without violating the optimality of the current solution is

$$\underline{lim}\Delta c_{Bk} = \max_j\{\bar{c}_j/y_{kj}, y_{kj} > 0\}. \quad (11.4.c)$$

Finally, if $y_{kj} < 0$, (11.3.1) holds if $\Delta c_{Bk} \leq 0$. Hence Δc_{Bk} may decrease without bound so that its lower limit is

$$\underline{lim}\Delta c_{Bk} = -\infty, \ y_{kj} < 0. \quad (11.5.a)$$

Thus the optimality criterion is affected by changes in c_{Bk} only when $\Delta c_{Bk} > 0$. In this instance (11.3.1) may be rewritten as $\bar{c}_j/y_{kj} \geq \Delta c_{Bk}, j=1,\ldots,n-m$, so that the upper limit to any change in c_{Bk} is

$$\overline{lim}\Delta c_{Bk} = \bar{c}_j/y_{kj}, y_{kj} < 0. \quad (11.5.b)$$

Hence x_{Rj} maintains its non-basic status provided that the increase in c_{Bk} does not exceed $\bar{c}_j/y_{kj}$. If $y_{kj} < 0$ for two or more non-basic variables, then (11.3.1) holds for all relevant j if the upper limit to which Δc_{Bk} can increase is specified as

$$\overline{lim}\Delta c_{Bk} = \min_j\{\bar{c}_j/y_{kj}, y_{kj} < 0\}. \quad (11.5.c)$$

Actually, the optimality criterion (11.3.1) will be maintained if both the lower and upper limits (11.4.c), (11.5.c) respectively are subsumed under the inequality

$$\underline{lim}\Delta c_{Bk} = \max_j\{\bar{c}_j/y_{kj}, y_{kj} > 0\} \leq \Delta c_{Bk} \leq \overline{lim}\Delta c_{Bk} = \min_j\{\bar{c}_j/y_{kj}, y_{kj} < 0\}. \quad (11.6)$$

Hence (11.6) provides the bounds within which Δc_{Bk} may decrease or increase without affecting the optimality of the current solution. As an alternative to the approach pursued in defining (11.4.c) and (11.5.c), we may express these limits in terms of the new level of c_{Bk}, call it $\hat{c}_{Bk}$, by noting that for $\Delta c_{Bk} = \hat{c}_{Bk} - c_{Bk}$, (11.4.c) becomes

$$\underline{lim}\hat{c}_{Bk} = \max_j\{c_{Bk}+\bar{c}_j/y_{kj}, \ y_{kj} > 0\}; \quad (11.4.c.1)$$

while (11.5.c) may be rewritten as

$$\overline{lim}\hat{c}_{Bk} = \min_j\{c_{Bk}+\bar{c}_j/y_{kj}, y_{kj} < 0\}. \tag{11.5.c.1}$$

Hence (11.4.c.1) ((11.5.c.1)) specifies the lower (upper) limit to which c_{Bk} can decrease (increase) without violating the optimality criterion (11.3.1). In view of this modification, (11.6) appears as

$$\underline{lim}\hat{c}_{Bk} = \max_j\{c_{Bk}+\bar{c}_j/y_{kj}, y_{kj} > 0\} \le \hat{c}_{Bk} \le \overline{lim}\hat{c}_{Bk} = \min_j\{c_{Bk}+\bar{c}_j/y_{kj}, y_{kj} < 0\}. \tag{11.6.1}$$

One final point is in order. If c_{Bk} changes in a fashion such that (11.6) and (11.6.1) are satisfied, then the current basic feasible solution remains optimal but the value of the objective function changes to $f = \mathbf{C}_B'\mathbf{B}^{-1}\mathbf{b} + \Delta c_{Bk}x_{Bk}$.

EXAMPLE 11.1. We found in chapter 7 that the optimal simplex matrix associated with the problem

$$\begin{aligned} \max f &= 2x_1 + 3x_2 + 4x_3 \quad \text{s.t.} \\ 2x_1 + x_2 + 4x_3 &\le 100 \\ x_1 + 3x_2 + x_3 &\le 40 \\ x_1, x_2, x_3 &\ge 0 \end{aligned}$$

appears as

$$\begin{bmatrix} \frac{5}{11} & 0 & 1 & \frac{3}{11} & -\frac{1}{11} & 0 & 23.63 \\ \frac{2}{11} & 1 & 0 & -\frac{1}{11} & \frac{4}{11} & 0 & 5.46 \\ \frac{4}{11} & 0 & 0 & \frac{9}{11} & \frac{8}{11} & 1 & 110.91 \end{bmatrix}.$$

A glance at the objective function row of this matrix reveals that $-\bar{c}_1 = \frac{4}{11}$, $-\bar{c}_2 = \frac{9}{11}$, and $-\bar{c}_3 = \frac{8}{11}$. Hence (11.2.b) indicates that $x_{R1} = x_1$, $x_{R2} = x_4$, and $x_{R3} = x_5$ should each maintain their non-basic status provided that their objective function coefficients do not increase by more than $\Delta c_{R1} = \frac{4}{11}$, $\Delta c_{R2} = \frac{9}{11}$, and $\Delta c_{R3} = \frac{8}{11}$ units respectively. Additionally, (11.2.c) may be employed to depict the upper limits to which $c_{R1} = c_1$, $c_{R2} = c_4$, and $c_{R3} = c_5$ can individually increase without violating primal optimality (Table 11.1).

TABLE 11.1.

Non-basic Variables	Objective Function Coefficients	Lower Limit	Upper Limit
$x_{R1} = x_1$	$c_{R1} = c_1 = 2$	$-\infty$	$c_{R1} - \bar{c}_1 = \frac{26}{11}$
$x_{R2} = x_4$	$c_{R2} = c_4 = 0$	$-\infty$	$c_{R2} - \bar{c}_2 = \frac{9}{11}$
$x_{R3} = x_5$	$c_{R3} = c_5 = 0$	$-\infty$	$c_{R3} - \bar{c}_3 = \frac{8}{11}$

Suppose that $c_{R1} = c_1$ is increased to 4. Since $\overline{lim}\hat{c}_{R1} = \frac{26}{11} = c_{R1} - \bar{c}_1 = 4 - \bar{c}_1$, it follows that $-\bar{c}_1 = -\frac{18}{11}$, which in turn implies that the current solution is no longer optimal. In terms of the preceding simplex matrix, $\frac{4}{11}$ is replaced by $-\frac{18}{11}$ in the last row of the same. An additional round of calculations thus yields a new optimal (though not unique) basic feasible solution consisting of $x_1 = 30, x_3 = 10$, and $f = 160$ as evidenced by the following simplex matrix

$$\begin{bmatrix} 0 & -\frac{5}{2} & 1 & \frac{1}{2} & -1 & 0 & 10 \\ 1 & \frac{11}{2} & 0 & -\frac{1}{2} & 2 & 0 & 30 \\ 0 & 9 & 0 & 0 & 4 & 1 & 160 \end{bmatrix}.$$

Next, from (11.6) we see that the limits within which $\Delta c_{B1} = \Delta c_3$, $\Delta c_{B2} = \Delta c_2$ can vary without affecting the optimality of the current solution are

$$\underline{lim}\Delta c_{B1} = max\left\{\frac{\bar{c}_1}{y_{11}} = \frac{-4/11}{5/11} = -\frac{4}{5}, \frac{\bar{c}_2}{y_{12}} = \frac{-9/11}{3/11} = -3\right\}$$
$$= -\frac{4}{5} \leq \Delta c_{B1} \leq \overline{lim}\Delta c_{B1} = min\left\{\frac{\bar{c}_3}{y_{13}} = \frac{-8/11}{-1/11} = 8\right\} = 8;$$

$$\underline{lim}\Delta c_{B2} = max\left\{\frac{\bar{c}_1}{y_{21}} = \frac{-4/11}{2/11} = -2, \frac{\bar{c}_3}{y_{23}} = \frac{-8/11}{4/11} = -2\right\}$$
$$= -2 \leq \Delta c_{B2} \leq \overline{lim}\Delta c_{B2} = min\left\{\frac{\bar{c}_2}{y_{22}} = \frac{-9/11}{-1/11} = 9\right\} = 9.$$

Moreover, we may utilize (11.6.1) to specify the limits within which $c_{B1} = c_3$, $c_{B2} = c_2$ can individually decrease or increase while preserving optimality (Table 11.2).

TABLE 11.2.

Basic Variables	Objective Function Coefficients	Lower Limit	Upper Limit
$x_{B1} = x_3$	$c_{B1} = c_3 = 4$	$c_{B1} + \frac{\bar{c}_1}{y_{11}} = \frac{16}{5}$	$c_{B1} + \frac{\bar{c}_3}{y_{13}} = 12$
$x_{B2} = x_2$	$c_{B2} = c_2 = 3$	$c_{B2} + \frac{\bar{c}_2}{y_{21}} =$ $c_{B2} + \frac{\bar{c}_3}{y_{23}} = 1$	$c_{B2} + \frac{\bar{c}_2}{y_{22}} = 12$

Suppose that $c_{B1} = c_3$ increases to 14. Since changing c_{B1} affects the objective function coefficients on the non-basic variables as well as the optimal value of f itself, it follows that f increases by $\Delta f = \Delta c_{B1} x_{B1} = 10$ (23.63) $= 236.3$ to $\widehat{f} = f + \Delta f = 110.91 + 236.3 = 347.21$ while, by virtue of (11.3.1), $-\bar{c}_1 = \frac{4}{11}$, $-\bar{c}_2 = \frac{9}{11}$, and $-\bar{c}_3 = \frac{8}{11}$ change respectively to

$$
\begin{aligned}
-\widehat{\bar{c}}_1 &= -\bar{c}_1 + y_{11}\Delta c_{B1} = \tfrac{4}{11} + \tfrac{5}{11}10 = \tfrac{54}{11}, \\
-\widehat{\bar{c}}_2 &= -\bar{c}_2 + y_{12}\Delta c_{B1} = \tfrac{9}{11} + \tfrac{3}{11}10 = \tfrac{39}{11}, \\
-\widehat{\bar{c}}_3 &= -\bar{c}_3 + y_{13}\Delta c_{B1} = \tfrac{8}{11} - \tfrac{1}{11}10 = -\tfrac{2}{11}.
\end{aligned}
$$

A substitution of these values into the last row of the optimal simplex matrix yields, after an additional iteration,

$$
\begin{bmatrix}
\frac{5}{11} & 0 & 1 & \frac{3}{11} & -\frac{1}{11} & 0 & 23.63 \\
\frac{2}{11} & 1 & 0 & -\frac{1}{11} & \frac{4}{11} & 0 & 5.46 \\
\frac{54}{11} & 0 & 0 & \frac{39}{11} & -\frac{2}{11} & 1 & 347.21
\end{bmatrix}
\longrightarrow
$$

$$
\begin{bmatrix}
\frac{1}{2} & \frac{1}{4} & 1 & \frac{1}{4} & 0 & 0 & 25 \\
\frac{1}{2} & \frac{11}{4} & 0 & -\frac{1}{4} & 1 & 0 & 15 \\
5 & \frac{1}{2} & 0 & \frac{7}{2} & 0 & 1 & 350
\end{bmatrix},
$$

and thus a new optimal basic feasible solution consisting of $x_3 = 25$, $x_5 = 15$, and $f = 350$.

(2) CHANGING A COMPONENT OF THE REQUIREMENTS VECTOR. If we next change $\mathbf{b}$ to $\mathbf{b} + \Delta\mathbf{b}$, the new vector of basic variables becomes

$$\begin{aligned}\hat{\mathbf{X}}_B &= \mathbf{B}^{-1}(\mathbf{b}+\Delta\mathbf{b}) \\ &= \mathbf{B}^{-1}\mathbf{b} + \mathbf{B}^{-1}\Delta\mathbf{b}.\end{aligned} \tag{11.7}$$

If this new basic solution is primal feasible ($\hat{\mathbf{X}}_B \geqq \mathbf{0}$), then, since $\mathbf{X}_B$ is independent of $\mathbf{C}$, primal optimality or dual feasibility is also preserved. For $\Delta\mathbf{b}' = (\Delta b_1, \ldots, \Delta b_m)$, (11.7) becomes

$$\hat{\mathbf{X}}_B = \mathbf{B}^{-1}\mathbf{b} + \begin{bmatrix} \boldsymbol{\beta}_1 \Delta\mathbf{b} \\ \vdots \\ \boldsymbol{\beta}_m \Delta\mathbf{b} \end{bmatrix} = \mathbf{X}_B + \begin{bmatrix} \sum_{j=1}^{m} \beta_{ij}\Delta b_j \\ \vdots \\ \sum_{j=1}^{m} \beta_{mj}\Delta b_j \end{bmatrix} \text{ or} \tag{11.8}$$

$$\hat{x}_{Bi} = x_{Bi} + \sum_{j=1}^{m} \beta_{ij}\Delta b_j,\ i=1,\ldots,m, \tag{11.8.1}$$

where $\boldsymbol{\beta}_i = (\beta_{i1}, \ldots, \beta_{im})$, $i=1,\ldots,m$, denotes the ith row of $\mathbf{B}^{-1}$. Let us assume that b_k changes. In order for this change to insure the feasibility of the resulting basic solution, we require, by virtue of (11.8.1), that

$$x_{Bi} + \beta_{ik}\Delta b_k \geq 0,\ i=1,\ldots,m. \tag{11.8.2}$$

Under what conditions do changes in b_k preserve the sense of this inequality? First, if $\beta_{ik} = 0$, then (11.8.2) holds for any Δb_k. Second, for $\beta_{ik} > 0$, (11.8.2) is satisfied if $\Delta b_k \geq 0$, *i.e.*, Δb_k may increase without bound so that its upper limit is

$$\overline{lim}\Delta b_k = +\infty,\ \beta_{ik} > 0. \tag{11.9.a}$$

Thus the feasibility of the new solution is affected only when $\Delta b_k < 0$. Since (11.8.2) implies that $-x_{Bi}/\beta_{ik} \leq \Delta b_k$, $i=1,\ldots,m$, it follows that the lower limit to which Δb_k can decrease without violating primal feasibility is

$$\underline{lim}\Delta b_k = -x_{Bi}/\beta_{ik},\ \beta_{ik} > 0. \tag{11.9.b}$$

Hence x_{Bi} remains in the set of basic variables so long as b_k does not decrease by more than $-x_{Bi}/\beta_{ik}$. If $\beta_{ik} > 0$ for two or more basic variables x_{Bi}, then to insure that (11.8.2) is satisfied for all such i, the lower limit to which Δb_k can decrease without violating the feasibility of the new primal solution is

$$\underline{lim}\Delta b_k = \max_i\{-x_{Bi}/\beta_{ik}, \beta_{ik} > 0\}. \tag{11.9.c}$$

Finally, if $\beta_{ik} < 0$, (11.8.2) holds if $\Delta b_k \leq 0$. Hence Δb_k can decrease without bound so that its lower limit is

$$\underline{lim}\Delta b_k = -\infty, \ \beta_{ik} < 0. \tag{11.10.a}$$

Thus the feasibility of the new basic solution is affected by changes in b_k only when $\Delta b_k > 0$. In this instance (11.8.2) may be rewritten as $-x_{Bi}/\beta_{ik} \geq \Delta b_k$, $i=1,\ldots,m$, so that the upper limit to Δb_k is

$$\overline{lim}\Delta b_k = -x_{Bi}/\beta_{ik}, \ \beta_{ik} < 0. \tag{11.10.b}$$

Hence x_{Bi} maintains its basic status provided that b_k does not increase by more than $-x_{Bi}/\beta_{ik}$. And if $\beta_{ik} < 0$ for more than one basic variable x_{Bi}, then (11.8.2) holds for all relevant i if the upper limit to which Δb_k can increase is specified as

$$\overline{lim}\Delta b_k = \min_i\{-x_{Bi}/\beta_{ik}, \beta_{ik} < 0\}. \tag{11.10.c}$$

If we choose to combine the lower and upper limits (11.9.c), (11.10.c) respectively into a single expression, then (11.8.2) will be maintained if

$$\begin{aligned}\underline{lim}\Delta b_k = \max_i\{-x_{Bi}/\beta_{ik}, \beta_{ik} > 0\} \leq \Delta b_k \leq \overline{lim}\Delta b_k = \\ \min_i\{-x_{Bi}/\beta_{ik}, \beta_{ik} < 0\}.\end{aligned} \tag{11.11}$$

Hence (11.11) provides the bounds within which Δb_k may decrease or increase without affecting the feasibility of $\hat{\mathbf{X}}_B$. As an alternative to (11.11), we may express the limits contained therein in terms of the new level of b_k, call it $\hat{b}_k$, by noting that since $\Delta b_k = \hat{b}_k - b_k$,

$$\begin{aligned}\underline{lim}\Delta\hat{b}_k = \max_i\{b_k - x_{Bi}/\beta_{ik}, \beta_{ik} > 0\} \leq \Delta\hat{b}_k \leq \overline{lim}\Delta\hat{b}_k = \\ \min_i\{b_k - x_{Bi}/\beta_{ik}, \beta_{ik} < 0\}.\end{aligned} \tag{11.11.1}$$

What about the effect of a change in a component of the requirements vector on the optimal value of the objective function? By virtue of the preceding discussion, if (11.11) or (11.11.1) is satisfied, those variables which are currently basic remain so. However, a change in any of the requirements $b_i, i=1,\ldots,m$, alters the optimal values of these variables, which in turn affect the optimal value of f. To determine the magnitude of this change in f, let us form

$$\begin{aligned} f(\hat{\mathbf{X}}_B) &= \mathbf{C}'_B\hat{\mathbf{X}}_B = \mathbf{C}'_B(\mathbf{B}^{-1}\mathbf{b}+\mathbf{B}^{-1}\Delta\mathbf{b}) \\ &= f(\mathbf{X}_B) + \mathbf{C}'_B\mathbf{B}^{-1}\Delta\mathbf{b} \\ &= f(\mathbf{X}_B) + \sum_{i=1}^{m} c_{Bi}\left(\sum_{j=1}^{m}\beta_{ij}\Delta b_j\right). \end{aligned}$$

If again b_k alone changes, then

$$f(\hat{\mathbf{X}}_B) = f(\mathbf{X}_B) + \sum_{i=1}^{m} c_{Bi}\beta_{ik}\Delta b_k,$$

and thus

$$\Delta f = f(\hat{\mathbf{X}}_B) - f(\mathbf{X}_B) = \Delta b_k \sum_{i=1}^{m} c_{Bi}\beta_{ik}. \tag{11.12}$$

By virtue of our earlier specification of the dual variables and (11.12),

$$\frac{\Delta f}{\Delta b_k} = \sum_{i=1}^{m} c_{Bi}\beta_{ik} = u_k, \tag{11.12.1}$$

i.e., as noted in chapter 5, $\Delta f/\Delta b_k$ represents a measure of the amount by which the optimal value of the primal objective function changes per unit change in the kth requirement b_k. Hence the total change in f when the kth requirement changes by Δb_k units is

$$\Delta f = u_k \Delta b_k. \tag{11.12.2}$$

EXAMPLE 11.2. Let us again look to the optimal simplex matrix provided at the outset of the previous example problem. In the light of (7.10), if we examine the columns of this final primal simplex matrix which were, in the original simplex matrix, the columns of the identity matrix, we find that the ith column of $\mathbf{B}^{-1}$ appears under the column which was initially the ith column of $\mathbf{I}_2$. Hence the inverse of the optimal basis matrix is

$$\mathbf{B}^{-1} = \begin{bmatrix} \frac{3}{11} & -\frac{1}{11} \\ -\frac{1}{11} & \frac{4}{11} \end{bmatrix}.$$

By utilizing the components of this inverse we may determine the limits within which $\Delta b_1, \Delta b_2$ can vary without affecting the feasibility of the new vector of basic variables according to

$$\begin{aligned} \underline{lim}\Delta b_1 &= max\left\{ \frac{-x_{B1}}{\beta_{11}} = \frac{-23.63}{3/11} = -86.64 \right\} = -86.64 \leq \Delta b_1 \\ &\leq \overline{lim}\Delta b_1 = min\left\{ \frac{-x_{B2}}{\beta_{21}} = \frac{-5.46}{-1/11} = 60.06 \right\} = 60.06; \end{aligned}$$

$$\begin{aligned} \underline{lim}\Delta b_2 &= max\left\{ \frac{-x_{B2}}{\beta_{22}} = \frac{-5.46}{4/11} = -15.02 \right\} = -15.02 \leq \Delta b_2 \\ &\leq \overline{lim}\Delta b_2 = min\left\{ \frac{-x_{B1}}{\beta_{12}} = \frac{-23.63}{-1/11} = 259.93 \right\} = 259.93. \end{aligned}$$

Additionally, we may utilize (11.11.1) to specify the limits within which b_1, b_2 can individually decrease or increase without violating the feasibility of $\hat{\mathbf{X}}_B$ (Table 11.3).

TABLE 11.3.

Basic Variables	Original Requirements	Lower Limit	Upper Limit
$x_{B1} = x_3$	$b_1 = 100$	$b_1 - \frac{x_{B1}}{\beta_{11}} = 13.36$	$b_1 - \frac{x_{B2}}{\beta_{21}} = 160.06$
$x_{B2} = x_2$	$b_2 = 40$	$b_2 - \frac{x_{B2}}{\beta_{22}} = 24.98$	$b_2 - \frac{x_{B1}}{\beta_{12}} = 299.93$

If we increase the first requirement b_1 by 10 units so that $\Delta \mathbf{b}' = (10, 0)$ then, from (11.7),

$$\hat{\mathbf{X}}_B = \begin{bmatrix} 23.63 \\ 5.46 \end{bmatrix} + \begin{bmatrix} \frac{3}{11} & -\frac{1}{11} \\ -\frac{1}{11} & \frac{4}{11} \end{bmatrix} \begin{bmatrix} 10 \\ 0 \end{bmatrix} = \begin{bmatrix} 26.36 \\ 4.55 \end{bmatrix}.$$

(Note that the components of this vector may alternatively be obtained from (11.8.2).) Hence this new basic solution is primal feasible as well as primal optimal. Moreover, from (11.12.2), the optimal value of f increases by $\Delta f = \frac{9}{11}10 = 8.18$ units.

Let us now increase the second requirement b_2 by 300 units. In this instance $\Delta \mathbf{b}' = (0, 300)$. Again employing (11.7) we obtain

$$\hat{\mathbf{X}}_B = \begin{bmatrix} 23.63 \\ 5.46 \end{bmatrix} + \begin{bmatrix} \frac{3}{11} & -\frac{1}{11} \\ -\frac{1}{11} & \frac{4}{11} \end{bmatrix} \begin{bmatrix} 0 \\ 300 \end{bmatrix} = \begin{bmatrix} -3.64 \\ 114.55 \end{bmatrix}.$$

(Here too (11.8.2) may be utilized to determine the components of $\hat{\mathbf{X}}_B$.) In addition, from (11.12.2), f increases to $f + \Delta f = 110.91 + \frac{8}{11}300 = 329$. Upon substituting these values into the last column of the optimal simplex matrix we obtain, using the dual simplex routine,

$$\begin{bmatrix} \frac{5}{11} & 0 & 1 & \frac{3}{11} & -\frac{1}{11} & 0 & -3.64 \\ \frac{2}{11} & 1 & 0 & -\frac{1}{11} & \frac{4}{11} & 0 & 114.55 \\ \frac{4}{11} & 0 & 0 & \frac{9}{11} & \frac{8}{11} & 1 & 329 \end{bmatrix} \longrightarrow \begin{bmatrix} -5 & 0 & -11 & -3 & 1 & 0 & 40 \\ 2 & 1 & 4 & 1 & 0 & 0 & 100 \\ 4 & 0 & 8 & 3 & 0 & 1 & 300 \end{bmatrix},$$

Hence the new optimal basic feasible solution which emerges is $x_2 = 100$, $x_5 = 40$, and $f = 300$.

(3) CHANGING A COMPONENT OF THE COEFFICIENT MATRIX. We now look to the effect on the optimal solution of a change in a component of the coefficient matrix $\mathbf{A}$ associated with the structural constraint system $\mathbf{AX} = \mathbf{b}$. Since $\mathbf{A}$ was previously partitioned as $\mathbf{A} = [\mathbf{B}, \mathbf{R}]$, it may be the case that: some component r_{ij}, $i=1,\ldots,m$; $j=1,\ldots,n-m$, of a non-basic vector $\mathbf{r}_j$ changes; or a component b_{ik}, $i,k=1,\ldots,m$, of a basis vector $\mathbf{b}_i$ changes.

Let us begin by assuming that the ℓth component of the nonbasic vector $\mathbf{r}'_k = (r_{1k}, \ldots, r_{\ell k}, \ldots, r_{mk})$ changes to $\hat{r}_{\ell k} = r_{\ell k} + \Delta r_{\ell k}$ so that $\mathbf{r}_k$ is replaced by

$\hat{\mathbf{r}}_k = \mathbf{r}_k + \Delta r_{\ell k}\mathbf{e}_\ell$ in $\mathbf{R} = [\mathbf{r}_1,\ldots,\mathbf{r}_k,\ldots,\mathbf{r}_{n-m}]$. Hence the matrix of non-basic vectors may now be expressed as $\hat{\mathbf{R}} = [\mathbf{r}_1,\ldots,\hat{\mathbf{r}}_k,\ldots,\mathbf{r}_{n-m}]$. With $\hat{\mathbf{X}}_B = \mathbf{B}^{-1}\mathbf{b} - \mathbf{B}^{-1}\hat{\mathbf{R}}\mathbf{X}_{\hat{R}}$ independent of $\Delta r_{\ell k}$ at a basic feasible solution (since $\mathbf{X}_{\hat{R}} = \mathbf{0}$), it follows that this non-basic solution is primal feasible, while if $-\mathbf{C}'_{\hat{R}} + \mathbf{C}'_B\mathbf{B}^{-1}\hat{\mathbf{R}} = -\mathbf{C}'_{\hat{R}} + \mathbf{U}'\hat{\mathbf{R}} \geqq \mathbf{0}'$, dual feasibility or primal optimality is also preserved. Since

$$
\begin{aligned}
-\mathbf{C}'_{\hat{R}} + \mathbf{U}'\hat{\mathbf{R}} &= (-c_{R1}+\mathbf{U}'\mathbf{r}_1,\ldots,-c_{Rk}+\mathbf{U}'\mathbf{r}_k+u_\ell\Delta r_{\ell k},\ldots,-c_{R,n-m}+\mathbf{U}'\mathbf{r}_{n-m}) \\
&= (-\bar{c}_1,\ldots,-\bar{c}_k+u_\ell\Delta r_{\ell k},\ldots,-\bar{c}_{n-m})
\end{aligned}
$$

and $-\bar{c}_j \geq 0$, $j=1,\ldots,k,\ldots,n-m$, it is evident that primal optimality is maintained if

$$-\bar{c}_k + u_\ell\Delta r_{\ell k} \geq 0. \tag{11.13}$$

Under what circumstances will changes in $r_{\ell k}$ preserve the sense of this inequality? First, if $u_\ell = 0$, then (11.13) holds regardless of the direction or magnitude of the change in $r_{\ell k}$. Second, for $u_\ell > 0$, (11.13) is satisfied if $\Delta r_{\ell k} \geq 0$, *i.e.*, $\Delta r_{\ell k}$ may increase without bound so that its upper limit becomes

$$\overline{lim}\Delta r_{\ell k} = +\infty,\ u_\ell > 0. \tag{11.14.a}$$

Hence (11.13) is affected only when $r_{\ell k}$ decreases or $\Delta r_{\ell k} < 0$. Since (11.13) implies that $\bar{c}_k/u_\ell \leq \Delta r_{\ell k}$, it is evident that the lower limit to which $\Delta r_{\ell k}$ can decrease without violating optimality is

$$\underline{lim}\ \Delta r_{\ell k} = \bar{c}_k/u_\ell,\ u_\ell > 0. \tag{11.14.b}$$

Thus $\hat{\mathbf{r}}_k$ enters the basis (or $x_{\hat{R}k}$ enters the set of basic variables) only if the decrease in $r_{\ell k}$ is greater than $\bar{c}_k/u_\ell$. If we express this lower limit in terms of the new level of $r_{\ell k}$, $\hat{r}_{\ell k}$, then for $\Delta r_{\ell k} = \hat{r}_{\ell k} - r_{\ell k}$, (11.14.b) becomes

$$\underline{lim}\ \hat{r}_{\ell k} = r_{\ell k} + \bar{c}_k/u_\ell,\ u_\ell > 0. \tag{11.14.b.1}$$

Next, let us assume that the lth component of the basis vector $\mathbf{b}'_k = (b_{1k},\ldots,b_{\ell k},\ldots,b_{mk})$ changes to $\hat{b}_{\ell k} = b_{\ell k} + \Delta b_{\ell k}$ so that $\mathbf{b}_k$ is replaced in the basis matrix $\mathbf{B} = [\mathbf{b}_1,\ldots,\mathbf{b}_k,\ldots,\mathbf{b}_m]$ by

$$\widehat{\mathbf{b}}_k = \begin{bmatrix} b_{1k} \\ \vdots \\ \widehat{b}_{\ell k} \\ \vdots \\ b_{mk} \end{bmatrix} = \begin{bmatrix} b_{1k} \\ \vdots \\ b_{\ell k} + \Delta b_{\ell k} \\ \vdots \\ b_{mk} \end{bmatrix} = \mathbf{b}_k + \Delta b_{\ell k}\mathbf{e}_\ell.$$

Now, for the current optimal feasible solution the inverse of the basis matrix $\mathbf{B}$ is known. Hence our problem becomes one of replacing the kth column of $\mathbf{B}$ by $\widehat{\mathbf{b}}_k$ so as to obtain the inverse of the new basis matrix $\widehat{\mathbf{B}} = [\mathbf{b}_1, \ldots, \widehat{\mathbf{b}}_k, \ldots, \mathbf{b}_m]$. Since $\widehat{\mathbf{b}}_k$ is expressible as a linear combination of the vectors within the current basis matrix, *i.e.*,

$$\widehat{\mathbf{b}}_k = \sum_{i=1}^{m} y_{ik}\mathbf{b}_i = \mathbf{B}\mathbf{Y}_k,$$

it follows that

$$\mathbf{Y}_k = \mathbf{B}^{-1}\widehat{\mathbf{b}}_k = \begin{bmatrix} \beta_1 \\ \vdots \\ \beta_k \\ \vdots \\ \beta_m \end{bmatrix} \widehat{\mathbf{b}}_k = \begin{bmatrix} \beta_1\mathbf{b}_k + \beta_{1\ell}\Delta b_{\ell k} \\ \vdots \\ \beta_k\mathbf{b}_k + \beta_{k\ell}\Delta b_{\ell k} \\ \vdots \\ \beta_m\mathbf{b}_k + \beta_{m\ell}\Delta b_{\ell k} \end{bmatrix} = \begin{bmatrix} \beta_{1\ell}\Delta_{\ell k} \\ \vdots \\ 1 + \beta_{k\ell}\Delta b_{\ell k} \\ \vdots \\ \beta_{mk}\Delta b_{\ell k} \end{bmatrix},$$

where $\beta_{i\ell}$, $i = 1, \ldots, m$, is the ℓth component of β_i, the ith row of $\mathbf{B}^{-1}$. Then, by virtue of (A.1) of appendix A,

$$\widehat{\mathbf{I}}_m = \begin{bmatrix} 1 & \cdots & \frac{-\beta_{1\ell}\Delta b_{\ell k}}{1+\beta_{k\ell}\Delta b_{\ell k}} & \cdots & 0 \\ \vdots & & \vdots & & \vdots \\ 0 & \cdots & \frac{1}{1+\beta_{k\ell}\Delta b_{\ell k}} & \cdots & 0 \\ \vdots & & \vdots & & \vdots \\ 0 & \cdots & \frac{-\beta_{m\ell}\Delta b_{\ell k}}{1+\beta_{k\ell}\Delta b_{\ell k}} & \cdots & 1 \end{bmatrix}.$$

Hence from (A.2), $\widehat{\mathbf{B}}^{-1} = \widehat{\mathbf{I}}_m\mathbf{B}^{-1}$ and thus

$$\hat{\mathbf{X}}_B = \hat{\mathbf{B}}^{-1}\mathbf{b} = \hat{\mathbf{I}}_m\mathbf{B}^{-1}\mathbf{b} = \hat{\mathbf{I}}_m\mathbf{X}_B = \begin{bmatrix} x_{B1} - \dfrac{\beta_{1\ell}\Delta b_{\ell k}}{1+\beta_{k\ell}\Delta b_{\ell k}} x_{Bk} \\ \vdots \\ \dfrac{1}{1+\beta_{k\ell}\Delta b_{\ell k}} x_{Bk} \\ \vdots \\ x_{Bm} - \dfrac{\beta_{m\ell}\Delta b_{\ell k}}{1+\beta_{k\ell}\Delta b_{\ell k}} x_{Bk} \end{bmatrix}.$$

For $\hat{\mathbf{X}}_B$ primal feasible, we require that

$$1 + \beta_{k\ell}\Delta b_{\ell k} > 0,$$

$$x_{Bi} - \frac{\beta_{i\ell}\Delta b_{\ell k}}{1 + \beta_{k\ell}\Delta b_{\ell k}} x_{Bk} \geq 0,\ i{=}1,\ldots,m.$$

If $1 + \beta_{k\ell}\Delta b_{\ell k} > 0$, then, upon solving the second inequality for $\Delta b_{\ell k}$,

$$\begin{aligned} &\text{(a)}\ \frac{x_{Bi}}{\beta_{i\ell}x_{Bk} - \beta_{k\ell}x_{Bi}} \leq \Delta b_{\ell k} \text{ for } \beta_{i\ell}x_{Bk} - \beta_{k\ell}x_{Bi} < 0; \\ &\text{(b)}\ \frac{x_{Bi}}{\beta_{i\ell}x_{Bk} - \beta_{k\ell}x_{Bi}} \geq \Delta b_{\ell k} \text{ for } \beta_{i\ell}x_{Bk} - \beta_{k\ell}x_{Bi} > 0,\ i{=}1,\ldots,m. \end{aligned} \tag{11.15}$$

Since these inequalities must hold for all $i{\neq}k$ when $1 + \beta_{k\ell}\Delta b_{\ell k} > 0$, the lower and upper limits to which $\Delta b_{\ell k}$ can decrease or increase without violating primal feasibility may be expressed, from (11.15.a, b) respectively, as

$$\begin{aligned} \underline{lim}\Delta b_{\ell k} &= \max_{i\neq k}\left\{\frac{x_{Bi}}{\beta_{i\ell}x_{Bk} - \beta_{k\ell}x_{Bi}} \,\middle|\, \beta_{i\ell}x_{Bk} - \beta_{k\ell}x_{Bi} < 0\right\} \leq \Delta b_{\ell k} \\ &\leq \overline{lim}\Delta b_{\ell k} = \min_{i\neq k}\left\{\frac{x_{Bi}}{\beta_{i\ell}x_{Bk} - \beta_{k\ell}x_{Bi}} \,\middle|\, \beta_{i\ell}x_{Bk} - \beta_{k\ell}x_{Bi} > 0\right\}. \end{aligned} \tag{11.16}$$

If we express these limits in terms of the new level of $b_{\ell k}, \hat{b}_{\ell k}$, then for $\Delta b_{\ell k} = \hat{b}_{\ell k} - b_{\ell k}$, (11.16) becomes

$$\underline{lim}\,\hat{b}_{\ell k} = \max_{i \neq k}\left\{ b_{\ell k} + \frac{x_{Bi}}{\beta_{i\ell}x_{Bk} - \beta_{k\ell}x_{Bi}} \,\middle|\, \beta_{i\ell}x_{Bk} - \beta_{k\ell}x_{Bi} < 0 \right\} \leq \hat{b}_{\ell k} \tag{11.16.1}$$

$$\leq \overline{lim}\,\hat{b}_{\ell k} = \min_{i \neq k}\left\{ b_{\ell k} + \frac{x_{Bi}}{\beta_{i\ell}x_{Bk} - \beta_{k\ell}x_{Bi}} \,\middle|\, \beta_{i\ell}x_{Bk} - \beta_{k\ell}x_{Bi} > 0 \right\}.$$

What is the effect of a change in $b_{\ell k}$ upon the primal optimality (dual feasibility) requirement $\mathbf{C}'_{\hat{B}}\hat{\mathbf{B}}^{-1}\mathbf{R} - \mathbf{C}'_R \geqq \mathbf{0}'$? Now the jth compo nent of

$$\mathbf{C}'_{\hat{B}}\hat{\mathbf{I}}_m\mathbf{B}^{-1}\mathbf{R} - \mathbf{C}'_R = \mathbf{C}'_{\hat{B}}\,\hat{\mathbf{I}}_m[\mathbf{Y}_1, \ldots, \mathbf{Y}_{n-m}] - \mathbf{C}'_{\mathbf{R}} \geqq \mathbf{0}'$$

is expressible as

$$\sum_{i \neq k} c_{\hat{B}i}y_{ij} + \frac{c_{\hat{B}k}y_{kj}}{1 + \beta_{k\ell}\Delta b_{\ell k}}$$

$$- \sum_{i \neq k} \frac{c_{\hat{B}i}\beta_{i\ell}\Delta b_{\ell k}}{1 + \beta_{k\ell}\Delta b_{\ell k}} y_{kj} - c_{Rj} \geq 0,\ j=1,\ldots,n-m. \tag{11.17}$$

If we add

$$c_{\hat{B}k}y_{kj} - c_{\hat{B}k}y_{kj} + \frac{c_{\hat{B}k}\beta_{k\ell}\Delta b_{\ell k}}{1 + \beta_{k\ell}\Delta b_{\ell k}} y_{kj} - \frac{c_{\hat{B}k}\beta_{k\ell}\Delta b_{\ell k}}{1 + \beta_{k\ell}\Delta b_{\ell k}} y_{kj}$$

to the left-hand side of (11.17) we obtain

$$\sum_i c_{\hat{B}i}y_{ij} - \frac{\Delta b_{\ell k}y_{kj}\sum_i c_{\hat{B}i}\beta_{i\ell}}{1 + \beta_{k\ell}\Delta b_{\ell k}} - c_{\mathbf{R}j}$$

$$= \frac{-\Delta b_{\ell k}y_{kj}u_\ell}{1 + \beta_{k\ell}\Delta b_{\ell k}} - \bar{c}_j \geq 0,\ j=1,\ldots,n-m. \tag{11.17.1}$$

If $1 + \beta_{k\ell}\Delta b_{\ell k} > 0$, then, upon solving for $\Delta b_{\ell k}$,

$$\text{(a)}\ \frac{-\bar{c}_j}{y_{kj}u_\ell + \bar{c}_j\beta_{k\ell}} \leq \Delta b_{\ell k} \text{ for } y_{kj}u_\ell + \bar{c}_j\beta_{k\ell} < 0;$$

(11.18)

$$\text{(b)}\ \frac{-\bar{c}_j}{y_{kj}u_\ell + \bar{c}_j\beta_{k\ell}} > \Delta b_{\ell k} \text{ for } y_{kj}u_\ell + \bar{c}_j\beta_{k\ell} > 0,\ j=1,\ldots,n-m.$$

Since these inequalities must hold for all j=1,...,$n-m$ when $1+\beta_{k\ell}\Delta b_{\ell k}>0$, the lower and upper limits to which $\Delta b_{\ell k}$ can decrease or increase without violating primal optimality may be expressed, from (11.18.a, b) respectively, as

$$\begin{aligned} \underline{lim}\Delta b_{\ell k} &= \max_j\left\{\frac{-\bar{c}_j}{y_{kj}u_\ell+\bar{c}_j\beta_{k\ell}}\,\middle|\, y_{kj}u_\ell+\bar{c}_j\beta_{k\ell}<0\right\}\leq \Delta b_{\ell k} \\ &\leq \overline{lim}\Delta b_{\ell k}=\left\{\frac{-\bar{c}_j}{y_{k\ell}u_\ell+\bar{c}_j\beta_{k\ell}}\,\middle|\, y_{kj}u_\ell+\bar{c}_j\beta_{k\ell}>0\right\}. \end{aligned} \tag{11.19}$$

Alternatively, we may express these limits in terms of the new level of $b_{\ell k}$, namely $\hat{b}_{\ell k}$, since, if $\Delta b_{\ell k} = \hat{b}_{\ell k}-b_{\ell k}$, (11.19) becomes

$$\begin{aligned} \underline{lim}\,\hat{b}_{\ell k} &= \max_j\left\{b_{\ell k}+\frac{-\bar{c}_j}{y_{kj}u_\ell+\bar{c}_j\beta_{k\ell}}\,\middle|\, y_{kj}u_\ell+\bar{c}_j\beta_{k\ell}<0\right\}\leq \Delta b_{\ell k} \\ &\leq \overline{lim}\,\hat{b}_{\ell k}=\min_j\left\{b_{\ell k}+\frac{-\bar{c}_j}{y_{k\ell}u_\ell+\bar{c}_j\beta_{k\ell}}\,\middle|\, y_{kj}u_\ell+\bar{c}_j\beta_{k\ell}>0\right\}. \end{aligned} \tag{11.19.1}$$

We next determine the effect of a change in the element $b_{\ell k}$ upon the optimal value of the objective function. To this end let

$$\begin{aligned} f(\mathbf{X}_{\hat{B}}) &= \mathbf{C}'_{\hat{B}}\mathbf{X}_{\hat{B}} = \mathbf{C}'_{\hat{B}}\hat{\mathbf{I}}_m\mathbf{X}_B \\ &= \sum_{i\neq k} c_{\hat{B}i}x_{Bi}+\frac{c_{\hat{B}k}x_{Bk}}{1+\beta_{k\ell}\Delta b_{\ell k}}-\sum_{i\neq k}\frac{c_{\hat{B}i}\beta_{i\ell}\Delta b_{\ell k}}{1+\beta_{k\ell}\Delta b_{\ell k}}x_{Bk}. \end{aligned} \tag{11.20}$$

Adding

$$c_{\hat{B}k}x_{Bk}-c_{\hat{B}k}x_{Bk}+\frac{c_{\hat{B}k}\beta_{k\ell}\Delta b_{\ell k}}{1+\beta_{k\ell}\Delta b_{\ell k}}x_{Bk}-\frac{c_{\hat{B}k}\beta_{k\ell}\Delta b_{\ell k}}{1+\beta_{k\ell}\Delta b_{\ell k}}x_{Bk}$$

to the right-hand side of (11.20) yields

$$\begin{aligned} f(\mathbf{X}_{\hat{B}}) &= \sum_i c_{\hat{B}}x_{Bi}-\sum_i\frac{c_{\hat{B}i}\beta_{i\ell}\Delta b_{\ell k}}{1+\beta_{k\ell}\Delta b_{\ell k}}x_{Bk} \\ &= f(\mathbf{X}_B)-\frac{\Delta b_{\ell k}x_{Bk}}{1+\beta_{k\ell}\Delta b_{\ell k}}\sum_i c_{\hat{B}i}\beta_{i\ell} = f(\mathbf{X}_B)-\frac{\Delta b_{\ell k}x_{Bk}u_\ell}{1+\beta_{k\ell}\Delta b_{\ell k}}. \end{aligned} \tag{11.20.1}$$

EXAMPLE 11.3. It was indicated at the beginning of example 11.1 that the optimal simplex matrix corresponding to the problem

$$\begin{aligned} max\ f &= 2x_1 + 3x_2 + 4x_3 \quad \text{s.t.} \\ 2x_1 + x_2 + 4x_3 &\le 100 \\ x_1 + 3x_2 + x_3 &\le 40 \\ x_1, x_2, x_3 &\ge 0 \end{aligned}$$

is

$$\begin{bmatrix} \frac{5}{11} & 0 & 1 & \frac{3}{11} & -\frac{1}{11} & 0 & 23.63 \\ \frac{2}{11} & 1 & 0 & -\frac{1}{11} & \frac{4}{11} & 0 & 5.46 \\ \frac{4}{11} & 0 & 0 & \frac{9}{11} & \frac{8}{11} & 1 & 110.91 \end{bmatrix}.$$

Moreover, since x_2, x_3 are basic variables,

$$\mathbf{B} = \begin{bmatrix} b_{11} & b_{12} \\ b_{21} & b_{22} \end{bmatrix} = \begin{bmatrix} 4 & 1 \\ 1 & 3 \end{bmatrix}, \ \mathbf{B}^{-1} = \begin{bmatrix} \beta_{11} & \beta_{12} \\ \beta_{21} & \beta_{22} \end{bmatrix} = \begin{bmatrix} \frac{3}{11} & -\frac{1}{11} \\ -\frac{1}{11} & \frac{4}{11} \end{bmatrix},$$

$$\mathbf{R} = \begin{bmatrix} r_{11} & r_{12} & r_{13} \\ r_{21} & r_{22} & r_{23} \end{bmatrix} = \begin{bmatrix} 2 & 1 & 0 \\ 1 & 0 & 1 \end{bmatrix}.$$

From the objective function row of the optimal simplex matrix we find, for instance, that $\bar{c}_1 = -\frac{4}{11}$. Hence (11.14.b) indicates that $x_{R1} = x_1$ should maintain its non-basic status provided that r_{11}, r_{21} do not decrease by more than $\Delta r_{11} = -\frac{4}{9}$, $\Delta r_{21} = -\frac{1}{2}$ respectively. In addition, (11.14.b.1) may be utilized to depict the lower limits to which r_{11}, r_{21} can individually decrease without violating primal optimality (Table 11.4).

Table 11.4.

Component of $\mathbf{r}_1$	Lower Limit	Upper Limit
$r_{11} = 2$	$r_{11} + \frac{\bar{c}_1}{u_1} = \frac{14}{9}$	$+\infty$
$r_{21} = 1$	$r_{21} + \frac{\bar{c}_1}{u_2} = \frac{1}{2}$	$+\infty$

Suppose that r_{11} is decreased to $\frac{1}{2}$. Then $\underline{lim}\,\hat{r}_{11} = \frac{14}{9} = r_{11}+\bar{c}_1/u_1 = \frac{1}{2}+\frac{11}{9}\bar{c}_1$ or $-\bar{c}_1 = -\frac{19}{22}$, which implies that the current solution is no longer optimal. In terms of the preceding simplex matrix, $\frac{4}{11}$ is replaced by $-\frac{19}{22}$ in the last row of the same. Furthermore, an additional adjustment in the optimal simplex matrix is in order. Since $\mathbf{X}_B = \mathbf{B}^{-1}\mathbf{b} - \mathbf{B}^{-1}\mathbf{R}\mathbf{X}_R = \mathbf{B}^{-1}\mathbf{b} - [\mathbf{Y}_1,\ldots,\mathbf{Y}_{n-m}]\mathbf{X}_R$, where $\mathbf{Y}_j = \mathbf{B}^{-1}\mathbf{r}_j$, $j=1,\ldots,n-m$, it follows that if $\mathbf{r}_k$ is replaced in $\mathbf{R}$ by $\hat{\mathbf{r}}_k = \mathbf{r}_k + \Delta r_{\ell k}\mathbf{e}_\ell$, then $\mathbf{Y}_k$ is replaced in $\mathbf{B}^{-1}\mathbf{R}$ by

$$\hat{\mathbf{Y}}_k = \mathbf{B}^{-1}\hat{\mathbf{r}}_k = \mathbf{B}^{-1}r_k + \Delta r_{\ell k}\begin{bmatrix}\beta_{1\ell}\\ \vdots \\ \beta_{k\ell}\\ \vdots \\ \beta_{m\ell}\end{bmatrix} = \mathbf{Y}_k + \Delta r_{\ell k}\begin{bmatrix}\beta_{1\ell}\\ \vdots \\ \beta_{k\ell}\\ \vdots \\ \beta_{m\ell}\end{bmatrix}.$$

Since the ith component of $\hat{\mathbf{Y}}_k$ is $\hat{y}_{ik} = y_{ik} + \Delta r_{\ell k}\beta_{i\ell}$, $y_{11} = \frac{5}{11}$ and $y_{21} = \frac{2}{11}$ are replaced by $\hat{y}_{11} = y_{11} + \Delta r_{11}\beta_{11} = \frac{1}{22}$ and $\hat{y}_{21} = y_{21} + \Delta r_{11}\beta_{21} = \frac{7}{22}$ respectively in the optimal simplex matrix. Hence an additional iteration yields the new optimal simplex matrix

$$\begin{bmatrix} \frac{1}{22} & 0 & 1 & \frac{3}{11} & -\frac{1}{11} & 0 & 23.63 \\ \frac{7}{22} & 1 & 0 & -\frac{1}{11} & \frac{4}{11} & 0 & 5.46 \\ -\frac{19}{22} & 0 & 0 & \frac{9}{11} & \frac{8}{11} & 1 & 110.91 \end{bmatrix} \longrightarrow$$

$$\begin{bmatrix} 0 & -\frac{1}{7} & 1 & \frac{22}{77} & -\frac{11}{77} & 0 & 22.85 \\ 1 & \frac{22}{7} & 0 & -\frac{2}{7} & \frac{8}{7} & 0 & 17.16 \\ 0 & \frac{19}{7} & 0 & \frac{44}{7} & \frac{132}{77} & 1 & 125.73 \end{bmatrix}$$

and thus a new optimal basic feasible solution consisting of $x_1 = 17.16$, $x_3 = 22.85$, and $f = 125.73$.

Next, from (11.16) we see that the limits within which the components of the basis vectors $\mathbf{b}_1' = (b_{11}, b_{21})$, $\mathbf{b}_2' = (b_{12}, b_{22})$ can vary without violating primal feasibility are:

$$\underline{lim}\,\Delta b_{11} = \max_{i\neq 1}\left\{\frac{x_{B2}}{\beta_{21}x_{B1} - \beta_{11}x_{B2}} = \frac{5.46}{\left(-\frac{1}{11}\right)(23.63) - \left(\frac{3}{11}\right)(5.46)}\right\}$$

$$= -1.50 \leq \Delta b_{11} \leq \overline{lim}\Delta b_{11} = +\infty;$$

$$\underline{lim}\Delta b_{21} = -\infty \leq \Delta b_{21} \leq \overline{lim}\Delta b_{21}$$

$$= \min_{i\neq 1}\left\{\frac{x_{B2}}{\beta_{22}x_{B1} - \beta_{12}x_{B2}} = \frac{5.46}{\left(\frac{4}{11}\right)(23.63) - \left(-\frac{1}{11}\right)(5.46)}\right\}$$

$$= 0.60;$$

$$\underline{lim}\Delta b_{12} = -\infty \leq \Delta b_{12} \leq \overline{lim}\Delta b_{12}$$

$$= \min_{i\neq 2}\left\{\frac{x_{B1}}{\beta_{11}x_{B2} - \beta_{12}x_{B1}} = \frac{23.63}{\left(\frac{3}{11}\right)(5.46) - \left(-\frac{1}{11}\right)(23.63)}\right\}$$

$$= 6.51;$$

$$\underline{lim}\,\Delta b_{22} = \max_{i\neq 2}\left\{\frac{x_{B1}}{\beta_{12}x_{B2} - \beta_{22}x_{B1}} = \frac{23.63}{\left(-\frac{1}{11}\right)(5.46) - \left(\frac{4}{11}\right)(23.63)}\right\}$$

$$= -2.60 \leq \Delta b_{22} \leq \overline{lim}\Delta b_{22} = +\infty.$$

Moreover, we may employ (11.16.1) to specify the limits to which b_{11}, b_{21}, b_{12} and b_{22} can individually decrease or increase without affecting the feasibility of the current solution (Table 11.5).

TABLE 11.5.

Component of $\mathbf{b}_i, i=1,2$	Lower Limit	Upper Limit
$b_{11}=4$	$b_{11}+\frac{x_{B2}}{\beta_{21}x_{B1}-\beta_{11}x_{B2}}=2.50$	$+\infty$
$b_{21}=1$	$-\infty$	$b_{21}+\frac{x_{B2}}{\beta_{22}x_{B1}-\beta_{12}x_{B2}}=1.60$
$b_{12}=1$	$-\infty$	$b_{12}+\frac{x_{B1}}{\beta_{11}x_{B2}-\beta_{21}x_{B1}}=7.51$
$b_{22}=3$	$b_{22}+\frac{x_{B1}}{\beta_{12}x_{B2}-\beta_{22}x_{B1}}=0.40$	$+\infty$

We now turn to the specification of the limits within which the components of $\mathbf{b}_1, \mathbf{b}_2$ can vary without affecting primal optimality. From (11.19):

$$\underline{lim}\Delta b_{11} = max\left\{\frac{-\bar{c}_3}{y_{13}u_1+\bar{c}_3\beta_{11}} = \frac{8/11}{\left(-\frac{1}{11}\right)\left(\frac{9}{11}\right)-\left(\frac{8}{11}\right)\left(\frac{3}{11}\right)}\right\}$$

$$= -2.66 \le \Delta b_{11} \le \overline{lim}\Delta b_{11}$$

$$= min\left\{\frac{-\bar{c}_1}{y_{11}u_1+\bar{c}_1\beta_{11}} = \frac{4/11}{\left(\frac{5}{11}\right)\left(\frac{9}{11}\right)-\left(\frac{4}{11}\right)\left(-\frac{3}{11}\right)}\right\} = 1.33;$$

$$\underline{lim}\Delta b_{21} = -\infty \le \Delta b_{21} \le \overline{lim}\Delta b_{21}$$

$$= min\left\{\frac{-\bar{c}_1}{y_{11}u_2+\bar{c}_1\beta_{12}} = \frac{4/11}{\left(\frac{5}{11}\right)\left(\frac{8}{11}\right)-\left(\frac{4}{11}\right)\left(-\frac{1}{11}\right)} = 1, \quad \frac{-\bar{c}_2}{y_{12}u_2+\bar{c}_2\beta_{12}} = \frac{9/11}{\left(\frac{3}{11}\right)\left(\frac{8}{11}\right)-\left(\frac{9}{11}\right)\left(-\frac{1}{11}\right)} = 3\right\} = 1;$$

$$\underline{lim}\Delta b_{12} = -\infty \leq \Delta b_{12} \leq \overline{lim}\Delta b_{12}$$

$$= min \left\{ \frac{-\bar{c}_1}{y_{21}u_1 + \bar{c}_1\beta_{21}} = \frac{4/11}{\left(\frac{2}{11}\right)\left(\frac{9}{11}\right) - \left(\frac{4}{11}\right)\left(-\frac{1}{11}\right)} = 2, \quad \frac{-\bar{c}_3}{y_{23}u_1 + \bar{c}_3\beta_{21}} = \frac{8/11}{\left(\frac{4}{11}\right)\left(\frac{9}{11}\right) - \left(\frac{8}{11}\right)\left(-\frac{1}{11}\right)} = 2 \right\} = 2;$$

$$\underline{lim}\Delta b_{22} = max \left\{ \frac{-\bar{c}_2}{y_{22}u_2 + \bar{c}_2\beta_{22}} = \frac{9/11}{\left(-\frac{1}{11}\right)\left(\frac{9}{11}\right) - \left(\frac{9}{11}\right)\left(\frac{4}{11}\right)} \right\}$$

$$= -2.25 \leq \Delta b_{22} \leq \overline{lim}\Delta b_{22} = +\infty.$$

Additionally, from (11.19.1), we may determine the lower and upper limits to which b_{11}, b_{21}, b_{12}, and b_{22} can individually decrease or increase while still preserving primal optimality (Table 11.6).

TABLE 11.6.

Component of $\mathbf{b}_i, i=1,2$	Lower Limit	Upper Limit
$b_{11} = 4$	$b_{11} + \dfrac{-\bar{c}_3}{y_{13}u_1 + \bar{c}_3\beta_{11}} = 1.34$	$b_{11} + \dfrac{-\bar{c}_1}{y_{11}u_1 + \bar{c}_1\beta_{11}} = 5.33$
$b_{21} = 1$	$-\infty$	$b_{21} + \dfrac{-\bar{c}_1}{y_{11}u_2 + \bar{c}_1\beta_{12}} = 2$
$b_{12} = 1$	$-\infty$	$b_{12} + \dfrac{-\bar{c}_j}{y_{2j}u_1 + \bar{c}_j\beta_{21}} = 2, j=1,3$
$b_{22} = 3$	$b_{22} + \dfrac{-\bar{c}_2}{y_{22}u_2 + \bar{c}_2\beta_{22}} = 0.75$	$+\infty$

Since changes in the components of $\mathbf{b}_1, \mathbf{b}_2$ may affect both primal feasibility and optimality, we may combine Tables 11.5, 6 to obtain the lower and upper limits to which b_{11}, b_{21}, b_{12}, and b_{22} can individually decrease or increase without affecting either primal feasibility or optimality (Table 11.7).

TABLE 11.7.

Component of $\mathbf{b}_i, i=1,2$	Lower Limit	Upper Limit
b_{11}	2.50	5.33
b_{21}	$-\infty$	1.60
b_{12}	$-\infty$	2
b_{22}	0.75	$+\infty$

Suppose now that $b_{11} = 4$ is decreased to 2. As evidenced from Tables 11.5, 7, such a change violates primal feasibility. Before initiating this change, let us drive $\mathbf{b}_1$ out of the basis. This may be accomplished by using the standard primal simplex exit criterion to find a pivot that will remove x_3 from the set of basic variables so as to yield a primal feasible though non-optimal basic solution. If x_4 is chosen to become basic, we ultimately have a problem in which only the component $r_{12} = 4$ of the non-basic vector $\mathbf{r}_2' = (4,1)$ need change. But this is similar to the one just solved. In this regard, pivoting on the element $\frac{3}{11}$ yields

$$\begin{bmatrix} \frac{5}{11} & 0 & 1 & \frac{3}{11} & -\frac{1}{11} & 0 & 23.63 \\ \frac{2}{11} & 1 & 0 & -\frac{1}{11} & \frac{4}{11} & 0 & 5.46 \\ \frac{4}{11} & 0 & 0 & \frac{9}{11} & \frac{8}{11} & 1 & 110.91 \end{bmatrix} \longrightarrow$$

$$\begin{bmatrix} \frac{5}{3} & 0 & \frac{11}{3} & 1 & -\frac{1}{3} & 0 & 86.66 \\ \frac{1}{3} & 1 & \frac{1}{3} & 0 & \frac{1}{3} & 0 & 13.33 \\ -1 & 0 & -3 & 0 & 1 & 1 & 40 \end{bmatrix}.$$

Moreover, with

$$\mathbf{B} = \begin{bmatrix} b_{11} & b_{12} \\ b_{21} & b_{22} \end{bmatrix} = \begin{bmatrix} 1 & 1 \\ 0 & 3 \end{bmatrix}, \quad \mathbf{B}^{-1} = \begin{bmatrix} \beta_{11} & \beta_{12} \\ \beta_{21} & \beta_{22} \end{bmatrix} = \begin{bmatrix} 1 & -\frac{1}{3} \\ 0 & \frac{1}{3} \end{bmatrix},$$

$$\mathbf{R} = \begin{bmatrix} r_{11} & r_{12} & r_{13} \\ r_{21} & r_{22} & r_{23} \end{bmatrix} = \begin{bmatrix} 2 & 4 & 0 \\ 1 & 1 & 1 \end{bmatrix}, \text{ and}$$

$$\widehat{\mathbf{Y}}_2 = \mathbf{Y}_2 + \Delta r_{12} \begin{bmatrix} \beta_{11} \\ \beta_{21} \end{bmatrix} = \begin{bmatrix} \frac{11}{3} \\ \frac{1}{3} \end{bmatrix} + (-2) \begin{bmatrix} 1 \\ 0 \end{bmatrix} = \begin{bmatrix} \frac{5}{3} \\ \frac{1}{3} \end{bmatrix},$$

it follows that $y_{12} = \frac{11}{3}$ is replaced by $\widehat{y}_{12} = \frac{5}{3}$ in the previous simplex matrix with $\widehat{y}_{22} = y_{22} = \frac{1}{3}$. Then

$$\begin{bmatrix} \frac{5}{3} & 0 & \frac{5}{3} & 1 & -\frac{1}{3} & 0 & 86.66 \\ \frac{1}{3} & 1 & \frac{1}{3} & 0 & \frac{1}{3} & 0 & 13.33 \\ -1 & 0 & -3 & 0 & 1 & 1 & 40 \end{bmatrix} \longrightarrow$$

$$\begin{bmatrix} 0 & -5 & 0 & 1 & -2 & 0 & 20 \\ 1 & 3 & 1 & 0 & 1 & 0 & 40 \\ 2 & 9 & 0 & 0 & 4 & 1 & 160 \end{bmatrix}.$$

Hence the new optimal basic feasible solution consists of $x_3 = 40$, $x_4 = 20$, and $f = 160$.

Finally, let $b_{22} = 3$ decrease to $\frac{1}{2}$. In this instance Tables 11.6, 7 indicate that this change violates primal optimality. Proceeding as in the previous case, let us remove $\mathbf{b}_2$ from the basis (or x_2 from the set of basic variables) by pivoting on the element $\frac{2}{11}$ (x_1 is chosen to become basic) to obtain

$$\begin{bmatrix} \frac{5}{11} & 0 & 1 & \frac{3}{11} & -\frac{1}{11} & 0 & 23.63 \\ \frac{2}{11} & 1 & 0 & -\frac{1}{11} & \frac{4}{11} & 0 & 5.46 \\ \frac{4}{11} & 0 & 0 & \frac{9}{11} & \frac{8}{11} & 1 & 110.91 \end{bmatrix} \longrightarrow$$

$$\begin{bmatrix} 0 & -\frac{5}{2} & 1 & \frac{1}{2} & -1 & 0 & 10 \\ 1 & \frac{11}{2} & 0 & -\frac{1}{2} & 2 & 0 & 30 \\ 0 & -2 & 0 & 1 & 0 & 1 & 100 \end{bmatrix}.$$

Here the component $r_{21} = 3$ of the non-basic vector $\mathbf{r}_1' = (1,3)$ is to be changed. Since

$$\mathbf{B} = \begin{bmatrix} b_{11} & b_{12} \\ b_{21} & b_{22} \end{bmatrix} = \begin{bmatrix} 4 & 2 \\ 1 & 1 \end{bmatrix}, \quad \mathbf{B}^{-1} = \begin{bmatrix} \beta_{11} & \beta_{12} \\ \beta_{21} & \beta_{22} \end{bmatrix} = \begin{bmatrix} \frac{1}{2} & -1 \\ -\frac{1}{2} & 2 \end{bmatrix},$$

$$\mathbf{R} = \begin{bmatrix} r_{11} & r_{12} & r_{13} \\ r_{21} & r_{22} & r_{23} \end{bmatrix} = \begin{bmatrix} 1 & 1 & 0 \\ 3 & 0 & 1 \end{bmatrix}, \text{ and}$$

$$\widehat{\mathbf{Y}}_1 = \mathbf{Y}_1 + \Delta r_{21} \begin{bmatrix} \beta_{12} \\ \beta_{22} \end{bmatrix} = \begin{bmatrix} -\frac{5}{2} \\ \frac{11}{2} \end{bmatrix} + (-2.5) \begin{bmatrix} -1 \\ 2 \end{bmatrix} = \begin{bmatrix} 0 \\ \frac{1}{2} \end{bmatrix},$$

it follows that $y_{11} = -\frac{5}{2}$, $y_{21} = \frac{11}{2}$ are replaced by $\widehat{y}_{11} = 0$, $\widehat{y}_{21} = \frac{1}{2}$ respectively in the previous simplex matrix. Hence

$$\begin{bmatrix} 0 & 0 & 1 & \frac{1}{2} & -1 & 0 & 10 \\ 1 & \frac{1}{2} & 0 & -\frac{1}{2} & 2 & 0 & 30 \\ 0 & -2 & 0 & 1 & 0 & 1 & 100 \end{bmatrix} \longrightarrow \begin{bmatrix} 0 & 0 & 2 & 1 & -2 & 0 & 20 \\ 2 & 1 & 2 & 0 & 2 & 0 & 80 \\ 4 & 0 & 2 & 0 & 6 & 1 & 240 \end{bmatrix}$$

so that the new optimal basic feasible solution is $x_2 = 80$, $x_4 = 20$, and $f = 240$.

11.2. Structural Changes

(1) ADDITION OF A NEW VARIABLE. We now look to the effect on the optimal solution of a given linear programming problem of the introduction of a new variable x_{n+1}, with c_{n+1} and $\mathbf{a}_{n+1}$ representing its associated objective function coefficient and vector of structural constraint coefficients respectively. In particular, we are now cofronted with solving a problem of the form

$$\begin{aligned} max\, f(\mathbf{X}, x_{n+1}) &= \mathbf{C}'\mathbf{X} + c_{n+1}x_{n+1} \quad \text{s.t.} \\ \left[\mathbf{A}, \mathbf{a}_{n+1}\right]\begin{bmatrix}\mathbf{X} \\ x_{n+1}\end{bmatrix} &= \mathbf{b},\ \mathbf{X} \geqq \mathbf{0},\ x_{n+1} \geq 0. \end{aligned}$$

If $\mathbf{X}_B = \mathbf{B}^{-1}\mathbf{b}$ represents an optimal basic feasible solution to the problem

$$\begin{aligned} max\, f(\mathbf{X}) = \mathbf{C}'\mathbf{X} \quad &\text{s.t.} \\ \mathbf{AX} = \mathbf{b},\ \mathbf{X} \geqq \mathbf{0}&, \end{aligned}$$

then we may also view it as a basic feasible solution to the former problem with x_{n+1} deemed non-basic or zero. Moreover, this solution will be optimal if

$$\bar{c}_{n+1-m} = c_{R,\,n+1-m} - \mathbf{C}'_B\mathbf{B}^{-1}\mathbf{r}_{n+1-m} = c_{n+1} - \mathbf{U}'\mathbf{a}_{n+1} \leq 0.$$

Obviously then, if $\bar{c}_{n+1-m} > 0$, the current solution is no longer optimal so that $\mathbf{a}_{n+1}$ must be brought into the basis, in which case we must insert into the optimal simplex matrix an additional column of the form

$$\begin{bmatrix}\mathbf{Y}_{n+1-m} \\ -\bar{c}_{n+1-m}\end{bmatrix} = \begin{bmatrix}\mathbf{B}^{-1}\mathbf{a}_{n+1} \\ -\bar{c}_{n+1-m}\end{bmatrix},$$

where $-\bar{c}_{n+1-m} < 0$.

(2) ADDITION OF A NEW STRUCTURAL CONSTRAINT. Oftentimes one is faced with the problem of introducing an additional structural constraint into a linear programming problem for which an optimal basic feasible solution has already been obtained. Let us write this new ($m+1^{st}$) structural constraint as

$$\sum_{j=1}^{n} a_{m+1,\,j}x_j \leq,\ \geq,\ = b_{m+1} \quad \text{or} \quad \boldsymbol{\alpha}\mathbf{X} \leq,\ \geq,\ = b_{m+1},$$

where $\boldsymbol{\alpha} = (a_{m+1,\,1}, \ldots, a_{m+1,\,n})$ is a $(1 \times n)$ vector of coefficients on the variables x_j, $j=1,\ldots,n$.

First, for $\boldsymbol{\alpha}\mathbf{X} \leq b_{m+1}$, if x_{n+1} represents a non-negative slack variable, then we seek a solution to a problem of the form

$$\begin{aligned} max\, f(\mathbf{X}, x_{n+1}) &= \mathbf{C}'\mathbf{X} + x_{n+1} \quad \text{s.t.} \\ \mathbf{A}\mathbf{X} + x_{n+1}\mathbf{0} &= \mathbf{b} \\ \boldsymbol{\alpha}\mathbf{X} + x_{n+1} &= b_{m+1} \\ \mathbf{X} \geqq \mathbf{0},\ x_{n+1} &\geq 0. \end{aligned}$$

The simplex matrix associated with this problem may be written as

$$\begin{array}{c} \quad \mathbf{X}_B\ \mathbf{X}_R\ x_{n+1} \\ \begin{bmatrix} \mathbf{B} & \mathbf{R} & \mathbf{0} & \mathbf{0} & \mathbf{b} \\ \boldsymbol{\alpha}_B & \boldsymbol{\alpha}_R & 1 & 0 & b_{m+1} \\ -\mathbf{C}'_B & -\mathbf{C}'_R & 0 & 1 & 0 \end{bmatrix}, \end{array}$$

where $\boldsymbol{\alpha} = (\boldsymbol{\alpha}_B, \boldsymbol{\alpha}_R)$. Here $\boldsymbol{\alpha}_B$ is a $(1 \times m)$ vector containing those components $a_{m+1,j}$ which correspond to the columns of $\mathbf{B}$ and $\boldsymbol{\alpha}_R$ is a $(1 \times n-m)$ vector whose components $a_{m+1,j}$ correspond to the columns of $\mathbf{R}$. If this simplex matrix is premultiplied by the inverse of the $m+1^{th}$ order basis matrix

$$\bar{\mathbf{B}} = \begin{bmatrix} \mathbf{B} & \mathbf{0} & \mathbf{0} \\ \boldsymbol{\alpha}_B & 1 & 0 \\ -\mathbf{C}'_B & 0 & 1 \end{bmatrix}, \quad \bar{\mathbf{B}}^{-1} = \begin{bmatrix} \mathbf{B}^{-1} & \mathbf{0} & \mathbf{0} \\ -\boldsymbol{\alpha}_B\mathbf{B}^{-1} & 1 & 0 \\ -\mathbf{C}'_B\mathbf{B}^{-1} & 0 & 1 \end{bmatrix},$$

the resulting product

$$\begin{bmatrix} \mathbf{I}_m & \mathbf{B}^{-1}\mathbf{R} & \mathbf{0} & \mathbf{0} & \mathbf{B}^{-1}\mathbf{b} \\ \mathbf{0}' & \boldsymbol{\alpha}_R - \boldsymbol{\alpha}_B\mathbf{B}^{-1}\mathbf{R} & 1 & 0 & b_{m+1} - \boldsymbol{\alpha}_B\mathbf{B}^{-1}\mathbf{b} \\ \mathbf{0}' & \mathbf{C}'_B\mathbf{B}^{-1}\mathbf{R} - \mathbf{C}'_R & 0 & 1 & \mathbf{C}'_B\mathbf{B}^{-1}\mathbf{b} \end{bmatrix}$$

has, with $\mathbf{X}_R = \mathbf{0}$, the partitioned rows

$$\begin{aligned} \mathbf{X}_B &= \mathbf{B}^{-1}\mathbf{b}, \\ x_{n+1} &= b_{m+1} - \boldsymbol{\alpha}_B\mathbf{B}^{-1}\mathbf{b} = b_{m+1} - \boldsymbol{\alpha}_B\mathbf{X}_B, \text{ and} \\ f &= \mathbf{C}'_B\mathbf{B}^{-1}\mathbf{b} = \mathbf{C}'_B\mathbf{X}_B \end{aligned}$$

respectively, That is

$$\mathbf{X}_{\bar{B}} = \begin{bmatrix} \mathbf{X}_B \\ x_{n+1} \end{bmatrix} = \begin{bmatrix} \mathbf{X}_B \\ b_{m+1} - \boldsymbol{\alpha}_B \mathbf{X}_B \end{bmatrix} \text{ or } \begin{cases} x_{\bar{B}i} = x_{Bi},\ i=1,\ldots,m; \\ x_{\bar{B},m+1} = x_{n+1} - \boldsymbol{\alpha}_B \mathbf{X}_B, \end{cases}$$

$$f(\mathbf{X}_{\bar{B}}) = f(\mathbf{X}_B).$$

Now, if $x_{\bar{B},m+1} \geq 0$, this basic solution to the enlarged problem is feasible, *i.e.*, the optimal basic feasible solution to the original problem also satisfies the new structural constraint or $\boldsymbol{\alpha}\mathbf{X}_{\bar{B}} \leq b_{m+1}$. Moreover, if the original basic solution was optimal, then an examination of the last row of the preceding simplex matrix reveals that the solution to the extended problem is also optimal (since the optimality criterion for each is the same) with equality holding between the optimal values of the objective functions for the original and enlarged problems. If $x_{\bar{B},m+1} < 0$ (in which case $\boldsymbol{\alpha}\mathbf{X}_{\bar{B}} > b_{m+1}$) so that primal feasibility is violated, then, since the optimality criterion is satisfied, the dual simplex method may be employed to obtain a basic feasible solution to the modified problem. In this regard, it is obvious that the optimal objective value for the original problem provides us with an upper bound on the same for the enlarged problem.

Next if $\boldsymbol{\alpha}\mathbf{X} \geq b_{m+1}$, then x_{n+1} represents a nonnegative surplus variable. In this circumstance let us add $\boldsymbol{\alpha}\mathbf{X} - x_{n+1} = b_{m+1}$ or $\bar{\boldsymbol{\alpha}}\mathbf{X} + x_{n+1} = \bar{b}_{m+1}$ to the original problem, where $\bar{\boldsymbol{\alpha}} = -\boldsymbol{\alpha}$ and $\bar{b}_{m+1} = -b_{m+1}$. The treatment then proceeds exactly as above with $\bar{\boldsymbol{\alpha}}$ and $\bar{b}_{m+1}$ replacing $\boldsymbol{\alpha}$ and b_{m+1} respectively.

Finally, if $\boldsymbol{\alpha}\mathbf{X} = b_{m+1}$, then x_{n+1} will represent a nonnegative artificial variable. If in the solution to the enlarged problem we find that $x_{\bar{B},m+1} = x_{n+1} = b_{m+1} - \boldsymbol{\alpha}_B \mathbf{X}_B = 0$, then it is evident that the optimal basic feasible solution for the original problem exactly satisfies the new constraint, in which case the original solution is also the optimal solution for the enlarged problem. The implication of this degenerate solution to the enlarged problem is that the added structural constraint is redundant. If $x_{\bar{B},m+1} < 0$, then the artificial variable x_{n+1} may be treated as if it were a slack variable, in which case the dual simplex method applies. And if $x_{\bar{B},m+1} \geq 0$, then the artificial variable x_{n+1}

may be handled as if it represented a surplus variable. Again the dual simplex method may be applied.

(3) DELETION OF A VARIABLE. The deletion of a positive variable is tantamount to driving it to zero so that its contribution to optimality is withdrawn. This being the case, it is obvious that the deletion of a non-basic variable is a trivial process. To delete a positive basic variable, we may simply pivot its associated basis vector out of the basis, where the pivotal column is determined by the dual simplex entry criterion. This procedure will result in a basic solution which is primal optimal but not necessarily primal feasible, at which point the column corresponding to the outgoing basis vector may be deleted from the simplex matrix. If warranted, reoptimization of the primal objection function may then be carried out using the dual simplex routine.

(4) DELETION OF A STRUCTURAL CONSTRAINT. If a structural constraint is to be deleted from the optimal simplex matrix, consideration must first be given to whether or not that constraint is binding at the optimal basic feasible solution. In this regard, if the original structural constraint was of the " $\leq$ " or " $\geq$ " variety and its associated slack or surplus variable appears in the optimal set of basic variables at a positive level, thus indicating that the constraint is not binding at the optimal extreme point, then clearly the current solution will also be an optimal basic feasible solution to the modified problem.

If, on the one hand, the structural constraint to be deleted is of the " $\leq$ " type and its corresponding slack variable is zero at the optimal basic feasible solution, thus indicating that the constraint is binding there, then the constraint may be deleted from the optimal simplex matrix by subtracting a non-negative surplus variable from its left-hand side. To justify this operation let us assume that the rth structural constraint is binding at an optimal basic feasible solution so that $a_{r1}x_1 + \cdots + a_{rn}x_n = b_r$. Given that this constraint is to be deleted from the optimal simplex matrix, the introduction of the surplus variable x_{n+1} allows for the possibility that $a_{r1}x_1 + \cdots + a_{rn}x_n > b_r$ once this constraint is deleted, *i.e.*, $a_{r1}x_1 + \cdots + a_{rn}x_n - b_r = x_{n+1} \geq 0$. A glance back at the section involving the addition of a new variable to a given linear programming problem reveals that in this instance we seek to generate a solution to a problem of the form

$$max\, f(\mathbf{X},\ x_{n+1}) \;=\; \mathbf{C}'\mathbf{X} + 0x_{n+1} \qquad \text{s.t.}$$

$$\left[\mathbf{A},\ -\mathbf{e}_r\right]\begin{bmatrix}\mathbf{X}\\ x_{n+1}\end{bmatrix} \;=\; \mathbf{b},\ \mathbf{X} \geqq \mathbf{0},\ x_{n+1} \geq 0.$$

Since $\bar{c}_{n+1-m} \;=\; c_{R,\,n+1-m} - \mathbf{C}_B'\mathbf{B}^{-1}\mathbf{r}_{n+1-m} \;=\; \mathbf{U}'\mathbf{e}_r \;=\; u_r > 0$, the current solution is no longer optimal so that we must insert into the optimal simplex matrix an additional column of the form

$$\begin{bmatrix}\mathbf{Y}_{n+1-m}\\ -\bar{c}_{n+1-m}\end{bmatrix} = \begin{bmatrix}-\mathbf{B}^{-1}\mathbf{e}_r\\ -u_r\end{bmatrix}.$$

The primal simplex routine then yields an optimal basic feasible solution to the modified problem. On the other hand, if the structural constraint to be deleted is of the " $\geq$ " variety and its associated surplus variable is zero at the optimal basic feasible solution so that the constraint is binding there, then the constraint may be deleted from the optimal simplex matrix by adding to its left-hand side a non-negative slack variable. Again looking to the rth (binding) structural constraint, the introduction of the slack variable x_{n+1} provides for the likelihood that $a_{r1}x_1 + \cdots + a_{rn}x_n < b_r$ when the constraint is deleted, *i.e.*, $x_{n+1} = b_r - a_{r1}x_1 - \cdots - a_{rn}x_n \geq 0$. Proceeding as above, we wish to determine a solution to

$$max\, f(\mathbf{X},\ x_{n+1}) \;=\; \mathbf{C}'\mathbf{X} + 0x_{n+1} \qquad \text{s.t.}$$

$$\left[\mathbf{A}, \mathbf{e}_r\right]\begin{bmatrix}\mathbf{X}\\ x_{n+1}\end{bmatrix} \;=\; \mathbf{b},\ \mathbf{X} \geqq \mathbf{0},\ x_{n+1} \geq 0.$$

And if the (r^{th}) structural constraint to be deleted was initially a strict equality, then we may simultaneously add the subtract non-negative slack and surplus variables x_{n+1}, x_{n+2} respectively on its left-hand side, *i.e.*, we must now solve

$$max\, f(\mathbf{X},\ x_{n+1}, x_{n+2}) \;=\; \mathbf{C}'\mathbf{X} + 0x_{n+1} + 0x_{n+2} \qquad \text{s.t.}$$

$$\left[\mathbf{A}, \mathbf{e}_r,\ -\mathbf{e}_r\right]\begin{bmatrix}\mathbf{X}\\ x_{n+1}\\ x_{n+2}\end{bmatrix} \;=\; \mathbf{b},\ \mathbf{X} \geqq \mathbf{0},\ x_{n+1} \geq 0,\ x_{n+2} \geq 0.$$

12 POSTOPTIMALITY ANALYSIS II

In the preceding chapter we considered, in part, an assortment of postoptimality problems involving discrete changes in only selected components of the matrices **C**, **b**, or **A**. Therein emphasis was placed upon the extent to which a given problem may be modified without breaching its feasibility or optimality. We now wish to extend this sensitivity analysis a bit further to what is called **parametric analysis.** That is, instead of just determining the amount by which a few individual components of the aforementioned matrices may be altered in some particular way before the feasibility or optimality of the current solution is violated, let us generate a sequence of basic solutions which in turn become optimal, one after the other, as all of the components of **C**, **b**, or a column of **A** vary continuously in some prescribed direction. In this regard, the following parametric analysis will involve a marriage between sensitivity analysis and simplex pivoting.

12.1. Parametric Analysis

If we seek to

$$\begin{aligned} \max f(\mathbf{X}) &= \mathbf{C}'\mathbf{X} \quad \text{s.t.} \\ \mathbf{AX} &= \mathbf{b},\ \mathbf{X} \geqq \mathbf{0}, \end{aligned}$$

then an optimal basic feasible solution emerges if $\mathbf{X}_B = \mathbf{B}^{-1}\mathbf{b} \geqq \mathbf{0}$ and $\mathbf{C}'_R - \mathbf{C}'_B\mathbf{B}^{-1}\mathbf{R} \leqq \mathbf{0}'$ (or, in terms of the optimal simplex matrix, $\mathbf{C}'_B\mathbf{B}^{-1}\mathbf{R} - \mathbf{C}'_R \geqq \mathbf{0}'$) with $f(\mathbf{X}) = \mathbf{C}'_B\mathbf{X}_B = \mathbf{C}'_B\mathbf{B}^{-1}\mathbf{b}$. Given this result as our starting point, let us examine the process of

(1) PARAMETRIZING THE OBJECTIVE FUNCTION. Given the above optimal basic feasible solution, let us replace $\mathbf{C}$ by $\mathbf{C}^* = \mathbf{C} + \theta\mathbf{S}$, where θ is a nonnegative scalar parameter and $\mathbf{S}$ is a specified, albeit arbitrary, $(n \times 1)$ vector which determines a given direction of change in the coefficients $c_j, j=1,\ldots,n$. In this regard, the $c_j^*, j=1,\ldots,n$, are specified as linear functions of the parameter θ. Currently $-\mathbf{C}_R' + \mathbf{C}_B'\mathbf{B}^{-1}\mathbf{R} \geqq \mathbf{0}'$ or $-\bar{c}_j = -c_{Rj} + \mathbf{C}_B'\mathbf{Y}_j \geq 0,\ j=1,\ldots,n-m$. If $\mathbf{C}^*$ is partitioned as

$$\begin{bmatrix} \mathbf{C}_B^* \\ \mathbf{C}_R^* \end{bmatrix} = \begin{bmatrix} \mathbf{C}_B \\ \mathbf{C}_R \end{bmatrix} + \theta \begin{bmatrix} \mathbf{S}_B \\ \mathbf{S}_R \end{bmatrix},$$

where $\mathbf{S}_B(\mathbf{S}_R)$ contains the components of $\mathbf{S}$ corresponding to the components of $\mathbf{C}$ within $\mathbf{C}_B(\mathbf{C}_R)$, then when $\mathbf{C}^*$ replaces $\mathbf{C}$, the revised optimality condition becomes

$$\begin{aligned} -(\mathbf{C}_R^*)' + (\mathbf{C}_B^*)'\mathbf{B}^{-1}\mathbf{R} &= -(\mathbf{C}_R' + \theta\mathbf{S}_R') + (\mathbf{C}_B' + \theta\mathbf{S}_B')\mathbf{B}^{-1}\mathbf{R} \\ &= (-\mathbf{C}_R' + \mathbf{C}_B'\mathbf{B}^{-1}\mathbf{R}) + \theta(-\mathbf{S}_R' + \mathbf{S}_B'\mathbf{B}^{-1}\mathbf{R}) \geqq \mathbf{0}' \end{aligned} \tag{12.1}$$

or, in terms of the individual components of (12.1),

$$-\bar{c}_j^* = -\bar{c}_j + \theta(-s_{Rj} + \mathbf{S}_B'\mathbf{Y}_j) \geq 0,\ j=1,\ldots,n-m. \tag{12.1.1}$$

(Note that the parametrization of f affects only primal optimality and not primal feasibility since $\mathbf{X}_B$ is independent of $\mathbf{C}$). Let us now determine the largest value of θ (known as its **critical value**, θ_c) for which (12.1.1) holds. Upon examining this expression it is evident that the critical value of θ is that for which any increase in θ beyond θ_c makes at least one of the $-\bar{c}_j^*$ values negative, thus violating optimality.

How large of an increase in θ preserves optimality? First, if $-\mathbf{S}_R' + \mathbf{S}_B'\mathbf{B}^{-1}\mathbf{R} \geqq \mathbf{0}'$ or $-s_{Rj} + \mathbf{S}_B'\mathbf{Y}_j \geq 0$, then θ can be increased without bound while still maintaining the revised optimality criterion since, in this instance, (12.1.1) reveals that $-\bar{c}_j^* \geq -\bar{c}_j \geq 0,\ j=1,\ldots,n-m$. Next, if $= -s_{Rj} + \mathbf{S}_B'\mathbf{Y}_j < 0$ for some particular value of j, then $-\bar{c}_j^* \geq 0$ for

$$\theta \leq \frac{\bar{c}_j}{-s_{Rj} + \mathbf{S}_B'\mathbf{Y}_j} = \theta_c.$$

Hence the revised optimality criterion is violated when $\theta > \theta_c$ for some non-basic variable x_{Rj}. Moreover, if $-s_{Rj}+\mathbf{S}_B' \mathbf{Y}_j < 0$ for two or more non-basic variables, then

$$\theta_c = \min_j \left\{ \frac{\bar{c}_j}{-s_{Rj}+\mathbf{S}_B' \mathbf{Y}_j} \middle| -s_{Rj}+\mathbf{S}_B' \mathbf{Y}_j < 0 \right\}. \tag{12.2}$$

So of θ increases and the minimum in (12.2) in attained for $j=k$, it follows that $\theta_c = \bar{c}_k/(-s_{Rk}+\mathbf{S}_B' \mathbf{Y}_k)$ or $-\bar{c}_k^* = 0$. With $-\bar{c}_k^* = 0$, the case of multiple optimal basic feasible solutions obtains, *i.e.* the current basic feasible solution (call it $\mathbf{X}_B^{(1)}$) remains optimal and the alternative optimal basic feasible solution (denoted $\mathbf{X}_B^{(2)}$) emerges when $\mathbf{r}_k$ is pivoted into the basis (or x_{Rk} enters the set of basic variables). And as θ increases slightly beyond θ_c, $-\bar{c}_k^* < 0$ so that $\mathbf{X}_B^{(2)}$ becomes the unique optimal basic feasible solution.

What about the choice of the direction vector $\mathbf{S}$? As indicated above, $\mathbf{S}$ represents the direction in which the objective function coefficients are varied. So while the components of $\mathbf{S}$ are quite arbitrary (*e.g.*, some of the c_j values may be increased while others are decreased), it is often the case that only a single objective function coefficient c_j is to be changed, *i.e.*, $\mathbf{S} = \mathbf{e}_j$ (or $-\mathbf{e}_j$). Addressing ourselves to this latter case, if $\mathbf{C}^* = \mathbf{C} + \theta \mathbf{e}_k$ and x_k is the non-basic variable x_{Rk}, then (12.1.1) reduces to $-\bar{c}_j^* = -\bar{c}_j \geq 0$ for $j \neq k$, while for $j=k$, $-\bar{c}_k^* = -\bar{c}_k + \theta(-s_{Rk}) = -\bar{c}_k - \theta \geq 0$. A comparison of this expression with (11.1) reveals that here $\theta \equiv \Delta c_{Rk}$ (generally, $\Delta c_{Rk} \equiv \theta s_{Rk}$). If θ increases so that $\theta_c = -\bar{c}_k$, or $-\bar{c}_k^* = 0$, then the preceding discussion dictates that the column within the simplex matrix associated with x_k is to enter the basis, with no corresponding adjustment in the current value of f warranted. Next, if x_k is the basic variable x_{Bk}, (12.1.1) becomes $-\bar{c}_j^* = -\bar{c}_j + \theta s_{Bk} y_{kj} = -\bar{c}_j + \theta y_{kj} \geq 0$, $j=1,\ldots,n-m$. Upon comparing this expression with (11.3.1) we find that here $\theta \equiv \Delta c_{Bk}$ (in general, $\Delta c_{Bk} \equiv \theta s_{Bk}$). In this instance (12.2) may be rewritten as

$$\theta_c = \min_j \left\{ \frac{\bar{c}_j}{y_{kj}} \middle| y_{kj} < 0 \right\}, \tag{12.2.1}$$

which is thus equivalent to (11.5.c). If the minimum in (12.2.1) is attained for $j=r$, then $\theta_c = \frac{\overline{c}_r}{y_{kr}}$ or $-\overline{c}_r^* = 0$. Then $\mathbf{r}_r$ may be pivoted into the basis so that again the case of multiple optimal basic feasible solutions holds. Furthermore, when x_{Rr} is pivoted into the set of basic variables, the current value of f must be adjusted by an amount $\Delta f = \Delta c_{Bk} x_k = \theta_c x_{Bk}$ as determined in the preceding chapter.

Let us now examine the geometric interpretation underlying the process of parametrizing the objective function along with the subsequent generation of a sequence of optimal basic feasible solutions as θ increases through a succession of critical values. In Figure 12.1, let A represent the unique optimal extreme point associated with a given maximization problem, where f is the corresponding hyperplane with normal $\mathbf{C}$. Given $\mathbf{S}$, if $\theta_1\mathbf{S}$, $0 < \theta_1 < \theta_c = \theta_c^{(1)}$, is added to $\mathbf{C}$ to yield a new normal $\mathbf{C}_1^* = \mathbf{C} + \theta_1\mathbf{S}$ (be virtue of the parallelogram law for vector addition) we obtain a new hyperplane $f(\theta_1)$ which also represents a maximal

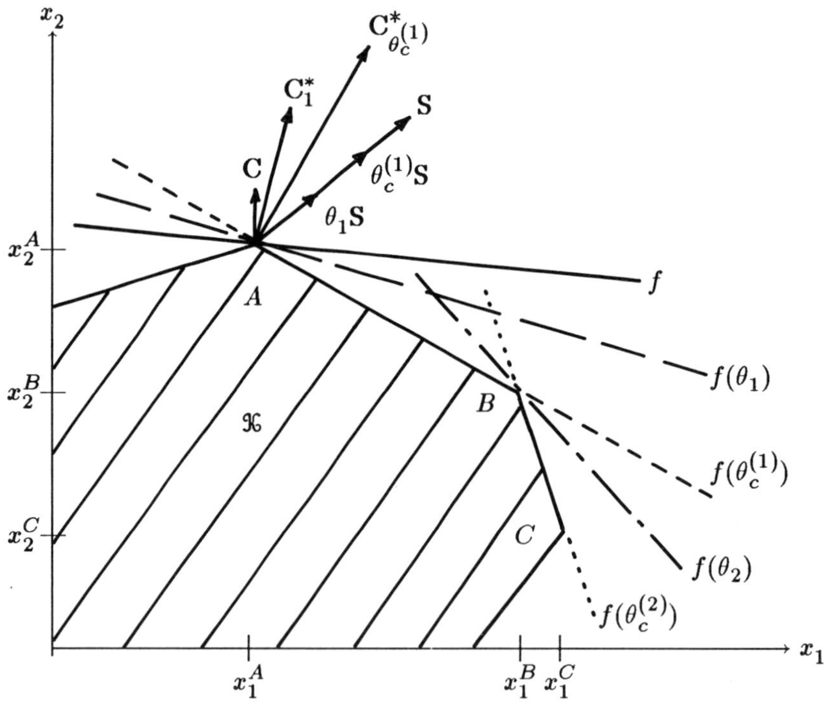

Figure 12.1

hyperplane since A continues to serve as the optimal extreme point. Thus A yields an optimal basic feasible solution to both the original and adjusted problems

$$\left.\begin{array}{rl} \max f(\mathbf{X}) = \mathbf{C}'\mathbf{X} & \text{s.t.} \\ \mathbf{AX} = \mathbf{b},\ \mathbf{X} \geqq \mathbf{0} & \end{array}\right\} \text{ and } \left\{\begin{array}{rl} \max f(\mathbf{X}) = (\mathbf{C}+\theta_1\mathbf{S})'\mathbf{X} & \text{s.t.} \\ \mathbf{AX} = \mathbf{b},\ \mathbf{X} \geqq \mathbf{0} & \end{array}\right.$$

respectively. If θ is next increased to $\theta_c = \theta_c^{(1)}$, the optimal hyperplane $f(\theta_c^{(1)})$ with normal $\mathbf{C}^*_{\theta_c^{(1)}} = \mathbf{C} + \theta_c^{(1)}\mathbf{S}$ is tangent to $\mathcal{K}$ along the edge from A to B so that the adjusted problem

$$\begin{aligned} \max f(\mathbf{X}) &= (\mathbf{C}+\theta_c^{(1)}\mathbf{S})'\mathbf{X} \quad \text{s.t.} \\ \mathbf{AX} &= \mathbf{b},\ \mathbf{X} \geqq \mathbf{0} \end{aligned}$$

has two optimal extreme point solutions: (1) point A; and (2) point B, with the latter emerging when the non-basic vector associated with the minimum in (12.2) is pivoted into the basis. (Note that the effect of parametrizing the objective function is to change its slope, *i.e.*, as θ increases from zero to $\theta_c^{(1)}$, the maximal hyperplane is rotated about extreme point A.) And if θ increases a bit beyond $\theta_c^{(1)}$ to, say, θ_2, with $\theta_c^{(1)} < \theta_2 < \theta_c^{(2)}$, B replaces A as the unique optimal extreme point solution. In this instance B represents an optimal basic feasible solution to the problem

$$\begin{aligned} \max f(\mathbf{X}) &= (\mathbf{C}+\theta_2\mathbf{S})'\mathbf{X} \quad \text{s.t.} \\ \mathbf{AX} &= \mathbf{b},\ \mathbf{X} \geqq \mathbf{0}. \end{aligned}$$

How much further can θ be increased beyond θ_2 without violating the optimality of extreme point B? Actually, the current basic feasible solution at B remains optimal for the problem

$$\begin{aligned} \max f(\mathbf{X}) &= [\mathbf{C} + (\theta_c^{(1)}+\theta)\mathbf{S}]'\mathbf{X} \quad \text{s.t.} \\ \mathbf{AX} &= \mathbf{b},\ \mathbf{X} \geqq \mathbf{0} \end{aligned}$$

as long as $\theta_c^{(1)} \leq \theta \leq \theta_c^{(2)}$, where $\theta_c^{(2)} = \theta_c^{(1)} + \theta_c$, with θ_c determined from (12.2) in conjunction with the appropriate information contained within the latest optimal simplex matrix. Moreover, if θ increases to $\theta_c^{(2)}$, the optimal hyperplane $f(\theta_c^{(2)})$ with normal $\mathbf{C}^*_{\theta_c^{(2)}} = \mathbf{C} + \theta_c^{(2)}\mathbf{S}$ is tangent to $\mathcal{K}$ along the edge from B to

C, with the adjusted problem

$$max\, f(\mathbf{X}) \;=\; (\mathbf{C}+(\theta_c^{(2)}\mathbf{S})'\mathbf{X} \quad \text{s.t.}$$
$$\mathbf{AX} \;=\; \mathbf{b},\ \mathbf{X} \geqq \mathbf{0}$$

again exhibiting two optimal extreme point solutions: (1) point B; and (2) point C, with only the latter surviving when the non-basic vector corresponding to the minimum in (12.2) enters the basis. And by increasing θ slightly above $\theta_c^{(2)}$, C replaces B as the unique optimal extreme point solution. Finally, with f bounded over $\mathcal{K}$, a basic feasible solution will ultimately be found which preserves optimality as θ approaches infinity.

Two final observations concerning the geometry of parametrizing the objective function are in order. First, the maximum of f (written $max\, f$) is a piecewise continuous linear function of θ. If the objective function associated with the hypothetical problem of Figure 12.1 appears as $f = c_1x_1 + c_2x_2$ and $\mathbf{S} = \mathbf{e}_1$, the parametrized objective function is $f = (c_1{+}\theta)x_1 + c_2x_2$. Then for $0 \leq \theta < \theta_c^{(1)}$, $max\, f = (c_1{+}\theta)x_1^A + c_2x_2^A$ is a linear function of θ with slope $d\, max\, f/d\theta = x_1^A$ (Figure 12.2.a). For $\theta_c^{(1)} \leq \theta \leq \theta_c^{(2)}$, $max\, f = \left(c_1 + (\theta_c^{(1)}{+}\theta)\right)x_1^B + c_2x_2^B$, with $d\, max\, f/d\theta = x_1^B$. And for $\theta_c^{(2)} \leq \theta < +\infty$, $max\, f = (c_1{+}\theta_c^{(2)})x_1^c + c_2x_2^c$, with $d\, max\, f/d\theta = x_1^c$. Second, both x_1, x_2 are discontinuous functions of θ since the values of these variables change abruptly between basic feasible solutions (Figures 12.2.b, c).

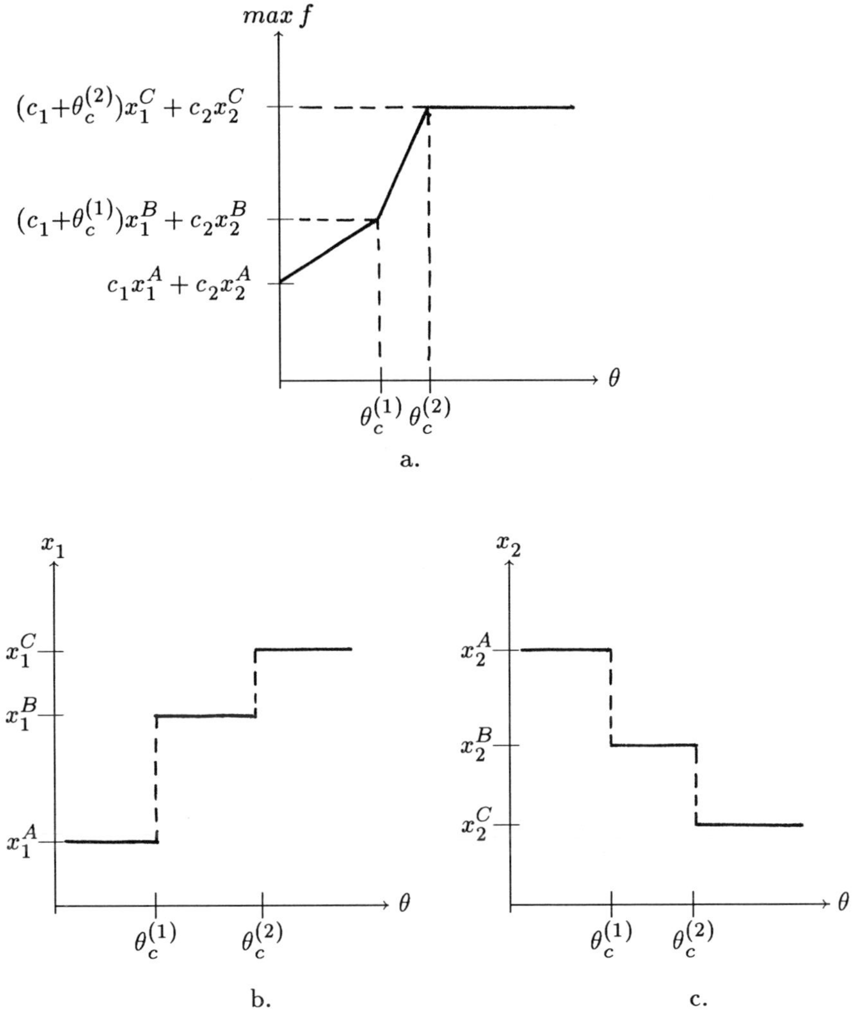

Figure 12.2

(2) PARAMETRIZING THE REQUIREMENTS VECTOR. Given that our starting point is again an optimal basic feasible solution to a linear programming problem, let us replace $\mathbf{b}$ by $\mathbf{b}^* = \mathbf{b} + \phi\mathbf{t}$, where ϕ is a non-negative scalar parameter and $\mathbf{t}$ is a given (arbitrary) $(m \times 1)$ vector which specifies the direction of change in the requirements $b_i, i{=}1,\ldots,m$. Hence the $b_i^*, i{=}1,\ldots,m$, vary linearly with the parameter ϕ. Currently primal feasibility holds with $\mathbf{X}_B = \mathbf{B}^{-1}\mathbf{b} \geqq \mathbf{0}$ or $x_{Bi} \geq 0,\ i{=}1,\ldots,m$. With $\mathbf{b}^*$ replacing $\mathbf{b}$, the revised primal feasibility requirement becomes

$$\begin{aligned} \mathbf{X}_B^* = \mathbf{B}^{-1}\mathbf{b}^* &= \mathbf{B}^{-1}(\mathbf{b}+\phi\mathbf{t}) \\ &= \mathbf{X}_B + \phi\mathbf{B}^{-1}\mathbf{t} \\ &= \mathbf{X}_B + \phi\begin{bmatrix} \boldsymbol{\beta}_1\mathbf{t} \\ \vdots \\ \boldsymbol{\beta}_m\mathbf{t} \end{bmatrix} \geqq \mathbf{0}, \end{aligned} \tag{12.3}$$

where $\boldsymbol{\beta}_i$ denotes the ith row of $\mathbf{B}^{-1}$. In terms of the individual components of (12.3),

$$x_{Bi}^* = x_{Bi} + \phi\boldsymbol{\beta}_i\mathbf{t} \geq 0,\ i=1,\ldots,m. \tag{12.3.1}$$

(Since the optimality criterion is independent of **b**, the parametrization of the latter affects only primal feasibility and not primal optimality.) What is the largest value of ϕ for which (12.3.1) holds? Clearly the implied critical value of ϕ, ϕ_c, is that for which any increase in ϕ beyond ϕ_c violates feasibility, *i.e.*, makes at least one of the x_{Bi}^* values negative.

To determine ϕ_c, let us first assume that $\boldsymbol{\beta}_i\mathbf{t} \geq 0$, $i=1,\ldots,m$. Then ϕ can be increased without bound while still maintaining the revised feasibility criterion since, in this instance, $x_{Bi}^* \geq x_{Bi} \geq 0$, $i=1,\ldots,m$. Next if $\boldsymbol{\beta}_i\mathbf{t} < 0$ for some i, $x_{Bi}^* \geq 0$ for

$$\phi \leq \frac{-x_{Bi}}{\boldsymbol{\beta}_i\mathbf{t}} = \phi_c.$$

Hence the revised feasibility criterion is violated when $\phi > \phi_c$ for some basic variable x_{Bi}. Moreover, if $\boldsymbol{\beta}_i\mathbf{t} < 0$ for two or more basic variables, then

$$\phi_c = \min_i\left\{ \frac{-x_{Bi}}{\boldsymbol{\beta}_i\mathbf{t}} \,\middle|\, \boldsymbol{\beta}_i\mathbf{t} < 0 \right\}. \tag{12.4}$$

Now, if ϕ increases and the minimum in (12.4) is attained for $i=k$, it follows that $\phi_c = -x_{Bk}/\boldsymbol{\beta}_k\mathbf{t}$ or $x_{Bk}^* = 0$. And as ϕ increases beyond ϕ_c, $x_{Bk}^* < 0$ so that $\mathbf{b}_k$ must be removed from the basis to preserve primal feasibility, with its replacement vector determined by the dual simplex entry criterion.

Although the specification of the direction vector **t** is arbitrary, it is frequently the case that only a single requirement b_i is to be changed, so that $\mathbf{t} = \mathbf{e}_i$ (or $-\mathbf{e}_i$). With respect to this particular selection of **t**, if $\mathbf{b}^* = \mathbf{b} + \phi\mathbf{e}_k$, then

(12.3.1) simplifies to $x^*_{Bk} = x_{Bi} + \phi\beta_{ik} \geq 0$, $i=1,\ldots,m$. A comparison of this expression with (11.8.2) indicates that here $\phi \equiv \Delta b_k$. For this choice of $\mathbf{t}$ then, (12.4) may be rewritten as

$$\phi_c = \min_i \left\{ \frac{-x_{Bi}}{\beta_{ik}} \middle| \beta_{ik} < 0 \right\}, \tag{12.4.1}$$

which is thus equivalent to (11.10.c). If the minimum in (12.4.1) is attained for $i = \ell$, $\phi_c = \dfrac{-x_{B\ell}}{\beta_{\ell k}}$ or $x^*_{B\ell} = 0$. If ϕ then increases above this critical value, $x^*_{B\ell}$ turns negative so that $\mathbf{b}_\ell$ is pivoted out of the basis to maintain primal feasibility. Moreover, as b_k changes, f must be modified by an amount $\Delta f = u_k \Delta b_k = u_k \phi_c$.

Let us now examine the geometry of parametrizing the requirements vector. For the current optimal basic feasible solution (with $\phi = 0$), $\mathbf{B}\mathbf{X}_B = \sum_{i=1}^{m} \mathbf{b}_i x_{Bi} = \mathbf{b}$, *i.e.*, $\mathbf{b}$ is a member of the convex polyhedral cone $\mathcal{C}_B = \{\mathbf{Y} \mid \mathbf{Y} = \mathbf{B}\mathbf{X}_B, \mathbf{X}_B \geqq \mathbf{0}\}$ spanned or generated by the columns of the basis matrix $\mathbf{B}$. (For instance, in Figure 12.3, let us assume that $\mathbf{B} = [\mathbf{a}_1, \mathbf{a}_2]$ so that $x_1\mathbf{a}_1 + x_2\mathbf{a}_2 = \mathbf{b}$, with both $x_1, x_2 > 0$ since $\mathbf{b} \in \mathcal{C}_B$ proper). Given $\mathbf{t} \in \mathcal{C} = \{\mathbf{Y} \mid \mathbf{Y} = \mathbf{A}\mathbf{X}, \mathbf{X} \geqq \mathbf{0}\}$, if $\phi_1\mathbf{t}$, $0 < \phi_1 < \phi_c < \phi_c^{(1)}$, is added to $\mathbf{b}$ to yield a new vector $\mathbf{b}^*_{\phi_1}$ (where $\mathbf{b}^*_{\phi_1}$ is

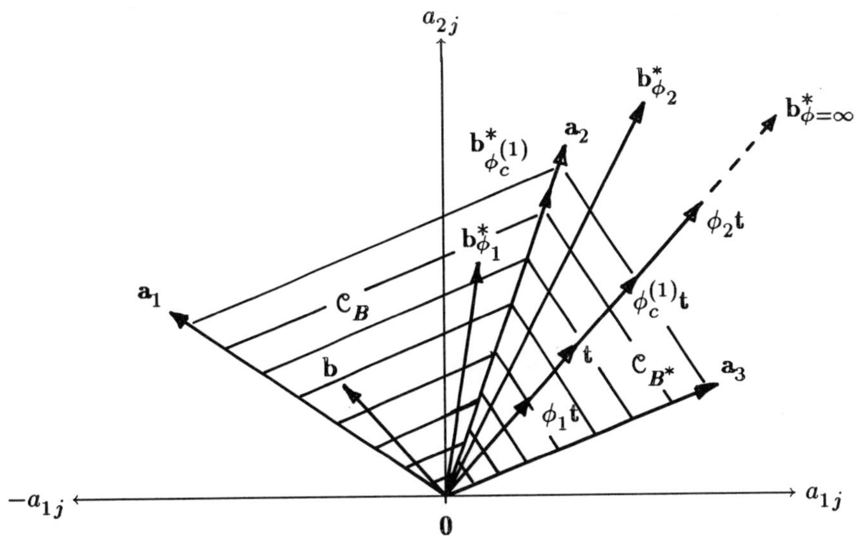

Figure 12.3

obtained via the parallelogram law for vector addition) and $\mathbf{b}^*_{\phi_1} \in \mathcal{C}_B$, then there exists an optimal basic feasible solution to the modified problem

$$\begin{aligned} max\, f(\mathbf{X}) &= \mathbf{C}'\mathbf{X} \quad \text{s.t.} \\ \mathbf{AX} &= \mathbf{b}^*_{\phi_1},\ \mathbf{X} \geqq \mathbf{0}, \end{aligned}$$

which may be written as

$$\mathbf{BX}^*_B = \sum_{i=1}^{m} \mathbf{b}_i x^*_{Bi} = \sum_{i=1}^{m} \mathbf{b}_i\,(x_{Bi}+\phi_1\beta_i\mathbf{t}) = \mathbf{b}^*_{\phi_1}. \tag{12.5}$$

If ϕ is next increased to $\phi_c = \phi_c^{(1)} = -\frac{x_{Bk}}{\beta_k\mathbf{t}}$, it follows that $x^*_{Bk} = 0$ so that (12.5) becomes

$$\sum_{\substack{i=1 \\ i\neq k}}^{m} \mathbf{b}_i(x_{Bi}+\phi_c^{(1)}\beta_i\mathbf{t}) = \mathbf{b}^*_{\phi_c^{(1)}}, \tag{12.6}$$

where $\mathbf{b}_k$ is to be pivoted out of the optimal basis. Moreover, once the dual simplex entry criterion chooses the vector to replace $\mathbf{b}_k$ in the basis, the incoming vector enters the basis at the zero level. If ϕ then increases beyond $\phi_c^{(1)}$, the incoming vector also appears in the basis at the zero level. In terms of Figure 12.3, if $\phi = \phi_1$, $\mathbf{b}^*_{\phi_1} \in \mathcal{C}_B$ so that $\mathbf{B}$ yields an optimal basic feasible solution to the adjusted problem obtained by replacing $\mathbf{b}$ by $\mathbf{b}^*_{\phi_1}$. And when $\phi_c = \phi_c^{(1)}$, $x^*_1 = 0$ so that $0\mathbf{a}_1 + x^*_2\mathbf{a}_2 = \mathbf{b}^*_{\phi_c^{(1)}}$. In this instance $\mathbf{b}^*_{\phi_c^{(1)}}$ lies along an edge of $\mathcal{C}_B$ so that the associated optimal basic feasible solution is degenerate as expected. Hence $\mathbf{a}_1$ may be deleted from the current basis and replaced by $\mathbf{a}_3$, where the new basis matrix is $\mathbf{B}^* = [\mathbf{a}_3, \mathbf{a}_2]$ and $\mathcal{C}_{B^*}$ is the convex polyhedral cone spanned by $\mathbf{a}_2$, $\mathbf{a}_3$. Since $\mathbf{b}^*_{\phi_c^{(1)}}$ also lies along an edge of $\mathcal{C}_{B^*}$, $0\mathbf{a}_3 + x^*_2\mathbf{a}_2 = \mathbf{b}^*_{\phi_c^{(1)}}$ so that $\mathbf{a}_3$ enters the basis at the zero level. And if θ increases a bit above $\theta_c^{(1)}$, to, say, ϕ_2, $\mathbf{b}^*_{\phi_2} \in \mathcal{C}_{B^*}$ proper with $x^*_3\mathbf{a}_3 + x^*_2\mathbf{a}_2 = \mathbf{b}^*_{\phi_2}$ and $x^*_3, x^*_2 > 0$. (Note that the process of parametrizing $\mathbf{b}$ involves rotating $\mathbf{b}$ about the origin.)

In general, as ϕ increases without bound, we obtain a series of critical values and a corresponding sequence of optimal basic feasible solutions. And with $\mathbf{t} \in \mathcal{C}$, $\mathbf{b}^*_\phi$ points in the direction of $\mathbf{t}$ as ϕ approaches infinity so that for the final basis,

t itself serves as the requirements vector (see Figure 12.3). For $\mathbf{t} \notin \mathcal{C}$, ϕ cannot become arbitrarily large without violating primal feasibility since, as Figure 12.4 indicates, $\bar{\phi}$ is the largest value of ϕ which restricts $\mathbf{b}^*$ to $\mathcal{C}$ (again the parallelogram law for vector addition applies). So for $\hat{\phi} > \bar{\phi}$, $\mathbf{b}^*_{\hat{\phi}} \notin \mathcal{C}$, *i.e.*, once ϕ increases beyond $\bar{\phi}$ and (12.4) determines the vector to leave the optimal basis, the dual simplex entry criterion cannot isolate a pivot element since in this case all elements within the pivot row of the simplex matrix are non-negative. Further increases in ϕ subsequently violate primal feasibility.

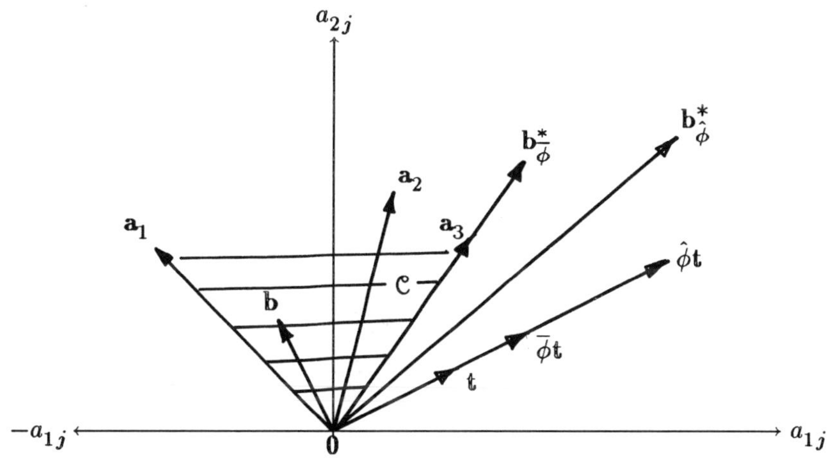

Figure 12.4

One final point is in order. In the preceding discussion we parametrized **b** directly. However, if we consider the associated dual problem, then parametrizing **b** can be handled in the same fashion as parametrizing the objective function.

(3) PARAMETRIZING AN ACTIVITY VECTOR. Given that we have obtained an optimal basic feasible solution to the problem

$$\begin{aligned} max\, f(\mathbf{X}) &= \mathbf{C}'\mathbf{X} \quad \text{s.t.} \\ \mathbf{AX} &= \mathbf{b},\ \mathbf{X} \geqq \mathbf{0}, \end{aligned}$$

let us undertake the parametrization of a column of the activity matrix **A**. With **A** previously partitioned as $\mathbf{A} = [\mathbf{B}, \mathbf{R}]$, it may be the case that some non-basic vector $\mathbf{r}_j$ is replaced by $\mathbf{r}_j^* = \mathbf{r}_j + \tau\mathbf{V}$, $j=1,\ldots,n-m$; or a basic vector $\mathbf{b}_i$ is replaced by $\mathbf{b}_i^* = \mathbf{b}_i + \tau\mathbf{V}$, $i=1,\ldots,m$, where τ represents a non-negative scalar parameter and $\mathbf{V}$ is an arbitrary $(m \times 1)$ vector which determines the direction of change in the components of either $\mathbf{r}_j$ or $\mathbf{b}_i$. In each individual case, the said components are expressed as linear functions of τ.

Let us initially assume that $\mathbf{r}_k$ is replaced by $\mathbf{r}_k^* = \mathbf{r}_k + \tau\mathbf{V}$ in $\mathbf{R} = [\mathbf{r}_1,\ldots,\mathbf{r}_k,\ldots,\mathbf{r}_{n-m}]$ so that the new matrix of non-basic vectors appears as $\mathbf{R}^* = [\mathbf{r}_1,\ldots,\mathbf{r}_k^*,\ldots,\mathbf{r}_{n-m}]$. Since the elements within the basis matrix $\mathbf{B}$ are independent of the parametrization of $\mathbf{r}_k$, primal feasibility is unaffected (we still have $\mathbf{X}_B = \mathbf{B}^{-1}\mathbf{b} \geqq \mathbf{0}$) but primal optimality may be compromised. In this regard, for $\tau = 0$, $\mathbf{C}_B'\mathbf{B}^{-1}\mathbf{R} - \mathbf{C}_R' \geqq \mathbf{0}'$ or $-\bar{c}_j = \mathbf{C}_B'\mathbf{Y}_j - c_{Rj} \geq 0$, $j=1,\ldots,n-m$. Upon parametrizing $\mathbf{r}_k$, the revised optimality criterion becomes

$$\mathbf{C}_B'\mathbf{B}^{-1}\mathbf{R}^* - \mathbf{C}_R' = (-\bar{c}_1,\ldots,-\bar{c}_k^*,\ldots,-\bar{c}_{n-m}) \geqq \mathbf{0}',$$

where

$$\begin{aligned} -\bar{c}_k^* &= \mathbf{C}_B'\mathbf{B}^{-1}\mathbf{r}_k^* - c_{Rk} = \mathbf{C}_B'\mathbf{B}^{-1}(\mathbf{r}_k + \tau\mathbf{V}) - c_{Rk} \\ &= -\bar{c}_k + \tau\mathbf{C}_B'\mathbf{B}^{-1}\mathbf{V} = -\bar{c}_k + \tau\mathbf{U}'\mathbf{V} \geq 0 \end{aligned} \tag{12.7}$$

and $\mathbf{U}' = \mathbf{C}_B'\mathbf{B}^{-1}$ is a $(1 \times m)$ vector of non-negative dual variables. Since $-\bar{c}_j \geq 0$, $j=1,\ldots,k,\ldots,n-m$, primal optimality will be preserved if (12.7) holds. Now as τ increases, primal optimality must be maintained. Given this restriction, what is the largest value of τ (its critical value, τ_c) for which (12.7) holds?

To find τ_c, we shall first assume that $\mathbf{U}'\mathbf{V} \geq 0$. Then τ can be increased indefinitely while still maintaining the revised optimality criterion. Next, if $\mathbf{U}'\mathbf{V} < 0$, then $-\bar{c}_k^* \geq 0$ for

$$\tau \leq \frac{\bar{c}_k}{\mathbf{U}'\mathbf{V}} = \tau_c. \tag{12.8}$$

In this regard, as τ increases to τ_c, it follows that $-\bar{c}_k^* = 0$ so that the case of multiple optimal solutions emerges. That is, the current basic feasible solution (call it $\mathbf{X}_B^{(1)}$) remains optimal and an alternative optimal basic feasible solution $\mathbf{X}_B^{(2)}$ obtains when $\mathbf{r}_k$ is pivoted into the basis, with the primal simplex exit criterion determining the vector to be removed from the current basis. And as τ increases a bit beyond τ_c, it follows that $-\overline{c}_k^* < 0$ so that $\mathbf{X}_B^{(2)}$ becomes the unique optimal basic feasible solution.

As indicated above, the choice of the direction vector $\mathbf{V}$ is completely arbitrary. However, if only a single component r_{ik} of the activity vector $\mathbf{r}_k$ is to change, then $\mathbf{V} = \mathbf{e}_i$ (or $-\mathbf{e}_i$). In this regard, if $\mathbf{r}_k^* = \mathbf{r}_k + \tau \mathbf{e}_\ell$, then (12.7) becomes $-\overline{c}_k^* = -\overline{c}_k + \tau u_\ell \geq 0$. A comparison of this inequality with (11.13) indicates that here $\tau \equiv \Delta r_{\ell k}$. So for this choice of $\mathbf{V}$, with $u_\ell \geq 0, \tau$ may be increased without bound while still preserving primal optimality. Next, if $\mathbf{r}_k^* = \mathbf{r}_k - \tau \mathbf{e}_\ell$, (12.7) simplifies to $-\overline{c}_k^* = -\overline{c}_k - \tau u_\ell \geq 0$. In this instance (12.8) may be rewritten as

$$\tau \leq \frac{-\overline{c}_k}{u_\ell} = \tau_c. \tag{12.8.1}$$

If τ then increases above this critical value, $-\overline{c}_k^*$ turns negative so that $\mathbf{r}_k$ is pivoted into the basis to maintain primal optimality.

When τ is increased slightly above $\tau = \tau_c^{(1)}$, it may be the case that: τ may be increased without bound while still preserving primal optimality; or there exists a second critical value of τ beyond which the new basis ceases to yield an optimal feasible solution. Relative to this latter instance, since $\mathbf{r}_k$ has entered the optimal basis, we now have to determine a procedure for parametrizing a basis vector.

To parametrize a vector within the optimal basis, let us replace the kth column $\mathbf{b}_k$ of the basis matrix $\mathbf{B} = [\mathbf{b}_1, \ldots, \mathbf{b}_k, \ldots, \mathbf{b}_m]$ by $\mathbf{b}_k^* = \mathbf{b}_k + \tau \mathbf{V}$ so as to obtain $\mathbf{B}^* = [\mathbf{b}_1, \ldots, \mathbf{b}_k^*, \ldots, \mathbf{b}_m]$. Since $\mathbf{b}_k^* = \sum_{i=1}^{m} y_{ik} \mathbf{b}_i = \mathbf{B}\mathbf{Y}_k^*$, it follows that

$$\mathbf{Y}_k^* = \mathbf{B}^{-1}\mathbf{b}_k^* = \begin{bmatrix} \beta_1 \\ \vdots \\ \beta_k \\ \vdots \\ \beta_m \end{bmatrix} \mathbf{b}_k^* = \begin{bmatrix} \beta_1 \mathbf{b}_k + \tau \beta_1 \mathbf{V} \\ \vdots \\ \beta_k \mathbf{b}_k + \tau \beta_k \mathbf{V} \\ \vdots \\ \beta_m \mathbf{b}_k + \tau \beta_m \mathbf{V} \end{bmatrix} = \begin{bmatrix} \tau \beta_1 \mathbf{V} \\ \vdots \\ 1 + \tau \beta_k \mathbf{V} \\ \vdots \\ \tau \beta_m \mathbf{V} \end{bmatrix}.$$

From (A.1) of appendix A,

$$\mathbf{I}_m^* = \begin{bmatrix} 1 & \cdots & \frac{-\tau\beta_1\mathbf{V}}{1+\tau\beta_k\mathbf{V}} & \cdots & 0 \\ \vdots & & \vdots & & \vdots \\ 0 & \cdots & \frac{1}{1+\tau\beta_k\mathbf{V}} & \cdots & 0 \\ \vdots & & \vdots & & \vdots \\ 0 & \cdots & \frac{-\tau\beta_m\mathbf{V}}{1+\tau\beta_k\mathbf{V}} & \cdots & 1 \end{bmatrix}.$$

And from (A.2), $(\mathbf{B}^*)^{-1} = \mathbf{I}_m^*\mathbf{B}^{-1}$ and thus

$$\mathbf{X}_B^* = (\mathbf{B}^*)^{-1} = \mathbf{I}_m^*\mathbf{B}^{-1}\mathbf{b} = \mathbf{I}_m^*\mathbf{X}_B$$

$$= \begin{bmatrix} x_{B1} - \frac{\tau\beta_1\mathbf{V}}{1+\tau\beta_k\mathbf{V}}x_{Bk} \\ \vdots \\ \frac{1}{1+\tau\beta_k\mathbf{V}}x_{Bk} \\ \vdots \\ x_{Bm} - \frac{\tau\beta_m\mathbf{V}}{1+\tau\beta_k\mathbf{V}}x_{Bk} \end{bmatrix}.$$

Since $\mathbf{X}_B^*$ must be primal feasible, we require that

$$1 + \tau\beta_k\mathbf{V} > 0,$$

$$x_{Bi} - \frac{\tau\beta_i\mathbf{V}}{1+\tau\beta_k\mathbf{V}}x_{Bk} \geq 0,\ i=1,\ldots,m.$$

If $1 + \tau\beta_k\mathbf{V} > 0$, then, upon solving the second inequality for τ and remembering that τ must be strictly positive,

$$\tau \leq \frac{x_{Bi}}{\beta_i\mathbf{V}x_{Bk} - \beta_k\mathbf{V}x_{Bi}},\ \beta_i\mathbf{V}x_{Bk} - \beta_k\mathbf{V}x_{Bi} > 0,\ i=1,\ldots,m. \tag{12.9}$$

Since (12.9) must hold for all $i \neq k$ when $1 + \tau\beta_k\mathbf{V} > 0$, it follows that the upper limit or critical value to which τ can increase without violating primal feasibility is

$$\tau_c' = \min_{i \neq k} \left\{ \frac{x_{Bi}}{\beta_i \mathbf{V} x_{Bk} - \beta_k \mathbf{V} x_{Bi}} \,\middle|\, \beta_i \mathbf{V} x_{Bk} - \beta_k \mathbf{V} x_{Bi} > 0 \right\}. \tag{12.10}$$

To determine the effect of parametrizing the basis vector upon primal optimality, let us express the revised optimality criterion as

$$\begin{aligned} \mathbf{C}'_{B^*}(\mathbf{B}^*)^{-1}\mathbf{R} - \mathbf{C}'_R &= \mathbf{C}'_{B^*}\mathbf{I}^*_m \mathbf{B}^{-1}\mathbf{R} - \mathbf{C}'_R \\ &= \mathbf{C}'_{B^*}\mathbf{I}^*_m[\mathbf{Y}_1, \ldots, \mathbf{Y}_{n-m}] - \mathbf{C}'_R \geqq \mathbf{0}' \end{aligned}$$

or, in terms of its components,

$$\begin{aligned} -\bar{c}_j^* &= \sum_{i \neq k} c_{B^*i}\, y_{ij} + \frac{c_{B^*k} y_{kj}}{1 + \tau \beta_k \mathbf{V}} - \sum_{i \neq k} \frac{\tau c_{B^*i} \beta_i \mathbf{V}}{1 + \tau \beta_k \mathbf{V}} y_{kj} \\ &\qquad - c_{Rj} \geq 0,\ j = 1, \ldots, n-m. \end{aligned} \tag{12.11}$$

If we now add and subtract the term

$$c_{B^*k} y_{kj} + \frac{\tau c_{B^*k} \beta_k \mathbf{V}}{1 + \tau \beta_k \mathbf{V}} y_{kj}$$

on the left-hand side of (12.11), the latter simplifies to

$$\begin{aligned} -\bar{c}_j^* &= \sum_i c_{B^*i} y_{ij} - \frac{\tau y_{kj} \sum_i c_{B^*i} \beta_i \mathbf{V}}{1 + \tau \beta_k \mathbf{V}} - c_{Rj} \\ &= -\frac{\tau y_{kj} \mathbf{U}'\mathbf{V}}{1 + \tau \beta_k \mathbf{V}} - \bar{c}_j \geq 0,\ j = 1, \ldots, n-m. \end{aligned} \tag{12.11.1}$$

If $1 + \tau \beta_k \mathbf{V} > 0$, and again remembering that τ must be strictly positive, we may solve (12.11.1) for τ as

$$\tau \leq \frac{-\bar{c}_j}{y_{kj} \mathbf{U}'\mathbf{V} + \bar{c}_j \beta_k \mathbf{V}},\ y_{kj} \mathbf{U}'\mathbf{V} + \bar{c}_j \beta_k \mathbf{V} > 0,\ j = 1, \ldots, n-m. \tag{12.12}$$

Since (12.12) is required to hold for all $j = 1, \ldots, n-m$ when $1 + \tau \beta_k \mathbf{V} > 0$, we may write the critical value to which τ can increase while still preserving primal optimality as

$$\tau_c'' = \min_j \left\{ \frac{-\bar{c}_j}{y_{kj} \mathbf{U}'\mathbf{V} + \bar{c}_j \beta_k \mathbf{V}} \,\middle|\, y_{kj} \mathbf{U}'\mathbf{V} + \bar{c}_j \beta_k \mathbf{V} > 0 \right\}. \tag{12.13}$$

As we have just seen, the parametrization of a basis vector may affect both the feasibility and optimality of the current solution. This being the case, the critical value of τ which preserves both primal feasibility and optimality simultaneously is simply

$$\tau_c = min\{\tau_c', \tau_c''\}. \tag{12.14}$$

Finally, to ascertain the effect on the optimal value of f of parametrizing $\mathbf{b}_k$, let us find

$$\begin{aligned} f(\mathbf{X}_{B^*}) &= \mathbf{C}_B' \mathbf{X}_{B^*} = \mathbf{C}_{B^*}' \mathbf{I}_m^* \mathbf{X}_B \\ &= \sum_{i \neq k} c_{B^*_i} x_{Bi} + \frac{c_{B^*_k} x_{Bk}}{1+\tau\beta_k\mathbf{V}} - \sum_{i\neq k} \frac{\tau c_{B^*_i}\beta_i\mathbf{V}}{1+\tau\beta_k\mathbf{V}} x_{Bk}. \end{aligned} \tag{12.15}$$

If the term

$$c_{B^*_k} x_{Bk} + \frac{\tau c_{B^*_k}\beta_k\mathbf{V}}{1+\tau\beta_k\mathbf{V}} x_{Bk}$$

is added and subtracted on the left-hand side of (12.15), we obtain

$$\begin{aligned} f(\mathbf{X}_{B^*}) &= \sum_i c_{B^*_i} x_{Bi} - \frac{\tau x_{Bk} \sum_i c_{B^*_i}\beta_i\mathbf{V}}{1+\tau\beta_k\mathbf{V}} \\ &= f(\mathbf{X}_B) - \frac{\tau x_{Bk}\mathbf{U}'\mathbf{V}}{1+\tau\beta_k\mathbf{V}}. \end{aligned} \tag{12.15.1}$$

As far as the specification of the arbitrary direction vector $\mathbf{V}$ is concerned, it is sometimes the case that only a single component of $\mathbf{b}_i$ is to change, *i.e.*, $\mathbf{V} = \mathbf{e}_i$ (or $-\mathbf{e}_i$). Then if $\mathbf{b}_k^* = \mathbf{b}_k + \tau\mathbf{e}_\ell$, (12.9) reduces to (11.15.b) (here $\tau \equiv \Delta b_{\ell k}$) so that (12.10) is equivalent to the right-hand inequality of (11.19). Finally, (12.15.1), (11.20.1) are one and the same for $\mathbf{V} = \mathbf{e}_\ell$.

12.2. The Primal-Dual Method Revisited

In this section we shall consider what may be termed a **parametric method** for solving a linear programming problem. While the technique to be presented shortly is not a post-optimality routine proper, it utilizes the basic approach of parametrizing the objective function discussed at the beginning of this chapter.

As indicated in chapter eight, when artificial variables appear in a linear programming problem, possibly the one-phase (M-penalty) or two-phase method may be employed to drive the artificial variables to zero so as to obtain an initial basic feasible solution to the original problem. Once such a solution emerges, the standard primal simplex pivot process can be used to generate a solution which is also optimal. However, at the end of phase I (or when all artificial vectors have turned non-basic in the M-penalty technique), the resulting solution may require a considerable number of additional iterations to achieve optimality, *i.e.*, both the one- and two-phase methods ignore optimality in their search for an initial basic feasible solution to the original problem.

It is this last observation then that has prompted the development of a method that simultaneously works towards both feasibility and optimality of the original problem. In this regard, when all artificial variables have been driven to zero, the resulting solution is both primal feasible and optimal, *e.g.*, phase I terminates with an optimal basic feasible solution to the given problem.

While the original formulation of the primal-dual method (chapter 10) uses alternatively the primal and dual problems, a modification of the standard primal-dual routine (van de Panne, Whinston(1968); van de Panne (1975)) will now be presented wherein a parametrized version of only the primal problem is utilized. As demonstrated in the preceding two references, the original version of the primal-dual method and its parametric analog are equivalent in the sense that each gives rise to the same sequence of basic feasible solutions.

Let us assume that the primal problem appears as

$$\begin{aligned} min\, f(\mathbf{X}) &= \mathbf{C}'\mathbf{X} \quad \text{s.t.} \\ \mathbf{AX} = \mathbf{b}, &\quad \text{with} \quad \mathbf{C}, \mathbf{X} \geqq \mathbf{0}. \end{aligned}$$

Then if we introduce into the structural constraint system an appropriate number of non-negative artificial variables (here we shall assume that m such variables are required), we obtain the **artificial problem**

$$\begin{aligned} min\, f(\mathbf{X},\, \mathbf{X}_a) &= \mathbf{C}'\mathbf{X} + \mathbf{0}'\mathbf{X}_a \quad \text{s.t.} \\ \mathbf{A}\mathbf{X} + \mathbf{I}_m\mathbf{X}_a &= \mathbf{b}, \quad \text{with } \mathbf{C},\, \mathbf{X},\, \mathbf{X}_a \geqq \mathbf{0} \text{ or} \\ max\{h(\mathbf{X},\mathbf{X}_a) &= -f(\mathbf{X},\mathbf{X}_a) = -\mathbf{C}'\mathbf{X} - \mathbf{0}'\mathbf{X}_a\} \quad \text{s.t.} \\ \mathbf{A}\mathbf{X} + \mathbf{I}_m\mathbf{X}_a &= \mathbf{b}, \text{ with } \mathbf{C},\, \mathbf{X},\, \mathbf{X}_a \geqq \mathbf{0}, \end{aligned}$$

where $\mathbf{X}_a$ is an $(m \times 1)$ vector of non-negative artificial variables. To consolidate our notation a bit, let us

$$\begin{aligned} max\, h(\bar{\mathbf{X}}) &= \bar{\mathbf{C}}'\bar{\mathbf{X}} \quad \text{s.t.} \\ \bar{\mathbf{A}}\bar{\mathbf{X}} &= \mathbf{b}, \quad \text{with } \bar{\mathbf{C}} \leqq \mathbf{0},\, \bar{\mathbf{X}} \geqq \mathbf{0}, \end{aligned}$$

where

$$\bar{\mathbf{C}} = \begin{bmatrix} -\mathbf{C} \\ \mathbf{0} \end{bmatrix},\ \bar{\mathbf{X}} = \begin{bmatrix} \mathbf{X} \\ \mathbf{X}_a \end{bmatrix},\ \text{and } \bar{\mathbf{A}} = [\mathbf{A},\, \mathbf{I}_m]$$

are respectively of order $(n+m \times 1)$, $(n+m \times 1)$, and $(m \times n+m)$. If we replace $\bar{\mathbf{C}}$ by

$$\bar{\mathbf{C}}^* = \bar{\mathbf{C}} + \lambda\mathbf{S} = \begin{bmatrix} -\mathbf{C} \\ \mathbf{0} \end{bmatrix} + \lambda \begin{bmatrix} \mathbf{0} \\ -\mathbf{1} \end{bmatrix} = \begin{bmatrix} -\mathbf{C} \\ -\lambda\mathbf{1} \end{bmatrix},$$

where λ is a non-negative scalar parameter and $\mathbf{S}$ is an $(n+m \times 1)$ vector which determines the direction of change in the coefficients $\bar{c}_j^*$, $j=1,\ldots,n-m$, then we ultimately seek to solve the **extended artificial problem**

$$\begin{aligned} max\{h(\bar{\mathbf{X}}) &= (\bar{\mathbf{C}}^*)'\bar{\mathbf{X}} = -\mathbf{C}'\mathbf{X} - \lambda\mathbf{1}'\mathbf{X}_a\} \quad \text{s.t.} \\ \mathbf{A}\mathbf{X} + \mathbf{I}_m\mathbf{X}_a &= \mathbf{b}, \quad \text{with } \mathbf{C},\, \mathbf{X},\, \mathbf{X}_a \geqq \mathbf{0}. \end{aligned}$$

The simplex matrix associated with this problem is

$$\begin{bmatrix} \mathbf{A} & \mathbf{I}_m & \mathbf{0} & \mathbf{b} \\ \mathbf{C}' & \lambda\mathbf{1}' & 1 & 0 \end{bmatrix}. \tag{12.16}$$

If we multiply the first m rows of this matrix by $-\lambda$ and add their sum to the last row of (12.16), we obtain

$$\begin{bmatrix} \mathbf{A} & \mathbf{I}_m & \mathbf{0} & \mathbf{b} \\ \mathbf{C}' - \lambda\mathbf{1}'\mathbf{A} & \mathbf{0}' & 1 & -\lambda\mathbf{1}'\mathbf{b} \end{bmatrix}. \tag{12.16.1}$$

Since an $(m+1 \times m+1)$ identity matrix appears in this latter simplex matrix, the $m+1$ basic variables are the m artificial variables and f. With $\lambda = 0$, $\mathbf{X} = \mathbf{0}$, it follows that $\mathbf{X}_a = \mathbf{b}$, $f = 0$ represents an initial optimal basic feasible solutions to the extended artificial problem since $\mathbf{C} \geqq \mathbf{0}$. The motivation underlying the primal-dual method then is to increase λ parametrically above its current zero level (*i.e.*, optimality is preserved from one basic feasible solution to the next) so that for each successive iteration an artificial vector is replaced by a legitimate one in the optimal basis. And as the parametrization process increases λ through a succession of critical values, the sequence of optimal basic feasible solutions to the extended artificial problem converges to an optimal basic feasible solution to the original problem.

In actual practice, the implementation of the primal-dual method mirrors the procedure employed above in parametrizing the objective function. To see this let us partition $\bar{\mathbf{C}}^*$ as

$$\bar{\mathbf{C}}^* = \begin{bmatrix} \mathbf{C}_R \\ \mathbf{C}_B \end{bmatrix} + \lambda \begin{bmatrix} \mathbf{S}_R \\ \mathbf{S}_B \end{bmatrix} = \begin{bmatrix} -\mathbf{C} \\ \mathbf{0} \end{bmatrix} + \lambda \begin{bmatrix} \mathbf{0} \\ -\mathbf{1} \end{bmatrix}.$$

With $\mathbf{B} = \mathbf{I}_m = \mathbf{B}^{-1}$, $\mathbf{R} = \mathbf{A}$, (12.16.1) may be rewritten as

$$\begin{bmatrix} \mathbf{B}^{-1}\mathbf{R} & \mathbf{I}_m & \mathbf{0} & \mathbf{b} \\ -\mathbf{C}_R+\lambda\mathbf{S}_B'\mathbf{B}^{-1}\mathbf{R} & \mathbf{0}' & 1 & -\lambda\mathbf{1}'\mathbf{b} \end{bmatrix} = \tag{12.16.2}$$

$$\begin{bmatrix} \mathbf{Y}_1 & \mathbf{Y}_2 & \cdots & \mathbf{Y}_n & \mathbf{I}_m & \mathbf{0} & \mathbf{b} \\ -c_R+\lambda\mathbf{S}_B'\mathbf{Y}_1 & -c_{R2}+\lambda\mathbf{S}_B'\mathbf{Y}_2 & \cdots & -c_{Rn}+\lambda\mathbf{S}_B'\mathbf{Y}_n & \mathbf{0}' & 1 & 1-\lambda\mathbf{1}'\mathbf{b} \end{bmatrix},$$

where a typical term taken from the first n of those appearing within the last row of this matrix is

$$-c_{Rj} + \lambda\mathbf{S}_B'\mathbf{Y}_j \geq 0,\ j=1,\ldots,n. \tag{12.17}$$

A careful examination of (12.17) reveals that it is equivalent to (12.1.1).

How large of an increase in λ above zero preserves primal optimality (dual feasibility)? First, if $\mathbf{S}'_B\mathbf{Y}_j \geq 0$ for all j, then λ can be increased without bound while still maintaining the optimality of the extended artificial problem. Next, if $\mathbf{S}'_B\mathbf{Y}_j < 0$ for at least one value of j, then the critical value of λ is

$$\lambda_c = \min_j\left\{\frac{c_{Rj}}{\mathbf{S}'_B\mathbf{Y}_j}\,\middle|\, \mathbf{S}'_B\mathbf{Y}_j < 0\right\}. \tag{12.18}$$

An examination of this expression indicates that it is equivalent to (12.2) So if λ increases and the minimum in (12.18) is attained for $j = k$, the kth non-basic column replaces an artificial vector in the basis, with the latter determined by the standard primal simplex exit criterion.

EXAMPLE 12.1. Using the parametric primal-dual technique, find

$$\begin{aligned} \min f &= x_1 + x_2 + x_3 + 2x_4 \quad \text{s.t.} \\ x_1 + 2x_2 \quad\quad &+ 3x_4 = 15 \\ x_2 + x_3 &+ 2x_4 = 10 \\ x_1, \ldots, x_4 &\geq 0. \end{aligned}$$

If x_5, x_6 represent artificial variables, the above problem may be rewritten in extended artificial form as

$$\begin{aligned} \max h(x_1, \ldots, x_6) &= -x_1 - x_2 - x_3 - 2x_4 - \lambda x_5 - \lambda x_6 \quad \text{s.t.} \\ x_1 + 2x_2 \quad\quad + 3x_4 + x_5 \quad &= 15 \\ x_2 + x_3 + 2x_4 \quad\quad + x_6 &= 10 \\ x_1, \ldots, x_6 &\geq 0, \end{aligned}$$

with associated simplex matrix

$$\begin{bmatrix} 1 & 2 & 0 & 3 & 1 & 0 & 0 & 15 \\ 0 & 1 & 1 & 2 & 0 & 1 & 0 & 10 \\ 1 & 1 & 1 & 2 & \lambda & \lambda & 1 & 0 \end{bmatrix} \longrightarrow \begin{bmatrix} 1 & 2 & 0 & 3 & 1 & 0 & 0 & 15 \\ 0 & 1 & 1 & 2 & 0 & 1 & 0 & 10 \\ 1 & 1 & 1 & 2 & 0 & 0 & 1 & 0 \\ -\lambda & -3\lambda & -\lambda & -5\lambda & 0 & 0 & 0 & -25\lambda \end{bmatrix}.$$

To facilitate the execution of subsequent rounds of the primal-dual method, each entry in the objective function row has been separated into two parts, one which does not involve λ, and one appearing directly beneath it which does. Hence this "double-row" representation actually depicts the "single" objective function row of the simplex matrix.

For $\lambda = 0$, we obtain an initial basic feasible solution to the extended artificial problem consisting of $x_5 = 15$, $x_6 = 10$, and $h = 0$. Let λ now increase in value above its current zero level. From (12.18),

$$\lambda_c^{(1)} = \frac{1}{3} = \min_j\left\{\frac{-1}{-1}, \frac{-1}{-3}, \frac{-1}{-1}, \frac{-2}{-5}\right\}$$

so that column two represents the pivot column. Upon performing a pivot operation we obtain

$$\begin{bmatrix} \frac{1}{2} & 1 & 0 & \frac{3}{2} & 0 & 0 & \frac{15}{2} \\ -\frac{1}{2} & 0 & 1 & \frac{1}{2} & 1 & 0 & \frac{5}{2} \\ \frac{1}{2} & 0 & 1 & \frac{1}{2} & 0 & 1 & -\frac{15}{2} \\ \frac{1}{2}\lambda & 0 & -\lambda & -\frac{1}{2}\lambda & 0 & 0 & -\frac{5}{2}\lambda \end{bmatrix},$$

where the column associated with the artificial variable x_5 has been deleted since x_5 has turned non-basic. In this instance the optimal basic feasible solution to the extended artificial problem is $x_2 = \frac{15}{2}$, $x_6 = \frac{5}{2}$, and $h = -\frac{15}{2} - \frac{5}{2}\lambda$. If we increase λ a bit further we may again employ (12.18) to determine

$$\lambda_c^{(2)} = 1 = \min_j\left\{\frac{-1}{-1}, \frac{-\frac{1}{2}}{-\frac{1}{2}}\right\}.$$

Since a tie for the minimum occurs, let us arbitrarily select column three of the preceding simplex matrix as the pivot column. Upon pivoting we have

$$\begin{bmatrix} \frac{1}{2} & 1 & 0 & \frac{3}{2} & 0 & \frac{15}{2} \\ -\frac{1}{2} & 0 & 1 & \frac{1}{2} & 0 & \frac{5}{2} \\ 1 & 0 & 0 & 0 & 1 & -10 \end{bmatrix},$$

where this time the column associated with the artificial variable x_6 has been deleted along with the bottom "λ row" since the latter contains only zero elements. Here $x_2 = \frac{15}{2}$, $x_3 = \frac{5}{2}$, and $h = -f = -10$ or $f = 10$ represents an optimal basic feasible solution to the original problem. Moreover, the optimal simplex matrix indicates that there exists multiple optimal basic feasible solutions to the original problem, as the reader may easily verify.

13 INTERIOR POINT METHODS

13.1. Optimization Over a Sphere

It was mentioned earlier that the standard simplex method searches for an optimum to a linear program by moving along the surface of a convex polyhedron from one extreme point to an adjacent extreme point in a fashion such that the objective value is nondecreasing between successive basic feasible solutions. The simplex process then either terminates, in a finite number of steps, at a finite optimal solution or reveals that no such solution exists. In this chapter we shall examine a set of ***interior point methods*** for solving a linear program (*e.g.*, projective and affine potential reduction, primal and dual affine scaling, and path following), so named because they converge to an optimal point (if one exists) by passing through the interior of a convex polytope in a fashion such that, at each iteration, a weighted least squares problem is solved.

The first serious challenge to the computational efficiency of the simplex method came in 1984 when N. Karmarkar (1984) introduced his projective scaling algorithm. Karmarkar's method is used to solve problems in what can be termed ***Karmarkar standard form***:

$$\begin{aligned} &min\ \mathbf{C}'\mathbf{X} \quad \text{s.t.} \\ &\mathbf{AX} = \mathbf{0},\ \mathbf{1}'\mathbf{X} = 1,\ \mathbf{X} \geq \mathbf{0}, \end{aligned} \tag{13.1.a}$$

where $\mathbf{C}$, $\mathbf{X} \epsilon \mathbf{R}^n$, $\mathbf{A}$ is $(m \times n)$, with $\rho(\mathbf{A}) = m$ and $n \geq 2$, and both $\mathbf{C}$, $\mathbf{A}$ are integer matrices. (Section 13.6 demonstrates that *any* linear program can be put in this form.) It is further assumed that:

(a) $\mathbf{A1} = \mathbf{0}$, *i.e.*, $\mathbf{X}_0 = (\frac{1}{n})\mathbf{1}$ is feasible; and

(b) the optimal objective value for (13.1.a) is zero.

Based upon these considerations, it is evident that the problem is feasible, bounded, and thus has an optimum. Moreover, the dual of (13.1.a) can be written

$$\begin{aligned} &max\ \mathbf{e}'_{m+1}\bar{\mathbf{U}} \quad \text{s.t.} \\ &[\mathbf{A}', \mathbf{1}]\ \bar{\mathbf{U}} \leqq \mathbf{C} \\ &\bar{\mathbf{U}} \text{ unrestricted.} \end{aligned}$$

If we let $\bar{\mathbf{U}}' = (\mathbf{U}', z)$, where $\mathbf{U} \in \mathbf{R}^m$ and $z \in \mathbf{R}$, then the preceding problem appears as

$$\begin{aligned} &max\ z \quad \text{s.t.} \\ &\mathbf{A}'\mathbf{U} + z\ \mathbf{1} \leqq \mathbf{C} \qquad\qquad (13.1.b) \\ &\mathbf{U}, z \text{ unrestricted.} \end{aligned}$$

This dual is also feasible. That is, for any $\widehat{\mathbf{U}} \in \mathbf{R}^m$, one can choose $\widehat{z} = \underset{j}{min}(\mathbf{C} - \mathbf{A}'\widehat{\mathbf{U}})_j$ so that the point $(\widehat{\mathbf{U}}, \widehat{z})$ is feasible in (13.1.b).

A glance at the constraints in (13.1.a) reveals that the feasible region $\mathcal{K}$ is formed as the intersection of the $(n-m)$-dimensional linear subspace $\mathcal{S} = \{\mathbf{X} \mid \mathbf{AX} = \mathbf{0},\ \ \mathbf{X} \in \mathbf{R}^n\}$ and the $(n-1)$-dimensional unit simplex $\sigma = \{\mathbf{X} \mid \mathbf{1}'\mathbf{X} = 1,\ \mathbf{X} \geqq \mathbf{0},\ \mathbf{X} \in \mathbf{R}^n\}$, *i.e.*, the polytope $\mathcal{K} = \mathcal{S} \cap \sigma$. Furthermore, the $m+1$ structural constraints in (13.1.a) are taken to be linearly independent, $\mathbf{X}_0 \in \mathcal{K}$, and $\mathbf{C}'\mathbf{X} \geq 0$ for each $\mathbf{X} \in \mathcal{K}$. Hence we seek to

$$min\ \mathbf{C}'\mathbf{X} \quad \text{s.t.} \quad \mathbf{X} \in \mathcal{K}. \qquad (13.1.1)$$

In contrast to the simplex method, the projective transformation routine generates a sequence of points inside $\mathcal{K}$ whose associated objective values (***costs*** as they are usually called in a minimization problem) approach zero. At termination, it seeks out a vertex of no greater cost, which is then deemed optimal.

To simplify the solution to (13.1.1), let us replace the constraint $\mathbf{X} \in \sigma$ by the ***roundness restriction*** $\mathbf{X} \in B(\mathbf{X}_0, r)$, where $B(\mathbf{X}_0, r) = \{\mathbf{X} \mid \|\mathbf{X} - \mathbf{X}_0\| < r\}$ is the largest sphere with center $\mathbf{X}_0$ and radius r that is contained in σ. Under this restriction, the new feasible region (corresponding to the intersection of the subspace $\mathcal{S}$ and a sphere centered in it) is a lower-dimensional sphere $\bar{B}(\mathbf{X}_0, r)$ of the same radius inside $\mathcal{S}$, *i.e.*, $\bar{B}(\mathbf{X}_0, r) = \mathcal{S} \cap B(\mathbf{X}_0, r)$, where $\mathbf{A}\mathbf{X}_0 = \mathbf{0}$. If $\mathbf{C}_p$ denotes the orthogonal projection of $\mathbf{C}$ onto $\mathcal{S}$, then we ultimately desire to

$$min\ \mathbf{C}_p'\mathbf{X} \quad \text{s.t.} \quad \mathbf{X} \in \bar{B}(\mathbf{X}_0, r). \tag{13.1.2}$$

To solve this problem, we need only start at $\mathbf{X}_0$ and take a step of length r in the direction of steepest descent $-\mathbf{C}_p$, thereby reaching a point $\mathbf{Z}$ on the surface of $\bar{B}(\mathbf{X}_0, r)$. The point $\mathbf{Z}$ so obtained minimizes the objective in (13.1.2). If $\bar{B}(\mathbf{X}_0, r)$ isn't much smaller than $\mathcal{K}$, then $\mathbf{C}_p'\mathbf{Z}$ should give a good approximation to the minimum of the original objective function in (13.1.a).

13.2. An Overview of Karmarkar's Algorithm

In view of (13.1.2), Karmarkar's algorithm creates a sequence of feasible points $\mathbf{X}_0, \mathbf{X}_1, \ldots, \mathbf{X}_k$ whose corresponding costs approach zero (under assumption (b) above) by undertaking the following sequence of steps:

STEP 0. Initialize.

$\mathbf{X}_0 = (\frac{1}{n})\mathbf{1} =$ center of σ.

STEP 1. Compute the next point in the sequence

$\mathbf{X}_{k+1} = \zeta(\mathbf{X}_k),\ k=0,1,\ldots.$

To determine the role of the function ζ, let us assume that the current solution point is $\mathbf{X}_k$ with $\mathbf{C}'\mathbf{X}_k > 0$. To obtain the next point, apply the projective transformation $T(\mathbf{X})$ of simplex σ that maps the input point $\mathbf{X}_k$ to the center $\mathbf{X}_0$, *i.e.*, $\mathbf{X}_k$ is replaced by $\mathbf{X}_0$ under this *centering transformation.* (Each iteration *always* starts at the center of σ since the improvement in the objective value obtained by moving from the center of the polytope in the direction of steepest descent is typically much greater than the improvement resulting by moving in the same direction from a point near the boundary of the polytope.) The feasible region $\bar{\mathcal{K}}$ of the new problem is formed, as was $\mathcal{K}$, by intersecting the simplex σ and a linear space. σ is then *rounded about* $\mathbf{X}_0$ by replacing it by the largest sphere $\bar{B}(\mathbf{X}_0, r)$ inscribed within $\mathfrak{K}$. By minimizing a transformed objective function over $\bar{B}(\mathbf{X}_0, r)$, we obtain a new point $\mathbf{Y}_{k+1}$ in transformed space whose cost in the transformed problem is smaller than that of $\mathbf{X}_0$. If the inverse of the transformation T is applied to $\mathbf{Y}_{k+1}$, we obtain the output or new solution point $\mathbf{X}_{k+1} = T^{-1}(\mathbf{Y}_{k+1})$ in our original or true space.

Because the cost of $\mathbf{Y}_{k+1}$ in the transformed problem is less than the cost of $\mathbf{X}_0$, the cost of $\mathbf{X}_{k+1}$ in the original problem will be less than the cost of $\mathbf{X}_k$. Repeated application of this procedure renders a sequence of points whose costs approach zero.

STEP 2. Check for infeasibility.

Since the transformed objective function is inherently nonlinear, Karmarkar introduces a ***potential function***

$$f(\mathbf{X}) = \sum_{i=1}^{n} \ln\left(\frac{\mathbf{C}'\mathbf{X}}{x_i}\right) \approx \mathbf{C}'\mathbf{X} \tag{13.2}$$

in true space to keep track of the progress of the objective function towards optimality. (This potential function is a surrogate for the original linear objective function since it tends to be much better behaved under the projective transformation.) An improvement δ in the potential function is expected at each step of the algorithm, *i.e.*, $f(\mathbf{X}_{k+1}) \leq f(\mathbf{X}_k) - \delta$. If the expected improvement in $f(\mathbf{X})$ does not materialize, then we terminate the process and conclude that the minimum value of the objective function is strictly positive. Hence the original problem does not possess a finite optimum, *i.e.*, the original problem is either infeasible or unbounded.

STEP 3. check for optimality.

This involves going from the current point in true space to the nearest extreme point in σ without increasing the value of the objective function and testing the extreme point for optimality.

Let us now look to the particulars of Karmarkar's algorithm. Specifically, to examine the preceding steps in detail, we need to further scrutinize: the projective transformation itself; the transformed problem; the notion of optimization over a sphere; requirements for potential function improvement; the test for optimality; and algorithm complexity.

13.3. The Projective Transformation $T(\mathbf{X})$

We noted above that to solve (3.1.2) we should move from $\mathbf{X}_0 = (\frac{1}{n})\mathbf{1}$ to the boundary of the inscribed sphere along the radius vector of most rapid objective function improvement or descent. After a single such step, the solution is no longer at the center of σ. To produce an algorithm that *always* steps from the center of the simplex, use a projective transformation so that, at each iteration, calculations are performed in a transformed space, where the current solution is the center of σ (and thus of the inscribed sphere), and then mapped back to true space. How shall we form this recentering transformation?

Karmarkar's projective transformation is a one-to-one and onto vector-valued function T: $\sigma \rightarrow \sigma$. Specifically, for each interior point of $\sigma = \{\mathbf{X} \mid \mathbf{1}'\mathbf{X} = 1, \mathbf{X} \geqq \mathbf{0}\}$ there is a unique projective transformation T of $\mathbf{1}'\mathbf{X} = 1$ fixing all vectors of σ and sending $\mathbf{X}$ to the center $\mathbf{X}_0 = (\frac{1}{n})\mathbf{1}$. Hence we have a transformation that converts the current problem into a new one in which the current feasible solution $\mathbf{X}_k > \mathbf{0}$ is replaced by $\mathbf{X}_0$.

Since the image of every $\mathbf{X} \in \sigma$ must also lie in σ, the sum of the coordinates of $T(\mathbf{X})$ must be one. Thus a transformation which maps $\mathbf{X}_k$ to $\mathbf{X}_0$ is to map the point $(x_1, \ldots, x_n)$ to $(x_1/x_{k1}, \ldots, x_n/x_{kn})$ and then divide by the sum of the coordinates, *i.e.*,

$$T: \mathbf{X} \rightarrow \mathbf{Y}, \text{ where } y_j = T_j(\mathbf{X}) = \frac{(x_j/x_{kj})}{\sum_{\ell=1}^{n}(x_\ell/x_{k\ell})}, \; j=1,\ldots,n. \tag{13.3}$$

In matrix terms, if

$$\mathbf{C}_k = \begin{bmatrix} x_{k1}^{-1} & 0 & \cdots & 0 \\ 0 & x_{k2}^{-1} & \cdots & 0 \\ \vdots & \vdots & & \vdots \\ 0 & 0 & \cdots & x_{kn}^{-1} \end{bmatrix},$$

then (13.3) becomes

$$Y = T(X) = \frac{C_k X}{1'C_k X}. \tag{13.3.1}$$

Alternatively, we may, at times, find it convenient to rewrite this expression as follows. If

$$D_k = C_k^{-1} = \begin{bmatrix} x_{k1} & 0 & \cdots & 0 \\ 0 & x_{k2} & \cdots & 0 \\ \vdots & \vdots & & \vdots \\ 0 & 0 & \cdots & x_{kn} \end{bmatrix},$$

then (13.3.1) is replaced by

$$Y = T(X) = \frac{D_k^{-1} X}{1'D_k^{-1} X}. \tag{13.3.2}$$

Under this transformation any point $X \in \sigma$ is transformed into $Y \in \sigma$ since $1'Y = 1'T(X) = 1$, $Y \geqq 0$. Thus the current point X_k is transformed into $Y_0 = X_0 = (\frac{1}{n})1$.

Since T is one-to-one and onto, it has a unique inverse given by

$$T^{-1}: Y \rightarrow X, \text{ where } x_j = T_j^{-1}(Y) = \frac{x_{kj} y_j}{\sum_{\ell=1}^{n} x_{k\ell} y_\ell}, j=1,\ldots,n, \tag{13.4}$$

or

$$X = T^{-1}(Y) = \frac{C_k^{-1} Y}{1'C_k^{-1} Y} = \frac{D_k Y}{1'D_k Y}.^{1} \tag{13.4.1}$$

Looking to the properties of the projective transformation T we have:

(a) $T(X_k) = (\frac{1}{n})1$.

(b) T is a one-to-one mapping from σ onto σ with T^{-1} given by (13.4).

(c) For $X \in \sigma$, let $Y = T(X)$. Then $\mathcal{S} = \{X \mid AX=0\} = \{T^{-1}(Y) \mid AD_k Y=0\}$.

(d) If $\bar{\mathcal{S}} = \{Y \mid AD_k Y=0\}$, then $T(\mathcal{S}) = \bar{\mathcal{S}}$. Hence if $X_k \in \mathcal{S}$ then $X_0 = Y_0 = (\frac{1}{n})1 \in \bar{\mathcal{S}}$.

---

[1] From (13.3.2), $X = (D_k Y)(1'D_k^{-1} Y)$. Since $1'X = 1$, the preceding expression for X becomes $1 = (1'D_k Y)(1'D_k^{-1} X)$ or $1'D_k^{-1} X = 1/1'D_k Y$. Hence

$$X = (D_k Y)(1'D_k^{-1} X) = \frac{D_k Y}{1'D_k Y}.$$

(e) Since $T(\sigma) = \sigma$ and $T(\mathcal{S}) = \bar{\mathcal{S}}$, it follows that
$T(\mathcal{K}) = T(\sigma \cap \mathcal{S}) = \sigma \cap \bar{\mathcal{S}} = \bar{\mathcal{K}}$.

For instance, to verify (a), it is easily seen, from (13.3), that

$$T_j(\mathbf{X}_k) = \frac{(x_{kj}/x_{kj})}{\Sigma_{\ell=1}^{n}(x_{k\ell}/x_{k\ell})} = \frac{1}{n}.$$

Hence $T(\mathbf{X}_k) = (\frac{1}{n})\mathbf{1}$. That (c) holds follows from the notion that $\mathbf{AX} = \mathbf{0}$ if and only if

$$\frac{\mathbf{AD}_k\mathbf{Y}}{\mathbf{1}'\mathbf{D}_k\mathbf{Y}} = \mathbf{0} \quad \text{or} \quad \mathbf{AD}_k\mathbf{Y} = \mathbf{0}.$$

Geometrically, Figure 13.1 illustrates the projective transformation of a point $\mathbf{X}_k$ to $\mathbf{Y}_0$.

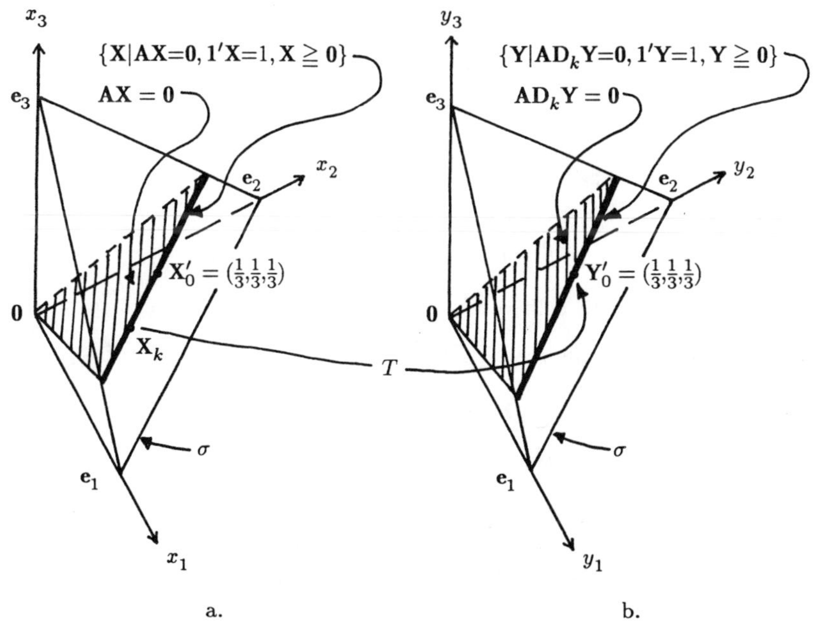

Projective Transformation of the

Feasible Region : $\mathbf{X}_k \in \{\mathbf{X} \mid \mathbf{AX}=\mathbf{0},\ \mathbf{1}'\mathbf{X}=1,\ \mathbf{X} \geqq \mathbf{0}\}$

is transformed under T to

$\mathbf{Y}_0 \in \{\mathbf{Y} \mid \mathbf{AD}_k\mathbf{Y} = \mathbf{0},\ \mathbf{1}'\mathbf{Y}=1,\ \mathbf{Y} \geqq \mathbf{0}\}$.

Figure 13.1

13.4. The Transformed Problem

Under (13.4.1), the original problem (13.1.a) can be transformed into a linear fractional program in $\mathbf{Y}$ over a subset of $\{\mathbf{Y} \mid \mathbf{1}'\mathbf{Y}=1,\ \mathbf{Y} \geqq \mathbf{0}\}$ as

$$\min \left\{ \frac{\mathbf{C}'\mathbf{D}_k\mathbf{Y}}{\mathbf{1}'\mathbf{D}_k\mathbf{Y}} \;\middle|\; \mathbf{A}\mathbf{D}_k\mathbf{Y} = \mathbf{0},\ \mathbf{1}'\mathbf{Y} = 1,\ \mathbf{Y} \geqq \mathbf{0} \right\}, \tag{13.5}$$

i.e., the projective transformation T carries σ into itself, $\mathbf{X}_k$ to $\mathbf{Y}_0 = (\frac{1}{n})\mathbf{1}$, the feasible region $\{\mathbf{X} \mid \mathbf{A}\mathbf{X}=\mathbf{0},\ \mathbf{1}'\mathbf{X}=1,\ \mathbf{X} \geqq \mathbf{0}\}$ into $\{\mathbf{Y} \mid \mathbf{A}\mathbf{D}_k\mathbf{Y}=\mathbf{0},\ \mathbf{1}'\mathbf{Y}=1, \mathbf{Y} \geqq \mathbf{0}\}$, and $\mathbf{C}'\mathbf{X}$ into the fractional objective $\mathbf{C}'\mathbf{D}_k\mathbf{Y}/\mathbf{1}'\mathbf{D}_k\mathbf{Y}$. By virtue of assumption (b) above, we need only minimize the numerator in (13.5) since the denominator is different from zero for all $\mathbf{Y}$ for which $\mathbf{1}'\mathbf{Y}=1$, $\mathbf{Y} \geqq \mathbf{0}$. Hence the preceding problem becomes the ***scaled linear program***

$$\begin{aligned} &\min \bar{\mathbf{C}}'\mathbf{Y} \quad \text{s.t.} \\ &\mathbf{P}\mathbf{Y} = \mathbf{P}_0,\ \mathbf{Y} \geqq \mathbf{0}, \end{aligned} \tag{13.5.1}$$

where

$$\bar{\mathbf{C}} = \mathbf{D}_k\mathbf{C},\quad \mathbf{P} = \begin{bmatrix} \mathbf{A}\mathbf{D}_k \\ \mathbf{1}' \end{bmatrix},\quad \mathbf{P}_0 = \begin{bmatrix} \mathbf{0} \\ 1 \end{bmatrix}.$$

As discussed earlier, instead of solving (13.5.1) directly, we shall solve its restriction to a sphere inscribed within σ. If $B(\mathbf{Y}_0, r)$ is the largest sphere with center $\mathbf{Y}_0 = (\frac{1}{n})\mathbf{1}$ and radius r contained in σ, then the said ***restriction*** of (13.5.1) is

$$\begin{aligned} &\min \bar{\mathbf{C}}'\mathbf{Y} \quad \text{s.t.} \\ &\mathbf{Y} \in \{\mathbf{Y} \mid \mathbf{P}\mathbf{Y}=\mathbf{P}_0\} \cap B(\mathbf{Y}_0, r),\ \mathbf{Y} \geqq \mathbf{0}, \end{aligned} \tag{13.5.2}$$

Here the feasible region is formed as the intersection of the $(n-m-1)$-dimensional affine space $\{\mathbf{Y} \mid \mathbf{P}\mathbf{Y}=\mathbf{P}_0\}$ and the sphere $B(\mathbf{Y}_0, r)$ and amounts to a $(n-m-1)$-dimensional sphere centered at $\mathbf{Y}_0$ with radius r. If $\mathbf{C}_p$ denotes the orthogonal projection of $\bar{\mathbf{C}}$ onto the null-space of $\mathbf{P}$ or onto $\{\mathbf{Y} \mid \mathbf{P}\mathbf{Y}=\mathbf{0}\}$ (so that $\mathbf{C}_p$ lies on the constraint surface of $\mathbf{P}\mathbf{Y}=\mathbf{P}_0$), then we ultimately desire to

$$\begin{aligned} &\min \mathbf{C}_p'\mathbf{Y} \quad \text{s.t.} \\ &\mathbf{Y} \in \{\mathbf{Y} \mid \mathbf{P}\mathbf{Y}=\mathbf{P}_0\} \cap B(\mathbf{Y}_0, r),\ \mathbf{Y} \geqq \mathbf{0}, \end{aligned} \tag{13.5.3}$$

where

$$\mathbf{C}_p = \left[\mathbf{I}_n - (\mathbf{P}^\dagger \mathbf{P})'\right]\bar{\mathbf{C}} = \left[\mathbf{I}_n - \mathbf{P}'(\mathbf{P}\mathbf{P}')^{-1}\mathbf{P}\right]\bar{\mathbf{C}}.^2 \tag{13.6}$$

This can be accomplished by moving from $\mathbf{Y}_0$ in the direction of steepest descent $-\mathbf{C}_p$ to the boundary of the sphere.

Let the direction of $\mathbf{C}_p \neq \mathbf{0}$ be $\widehat{\mathbf{C}}_p = \mathbf{C}_p / \|\mathbf{C}_p\|$. Then using a ***step parameter*** α, $0 < \alpha < 1$, and taking a step of length αr in the projected steepest descent direction $-\mathbf{C}_p$ renders the new point

$$\mathbf{Y}_{k+1} = \mathbf{Y}_0 - \alpha r \widehat{\mathbf{C}}_p \tag{13.7}$$

in transformed space. To generate the corresponding point in true space we need only employ (13.4.1) to obtain $\mathbf{X}_{k+1} = T^{-1}(\mathbf{Y}_{k+1})$ or

$$\mathbf{X}_{k+1} = \frac{\mathbf{D}_k \mathbf{Y}_{k+1}}{\mathbf{1}'\mathbf{D}_k \mathbf{Y}_{k+1}}. \tag{13.8}$$

Given that $\mathbf{Y}_{k+1} > \mathbf{0}$ (it lies in the interior of $B(\mathbf{Y}_0, \alpha r)$), it follows that $\mathbf{X}_{k+1} > \mathbf{0}$.

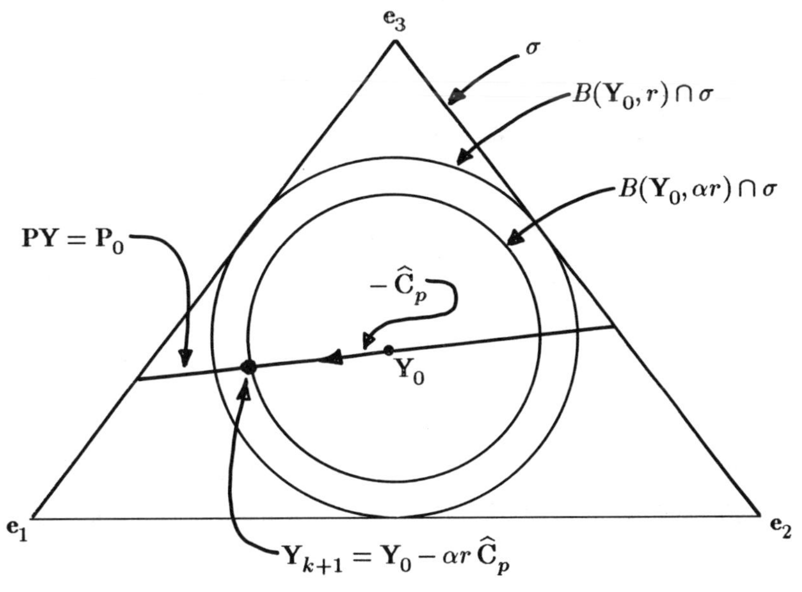

Figure 13.2

[2]See Appendix C for details regarding the derivation of this projection vector.

Looking to the geometry of (step 3 of) Karmarkar's algorithm, Figure 13.2 depicts the intersection of a two-dimensional simplex $\sigma = \{\mathbf{Y} \mid \mathbf{1}'\mathbf{Y}=1, \mathbf{Y} \geqq \mathbf{0}\}$ and a three-dimensional sphere with center $\mathbf{Y}_0$ and radius r chosen so that the said intersection is a two-dimensional sphere or circle inscribed within the simplex. Starting at $\mathbf{Y}_0$, the algorithm moves us a distance αr along the constraint surface $\{\mathbf{Y} \mid \mathbf{P}\mathbf{Y}=\mathbf{P}_0\}$ in the direction of $-\mathbf{C}_p$ to the boundary of the sphere $B(\mathbf{Y}_0, \alpha r)$.

We noted earlier that the approximation to the true cost objective $\mathbf{C}'\mathbf{X}$ in true space was given by the potential function (13.2) as

$$f(\mathbf{X}) = \sum_{j=1}^{n} \ln\left(\frac{\mathbf{C}'\mathbf{X}}{x_j}\right) = n\ln(\mathbf{C}'\mathbf{X}) - \sum_{j=1}^{n} \ln(x_j). \tag{13.2.1}$$

By virtue of (13.4) and (13.4.1), $\mathbf{C}'\mathbf{X}/x_j = \mathbf{C}'\mathbf{D}_k\mathbf{Y}/x_{kj}y_j = \bar{\mathbf{C}}'\mathbf{Y}/x_{kj}y_j$ so that the approximation to transformed cost in transformed space is provided by the ***transformed potential function***

$$\begin{aligned} g(\mathbf{Y}) &= \sum_{j=1}^{n} \ln\left(\frac{\bar{\mathbf{C}}'\mathbf{Y}}{y_j}\right) - \sum_{j=1}^{n} \ln(x_{kj}) \\ &= n\ln(\bar{\mathbf{C}}'\mathbf{Y}) - \sum_{j=1}^{n} \ln(y_j) - \sum_{j=1}^{n} \ln(x_{kj}). \end{aligned} \tag{13.9}$$

Direct substitution easily establishes that $f(\mathbf{X}) = g(T(\mathbf{X}))$, *i.e.*, f measures the same progress towards optimality in true space that g measures in transformed space.

In sum, Karmarkar's algorithm is a projected gradient method designed to minimize the potential function f, where each iteration is performed in transformed space and the transformation moves the current point $\mathbf{X}_k \epsilon \sigma$ to the center of the simplex while keeping the form of f invariant. Under repeated projective transformations, each followed by optimization over a sphere inscribed within the feasible region, the algorithm generates a sequence of interior points converging to the optimal solution.

13.5. Potential Function Improvement and Computational Complexity

We previously denoted the radius of the largest sphere with center $\mathbf{Y}_0 = (\frac{1}{n})\mathbf{1}$ that can be inscribed within the simplex $\{\mathbf{Y} \mid \mathbf{1}'\mathbf{Y}=1, \mathbf{Y} \geqq \mathbf{0}\}$ as r. If we view r as the distance from the center of this simplex $\mathbf{Y}_0' = (\frac{1}{n}, \ldots, \frac{1}{n})$ to one of its facets, where a ***facet*** is actually a simplex of one fewer dimension, then choosing, say, the facet at which the component $y_1 = 0$, it is easily verified that r is the distance from $(\frac{1}{n}, \ldots, \frac{1}{n})$ to $\left(0, \frac{1}{n-1}, \ldots, \frac{1}{n-1}\right)$ or $r = \frac{1}{\sqrt{n(n-1)}}$.

For computational purposes, instead of optimizing over $B(\mathbf{Y}_0, r)$, the algorithm is designed to optimize over the smaller sphere $B(\mathbf{Y}_0, \alpha r)$, $0 < \alpha < 1$. This enables the optimum of the transformed potential function to be approximated very closely by the optimum of the linear objective in (13.5.3).

Karmarkar's analysis of how large a step αr to take is treated in terms of potential functions. Specifically, Karmarkar (1984, pp. 379-381) addresses the problem of minimizing $g(\mathbf{Y})$ for $\mathbf{Y} \epsilon \{\mathbf{Y} \mid \mathbf{P}\mathbf{Y}=\mathbf{P}_0\} \cap B(\mathbf{Y}_0, \alpha r) = \overline{\overline{\mathcal{K}}}$. He establishes that there exists a point that achieves a constant reduction in this transformed potential function, *i.e.*, there exists a point $\overline{\mathbf{Y}} \epsilon \overline{\overline{\mathcal{K}}}$ such that $g(\overline{\mathbf{Y}}) \leq g(\mathbf{Y}_0) - \delta$. And since $min\, g(\mathbf{Y})$ can be approximated by $min\, \overline{\mathbf{C}}'\mathbf{Y}$ (*i.e.*, the potential function is *almost invariant* under $T(\mathbf{X})$ or $g(T(\mathbf{X})) - f(\mathbf{X})$ is independent of $\mathbf{X}$), Karmarkar next proves that if $\overline{\mathbf{Y}}$ minimizes $\overline{\mathbf{C}}'\mathbf{Y}$ over $\overline{\overline{\mathcal{K}}}$, then $g(\overline{\mathbf{Y}}) \leq g(\mathbf{Y}_0) - \delta$. How can we find the $\overline{\mathbf{Y}}$ that minimizes $\overline{\mathbf{C}}'\mathbf{Y}$? As Karmarkar easily demonstrates, the point $\overline{\mathbf{Y}}$ generated by the projective scaling algorithm minimizes $\overline{\mathbf{C}}'\mathbf{Y}$ over $\overline{\overline{\mathcal{K}}}$.

Based upon the preceding discussion we may keep track of any potential function improvement by stating the following

13.1. LEMMA. For $\mathbf{Y}_{k+1}$ given by (13.3.2), either

(a) $\overline{\mathbf{C}}'\mathbf{Y}_{k+1} = 0$ or

(b) $g(\mathbf{Y}_{k+1}) \leq g(\mathbf{Y}_0) - \delta$.

(Since $f(\mathbf{X}) = g(T(\mathbf{X}))$, it follows that if $g(\mathbf{Y}_{k+1}) - g(\mathbf{Y})_0) - \delta$, then $g(T(\mathbf{X}_{k+1})) \leq g(T(\mathbf{X}_k)) - \delta$ or $f(\mathbf{X}_{k+1}) \leq f(\mathbf{X}_k) - \delta$.) If case (a) obtains, then the inverse transformation (13.4.1) is used to return $\mathbf{X}_{k+1}$ and we stop. If case (b) holds, we again apply the inverse of T to get $\mathbf{X}_{k+1}$. The *almost invariance* of f implies that $f(\mathbf{X}_{k+1}) \leq f(\mathbf{X}_k) - \delta$ (provided that the minimum cost is zero). The procedure is repeated, with each iteration decreasing f by δ.

By virtue of the preceding discussion we should expect an improvement δ (which depends upon the α chosen) in the potential function at each step. If the expected improvement does not materialize, *i.e.*, if $f(\mathbf{X}_{k+1}) > f(\mathbf{X}_k) - \delta$, then we stop and conclude that the minimum value of the objective function must be strictly positive, in which case the original problem (13.1.a) has no solution.

The notion of the ***complexity of an algorithm*** addresses the issue of the computational resources expended in attaining a solution to a particular instance of a problem. More specifically, complexity theory categories problems in terms of the forms of the mathematical functions which express the amount of computation as a function of instance size. Moreover, computational complexity assessments are based on the worst-case computational performance of the algorithm. For instance, if an algorithm requires a number of computations that can be expressed as an rth order polynomial of the problem size n, then the algorithm is said to be ***of order*** $\boldsymbol{n^r}$ and written $O(n^r)$.

In general, computations can grow ***polynomially*** or ***exponentially.*** For instance, the standard simplex algorithm is termed *computationally inefficient* because of its worst-case behavior, *i.e.*, although it usually requires a low-order polynomial number of iterations to reach an optional solution (provided one exists), it exhibits an exponential growth in computational effort on certain types of problems as problem size increases (Klee, Minty (1972) construct a class of problems in $m{=}n$ constraints and $2n$ nonnegative variables for which the simplex method requires $2^n - 1$ iterations). However, Karmarkar's algorithm is considered *computationally efficient* since it exhibits a growth in solution effort that is polynomial in the size of the problem. To see this let us consider an *instance* of Karmarkar's problem in standard form. The *size of this instance* is given by m, n, and L, where L, termed the ***input length*** of an instance of the problem, is the number of binary bits required to record all of the data of the problem. Given integer data matrices **A** and **C**, the length of input can be expressed as

$$L = \sum_{i=0}^{m} \sum_{j=1}^{n} \lceil \log(|a_{ij}|+1) \rceil + 1,$$

where $\lceil x \rceil$ denotes the *smallest integer* $\geq x$ and $a_{0j} \equiv c_j$ (Goldfarb, Todd, (1989), p. 142).

In view of the preceding discussion, Karmarkar's algorithm generates a sequence of strictly feasible solutions, starting from $\mathbf{X}_0 = (\frac{1}{n})\mathbf{1}$, that satisfy, for $\bar{\alpha} = \frac{1}{3}$ (a value that seems to work well), $g(\mathbf{Y}_{k+1}) - g(\mathbf{Y}_0) \leq -\frac{1}{5}$.[3] So when $\alpha \approx \frac{1}{3}$, g, and thus f, falls by about $\delta \approx \frac{1}{5}$ after every iteration. Hence over k iterations, $f(\mathbf{X}_{k+1}) - f(\mathbf{X}_k) \leq -\frac{k}{5}$. From (13.2.1),

$$f(\mathbf{X}_k) - f(\mathbf{X}_0) = n \ln\left[\frac{\mathbf{C}'\mathbf{X}_k}{\mathbf{C}'\mathbf{X}_0}\right] - \sum_{j=1}^{n} \ln(n x_{kj}).$$

Using this expression it can be shown (Bazaraa, *et al.*, *op. cit.*) that

$$n \ln\left[\frac{\mathbf{C}'\mathbf{X}_k}{\mathbf{C}'\mathbf{X}_0}\right] \leq -\frac{k}{5} \quad \text{or}$$

$$\mathbf{C}'\mathbf{X}_k \leq e^{-k/5n}\, \mathbf{C}'\mathbf{X}_0, \; k=0,1,2,\dots .$$

For $k = 10nL$, this performance guarantee renders $\mathbf{C}'\mathbf{X}_k < 2^{-L}$, *i.e.*,

$$\mathbf{C}'\mathbf{X}_k \leq e^{-k/5n}\, \mathbf{C}'\mathbf{X}_0 = e^{-2L}\mathbf{C}'(\tfrac{1}{n}\mathbf{1}) < 2^{-2L}(2^L) = 2^{-L}.$$

Hence Karmarkar's algorithm produces a solution with objective value less than 2^{-L} within $O(nL)$ iterations, and with an overall effort (in terms of arithmetic operations) of polynomial complexity $O(n^{3.5}L)$ in the worst case (Karmarkar, (1984)).

---

[3] Using (13.9) it can be shown that

$$g(\mathbf{Y}_{k+1}) - g(\mathbf{Y}_0) \leq -\bar{\alpha} + \frac{\bar{\alpha}^2}{2(1-\bar{\alpha})^2} = -\delta,$$

where $\bar{\alpha} = n\alpha/(n-1)$, $0<\alpha<1$ (Bazaraa, *et al.* (1990), pp. 394-398). A simplified version of this argument sets $\delta = \alpha - \alpha^2/(1-\alpha)$, $0<\alpha<1/2$.

13.6. A Summary of Karmarkar's Algorithm

We may summarize the sequence of steps involved in implementing Karmarkar's projective scaling algorithm as follows.

13.2. KARMARKAR'S ALGORITHM. Quite generally, the algorithm creates a sequence of points $\mathbf{X}_0, \mathbf{X}_1, \ldots, \mathbf{X}_k$ from the steps:

STEP 0. Initialize.

Choose $\mathbf{X}_0 = (\frac{1}{n})\mathbf{1}$ as the center of the simplex σ.

STEP 1. Compute the next point in the sequence.

a. Perform the projective transformation $T(\mathbf{X})$ of the simplex σ that maps the input point $\mathbf{X} \in \sigma$ to the center $\mathbf{Y}_0 \in \sigma$.

b. Optimize (approximately) the transformed objective function over an inscribed sphere to find a point $\mathbf{Y}_{k+1}$.

c. Apply the inverse of T to $\mathbf{Y}_{k+1}$ to obtain the output point $\mathbf{X}_{k+1}$.

STEP 2. Check for infeasibility.

If $\mathbf{C}'\mathbf{X}_{k+1} = 0$, stop; if $f(\mathbf{X}_{k+1}) > f(\mathbf{X}_k) - \delta$, conclude the optimal cost value is positive so that the problem has no feasible solution.

STEP 3. Check for optimality.

Find a vertex $\bar{\mathbf{X}}$ of cost no greater that $\mathbf{C}'\mathbf{X}_{k+1}$. Then take $\mathbf{C}'\bar{\mathbf{X}} = 0$.

More specifically, for problem (13.1.a) with $\mathbf{A}$, $\mathbf{C}$ integer matrices and $\rho(\mathbf{A}) = m$, let $\mathbf{X}_0 = (\frac{1}{n})\mathbf{1}$ be feasible and let the optimal objective value be zero. Execute:

STEP 0. Initialization.

Choose $\mathbf{X}_0 = (\frac{1}{n})\mathbf{1}$, $r = 1 \big/ \sqrt{n(n-1)}$, $\alpha = (n-1)/3n$ and fix the tolerance L.

Set $k = 0$.

STEP 1. Stopping.

If $\mathbf{C}'\mathbf{X}_k < 2^{-L}$, stop. $\mathbf{X}_k$ may be rounded to an optimal solution. Otherwise proceed to

STEP 2. Projection.

a. Let

$$\mathbf{D}_k = diag(x_{k1},\ldots,x_{kn})\,,\ \bar{\mathbf{C}} = \mathbf{D}_k\mathbf{C},\ \mathbf{P} = \begin{bmatrix} \mathbf{AD}_k \\ \mathbf{1}' \end{bmatrix}.$$

b. Let the projection of $\bar{\mathbf{C}}$ onto $\mathbf{P}$'s null-space be

$$\mathbf{C}_p = [\mathbf{I}_n - \mathbf{P}'(\mathbf{PP}')^{-1}\mathbf{P}]\,\bar{\mathbf{C}}.$$

c. Let $\widehat{\mathbf{C}}_p = \mathbf{C}_p \big/ \|\mathbf{C}_p\|$. If $\mathbf{C}_p = \mathbf{0}$, let $\widehat{\mathbf{C}}_p = \mathbf{0}$.

STEP 3. New Point Update in Transformed Space.

Take a step of length αr in the direction of $-\mathbf{C}_p$ by forming

$$\mathbf{Y}_{k+1} = \mathbf{Y}_0 - \alpha r\widehat{\mathbf{C}}_p,\ 0 < \alpha < 1.$$

STEP 4. New Point Update in True Space.

$$\mathbf{X}_{k+1} = T^{-1}(\mathbf{Y}_{k+1}) = \frac{\mathbf{D}_k\mathbf{Y}_{k+1}}{\mathbf{1}'\mathbf{D}_k\mathbf{Y}_{k+1}}.$$

STEP 5. Advance k to $k+1$.

Return to step 1.

In step 1 immediately above it was indicated that if stopping is in order, then $\mathbf{X}_k$ may be rounded to an optimal solution. If n linearly independent structural constraints are binding at $\mathbf{X}_k$, then clearly $\mathbf{X}_k$ constitutes an optimal basic feasible solution to the linear program. Otherwise, there exists a feasible direction $\mathbf{d} \neq \mathbf{0}$, $\|\mathbf{d}\| = 1$, satisfying $\mathbf{1}'\mathbf{X} = 1$ and the binding constraints within $\mathbf{AX} = \mathbf{0}$. If $\mathbf{C}'\mathbf{d} < 0$, we use the algorithm to move along the direction $\mathbf{d}$ (or along $-\mathbf{d}$ if $\mathbf{C}'\mathbf{d} > 0$) until we reach some constraint boundary. Hence at least one additional linearly independent structural constraint is binding at the new solution point, where the new objective value is no greater than $\mathbf{C}'\mathbf{X}_k$. This ***purification process*** is continued until a basic feasible solution $\bar{\mathbf{X}}$ to (13.1.a) emerges with $\mathbf{C}'\bar{\mathbf{X}} < 2^{-L}$. Thus $\bar{\mathbf{X}}$ is optimal with $\mathbf{C}'\bar{\mathbf{X}} = 0$.

13.7. Transforming a General Linear Program to Karmarkar Standard Form

There are a few ways in which a given linear programming problem can be converted to the form depicted by (13.1.a). First, let us

$$\begin{aligned} &min\, \mathbf{C}'\mathbf{X} \quad \text{s.t.} \\ &\mathbf{AX} = \mathbf{b},\ \mathbf{X} \geqq \mathbf{0}, \end{aligned} \tag{13.10}$$

where $\mathbf{A}$ is $(m \times n)$ and $\rho(\mathbf{A}) = m$. Next, we introduce the ***compactifying (bounding) constraint*** $\mathbf{1}'\mathbf{X} \leq nQ$, where Q is chosen large enough so that any feasible solution $\mathbf{X}$ satisfies $\mathbf{1}'\mathbf{X} < nQ$ unless (13.10) is unbounded. Taking x_{n+1} as a nonnegative slack variable, (13.10) becomes

$$\begin{aligned} &min\, \mathbf{C}'\mathbf{X} \quad \text{s.t.} \\ &\mathbf{AX} = \mathbf{b} \\ &\mathbf{1}'\mathbf{X} + x_{n+1} \;=\; nQ \\ &\mathbf{X} \geqq \mathbf{0},\ x_{n+1} \geq 0. \end{aligned} \tag{13.10.1}$$

If we set

$$\begin{aligned} &\bar{\mathbf{X}} = (nQ)^{-1}\mathbf{X}, \\ &\bar{x}_{n+1} = (nQ)^{-1}\ x_{n+1}, \\ &\bar{\mathbf{b}} = (nQ)^{-1}\mathbf{b}, \end{aligned}$$

then we may rewrite (13.10.1) as

$$\begin{aligned} &min\, \mathbf{C}'\bar{\mathbf{X}} \quad \text{s.t.} \\ &\mathbf{A}\bar{\mathbf{X}} = \bar{\mathbf{b}} \\ &\mathbf{1}'\bar{\mathbf{X}} + \bar{x}_{n+1} \;=\; 1 \\ &\bar{\mathbf{X}} \geqq \mathbf{0},\ \bar{x}_{n+1} \geq 0. \end{aligned} \tag{13.10.2}$$

Subtracting $\bar{\mathbf{b}}\mathbf{1}'\bar{\mathbf{X}} + \bar{\mathbf{b}}\bar{x}_{n+1} = \bar{\mathbf{b}}$ from $\mathbf{A}\bar{\mathbf{X}} = \bar{\mathbf{b}}$ leaves us with the problem

$$\begin{aligned} &min\, \mathbf{C}'\bar{\mathbf{X}} \quad \text{s.t.} \\ &(\mathbf{A} - \bar{\mathbf{b}}\mathbf{1}')\bar{\mathbf{X}} - \bar{\mathbf{b}}\bar{x}_{n+1} = \mathbf{0} \\ &\mathbf{1}'\bar{\mathbf{X}} + \bar{x}_{n+1} \;=\; 1 \\ &\bar{\mathbf{X}} \geqq \mathbf{0},\ \bar{x}_{n+1} \geq 0. \end{aligned} \tag{13.10.3}$$

Let us now introduce an artificial variable $\bar{x}_{n+2} \geq 0$ into (13.10.3) in order to render $(\bar{x}_1, \ldots, \bar{x}_n,\ \bar{x}_{n+1},\ \bar{x}_{n+2}) = \ (1/n+2, \ldots, 1/n+2)$ an initial feasible solution. To this end let the objective function coefficient on $\bar{x}_{n+2}$ be M and, within the structural constraints, let the coefficients on each of the resulting homogeneous equations sum to zero and let the coefficient on $\bar{x}_{n+2}$ within the final constraint equation be one. Hence (13.10.3) becomes

$$\begin{aligned} & min\, \mathbf{C}'\bar{\mathbf{X}} + M\bar{x}_{n+2} \quad \text{s.t.} \\ & (\mathbf{A}-\bar{\mathbf{b}}\mathbf{1}')\bar{\mathbf{X}} - \bar{\mathbf{b}}\bar{x}_{n+1} - \left[(\mathbf{A}-\bar{\mathbf{b}}\mathbf{1}')\mathbf{1} - \bar{\mathbf{b}}\right]\bar{x}_{n+2} = \mathbf{0} \\ & \mathbf{1}'\bar{\mathbf{X}} + \bar{x}_{n+1} + \bar{x}_{n+2} = 1 \\ & \bar{\mathbf{X}} \geqq \mathbf{0},\ \bar{x}_{n+1} \geq 0,\ \bar{x}_{n+2} \geq 0. \end{aligned} \tag{13.10.4}$$

Program (13.10) has been transformed to Karmarkar standard form (13.1.a) in $n+2$ variables with $(\frac{1}{n+2})\mathbf{1}$ a feasible solution (assumption (a) above). If the optimal objective value for (13.10.4) is zero (assumption (b)), then Karmarkar's algorithm applies. (If in solving (13.10.4) we find that $\bar{x}_{n+2} > 0$, then (13.10) is infeasible. And if $\bar{x}_{n+1} = 0$, then (13.10) is unbounded.) If a problem is in Karmarkar standard form (13.1.a) and has a known objective function value of v^*, then to construct a new objective function with a minimum of zero as required by Karmarkar's algorithm, simply form $\bar{f} = f - v^* = \mathbf{C}'\mathbf{X} - v^*\mathbf{1}'\mathbf{X} = (\mathbf{C} - v^*\mathbf{1})'\mathbf{X}$. Hence Karmarkar's algorithm can be applied using $\bar{f}$ instead of f.

Next, to obtain a transformed problem which satisfies both assumptions (a), (b) stated earlier, let us apply the preceding procedure to the combined primal and dual problems (Karmarkar (1984); Goldfarb, Mehrotra (1988)). For example, from the primal-dual pair of problems

Primal	Dual
$min\, \mathbf{C}'\mathbf{X}$ s.t.	$max\, \mathbf{b}'\mathbf{U}$ s.t.
$\mathbf{AX} \geqq \mathbf{b},\ \mathbf{X} \geqq \mathbf{0}$	$\mathbf{A}'\mathbf{U} \leqq \mathbf{C},\ \mathbf{U} \geqq \mathbf{0}$

we may form the ***combined primal-dual problem***

$$\begin{aligned} & \mathbf{AX} \geqq \mathbf{b} \\ & \mathbf{A}'\mathbf{U} \leqq \mathbf{C} \\ & \mathbf{C}'\mathbf{X} - \mathbf{b}'\mathbf{U} = 0 \\ & \mathbf{X} \geqq \mathbf{0},\ \mathbf{U} \geqq \mathbf{0}, \end{aligned} \tag{13.11}$$

where $\mathbf{A}$ is $(m \times n)$ and $\rho(\mathbf{A}) = m$. Clearly this combined problem is feasible if and only if the original primal problem has a finite optimal solution.

Upon inserting vectors of surplus and slack variables $\mathbf{X}_s$, $\mathbf{U}_s$ respectively into the structural constraint inequalities of (13.11) and introducing a compactifying constraint on the sum of the variables, (13.11) becomes

$$\begin{aligned}
&\mathbf{A}_1\mathbf{X}_1 = \mathbf{b}\\
&\mathbf{A}_2\mathbf{U}_1 = \mathbf{C}\\
&(\mathbf{C}', \mathbf{0}')\mathbf{X}_1 - (\mathbf{b}', \mathbf{0}')\mathbf{U}_1 = 0\\
&\mathbf{1}'\mathbf{X}_1 + \mathbf{1}'\mathbf{U}_1 + x_{n+1} = R\\
&\mathbf{X}_1, \mathbf{U}_1 \geqq \mathbf{0};\ x_{n+1} \geq 0,
\end{aligned} \tag{13.11.1}$$

where

$$\mathbf{X}_1 = \begin{bmatrix} \mathbf{X} \\ \mathbf{X}_s \end{bmatrix},\ \mathbf{U}_1 = \begin{bmatrix} \mathbf{U} \\ \mathbf{U}_s \end{bmatrix},\ \mathbf{A}_1 = [\mathbf{A}, \mathbf{I}_m],\ \mathbf{A}_2 = [\mathbf{A}', -\mathbf{I}_n],$$

x_{n+1} represents a slack variable, and R depicts an integer upper bound.

Next, setting

$$\begin{aligned}
\bar{\mathbf{X}}_1 &= R^{-1}\mathbf{X}_1,\ \bar{\mathbf{U}}_1 = R^{-1}\mathbf{U}_1;\\
\bar{\mathbf{b}} &= R^{-1}\mathbf{b},\ \bar{\mathbf{C}} = R^{-1}\mathbf{C};\\
\bar{x}_{n+1} &= R^{-1}x_{n+1}
\end{aligned}$$

and subtracting $\bar{\mathbf{b}}(\mathbf{1}'\bar{\mathbf{X}}_1+\mathbf{1}'\bar{\mathbf{U}}_1+\bar{x}_{n+1}) = \bar{\mathbf{b}}$ from $\mathbf{A}_1\bar{\mathbf{X}}_1 = \bar{\mathbf{b}}$ and $\bar{\mathbf{C}}(\mathbf{1}'\bar{\mathbf{X}}_1+\mathbf{1}'\bar{\mathbf{U}}_1+\bar{x}_{n+1}) = \bar{\mathbf{C}}$ from $\mathbf{A}_2\bar{\mathbf{U}}_1 = \bar{\mathbf{C}}$ enables us to transform (13.11.1) to

$$\begin{aligned}
&(\mathbf{A}_1-\bar{\mathbf{b}}\mathbf{1}')\bar{\mathbf{X}}_1 - \bar{\mathbf{b}}\mathbf{1}'\bar{\mathbf{U}}_1 - \bar{\mathbf{b}}\bar{x}_{n+1} = \mathbf{0}\\
&-\bar{\mathbf{C}}\mathbf{1}'\bar{\mathbf{X}}_1 + (\mathbf{A}_2-\bar{\mathbf{C}}\mathbf{1}')\bar{\mathbf{U}}_1 - \bar{\mathbf{C}}\bar{x}_{n+1} = \mathbf{0}\\
&(\bar{\mathbf{C}}', \mathbf{0}')\bar{\mathbf{X}}_1 - (\bar{\mathbf{b}}', \mathbf{0}')\bar{\mathbf{U}}_1 = 0\\
&\mathbf{1}'\bar{\mathbf{X}}_1 + \mathbf{1}'\bar{\mathbf{U}}_1 + \bar{x}_{n+1} = 1\\
&\bar{\mathbf{X}}_1, \bar{\mathbf{U}}_1 \geqq \mathbf{0};\ \bar{x}_{n+1} \geq 0.
\end{aligned} \tag{13.11.2}$$

Finally, let us modify (13.11.2) slightly by introducing $\bar{x}_{n+2} \geq 0$ as an artificial variable whose value must be driven to zero and whose structural constraint coefficients are chosen so that within each homogeneous equation the sum of the coefficients is zero and within the last equality constraint the x_{n+2} coefficient is unity. We thus obtain

$$
\begin{aligned}
&\mathit{min}\, \bar{x}_{n+2} \quad \text{s.t.} \\
&(\mathbf{A}_1 - \bar{\mathbf{b}}\mathbf{1}')\bar{\mathbf{X}}_1 - \bar{\mathbf{b}}\mathbf{1}'\bar{\mathbf{U}}_1 - \bar{\mathbf{b}}\bar{x}_{n+1} - \left[(\mathbf{A}_1 - 2\bar{\mathbf{b}}\mathbf{1}')\mathbf{1} - \bar{\mathbf{b}}\right]\bar{x}_{n+2} = \mathbf{0} \\
&-\bar{\mathbf{C}}\mathbf{1}'\bar{\mathbf{X}}_1 + (\mathbf{A}_2 - \bar{\mathbf{C}}'\mathbf{1})\bar{\mathbf{U}}_1 - \bar{\mathbf{C}}\bar{x}_{n+1} - \left[(\mathbf{A}_2 - 2\bar{\mathbf{C}}\mathbf{1}')\mathbf{1} - \bar{\mathbf{C}}\right]\bar{x}_{n+2} = \mathbf{0} \\
&(\bar{\mathbf{C}}', \mathbf{0}')\bar{\mathbf{X}}_1 - (\bar{\mathbf{b}}', \mathbf{0}')\bar{\mathbf{U}}_1 - (\bar{\mathbf{C}}'\mathbf{1} - \bar{\mathbf{b}}'\mathbf{1})\bar{x}_{n+2} = 0 \qquad (13.11.3) \\
&\mathbf{1}'\bar{\mathbf{X}}_1 + \mathbf{1}'\bar{\mathbf{U}}_1 + \bar{x}_{n+1} + \bar{x}_{n+2} = 1 \\
&\qquad \bar{\mathbf{X}}_1, \bar{\mathbf{U}}_1 \geqq \mathbf{0};\ \bar{x}_{n+1}, \bar{x}_{n+2} \geq 0.
\end{aligned}
$$

Here (13.11.3) is of the requisite form (13.1) and also satisfies assumptions (a), (b) above. Solution by Karmarkar's algorithm then yields an optimal pair of primal-dual solutions(provided they exist).

13.8. Extensions and Modifications of Karmarkar's Algorithm

Discussions pertaining to the *largest* decrease in the potential function f via selecting the best value of the step parameter α at each iteration have been offered by Blair (1986), Padberg (1985), Shanno (1988), and Todd and Burrell (1986). While specific values of α have been suggested as *optimal,* Todd and Burrell (1986) note that it is worthwhile to conduct a line search on the potential in the direction $-\mathbf{C}_p$ from $\mathbf{X}_k$ rather than choosing a fixed α (since along any such line f has a unique minimum).

We noted at the outset of this chapter that the optimal value of the objective function in (13.1.a) is zero (assumption (b)). Attempts at relaxing this restriction have been made by Anstreicher (1986), Gay (1987), Todd and Burrell (1986), and Ye and Kojima (1987). For instance, Todd and Burrell (1986) demonstrate that, in transformed space, the objective value moves toward a lower bound given by duality, rather than toward zero. Thus their technique, which is essentially a primal projective algorithm using dual variables, applies even when the optimal value of the objective is not zero or when it is actually unknown. Moreover, it applies even in degenerate cases and generates both primal and dual interior solutions with lower bound objective values which converge (from below) to the common optimal primal and dual solutions. (Note that the Todd and Burrell extension of Karmarkar's algorithm provides us with an optimal dual solution — a feature absent from Karmarkar's original formulation.)

To see how the Todd-Burrell method works let us first drop the assumption that the optimal objective value for (13.1.a) is zero and rewrite its dual ((13.1.b)) as

$$\begin{gathered} max\ z \quad \text{s.t.} \\ \mathbf{A}'\mathbf{U} + z\mathbf{1} \leqq \mathbf{C}. \end{gathered} \tag{13.12}$$

Let the optimal solution for the primal objective (13.1.a) be z^* (a part of the optimal solution of (13.12)). Since z^* is unknown, we may use lower bounds z_k on z^*. To this end let us replace $\mathbf{C}$ by $\mathbf{C}(z_k) = \mathbf{C} - z_k\mathbf{1}$ at the kth iteration. Todd and Burrell (1986) then demonstrate that Karmarkar's algorithm can be modified to generate sequences $\{\mathbf{X}_k\}$, $\{(\mathbf{U}_k, z_k)\}$ satisfying:

(a) $\mathbf{X}_k \in \{\mathbf{X} \mid \mathbf{A}\mathbf{X} = \mathbf{0},\ (\frac{1}{n})\mathbf{1}'\mathbf{X} = 1,\ \mathbf{X} > \mathbf{0}\}$,

$$\begin{aligned} &\text{(b)}\ \mathbf{U}_k = (\mathbf{A}\mathbf{D}_k^2\mathbf{A}')^{-1}\mathbf{A}\mathbf{D}_k^2\mathbf{C} \in R^m, \\ &\text{(c)}\ z_k = \min_j (\mathbf{C} - \mathbf{A}'\mathbf{U}_k)_j \end{aligned} \tag{13.13}$$

such that

(d) $f(\mathbf{X}_0;\ \mathbf{C}(z_k)) - f(\mathbf{X}_k;\ \mathbf{C}(z_k)) \geq \frac{k}{5n}$ for all k.

Here $\mathbf{U}_k$ in (13.13.b) results as the solution of the system of normal equations $(\mathbf{A}\mathbf{D}_k^2\mathbf{A}')\mathbf{U} = \mathbf{A}\mathbf{D}_k^2\mathbf{C}$ for the weighted least squares problem $\min_{\mathbf{U}} \| \mathbf{D}_k(\mathbf{A}'\mathbf{U} - \mathbf{C}) \|^2$ (Appendix C). (For $k = 0$ choose $\mathbf{X}_0 = (\frac{1}{n})\mathbf{1}$, $\mathbf{U}_0' = (\mathbf{A}\mathbf{A}')^{-1}\mathbf{A}\mathbf{C}$, and the corresponding z_0. Then (13.13.a-d) hold trivially.)

Based upon these considerations we now state the Todd-Burrell

13.3. ALGORITHM.

(1) At iteration k we have $\mathbf{X}_k$, $\mathbf{U}_k$, z_k. Compute, for $\mathbf{A}_k = \mathbf{A}\mathbf{D}_k$,

$$\mathbf{W} = \mathbf{P}_{\mathbf{A}_k}\mathbf{D}_k\mathbf{C} \quad \text{and} \quad \mathbf{V} = \mathbf{P}_{\mathbf{A}_k}\mathbf{X}_k,$$

where

$$\mathbf{P}_{\mathbf{A}_k} = \mathbf{I}_n - \mathbf{A}_k'(\mathbf{A}_k\mathbf{A}_k')^{-1}\mathbf{A}_k.$$

is the projection onto the null-space of $\mathbf{A}_k$.

(2) If

$$\min_j(\mathbf{W}-z_k\mathbf{V})_j \leq 0$$

(we have not improved our previous lower bound z^k on z^*), then set $\mathbf{U}_{k+1} = \mathbf{U}_k$, $z_{k+1} = z_k$. Otherwise find $z_{k+1} > z_k$ with

$$\min_j\left[\mathbf{D}_k(\mathbf{C}(z_{k+1}) - \mathbf{A}'\mathbf{U}_{k+1})\right] = \min_j(\mathbf{W}-z_{k+1}\mathbf{V})_j = 0$$

(we apply a minimum ratio test to obtain z_{k+1}) and set

$$\mathbf{U}_{k+1} = (\mathbf{A}_k')^{\dagger}\mathbf{X}_k\mathbf{C}(z_{k+1}) = (\mathbf{A}_k\mathbf{A}_k')^{-1}\mathbf{A}_k\mathbf{D}_k\mathbf{C}(z_{k+1}).$$

(3) Compute, for $\mathbf{P} = \mathbf{I}_n - \mathbf{1}\mathbf{1}'/n$ (the matrix for a projection onto $\{\mathbf{Y} \mid \mathbf{1}'\mathbf{Y}=0\} \epsilon \mathbf{R}^n$),

$$\hat{\mathbf{d}}_k = -\mathbf{P}(\mathbf{W}-z_{k+1}\mathbf{V}),\ \mathbf{d}_k = \mathbf{D}_k\hat{\mathbf{d}}_k,$$

and

$$\mathbf{Y}_{k+1} = \mathbf{X}_k + \tfrac{1}{3}\mathbf{d}_k \Big/ \left\|\hat{\mathbf{d}}_k\right\|.$$

(4) Set

$$\mathbf{X}_{k+1} = \mathbf{Y}_{k+1} \Big/ \mathbf{1}'\mathbf{Y}_{k+1}.$$

On the basis of this routine we are thus able to conclude that

13.4. THEOREM [Todd, Burrell (1986)]. The preceding algorithm generates a sequence $\{\mathbf{X}_k\}$ of primal feasible and a sequence $\{(\mathbf{U}_k, z_k)\}$ of dual feasible solutions, with $\mathbf{C}'\mathbf{X}_k$ and z_k approaching z^*. Indeed,

$$(\mathbf{C}'\mathbf{X}_k - z^*) \leq exp(-k/5n)(\mathbf{C}'\mathbf{X}_0 - z^*).$$

As was the case with Karmarkar's basic projection algorithm, the Todd-Burrell mod-ification also converges in $O(nL)$ iterations and $O(n^{3.5}L)$ arithmetic operations.

Variants of Karmarkar's algorithm which apply to the standard form of a linear program

$$\begin{aligned} &min\, \mathbf{C}'\mathbf{X} \quad \text{s.t.} \\ &\mathbf{AX} = \mathbf{b},\ \mathbf{X} \geqq \mathbf{0} \end{aligned} \tag{13.14}$$

rather than to Karmarkar's standard form (13.1.a) and which do not impose any *a priori* restriction on the optimal value of the objective function have been introduced by Barnes (1986), Anstreicher (1986), de Ghellinck and Vial (1986), Gay (1987), Ye and Kojima (1987), Todd (1988a), and Gonzaga (1989a). Such methods are referred to as ***standard form projective algorithms.*** Common to all of these techniques is a procedure for maintaining lower bounds on the optimal objective value (typically the one advanced by Todd and Burrell is used although an alternative procedure based on geometry has been proposed by Anstreicher (1986)) and the reduction of a potential function which drives the gap between the objective value and the lower bound to zero. For instance, we may rewrite (13.14) as

$$min\,(\mathbf{C}', 0)\begin{bmatrix}\mathbf{X}\\ \eta\end{bmatrix} \quad \text{s.t.}$$
$$(\mathbf{A},-\mathbf{b})\begin{bmatrix}\mathbf{X}\\ \eta\end{bmatrix} = \mathbf{0},\ (\mathbf{0}', 1)\begin{bmatrix}\mathbf{X}\\ \eta\end{bmatrix} = 1,\ \mathbf{X} \geqq \mathbf{0},\ \eta \geq 0 \tag{13.15}$$

or

$$\begin{aligned} &min\,\widehat{\mathbf{C}}'\widehat{\mathbf{X}} \quad \text{s.t.} \\ &\widehat{\mathbf{A}}\widehat{\mathbf{X}} = \mathbf{0},\ \mathbf{d}'\widehat{\mathbf{X}} = 1,\ \widehat{\mathbf{X}} \geqq \mathbf{0}, \end{aligned} \tag{13.15.1}$$

where

$$\widehat{\mathbf{C}} = \begin{bmatrix}\mathbf{C}\\ 0\end{bmatrix},\ \widehat{\mathbf{X}} = \begin{bmatrix}\mathbf{X}\\ \eta\end{bmatrix},\ \widehat{\mathbf{A}} = (\mathbf{A}, -\mathbf{b}),\ \mathbf{d} = \begin{bmatrix}0\\ 1\end{bmatrix}.$$

Clearly optimal primal and dual solutions to (13.14) may be obtained directly from the same for (13.15.1) and conversely.

As indicated by Todd (1988a), if $\mathbf{X}_*$ is a strictly feasible solution to (13.15.1), then a scalar multiple of $\mathbf{X}_*$ is strictly feasible for

$$\begin{aligned} &min\,(\widehat{\mathbf{C}}-z\mathbf{d})'\widehat{\mathbf{X}} \quad \text{s.t.} \\ &\widehat{\mathbf{A}}\widehat{\mathbf{X}} = \mathbf{0},\ \mathbf{1}'\widehat{\mathbf{X}} = 1,\ \widehat{\mathbf{X}} \geqq \mathbf{0}. \end{aligned} \tag{13.16}$$

Moreover, if z^* is the optimal objective value for (13.15.1), then a scalar multiple of z^* is optimal in (13.16). Based upon these considerations, if at iteration k we have a strictly positive feasible solution $\mathbf{X}_k$ for (13.16) and z_k depicts a lower bound on the optimal objective value for the same, then z_k is also a lower bound on the optimal objective value for (13.15.1) under dual feasibility. If the feasible region for (13.15.1) is bounded, then we may apply the preceding algorithm advanced by Todd and Burrell and given immediately above to (13.16) so as to generate a solution to (13.15.1) and thus to (13.14). The above theorem accompanying their algorithm applies here also.

We noted in the preceding section that Karmarkar's projective algorithm can be applied to a combined primal-dual linear program since this formulation conveniently has a feasible interior solution on the simplex. However, applying the projective algorithm in *primal-only* form requires that an initial feasible solution be at hand.

While the problem of obtaining an initial interior solution has typically been addressed by simply assuming that one exists or by introducing an artificial variable and its attendant M-penalty into the problem objective, Anstreicher (1986, 1989) has approached this problem by including an explicit constraint that restricts an artificial variable to zero at optimality, thus avoiding any M-penalty term in the standard form linear program. That is, Anstreicher deals with the artificial variable $\tau \geq 0$ directly by imposing a constraint (satisfied in the limit) that it be zero, thus leading to the problem

$$min(\widehat{\mathbf{C}}', 0, 0)\begin{bmatrix}\widehat{\mathbf{X}}\\ \eta\\ \tau\end{bmatrix} \quad \text{s.t.}$$

$$(\widehat{\mathbf{A}}, -\widehat{\mathbf{b}}, \widehat{\mathbf{b}} - \widehat{\mathbf{A}}\mathbf{1})\begin{bmatrix}\widehat{\mathbf{X}}\\ \eta\\ \tau\end{bmatrix} = \mathbf{0}$$

$$(\mathbf{0}', 1, 0)\begin{bmatrix}\widehat{\mathbf{X}}\\ \eta\\ \tau\end{bmatrix} = 1 \qquad (13.15.2)$$

$$(\mathbf{0}', 0, 1)\begin{bmatrix}\widehat{\mathbf{X}}\\ \eta\\ \tau\end{bmatrix} = 0$$

$$\begin{bmatrix}\widehat{\mathbf{X}}\\ \eta\\ \tau\end{bmatrix} \geqq \mathbf{0}.$$

The procedure for updating lower bounds on the primal objective value z^* also renders dual feasible solutions and his variant of Karmarkar's projective algorithm generates a sequence of strictly positive points satisfying all constraints save $\tau = 0$. Starting from $\mathbf{1}$, if (13.15.2) has a nonempty bounded set of optimal solutions,[4] the method approaches feasibility and optimality simultaneously. Once a lower bound is found, the next iteration always begins at the initial solution $\mathbf{X}_0 = \mathbf{1}$. Moreover, the algorithm either detects an infeasibility in a finite number of iterations or produces a sequence of lower bounds for z^* which diverges to $-\infty$.

A modification of Anstreicher's standard form projective algorithm has been offered by Todd (1992). Starting from a problem of the form

$$\begin{aligned} &min\, \mathbf{C}'\mathbf{X} \quad \text{s.t.} \\ &\qquad \mathbf{AX} = \mathbf{0} \\ &\qquad \mathbf{d}'\mathbf{X} = 1 \\ &\qquad \xi'\mathbf{X} = 0 \\ &\qquad \mathbf{X} \geqq \mathbf{0} \end{aligned} \tag{13.15.3}$$

and given that a strictly positive solution satisfying all but one constraint (the exception is $\xi'\mathbf{X} = \mathbf{0}$) is known, Todd develops improved search directions by solving a two-constraint linear program (essentially a relaxation of a scaled version of (13.15.3)), the optimum value of which, if finite, serves to provide a lower bound on the objective value z^* and signals the existence of *good* search directions. These modified directions are generated by a norm-minimizing subproblem, where the solution of the relaxed problem provides a feasible solution of the direction-finding subproblem. The solution of this subproblem is a search direction that assures progress towards feasibility, optimality, and, if possible, monotonicity. In Todd's approach, when a lower bound is found, the next iterate is derived from the current one and not from $\mathbf{X}_0 = \mathbf{1}$ and $z_0 = z_k$ as in Anstreicher's algorithm.

[4] If we set $\mathbf{C}' = (\widehat{\mathbf{C}}', 0, 0)$, $\mathbf{A} = (\widehat{\mathbf{A}}, -\widehat{\mathbf{b}}, \widehat{\mathbf{b}} - \widehat{\mathbf{A}}\mathbf{1})$, $\mathbf{d}' = (\mathbf{0}', 1, 0)$ and $\mathbf{X}' = (\widehat{\mathbf{X}}', \eta, \tau)$ and if (13.15.2) is feasible with $\{\mathbf{X} \mid \mathbf{C}'\mathbf{X} \leq 0,\ \mathbf{AX} = \mathbf{0}, \mathbf{d}'\mathbf{X} = 0,\ \mathbf{X} \geqq \mathbf{0}\} = \{\mathbf{0}\}$, then (13.15.2) has a bounded set of optimal solutions.

Additional standard form primal projective algorithms which are also combined phase I–phase II methods (*i.e.*, they do not require an initial feasible solution) are those of de Ghellinck and Vial (1986), Fraley (1990), Shaw and Goldfarb (1991), and Anstreicher (1993). De Ghellinck and Vial develop an interior point method that produces a sequence of not necessarily feasible iterates that, like those in the algorithms of Anstreicher (1986, 1989) and Todd (1992), approach feasibility and optimality simultaneously. However, unlike Anstreicher's and Todd's routines, their method does not explicitly incorporate an artificial variable in the constraint system.

The lower-bound updating procedure of de Ghellinck and Vial (which was modified by Fraley) is based on solving n quadratic equations and their algorithm has its foundation in their 1987 feasibility method for solving a linear system of homogeneous equations. A study by Todd and Wang (1993) which compares search directions and lower-bound updates in the algorithms of Anstreicher (1986, 1989), Todd (1992), and de Ghellinck and Vial (1986) concludes that Todd's direction (obtained from his direction-finding subproblem) and Fraley's lower-bound update have modest computational advantages over those of the others.

The version of the standard form projective algorithm advanced by Shaw and Goldfarb (1991), using a constant, fixed steplength, achieves $O(\sqrt{n}L)$-step complexity rather than the $O(nL)$-step complexity originally demonstrated for the projective algorithm. A modification of their approach which utilizes a line search of the potential function on each iteration to accelerate convergence is offered by Anstreicher (1993). In particular, Anstreicher further weakens the *weakened ball* lower bound update of Shaw and Goldfarb so as to control the size of the lower bound adjustment when the bound is updated and adds a single *simple constraint* to the *ball minimization problem* on which the lower bounding and direction finding logic of the algorithm is based. In this way Anstreicher derives a new, strictly monotone (in the presence of the *simple constraint*) version of the standard form projective algorithm which retains the $O(\sqrt{n}L)$-step complexity property.

An algorithm which is an analog of Karmarkar's algorithm and which works directly in the space of the linear structural constraint inequalities $\mathbf{AX} \leqq \mathbf{b}$ and not in the space of feasible solutions to $\mathbf{AX} = \mathbf{0}$, $\mathbf{1}'\mathbf{X} = 1$, $\mathbf{X} \geqq \mathbf{0}$ has been proposed by Freund (1988). The algorithm performs a simple projective transformation of the feasible region that maps the current iterate $\mathbf{X}$ to the analytic center of the transformed feasible region. Freund's algorithm employs a set of assumptions similar to those used by Karmarkar and also utilizes a potential function to chart the progress of the algorithm.

Finally, given the primal-dual pair of problems (13.14) and (13.17.1) (*infra*), Ye (1990c) presents, in the tradition of Karmarkar (1984), a class of projective transformations under which potential functions (such as the primal-dual potential function of Todd and Ye (1990)

$$\phi(\mathbf{X},\mathbf{S}) = \rho \ln \mathbf{X}'\mathbf{S} - \sum_{j=1}^{n} \ln(x_j s_j),$$

with $n \leq \rho < +\infty$) are invariant and which leads to a new projective algorithm which converges in $O(\sqrt{n}\,L)$ iterations. The algorithm is not central path following and assumes only that: (a) the interiors of the feasible regions of the primal and dual problems are nonempty; and (b) the coefficient matrix $\mathbf{A}$ for the primal structural constraints has full row rank m.

For r a positive integer and $\mathbf{X}_0 \in \mathcal{K} = \{\mathbf{X} \in \mathbf{R}^n \mid \mathbf{AX} = \mathbf{b},\ \mathbf{X} \geqq \mathbf{0}\}$ an initial feasible solution, Ye defines a one-to-one projective transformation from $\mathcal{K}$ in $\mathbf{R}^n$ onto $\widehat{\mathcal{K}}$ in $\mathbf{R}^{n+r}$ as

$$T: \mathbf{X} \rightarrow \mathbf{Y} = \begin{bmatrix} \mathbf{Y}_{1:n} \\ y_{n+1} \\ \vdots \\ y_{n+r} \end{bmatrix},$$

where

$$\mathbf{Y}_{1:n} = \frac{(n+r)\bar{\mathbf{X}}_0^{-1}\mathbf{X}}{r + \mathbf{1}'\bar{\mathbf{X}}_0^{-1}\mathbf{X}}$$

contains the first n components of $\mathbf{Y} \in \mathbf{R}^{n+r}$,

$$y_j = \frac{n+r}{r + \mathbf{1}'\bar{\mathbf{X}}_0{}^{-1}\mathbf{X}},\ j = n+1,\ldots,n+r,$$

$\bar{\mathbf{X}}_0 = diag(\mathbf{X}_0)$, and the inverse transformation is given by

$$\mathbf{X} = T^{-1}(\mathbf{Y}) = \frac{\bar{\mathbf{X}}_0 \mathbf{Y}_{1:n}}{\sum_{j=n+1}^{n+r} (y_j/r)}.$$

Since $\mathbf{1}'\mathbf{Y} = n+r$, $\widehat{\mathcal{K}}$ is contained in the $(n+r)$-simplex $\sigma = \{\mathbf{Y} \in \mathbf{R}^{n+r} \mid \mathbf{1}'\mathbf{Y} = n+r, \mathbf{Y} \geqq \mathbf{0}\}$. In particular, $T(\mathbf{X}_0) = \mathbf{1}$ (the center of σ).

If we write the primal objective function and structural constraint system as $\mathbf{C}'\mathbf{X} - z \geq 0$ (where $z = z_*$ is the minimal objective value for the primal problem) and $\mathbf{AX} - \mathbf{b} = \mathbf{0}$ respectively, then under T above, the primal problem corresponds to the program

$$\begin{aligned} &min\, \widehat{\mathbf{C}}(z)'\mathbf{Y} \quad \text{s.t.} \\ &\mathbf{Y} \in \widehat{\mathcal{K}} = \{\mathbf{Y} \in \mathbf{R}^{n+r} \mid \widehat{\mathbf{A}}\mathbf{Y} = \mathbf{0},\ \mathbf{1}'\mathbf{Y} = n+r,\ \mathbf{Y} \geqq \mathbf{0}\}, \end{aligned}$$

where

$$\begin{aligned} \widehat{\mathbf{C}}(z)' &= (\mathbf{C}'\bar{\mathbf{X}}_0,\ -\tfrac{z}{r}, \ldots, -\tfrac{z}{r}), \\ \widehat{\mathbf{A}} &= (\mathbf{A}\bar{\mathbf{X}}_0,\ -\tfrac{\mathbf{b}}{r}, \ldots, -\tfrac{\mathbf{b}}{r}). \end{aligned}$$

(Note that $\widehat{\mathbf{C}}(z)'\mathbf{1} = \mathbf{C}'\mathbf{X}_0 - z$.)

Setting $z_0 = \mathbf{b}'\mathbf{U}_0$ and $\mathbf{S}_0 = \mathbf{C} - \mathbf{A}'\mathbf{U}_0 > \mathbf{0}$ we may form the ***ball optimization problem***

$$\begin{aligned} &min\, \widehat{\mathbf{C}}(z)'\mathbf{Y} \quad \text{s.t.} \\ &\widehat{\mathbf{A}}\mathbf{Y} = \mathbf{0} \\ &\mathbf{1}'\mathbf{Y} = n+r \\ &\|\mathbf{Y} - \mathbf{1}\|^2 \leq \beta^2,\ 0 < \beta < 1, \end{aligned}$$

for which the minimal solution can be expressed as

$$\mathbf{Y}(z) = \mathbf{1} - \beta \frac{\mathbf{p}(z)}{\|\mathbf{p}(z)\|},$$

where $\mathbf{p}(z) = \widehat{\mathbf{P}}\widehat{\mathbf{C}}(z)$ and $\widehat{\mathbf{P}}$ is the projection matrix onto the null-space of $\widehat{\mathbf{A}}\mathbf{Y} = \mathbf{0}$, $\mathbf{1}'\mathbf{Y} = 0$.

Let us rewrite $\mathbf{p}(z)$ as

$$\mathbf{p}(z) = \widehat{\mathbf{C}}(z) - \widehat{\mathbf{A}}'\widehat{\mathbf{U}}(z) - \frac{\widehat{\mathbf{C}}(z)'\mathbf{1}}{n+r}\,\mathbf{1},$$

where $\widehat{\mathbf{U}}(z) = (\widehat{\mathbf{A}}\widehat{\mathbf{A}}')^{-1}\widehat{\mathbf{A}}\widehat{\mathbf{C}}(z)$. Then in terms of the coefficients and matrices of the original primal and dual problems (13.14) and (13.17.1) respectively, $\mathbf{p}(z)$ can be expressed as

$$\mathbf{p}(z) = \begin{bmatrix} \bar{\mathbf{X}}_0(\mathbf{C}-\mathbf{A}'\widehat{\mathbf{U}}(z)) \\ (\mathbf{b}'\widehat{\mathbf{U}}(z) - z)/r \\ \vdots \\ (\mathbf{b}'\widehat{\mathbf{U}}(z) - z)/r \end{bmatrix} - \frac{\mathbf{C}'\mathbf{X}_0 - z}{n+r}\mathbf{1}\,.$$

On the basis of a potential reduction theorem, Ye (1990c, p. 462) establishes the fact that the primal-dual potential function ϕ can be reduced by a constant by solving the above ball optimization problem on the interior of the primal and dual feasible regions for arbitrary $\mathbf{X}_0$, $\mathbf{U}_0$. In addition, in the formal statement of the algorithm (Ye (1990c, p. 463)), a line search is undertaken to minimize ϕ, *i.e.*, at the kth iteration,

$$\mathbf{Y}(z_k) = \mathbf{1} - \bar{\beta}\mathbf{p}(z_k),\ \bar{\beta} = \underset{\beta \geq 0}{arg\ min}\ \phi(\mathbf{1}-\beta\mathbf{p}(z_k), \mathbf{S}_k).$$

This primal projective algorithm can start from any interior point $\mathbf{X}_0$ and a strictly lower bound $z_0 < z_*$ and its performance is such that, if $\phi(\mathbf{X}_0, \mathbf{S}_0) \leq O(\sqrt{n}\,L)$, then it terminates in $O(\sqrt{n}\,L)$ iterations, with each iteration requiring $O(n^3)$ arithmetic operations.

13.9. Methods Related to Karmarkar's Routine: Affine Scaling Algorithms

The following collection of solution techniques rely upon an affine rather than a projective transformation at the start of each iteration. Moreover, they operate directly on the problem in standard form and do not impose any *a priori* restriction on the optimal value of the objective function.

Let us represent the standard form of a linear program as (13.14); its corresponding dual problem may be written as

$$\begin{gathered} max\ \mathbf{b}'\mathbf{U} \quad \text{s.t.} \\ \mathbf{A}'\mathbf{U} \leq \mathbf{C},\ \mathbf{U} \text{ unrestricted.} \end{gathered} \tag{13.17}$$

Given these primal and dual structures, we first consider what is called the ***primal affine scaling algorithm.*** (Arbel (1993), Ašić, *et al.* (1990), Barnes (1986), Barnes, *et al.* (1988), Vanderbei, *et al.* (1986), and Vanderbei and Lagarias (1990)). This variant of Karmarkar's algorithm: (1) is formulated within the positive orthant rather than within the unit simplex; (2) assumes both primal and dual nondegeneracy;[5] (3) assumes that (13.14) is bounded and feasible; (4) employs a recentered projected gradient search for the optimum; and (5) assumes the existence of an ***interior feasible point*** ($\mathbf{X}_k$), *e.g.*, a point $\mathbf{X}_k > \mathbf{0}$ which satisfies $\mathbf{AX} = \mathbf{b}$.

As we shall soon see, the primal affine scaling algorithm starts with an interior feasible point (say $\mathbf{X}_0$) in the original or true space, centers it using a scaling operation, takes a step from the centered point $\mathbf{Y}_0$ in the direction of steepest descent in the scaled space, and obtains a new point (call it $\mathbf{Y}_1$) which is then transformed back to a point $\mathbf{X}_1$ in the original space via a rescaling operation.

To this end let $\mathbf{X}_k$ be an interior feasible point for (13.14). If $\mathbf{D}_k$ represents an nth order diagonal matrix containing the components of $\mathbf{X}_k$, then these components are scaled under the affine transformation $T: \mathbf{X} \rightarrow \mathbf{Y}$, where

$$\mathbf{Y} = T(\mathbf{X}) = \mathbf{D}_k^{-1}\mathbf{X}.$$

(Note that $T^{-1}: \mathbf{Y} \rightarrow \mathbf{X}$ yields $\mathbf{X} = \mathbf{D}_k\mathbf{Y}$.) Clearly $\mathbf{Y}_k = \mathbf{D}_k^{-1}\mathbf{X}_k = \mathbf{1}$. Under this change of units the scaled vector $\mathbf{Y}_k$ is placed equidistant from the walls of the polytope $\mathbf{AX} = \mathbf{b}$, $\mathbf{X} \geqq \mathbf{0}$. (Note that $\mathbf{D}_k^{-1}$ exists since $\mathbf{X}_k > \mathbf{0}$.) Hence the ***scaled linear program*** appears as

$$\begin{aligned} &min\ \bar{\mathbf{C}}'\mathbf{Y} \qquad \text{s.t.} \\ &\bar{\mathbf{A}}\mathbf{Y} = \mathbf{b},\ \mathbf{Y} \geqq \mathbf{0}, \end{aligned} \tag{13.18}$$

where

$$\bar{\mathbf{C}} = \mathbf{D}_k\mathbf{C},\ \bar{\mathbf{A}} = \mathbf{A}\mathbf{D}_k.$$

Under assumption (2) above, $\bar{\mathbf{A}}$ has full row rank for every feasible $\mathbf{X}$.

In order to obtain a steepest descent direction vector $\bar{\mathbf{d}}_k$ for the scaled problem (13.18), let us first determine the orthogonal projection of $\bar{\mathbf{C}}$ onto the

[5]The seminal work by Dikin (1967) which first explored an affine scaling procedure assumed only primal nondegenracy.

null-space of $\bar{\mathbf{A}}$ or

$$\mathbf{C}_P = \bar{\mathbf{P}}_r\bar{\mathbf{C}} = \left[\mathbf{I}_n - \bar{\mathbf{A}}'(\bar{\mathbf{A}}\bar{\mathbf{A}}')^{-1}\bar{\mathbf{A}}\right]\bar{\mathbf{C}}$$

$$= \mathbf{D}_k\left[\mathbf{C} - \mathbf{A}'(\mathbf{A}\mathbf{D}_k^2\mathbf{A}')^{-1}\mathbf{A}\mathbf{D}_k^2\mathbf{C}\right],^{6} \tag{13.19}$$

where the inverse of $\bar{\mathbf{A}}\bar{\mathbf{A}}'$ exists since $\bar{\mathbf{A}}$ has full row rank. If we let

$$\mathbf{U}_k = (\mathbf{A}\mathbf{D}_k^2\mathbf{A}')^{-1}\mathbf{A}\mathbf{D}_k^2\,\mathbf{C} \tag{13.20}$$

represent the m-vector of dual variables, then (13.19) can be rewritten as

$$\mathbf{C}_P = \mathbf{D}_k(\mathbf{C}-\mathbf{A}'\mathbf{U}_k) = \mathbf{D}_k\mathbf{r}_k, \tag{13.19.1}$$

where $\mathbf{r}_k$ is the vector of reduced costs corresponding to $\mathbf{X}_k$. (Note that if $\mathbf{C}_P \leqq \mathbf{0}$ for some feasible interior point of (13.14), then this problem is unbounded. Also, if for some feasible interior point of the same we have $\mathbf{C}_P = \mathbf{0}$, then every feasible point is optimal.) Then the direction of steepest descent in scaled space is

$$\bar{\mathbf{d}}_k = -\mathbf{C}_P = -\mathbf{D}_k\mathbf{r}_k, \tag{13.21}$$

i.e., the direction opposite to that of the projected gradient $\mathbf{C}_P$, where the latter is the projection of the scaled gradient $\bar{\mathbf{C}}$ onto the null-space of the scaled coefficient matrix $\bar{\mathbf{A}}$. If $\bar{\mathbf{d}}_k \leqq \mathbf{0}$, then $\mathbf{U}_k$ is dual feasible so that $\mathbf{b}'\mathbf{U}_k$ yields a valid lower bound on the primal objective value for the transformed problem.

Given $\bar{\mathbf{d}}_k$ in scaled space, its counterpart in true or unscaled space is the steepest descent direction

$$\mathbf{d}_k = \mathbf{D}_k\bar{\mathbf{d}}_k = -\mathbf{D}_k^2\mathbf{r}_k. \tag{13.22}$$

Taking a step in the $\mathbf{d}_k$ direction yields the next iterate of $\mathbf{X}$, namely

$$\mathbf{X}_{k+1} = \mathbf{X}_k + \alpha_k\mathbf{d}_k, \tag{13.23}$$

where $0 < \alpha_k \epsilon R$ is the steplength. To ensure that $\mathbf{X}_{k+1} \geqq \mathbf{0}$, α_k must be selected so that $x_j^{k+1} = x_j^k + \alpha_k d_j^k \geq 0$, j=1,...,n, or

$$\alpha_k = \min_j\left\{-x_j^k \Big/ d_j^k,\ d_j^k < 0,\ j{=}1,\ldots,n\right\}, \tag{13.24}$$

[6]See Appendix C to this chapter regarding the derivation of this expression.

where d_j^k is the jth component of the n-vector $\mathbf{d}_k$.

While (13.24) guarantees primal feasibility, it also forces at least one component of $\mathbf{X}_{k+1}$ to vanish at the boundary of the constraint polytope. When this happens $\mathbf{X}_{k+1}$ is not an interior point of the said polytope. So to guarantee that $\mathbf{X}_{k+1}$ represents an interior point of the constraint polytope, we need only take some fraction r (typically $r = 0.95$) of the steplength α_k, *i.e.*, (13.23) is modified as

$$\mathbf{X}_{k+1} = \mathbf{X}_k + r\alpha_k \mathbf{d}_k,\ 0 < r < 1. \tag{13.23.1}$$

At this point two comments concerning the steepest descent direction $\mathbf{d}_k$ are in order. First, $\mathbf{d}_k$ is in the null space of $\mathbf{A}$ since $\mathbf{AX}_{k+1} = \mathbf{AX}_k + r\alpha_k\mathbf{Ad}_k = \mathbf{b}$ or $\mathbf{Ad}_k = \mathbf{0}$. Second, $\mathbf{d}_k$ indeed takes us in a descent direction since $\mathbf{C}'\mathbf{d}_k = \mathbf{C}'\mathbf{D}_k\bar{\mathbf{d}}_k$ $= -\bar{\mathbf{C}}'\bar{\mathbf{P}}_r\bar{\mathbf{C}} = -\bar{\mathbf{C}}'\bar{\mathbf{P}}_r^2\bar{\mathbf{C}} = -\|\bar{\mathbf{P}}_r\bar{\mathbf{C}}\|^2 \leq 0.$

What about the determination of the dual variables $\mathbf{U}_k$ given by (13.20)? It can be shown that they are the solution of the normal equations

$$(\mathbf{AD}_k^2\mathbf{A}')\mathbf{U}_k = \mathbf{AD}_k^2\mathbf{C}$$

for the weighted or scaled least squares problem

$$\underset{\mathbf{U}_k}{min} \|\mathbf{D}_k(\mathbf{C}-\mathbf{A}'\mathbf{U}_k)\|^2.$$

For this problem the vector of residuals is the reduced cost vector $\mathbf{r}_k$ given in (13.19.1).

Let us next consider the limiting behavior of the primal affine scaling algorithm. The limit points of the sequence $\mathbf{X}_k$, k=0,1,... generated by (13.23.1) belong to the set $\mathcal{S} = \{\mathbf{X} \mid \mathbf{X} \geqq \mathbf{0},\ \mathbf{AX} = \mathbf{b},\ \mathbf{D}_k\mathbf{r}_k = \mathbf{0}\}$. Under the assumption that $\bar{\mathbf{A}}$ has full row rank and $\mathbf{r}_k$ has at most m zero components, it follows that $\mathcal{S}$ is the set of all vertices of the feasible region. And since $\mathbf{D}_k\mathbf{r}_k = \mathbf{0}$, $\mathcal{S}$ consists of those feasible points for which complementary slackness holds, *i.e.*, $x_j^k r_j^k = 0$, j=1,..., n. Based upon these considerations, it can be shown (Vanderbei, *et al.* (1986, pp. 399-402)) that

$$\lim_{k\to\infty} \mathbf{D}_{\mathbf{X}_k}\mathbf{r}(\mathbf{X}_k) = \mathbf{0}.$$

Moreover, for $\mathcal{T} = \{\mathbf{X} \mid \mathbf{X} \geqq \mathbf{0},\ \mathbf{AX} = \mathbf{b},\ \mathbf{r}_k \geqq \mathbf{0}\}$ the set of points for which dual feasibility holds, it follows that if $\mathbf{X} \in \mathcal{S} \cap \mathcal{T}$, then $\mathbf{X}$ is optimal. Finally, under this characterization of optimality, it can also be shown that the sequence $\{\mathbf{X}_k\}$ is convergent with $\mathbf{X}_\infty = \lim_{k \to \infty} \mathbf{X}_k$ optimal.

We may summarize the sequence of steps involved in executing the primal affine scaling algorithm as follows:

13.5 PRIMAL AFFINE SCALING ALGORITHM.

The algorithm generates a sequence of points $\mathbf{X}_0, \mathbf{X}_1, \ldots, \mathbf{X}_k$ from the steps:

STEP 0. Initialize.

Choose $\mathbf{X}_0 > \mathbf{0}$ as a feasible interior solution and set $\mathbf{X}_k = \mathbf{X}_0$.

STEP 1. Find the vector of dual variables.

The dual vector $\mathbf{U}_k$ is the solution to the system of normal equations

$$(\mathbf{AD}_k^2\mathbf{A}')\mathbf{U}_k = \mathbf{AD}_k^2\mathbf{C},$$

where $\mathbf{D}_k = diag(x_1^k, \ldots, x_n^k)$.

STEP 2. Find the vector of reduced costs.

The reduced cost vector is $\mathbf{r}_k = \mathbf{C} - \mathbf{A}'\mathbf{U}_k$.

If $\mathbf{r}_k \geqq \mathbf{0}$, then $\mathbf{U}_k$ is dual feasible.

STEP 3. Find the direction of steepest descent in true space.

The vector of steepest descent is

$$\mathbf{d}_k = -\mathbf{D}_k^2\mathbf{r}_k.$$

If $\mathbf{d}_k > \mathbf{0}$, stop; the original primal problem is unbounded.

STEP 4. Compute the next point in the sequence.

The new point update in true space is

$$\mathbf{X}_{k+1} = \mathbf{X}_k + r\alpha_k\mathbf{d}_k,\quad 0 < r < 1,$$

where

$$\alpha_k = \min_i \left\{ -x_j^k \Big/ d_j^k,\ d_j^k < 0,\ j=1,\ldots,n \right\}.$$

(Alternatively, set $r\alpha_k = r \Big/ \|\mathbf{d}_k\|$. This choice corresponds to a descent step that moves a proportion r of the way to the boundary of a sphere inscribed within the non-negative orthant and centered at $\mathbf{Y}_k = \mathbf{1}$ in scaled space.)

STEP 5. Check for optimality.

The sufficient optimality criterion informs us that $\mathbf{C}'\mathbf{X}_k = \mathbf{b}'\mathbf{U}_k$ holds only at optimality while the weak duality theorem states that the dual objective value provides us with a lower bound on the primal objective value (since we are minimizing in the primal problem), with $\mathbf{C}'\mathbf{X}_k - \mathbf{b}'\mathbf{U}_k$ serving as the duality gap. So for a small prespecified tolerance level $0 < \in \epsilon R$, the iterations stop when the ***relative duality gap***

$$\frac{|\mathbf{C}'\mathbf{X}_k - \mathbf{b}'\mathbf{U}_k|}{1 + |\mathbf{C}'\mathbf{X}_k|} \leq \in .$$

Otherwise, proceed to

STEP 6. Advance k to $k+1$.

Return to step 1.

We end this presentation of primal affine scaling with a few comments pertaining to the practical implementation of the primal affine scaling algorithm along with a discussion on how it compares with the standard simplex method.

First, any unrestricted primal variable is obviously excluded from the ratio test given in (13.24). If any such variable turns negative on any iteration of the primal affine scaling routine, simply negate its current value to make it positive. This will ensure that $\mathbf{D}_k$ has a strictly positive diagonal. In addition we also negate its associated column in $\mathbf{A}$ and component in $\mathbf{C}$. The algorithm is then continued.

The second point deals with starting the primal affine scaling algorithm. Given an interior point $\mathbf{X}_0 > \mathbf{0}$ which does not satisfy $\mathbf{AX} = \mathbf{b}$, let us form the ***residual vector*** $\mathbf{b} - \mathbf{AX}_0$ and include it as an $n+1^{st}$ column within the structural constraint matrix so as to obtain

$$\bar{\mathbf{A}}\bar{\mathbf{X}} = [\mathbf{A} \vdots \mathbf{b} - \mathbf{AX}_0]\begin{bmatrix}\mathbf{X} \\ x_{n+1}\end{bmatrix} = \mathbf{b},$$

where x_{n+1} is a nonnegative artificial variable corresponding to the last column of $\bar{\mathbf{A}}$. If $\bar{\mathbf{X}}_0' = (\mathbf{X}_0', 1)$, then

$$\bar{\mathbf{A}}\bar{\mathbf{X}}_0 = \left[\mathbf{A} \vdots \mathbf{b} - \mathbf{A}\mathbf{X}_0\right]\begin{bmatrix} \mathbf{X}_0 \\ 1 \end{bmatrix} = \mathbf{b}.$$

Hence $\bar{\mathbf{X}}_0$ serves as a new initial feasible solution vector to $\bar{\mathbf{A}}\bar{\mathbf{X}} = \mathbf{b}$ given any $\mathbf{X}_0 > \mathbf{0}$. Under these changes the modified primal problem is

$$min\,(\mathbf{C}, M)'\bar{\mathbf{X}} \quad \text{s.t.}$$
$$\bar{\mathbf{A}}\bar{\mathbf{X}} = \mathbf{b},\ \bar{\mathbf{X}} \geqq \mathbf{0},$$

where M is a large positive constant (penalty). Since M can be infinitely large, we are assured that x_{n+1} vanishes at the optimum given that the original primal problem is feasible. If x_{n+1} does not vanish at optimality, then the original primal problem is infeasible.

Recently Ishihara and Kojima (1993) presented a method for updating M in an affine scaling algorithm which is essentially a variant of the technique proposed by Kojima, Mizuno, and Yoshise (1993) for implementing primal-dual interior-point algorithms. The method:

(i) detects whether M should be increased;

(ii) allows M to be increased (if warranted) during an iteration of the affine scaling algorithm; and

(iii) provides information on the possible infeasibility of the primal and dual programs as M increases in value.

An interesting set of parallels with, and differences between, the primal affine scaling method and the simplex method has been highlighted by Vanderbei, *et al.* (1986). Specifically:

1. The simplex algorithm starts from an initial feasible vertex $\mathbf{X}' = (\mathbf{X}_B', \mathbf{0})$ while the affine scaling algorithm starts from a strictly feasible interior solution $\mathbf{X} > \mathbf{0}$.

2. Both algorithms calculate the vector of reduced costs as $\mathbf{r} = \mathbf{C} - \mathbf{A}'\mathbf{U} = \mathbf{C} - \mathbf{A}'(\mathbf{A}\mathbf{D}^2\mathbf{A}')^{-1}\mathbf{A}\mathbf{D}^2\mathbf{C}$. Furthermore, for the simplex algorithm,

$$\mathbf{r} = \begin{bmatrix} \mathbf{0} \\ \cdots\cdots\cdots \\ \mathbf{C}_R - (\mathbf{B}^{-1}\mathbf{R})'\mathbf{C}_B \end{bmatrix}.$$

3. The simplex algorithm stopping criterion for detecting an optimal $\mathbf{X}$ is $\mathbf{S} \geqq \mathbf{0}$; for the affine scaling algorithm no *exact* stopping rule exists.
4. Both algorithms use $\mathbf{r}$ to find a direction $\mathbf{d}$ within the null-space of $\mathbf{A}$ that will decrease the objective function. For the simplex algorithm the decision rule is to choose i such that $s_i < 0$ and find

$$\mathbf{d} = \mathbf{e}_i - \begin{bmatrix} \mathbf{B}^{-1}\mathbf{A}\mathbf{e}_i \\ \cdots\cdots\cdots \\ \mathbf{0} \end{bmatrix};$$

for the affine scaling algorithm we choose the steepest descent direction $\mathbf{d} = -\mathbf{D}^2\mathbf{r}$.
5. Both algorithms move from $\mathbf{X}$ to $\hat{\mathbf{X}} = \mathbf{X} + (\frac{r}{\alpha})\,\mathbf{d}$, where:
   (i) for the simplex algorithm $r = 1$, $\alpha = \max_i \{s_i/x_i, x_i > 0\}$, and $\hat{\mathbf{X}}$ is a new feasible vertex; and;
   (ii) for the affine scaling algorithm, $0 < r < 1$, $\alpha = \max_i\{s_i/x_i\}$, and $\hat{\mathbf{X}}$ is a new feasible interior point.

An important mode of implementation of the primal affine scaling technique is to apply it to the dual of (13.14) (see (13.17)) or to

$$\begin{gathered} \max\ \mathbf{b}'\mathbf{U} \quad \text{s.t.} \\ \mathbf{A}'\mathbf{U} + \mathbf{S} = \mathbf{C},\ \mathbf{U} \text{ unrestricted},\ \mathbf{S} \geqq \mathbf{0}, \end{gathered} \tag{13.17.1}$$

where $\mathbf{S}$ is an $(n \times 1)$ vector of dual slack variables (the reduced cost vector) and $\mathbf{A}$ is of full row rank m. The resulting computational routine has become known as the ***dual affine scaling algorithm***. (See, for instance, Arbel (1993), Marsten, *et al.* (1989), and Monma and Morton (1987).)

Let $\bar{\mathbf{U}}$, $\bar{\mathbf{S}}$ represent an initial feasible pair of solution vectors to (13.17.1), *i.e.*, $\bar{\mathbf{S}} \geqq \mathbf{0}$, $\mathbf{A}'\bar{\mathbf{U}} + \bar{\mathbf{S}} = \mathbf{C}$. For $\bar{\mathbf{S}} > \mathbf{0}$ we may apply an affine scaling transformation to $\mathbf{S}_k$ so as to obtain $\mathbf{t}_k = \mathbf{D}_k^{-1}\mathbf{S}_k$, where $\mathbf{D}_k = diag(\bar{s}_1, \ldots, \bar{s}_n)$. Hence (13.17.1) becomes

$$\begin{gathered} \max\ \mathbf{b}'\mathbf{U}_k \quad \text{s.t.} \\ \mathbf{D}_k^{-1}\mathbf{A}'\mathbf{U}_k + \mathbf{t}_k = \mathbf{D}_k^{-1}\mathbf{C},\ \mathbf{U}_k \text{ unrestricted},\ \mathbf{t}_k \geqq \mathbf{0}. \end{gathered} \tag{13.17.2}$$

Clearly $\bar{\mathbf{U}}$, $\bar{\mathbf{t}}$ are feasible for this program given that $\bar{\mathbf{U}}$, $\bar{\mathbf{S}}$ are feasible in (13.17.1).

If we multiply both sides of the constraint equality in (13.17.2) by $\mathbf{A}\mathbf{D}_k^{-1}$, then we may solve for

$$\mathbf{U}_k = \mathbf{R}_k\mathbf{D}_k^{-1}\mathbf{C} - \mathbf{R}_k\mathbf{t}_k, \tag{13.25}$$

where $\mathbf{R}_k = (\mathbf{A}\mathbf{D}_k^{-2}\mathbf{A}')^{-1}\mathbf{A}\mathbf{D}_k^{-1}$. Upon substituting (13.25) into (13.17.2) we may eliminate $\mathbf{U}_k$ so that the dual problem can be rewritten solely in terms of $\mathbf{t}_k$ as

$$\begin{aligned} max\ \mathbf{b}'\mathbf{U}_k &= min(-\mathbf{b}'\mathbf{U}_k) \\ &= min(\mathbf{b}'\mathbf{R}_k\mathbf{t}_k - \mathbf{b}'\mathbf{R}_k\mathbf{D}_k^{-1}\mathbf{C}) \quad \text{s.t.} \\ \bar{\mathbf{R}}_k\mathbf{t}_k &= \bar{\mathbf{R}}_k\mathbf{D}_k^{-1}\mathbf{C},\ \mathbf{t}_k \geqq \mathbf{0}, \end{aligned} \tag{13.17.3}$$

where $\bar{\mathbf{R}}_k = \mathbf{I}_n - \mathbf{D}_k^{-1}\mathbf{A}'\mathbf{R}_k$. Since (13.17.3) is of the form (13.14), we may apply the primal affine scaling algorithm to the former (with the constant term in the objective function suppressed).

From (13.19), the orthogonal projection of $\mathbf{b}'\mathbf{R}_k$ onto the null-space of $\bar{\mathbf{R}}_k$ is

$$\begin{aligned} \mathbf{C}_p &= \left[\mathbf{I}_n - \bar{\mathbf{R}}_k'(\bar{\mathbf{R}}_k\bar{\mathbf{R}}_k')^{-1}\bar{\mathbf{R}}_k\right]\mathbf{R}_k'\mathbf{b} \\ &= \mathbf{R}_k'\mathbf{b} = \mathbf{D}_k^{-1}\mathbf{A}'(\mathbf{A}\mathbf{D}_k^{-2}\mathbf{A}')^{-1}\mathbf{b} \end{aligned} \tag{13.26}$$

since $\bar{\mathbf{R}}_k\mathbf{R}_k' = \mathbf{0}$.

Let the new iterates for the dual problem be

$$\begin{aligned} \mathbf{U}_{k+1} &= \mathbf{U}_k + \alpha_k\mathbf{d}_k^U,\ \mathbf{U}_k \text{ unrestricted} \\ \mathbf{t}_{k+1} &= \mathbf{t}_k + \alpha_k\mathbf{d}_k^t,\ \mathbf{t}_k \geqq \mathbf{0},\ k{=}1,\ldots, \end{aligned}$$

where $\mathbf{d}_k^U \epsilon R^m$, $\mathbf{d}_k^t \epsilon R^n$ represent direction vectors for the $\mathbf{U}$, $\mathbf{t}$ variables respectively and α_k depicts the maximum allowable step size. To maintain dual feasibility, these iterates must satisfy the equality constraints in (13.17.2) or

$$\mathbf{D}_k^{-1}\mathbf{A}'(\mathbf{U}_k + \alpha_k\mathbf{d}_k^U) + \mathbf{t}_k + \alpha_k\mathbf{d}_k^t = \mathbf{D}_k^{-1}\mathbf{C}$$

and thus

$$\mathbf{D}_k^{-1}\mathbf{A}'\mathbf{d}_k^U + \mathbf{d}_k^t = \mathbf{0}. \tag{13.27}$$

Multiplying both sides of (13.27) by $\mathbf{A}\mathbf{D}_k^{-1}$ and solving for $\mathbf{d}_k^U$ yields

$$\mathbf{d}_k^U = -(\mathbf{A}\mathbf{D}_k^{-2}\mathbf{A}')^{-1}\mathbf{A}\mathbf{D}_k^{-1}\mathbf{d}_k^t. \tag{13.28}$$

Although (13.28) is valid for any direction vector $\mathbf{d}_k^t$, let us use the steepest descent direction $\mathbf{d}_k^t = -\mathbf{C}_p$ ((13.26)) so that the preceding expression becomes

$$\mathbf{d}_k^U = (\mathbf{A}\mathbf{D}_k^{-2}\mathbf{A}')^{-1}\mathbf{b}. \tag{13.29}$$

Given (13.29), the direction vector $\mathbf{d}_k^t$ can be obtained from (13.27) as

$$\mathbf{d}_k^t = -\mathbf{D}_k^{-1}\mathbf{A}'\mathbf{d}_k^U. \tag{13.30}$$

Applying the reverse affine transformation to $\mathbf{d}_k^t$ yields a direction vector for the original slack variables

$$\mathbf{d}_k^S = \mathbf{D}_k\mathbf{d}_k^t = -\mathbf{A}'\mathbf{d}_k^U. \tag{13.31}$$

Hence from (13.29), (13.31) we have a feasible direction vector $\mathbf{d}_k' = (\mathbf{d}_k^S, \mathbf{d}_k^U)'$. Our final set of iterates can thus be written as

$$\begin{aligned} \mathbf{U}_{k+1} &= \mathbf{U}_k + \alpha_k\mathbf{d}_k^U, \ \mathbf{U}_k \text{ unrestricted} \\ \mathbf{S}_{k+1} &= \mathbf{S}_k + \alpha_k\mathbf{d}_k^S, \ \mathbf{S}_k \geqq \mathbf{0}, \ k=0,1,\ldots. \end{aligned}$$

To preserve $\mathbf{S}_k \geqq \mathbf{0}$, let us choose α_k in a fashion such that $s_j^{k+1} = s_k^k + \alpha_k d_{kj}^s \geq 0$, $j=1,\ldots,n$, *i.e.*, according to the ratio test

$$\alpha_k = \min_j\left\{ -s_j^k \Big/ d_{kj}^s, d_{kj}^s \leq 0, \ j=1,\ldots,n\right\}, \tag{13.32}$$

where d_{kj}^s is the jth component of the n-vector $\mathbf{d}_k^S$. Moreover, by virtue of (13.32) and an argument similar to that underlying (13.23.1), we may form the modified $\mathbf{S}$ iterate as

$$\mathbf{S}_{k+1} = \mathbf{S}_k + r\alpha_k\mathbf{d}_k^S, \ 0 < r < 1.$$

Finally, one can easily obtain an estimate of the primal vector $\mathbf{X}$ from the current $\mathbf{U}$, $\mathbf{S}$ iterates according to an appropriate modification of (13.22), *i.e.*,

$$\mathbf{X}_{k+1} = -\mathbf{D}_k^{-2}\mathbf{d}_k^S = -\mathbf{D}_k^{-2}(-\mathbf{A}'\mathbf{d}_k^U) = \mathbf{D}_k^{-2}\mathbf{A}'(\mathbf{A}\mathbf{D}_k^{-2}\mathbf{A}')^{-1}\mathbf{b}.$$

Clearly $\mathbf{A}\mathbf{X}_{k+1} = \mathbf{b}$. If, in addition, $\mathbf{X}_{k+1} \geqq \mathbf{0}$, then we have a feasible solution to the primal problem. Moreover $\mathbf{C}'\mathbf{X}_{k+1}$ then provides us with a valid lower bound on the optimal objective value for the dual.

Let us now consolidate our results on the dual affine scaling routine by formally stating the:

13.6. DUAL AFFINE SCALING ALGORITHM.

The algorithm generates a sequence of pairs of points $(\mathbf{U}_0, \mathbf{S}_0)$, $(\mathbf{U}_1, \mathbf{S}_1), \ldots,$ $(\mathbf{U}_k, \mathbf{S}_k)$ from the steps:

STEP 0. Initialize.

Choose $\mathbf{U}_0$, $\mathbf{S}_0 > \mathbf{0}$ as a feasible interior solution and set $\mathbf{U}_k = \mathbf{U}_0$, $\mathbf{S}_k = \mathbf{S}_0$.

STEP 1. Find the direction vectors for the dual variables.

a. The direction vector $\mathbf{d}_k^U$ is the solution to the system of normal equations

$$(\mathbf{A}\mathbf{D}_k^{-2}\mathbf{A}')\mathbf{d}_k^U = \mathbf{b},$$

where $\mathbf{D}_k = diag(s_1^k, \ldots, s_n^k)$.

b. The direction vector $\mathbf{d}_k^S$ is determined from

$$\mathbf{d}_k^S = -\mathbf{A}'\mathbf{d}_k^U.$$

If $\mathbf{d}_k^S > \mathbf{0}$, stop; the dual problem is unbounded and thus the primal problem has no feasible solution.

STEP 2. Compute the next pair of points in the sequence.

The new point updates are

$$\mathbf{U}_{k+1} = \mathbf{U}_k + r\alpha_k \mathbf{d}_k^U,$$
$$\mathbf{S}_{k+1} = \mathbf{S}_k + r\alpha_k \mathbf{d}_k^S,$$

where

$$\alpha_k = \min_j \left\{ -s_j^k \Big/ d_{kj}^s, d_{kj}^s < 0,\ j=1,\ldots,n \right\},\quad 0 < r < 1.$$

STEP 3. Determine the primal solution.

The current primal variable estimates corresponding to the latest dual variable values are found using

$$\mathbf{X}_{k+1} = -\mathbf{D}_k^{-2}\mathbf{d}_k^S.$$

If $\mathbf{d}_k^S \leqq \mathbf{0}$, then $\mathbf{X}_{k+1}$ is feasible for the primal problem and thus $\mathbf{C}'\mathbf{X}_{k+1}$ serves as an upper bound on the optimal dual objective value.

STEP 4. Check for optimality.

For $0 \leq \in \epsilon R$ a small prespecified tolerance level, stop when the relative improvement in the dual objective

$$\frac{|\mathbf{b}'\mathbf{U}_{k+1} - \mathbf{b}'\mathbf{U}_k|}{|\mathbf{b}'\mathbf{U}_k|} < \in .$$

Otherwise proceed to

STEP 5. Advance k to $k+1$.

Return to step 1.

The preceding dual affine scaling algorithm has been extended by Marsten, *et al.* (1989) to the case where the variables are free as well as subject to simple upper bound constraints. In addition, these authors provide a discussion on ways of reducing problem size and on the effects of scaling on the performance of the algorithm.

We assumed at the outset of our discussion of the primal affine scaling algorithm that both primal and dual nondegeneracy must hold. However, new convergence results for affine scaling offered by Tseng and Luo (1992) and Tsuchiya (1990) do not require any nondegeneracy assumption. Specifically, it is demonstrated that affine scaling methods for linear programs are globally convergent without either primal or dual nondegeneracy assumptions if some restrictions are imposed on stepsize, *e.g.*, Tseng and Luo assume a very small stepsize of order $2^{-O(L)}$ while Tsuchiya requires that the stepsize be about one-eighth of the maximum allowable distance.

13.10. Methods Related to Karmarkar's Routine: A Path-Following Algorithm

In this section we examine a ***primal-dual interior-point algorithm*** which, at each iteration, preserves both primal and dual feasibility while following the so-called ***central path*** toward optimality. Along the said path the algorithm moves us in a set of directions which successively reduce the duality gap so that, in the limit, as the duality gap is closed and optimality is approached, complementary slackness holds. The process converges in $O(\sqrt{n}L)$ iterations and $O(n^3L)$ arithmetic operations.

As will be demonstrated below, the path-following algorithm uses a projected Newton method to trace out the central path of minimizers for the logarithmic barrier family of subproblems. Here the central path amounts to the classical barrier path under the restriction of primal-dual complementarity (Arbel (1993), Gill, *et al.* (1986), McShane, *et al.* (1989), Megiddo (1986, 1989b), Kojima, *et al.* (1989b), and Monteiro and Adler (1987)).

Our starting point is the set of primal and dual problems given by (13.14) and (13.17.1) respectively, where, as before, it is assumed that **A** has full row rank m. We may reformulate these problems by replacing the nonnegativity conditions within each by an appropriate barrier term so as to obtain:

Primal Barrier Problem

$$\begin{aligned} &min\, \mathbf{C}'\mathbf{X} - \mu \textstyle\sum_{j=1}^{n} \ln(x_j) \qquad \text{s.t.} \\ &\qquad \mathbf{AX} = \mathbf{b},\ \mu > 0. \end{aligned} \tag{13.33}$$

Dual Barrier Problem

$$\begin{aligned} &max\, \mathbf{b}'\mathbf{U} + \mu \textstyle\sum_{j=1}^{n} \ln(s_j) \qquad \text{s.t.} \\ &\qquad \mathbf{A}'\mathbf{U} + \mathbf{S} = \mathbf{C},\ \mu > 0, \end{aligned} \tag{13.34}$$

where μ(usually small and positive) is the ***barrier parameter***. Here the objectives in (13.33), (13.34) depict ***logarithmic barrier functions*** while the ***barrier term***, say $\sum_{j=1}^{n} \ln(x_j)$, keeps the solution centered within the feasible polytope and away from the boundary of the same, where a singularity occurs when $x_j = 0$, $j=1,\ldots,n$, *i.e.*, as some $x_j \rightarrow 0$, $\ln(x_j) \rightarrow \infty$ so that an infinite penalty is imposed for departing from the central path of interior solutions. At each iteration of the path-following algorithm (13.33), (13.34) represent individual subproblems that must be solved and where an updated μ is specified for each new subproblem.

For a given μ and for feasible interior vectors $\mathbf{X} > \mathbf{0}$, $\mathbf{U}$, and reduced costs $\mathbf{S} > \mathbf{0}$, the Lagrangians for (13.33), (13.34) are respectively

$$\mathcal{L}^P(\mathbf{X}, \mathbf{U}, \mu) = \mathbf{C}'\mathbf{X} - \mu \textstyle\sum_{j=1}^{n} \ln(x_j) + \mathbf{U}'(\mathbf{b} - \mathbf{AX}), \tag{13.35}$$

$$\mathcal{L}^D(\mathbf{U}, \mathbf{S}, \mu) = \mathbf{b}'\mathbf{U} + \mu \textstyle\sum_{j=1}^{n} \ln(s_j) + \mathbf{X}'(\mathbf{C} - \mathbf{A}'\mathbf{U} - \mathbf{S}). \tag{13.36}$$

Looking to the Karush-Kuhn-Tucker first-order conditions for an optimum we have, from (13.35),

$$\begin{aligned} &\text{(a)} \quad \partial \mathcal{L}^P / \partial \mathbf{X} = \mathbf{C} - \mu \bar{\mathbf{X}}^{-1}\mathbf{1} - \mathbf{A}'\mathbf{U} = \mathbf{0}, \\ &\text{(b)} \quad \partial \mathcal{L}^P / \partial \mathbf{U} = \mathbf{b} - \mathbf{A}\mathbf{X} = \mathbf{0}; \end{aligned} \tag{13.37}$$

and from (13.36),

$$\begin{aligned} &\text{(a)} \quad \partial \mathcal{L}^D / \partial \mathbf{X} = \mathbf{C} - \mathbf{A}'\mathbf{U} - \mathbf{S} = \mathbf{0}, \\ &\text{(b)} \quad \partial \mathcal{L}^D / \partial \mathbf{U} = \mathbf{b} - \mathbf{A}\mathbf{X} = \mathbf{0} \\ &\text{(c)} \quad \partial \mathcal{L}^D / \partial \mathbf{S} = \mu \bar{\mathbf{S}}^{-1}\mathbf{1} - \mathbf{X} = \mathbf{0}, \end{aligned} \tag{13.38}$$

where $\bar{\mathbf{X}}$, $\bar{\mathbf{S}}$ are the nth order diagonal matrices

$$\bar{\mathbf{X}} = \begin{bmatrix} x_1 & & \mathbf{0} \\ & \ddots & \\ \mathbf{0} & & x_n \end{bmatrix}, \quad \bar{\mathbf{S}} = \begin{bmatrix} s_1 & & \mathbf{0} \\ & \ddots & \\ \mathbf{0} & & s_n \end{bmatrix}.$$

(Note that $\mathbf{X} = \bar{\mathbf{X}}\mathbf{1}$, $\mathbf{S} = \bar{\mathbf{S}}\mathbf{1}$ and that $\bar{\mathbf{X}}^{-1}$ and $\bar{\mathbf{S}}^{-1}$ exist for interior $\mathbf{X}$ and $\mathbf{S}$ respectively.) Since the objective function in (13.33)((13.34)) is convex (concave), subproblem (13.33)((13.34)) has a unique global minimum (maximum) which is completely determined by (13.37)((13.38)).

Given the same μ in both (13.37) and (13.38), a substitution of $\mathbf{C}$ from (13.38.a) into (13.37.a) yields $\bar{\mathbf{X}}\bar{\mathbf{S}}\mathbf{1} = \mu\mathbf{1}$ or (13.38.c). In this regard, the system of equations characterizing the optimality for the primal-dual pair of problems with the same μ is

$$\begin{aligned} &\text{(a)} \quad \bar{\mathbf{X}}\bar{\mathbf{S}}\mathbf{1} = \mu\mathbf{1} \text{ or } x_j s_j = \mu,\ j=1,\ldots,n, \\ &\text{(b)} \quad \mathbf{A}\mathbf{X} = \mathbf{b}, \\ &\text{(c)} \quad \mathbf{A}'\mathbf{U} + \mathbf{S} = \mathbf{C}. \end{aligned} \tag{13.39}$$

Here (13.39.a) depicts the ***μ-complementarity relation***; (13.39.a, b) assure primal feasibility; (13.39.a, c) assure dual feasibility; and, as $\mu \to 0$, (13.39) itself guarantees optimality for the original primal and dual problems. (In fact, given that $\mathbf{X} > \mathbf{0}$, $\mathbf{U}$, and $\mathbf{S} > \mathbf{0}$ are feasible interior solutions to (13.33), (13.34) above, (13.39) provides a necessary and sufficient condition for each of these subproblems to attain an optimal solution.) Moreover, if either the primal or dual feasible region is bounded and has a nonempty interior, then (13.39) has a unique solution for $\mathbf{X} \geqq \mathbf{0}$, $\mathbf{S} \geqq \mathbf{0}$.

Based upon the preceding discussion we can now offer a more precise definition of the central path or trajectory. For a fixed μ, we can solve (13.39) for $\mathbf{X}$, $\mathbf{U}$, and $\mathbf{S}$ in terms of μ. In fact, for each such μ , we generate an entire family of solutions depending on μ or $\mathbf{X}(\mu)$, $\mathbf{U}(\mu)$, and $\mathbf{S}(\mu)$. Let

$$\begin{aligned} \mathcal{S} &= \{\mathbf{X} \mid \mathbf{AX} = \mathbf{b},\ \mathbf{X} > \mathbf{0}\} \in \mathbf{R}^n, \\ \mathcal{T} &= \{(\mathbf{U},\ \mathbf{S}) \mid \mathbf{A}'\mathbf{U} + \mathbf{S} = \mathbf{C},\ \mathbf{S} > \mathbf{0}\} \in \mathbf{R}^m \times \mathbf{R}^n, \text{ and} \\ \mathcal{W} &= \{\mathbf{W} = (\mathbf{X}, \mathbf{U}, \mathbf{S}) \mid \mathbf{X} \in \mathcal{S},\ (\mathbf{U},\ \mathbf{S}) \in \mathcal{T}\} \in \mathbf{R}^m \times \mathbf{R}^n \times \mathbf{R}^n. \end{aligned}$$

Then the central path $\Gamma(\mu)$ is defined as the set of vectors $\mathbf{X}(\mu)$, $\mathbf{U}(\mu)$, and $\mathbf{S}(\mu)$ satisfying (13.39) or

$$\Gamma(\mu) = \Big\{\mathbf{W}(\mu) = \big(\mathbf{X}(\mu),\ \mathbf{U}(\mu),\ \mathbf{S}(\mu)\big) \in \mathcal{W} \mid x_j s_j = \mu > 0,\ j=1,\ldots,n\Big\}. \tag{13.40}$$

As $\mu \to 0$, Γ leads to a pair of optimal solutions $\mathbf{X}_*$, $(\mathbf{U}_*,\ \mathbf{S}_*)$ to the primal and dual problems respectively which satisfy, in the limit, the complementary slackness conditions $x_j^* s_j^* = 0$, $j=1,\ldots,n$. Thus optimal solutions to the primal and dual problems can be approximated by tracing the trajectory $\Gamma(\mu)$ until the parameter μ becomes sufficiently small.

The following interior-point primal-dual algorithm traces out $\Gamma(\mu)$ on its route to optimality and thus constitutes a *path-following* routine. Furthermore, if $\mathbf{W}(\mu) \in \Gamma(\mu)$, then the gap between the primal objective value at $\mathbf{X}(\mu) \in \mathcal{S}(\mu)$ and the dual objective value at $(\mathbf{U}(\mu),\ \mathbf{S}(\mu)) \in \mathcal{T}(\mu)$ is

$$\begin{aligned} \mathbf{C}'\mathbf{X}(\mu) - \mathbf{b}'\mathbf{U}(\mu) &= \mathbf{C}'\mathbf{X}(\mu) - \mathbf{X}(\mu)'\mathbf{A}'\mathbf{U}(\mu) \\ &= \mathbf{X}(\mu)'\mathbf{S}(\mu) = \mathbf{1}'\overline{\mathbf{X}}(\mu)'\overline{\mathbf{S}}(\mu)\mathbf{1} \\ &= \mathbf{1}'(\mu\mathbf{1}) = n\mu. \end{aligned} \tag{13.41}$$

To initiate the (central) path-following algorithm we need a set of feasible interior solutions $(\mathbf{X}, \mathbf{U}, \mathbf{S}) \in \mathcal{S} \times \mathcal{T}$ (*i.e.*, (13.39.b,c) are satisfied but (13.39.a) is not). Following Kojima, *et al.* (1989b) and McShane, *et al.* (1989), let us derive the (projected) Newton direction of (13.39) at $(\mathbf{X}, \mathbf{U}, \mathbf{S})$.[7] That is, for a given μ, we seek to determine a set of step directions (*corrections* to the current estimates) $\mathbf{d}^X$, $\mathbf{d}^U$, and $\mathbf{d}^S$ which, for small movements along them, preserve feasibility while reducing the *errors* in (13.39.a). The resulting directions will yield a new solution point which satisfies (13.39).

From (13.42), with $\mathbf{V}(\mu) = \mu\mathbf{1} - \bar{\mathbf{X}}\bar{\mathbf{S}}\mathbf{1}$, we obtain

$$\begin{bmatrix} \bar{\mathbf{S}} & \mathbf{0} & \bar{\mathbf{X}} \\ \mathbf{A} & \mathbf{0} & \mathbf{0} \\ \mathbf{0} & \mathbf{A}' & \mathbf{I}_n \end{bmatrix} \begin{bmatrix} \mathbf{d}^X \\ \mathbf{d}^U \\ \mathbf{d}^S \end{bmatrix} = \begin{bmatrix} \mathbf{V}(\mu) \\ \mathbf{0} \\ \mathbf{0} \end{bmatrix} \quad \text{or} \quad \begin{array}{rcl} \bar{\mathbf{S}}\mathbf{d}^X \qquad + \bar{\mathbf{X}}\mathbf{d}^S & = & \mathbf{V}(\mu) \\ \mathbf{A}\mathbf{d}^X \qquad\qquad & = & \mathbf{0} \\ \mathbf{A}'\mathbf{d}^U + \ \mathbf{d}^S & = & \mathbf{0}. \end{array} \tag{13.43}$$

Since the solution to (13.43) depends linearly upon μ, we may express the same as

$$\begin{aligned} \mathbf{d}^X(\mu) &= [\bar{\mathbf{S}}^{-1} - \bar{\mathbf{S}}^{-1}\bar{\mathbf{X}}\mathbf{A}'(\mathbf{A}\bar{\mathbf{S}}^{-1}\bar{\mathbf{X}}\mathbf{A}')\mathbf{A}\bar{\mathbf{S}}^{-1}]\,\mathbf{V}(\mu). \\ \mathbf{d}^U(\mu) &= -(\mathbf{A}\bar{\mathbf{S}}^{-1}\bar{\mathbf{X}}\mathbf{A}')^{-1}\mathbf{A}\bar{\mathbf{S}}^{-1}\,\mathbf{V}(\mu), \\ \mathbf{d}^S(\mu) &= \mathbf{A}'(\mathbf{A}\bar{\mathbf{S}}^{-1}\bar{\mathbf{X}}\mathbf{A}')^{-1}\mathbf{A}\bar{\mathbf{S}}^{-1}\,\mathbf{V}(\mu) \end{aligned} \tag{13.44}$$

or

$$\begin{aligned} \mathbf{d}^U(\mu) &= -(\mathbf{A}\bar{\mathbf{S}}^{-1}\bar{\mathbf{X}}\mathbf{A}')^{-1}\mathbf{A}\bar{\mathbf{S}}^{-1}\,\mathbf{V}(\mu), \\ \mathbf{d}^S(\mu) &= -\mathbf{A}'\mathbf{d}^U(\mu), \\ \mathbf{d}^X(\mu) &= \bar{\mathbf{S}}^{-1}\mathbf{V}(\mu) - \bar{\mathbf{S}}^{-1}\bar{\mathbf{X}}\mathbf{d}^S(\mu). \end{aligned} \tag{13.44.1}$$

---

[7] To employ Newton's method, let us rewrite (13.39) as the vector-valued function

$$\mathbf{F}(\mathbf{X}, \mathbf{U}, \mathbf{S}) = \begin{bmatrix} f^1(\mathbf{X}, \mathbf{U}, \mathbf{S}) \\ f^2(\mathbf{X}, \mathbf{U}, \mathbf{S}) \\ f^3(\mathbf{X}, \mathbf{U}, \mathbf{S}) \end{bmatrix} = \begin{bmatrix} \bar{\mathbf{X}}\bar{\mathbf{S}}\mathbf{1} - \mu\,\mathbf{1} \\ \mathbf{A}\mathbf{X} - \mathbf{b} \\ \mathbf{A}'\mathbf{U} + \mathbf{S} - \mathbf{C} \end{bmatrix} = \mathbf{0}.$$

In addition, let $(\mathbf{X}_k, \mathbf{U}_k, \mathbf{S}_k)$ be a given point. By Taylor's expansion,

$$\mathbf{F}(\mathbf{X}, \mathbf{U}, \mathbf{S}) \approx \mathbf{F}(\mathbf{X}_k, \mathbf{U}_k, \mathbf{S}_k) + \mathbf{J}_F(\mathbf{X}_k, \mathbf{U}_k, \mathbf{S}_k)\,\mathbf{d}_k,$$

where $\mathbf{J}_F$ is the Jacobian matrix $\partial(f^1, f^2, f^3)/\partial(\mathbf{X}, \mathbf{U}, \mathbf{S})$ and

$$\mathbf{d}_k = \begin{bmatrix} \mathbf{X} - \mathbf{X}_k \\ \mathbf{U} - \mathbf{U}_k \\ \mathbf{S} - \mathbf{S}_k \end{bmatrix} = \begin{bmatrix} \mathbf{d}_k^X \\ \mathbf{d}_k^U \\ \mathbf{d}_k^S \end{bmatrix}.$$

Hence the original equations may be approximated by

$$\mathbf{F}(\mathbf{X}_k, \mathbf{U}_k, \mathbf{S}_k) + \mathbf{J}_F(\mathbf{X}_k, \mathbf{U}_k, \mathbf{S}_k)\,\mathbf{d}_k = \mathbf{0}. \tag{13.42}$$

Now, let us consider the diagonal matrix

$$\mathbf{D} = (\bar{\mathbf{X}}\bar{\mathbf{S}}^{-1})^{1/2} = diag\Big((x_1/s_1)^{1/2},\ldots,(x_n/s_n)^{1/2}\Big).$$

To rationalize the role of $\mathbf{D}$, let us note that we may view $\bar{\mathbf{X}}$ as the rescaling matrix for primal space while $\bar{\mathbf{S}}^{-1}$ has the same function in dual space. For the combined primal-dual algorithm, $\mathbf{D}$ is the common rescaling matrix (the geometric mean of $\bar{\mathbf{X}}$ and $\bar{\mathbf{S}}^{-1}$) that is applied simultaneously to the primal and dual spaces (Kojima, *et al.* (1989)). In addition, let $\mathbf{t}(\mu) = (\bar{\mathbf{X}}\bar{\mathbf{S}})^{1/2}\mathbf{1} - \mu(\bar{\mathbf{X}}\bar{\mathbf{S}})^{1/2}$. Finally, let $\mathbf{P} = \mathbf{DA}'(\mathbf{AD}^2\mathbf{A}')^{-1}\mathbf{AD}\,\epsilon\,\mathbf{R}^n$ be the orthogonal projection matrix onto the linear subspace $\mathcal{M} = \{\mathbf{Y}|\mathbf{DA}'\mathbf{U} = \mathbf{Y}\}$ while $\mathbf{I}_n - \mathbf{P}$ depicts the projection matrix operator orthogonal to $\mathcal{M}$. Then the projected Newton step directions can be expressed as

$$\begin{aligned} \mathbf{d}^X(\mu) &= \mathbf{D}(\mathbf{I}_n - \mathbf{P})\,\mathbf{t}(\mu), \\ \mathbf{d}^U(\mu) &= -(\mathbf{AD}^2\mathbf{A}')^{-1}\mathbf{ADt}(\mu), \\ \mathbf{d}^S(\mu) &= \mathbf{D}^{-1}\mathbf{Pt}(\mu). \end{aligned} \tag{13.45}$$

Given this set of direction vectors, the new iterates can be determined, for a given μ, as

$$\begin{aligned} \hat{\mathbf{X}} &= \mathbf{X} + \alpha^P\mathbf{d}^X,\ \hat{\mathbf{X}} \geqq \mathbf{0} \\ \hat{\mathbf{U}} &= \mathbf{U} + \alpha^D\mathbf{d}^U,\ \hat{\mathbf{U}}\text{ unrestricted}, \\ \hat{\mathbf{S}} &= \mathbf{S} + \alpha^D\mathbf{d}^S,\ \hat{\mathbf{S}} \geqq \mathbf{0}, \end{aligned} \tag{13.46}$$

where α^P, α^D must be selected so that $\hat{\mathbf{X}} \geqq \mathbf{0}$ and $\hat{\mathbf{S}} \geqq \mathbf{0}$ or

$$\begin{aligned} &\text{(a)}\ \alpha^P = \min_j\Big\{-x^j/dx_j,\ dx_j < 0, j=1,\ldots,n\Big\}, \\ &\text{(b)}\ \alpha^D = \min_j\{-s_j/ds_j, ds_j < 0, j=1,\ldots,n\}, \end{aligned} \tag{13.47}$$

where dx_j, ds_j are the jth components of the n-vectors $\mathbf{d}^X, \mathbf{d}^S$ respectively. And if we again employ an argument similar to that underlying (13.23.1), then we may form the modified $\mathbf{X}$, $\mathbf{S}$ iterates as

$$\hat{\mathbf{X}} = \mathbf{X} + r\alpha^P\mathbf{d}^X,\ 0 < r < 1,$$

$$\hat{\mathbf{S}} = \mathbf{S} + r\alpha^D\mathbf{d}^S,\ 0 < r < 1.$$

We may summarize the sequence of steps involved in executing the primal-dual path-following algorithm as follows:

13.7. PRIMAL-DUAL PATH-FOLLOWING ALGORITHM.

The algorithm generates a sequence of primal-dual triples $(\mathbf{X}_0, \mathbf{U}_0, \mathbf{S}_0)$, $(\mathbf{X}_1, \mathbf{U}_1, \mathbf{S}_1)$, ..., $(\mathbf{X}_k, \mathbf{U}_k, \mathbf{S}_k)$ from the steps:

STEP 0. Initialize.

Choose $\mathbf{X}_0 > \mathbf{0}$, $\mathbf{U}_0$, $\mathbf{S}_0 > \mathbf{0}$ as a feasible interior solution and set $\mathbf{X}_k = \mathbf{X}_0$, $\mathbf{U}_k = \mathbf{U}_0$, $\mathbf{S}_k = \mathbf{S}_0$.

STEP 1. Find the primal-dual scaling matrix.

Define

$$\bar{\mathbf{X}}_k = diag\left(x_1^k, \ldots, x_n^k\right),$$

$$\bar{\mathbf{S}}^{-1} = diag\left(1/s_1^k, \ldots, 1/s_n^k\right)$$

so that

$$\mathbf{D}_k^2 = \bar{\mathbf{S}}^{-1}\bar{\mathbf{X}} = diag(x_1^k/s_1^k, \ldots, x_n^k/s_n^k).$$

STEP 2. Evaluate the barrier parameter μ and the vector $\mathbf{V}(\mu)$. Let

$$\bar{\mu} = \frac{\mathbf{C}'\mathbf{X}_k - \mathbf{b}'\mathbf{U}_k}{n},$$

$$\mathbf{V}(\bar{\mu}) = \bar{\mu}\mathbf{1} - \bar{\mathbf{X}}\bar{\mathbf{S}}\mathbf{1}.$$

STEP 3. Find the direction vectors for the primal-dual set of variables.

a. The direction vector $\mathbf{d}_k^U$ is the solution to the system of normal equations

$$(\mathbf{A}\mathbf{D}_k^2\mathbf{A}')\,\mathbf{d}_k^U = -\mathbf{A}\bar{\mathbf{S}}^{-1}\mathbf{V}(\bar{\mu})$$

for the weighted least squares problem

$$\min_{\mathbf{d}_k^U} \left\| \mathbf{D}_k(\mathbf{A}'\mathbf{d}_k^U + \bar{\mathbf{X}}^{-1}\mathbf{V}(\bar{\mu})) \right\|^2.$$

b. The direction vector $\mathbf{d}_k^S$ is determined from

$$\mathbf{d}_k^S = -\mathbf{A}'\mathbf{d}_k^U.$$

c. The direction vector $\mathbf{d}_k^X$ is determined as

$$\mathbf{d}_k^X = \left[\bar{\mathbf{S}}^{-1} - \mathbf{D}_k^2\mathbf{A}'(\mathbf{A}\mathbf{D}_k^2\mathbf{A}')^{-1}\mathbf{A}\bar{\mathbf{S}}^{-1}\right]\mathbf{V}(\bar{\mu}).$$

STEP 4. Compute the next primal-dual triple in the sequence.

The new primal-dual variable updates are

$$\begin{aligned} \mathbf{X}_{k+1} &= \mathbf{X}_k + r\alpha_k^P \mathbf{d}_k^X, \\ \mathbf{U}_{k+1} &= \mathbf{U}_k + r\alpha_k^D \mathbf{d}_k^U, \\ \mathbf{S}_{k+1} &= \mathbf{S}_k + r\alpha_k^D \mathbf{d}_k^S, \end{aligned}$$

where

$$\begin{aligned} \alpha_k^P &= \min_j \left\{ -x_j^k / dx_j^k,\ dx_j^k < 0, j{=}1,\ldots,n \right\}, \\ \alpha_k^D &= \min_j \left\{ -s_j^k / ds_j^k,\ ds_j^k < 0, j{=}1,\ldots,n \right\}, \\ & 0 < r < 1. \end{aligned}$$

STEP 5. Check for optimality.

For $0 < \in \epsilon R$ a small prespecified tolerance level, stop when the ***relative duality gap***

$$\frac{\left| \mathbf{C}'\mathbf{X}_k - \mathbf{b}'\mathbf{U}_k \right|}{\left| \mathbf{C}'\mathbf{X}_k \right|} \leq \in .$$

Otherwise proceed to

STEP 6. Advance k to $k+1$.

Return to Step 1.

A study which offers a comprehensive implementation of this primal-dual algorithm has been provided by Lustig, Marsten, and Shanno (1991). Their approach allows for the easy handling of simple bounds on the primal variables and incorporates free variables into the calculations. These authors discuss in detail a variety of computational issues (*e.g.*, the role of barrier parameter, effects of preprocessing, scaling, removing variables which appear to be approaching zero as their optimal value, and reducing infeasibility by altering slack variables whenever possible) and provides an extensive set of numerical results for a variety of test problems.

At step 0 of the above algorithm we must find an $(\mathbf{X}, \mathbf{U}, \mathbf{S}) \epsilon \mathcal{W}$ so that the algorithm can be initialized. To this end let $(\mathbf{X}_0, \mathbf{U}_0, \mathbf{S}_0)$ be an initial point satisfying $\mathbf{X}_0 > \mathbf{0}$ and $\mathbf{S}_0 > \mathbf{0}$. In addition, let $x_{n+1}, x_{n+2} \geq 0$ represent primal artificial variables while $u_{m+1}, s_{n+1}, s_{n+2} \geqq 0$ are to serve as dual artificial variables. Then corresponding to the primal and dual problems (13.14), (13.17.1) respectively let us form the following artificial primal and dual programs (Kojima, *et al.* (1989b); McShane, *et al.* (1989)):

$$\begin{aligned} &min \mathbf{C}'\mathbf{X} + c_{n+1}x_{n+1} \quad \text{s.t.} \\ &\mathbf{AX} + (\mathbf{b}-\mathbf{AX}_0)x_{n+1} = \mathbf{b} \\ &(\mathbf{A}'\mathbf{U}_0+\mathbf{S}_0-\mathbf{C})'\mathbf{X} + x_{n+2} = b_{m+1} \\ &\mathbf{X} \geqq \mathbf{0};\ x_{n+1}, x_{n+2} \geqq 0, \end{aligned} \tag{13.48}$$

where c_{n+1}, b_{m+1} are taken to be sufficiently large positive numbers; and

$$\begin{aligned} &max \mathbf{b}'\mathbf{U} + b_{m+1}u_{m+1} \quad \text{s.t.} \\ &\mathbf{A}'\mathbf{U} + (\mathbf{A}'\mathbf{U}_0+\mathbf{S}_0-\mathbf{C})u_{m+1} + \mathbf{S} = \mathbf{C} \\ &(\mathbf{b}-\mathbf{AX}_0)'\mathbf{U} + s_{n+1} = c_{n+1} \\ &u_{m+1} + s_{n+2} = 0 \\ &\mathbf{U}, \mathbf{S} \geqq \mathbf{0};\ u_{m+1}, s_{n+1}, s_{n+2} \geqq 0. \end{aligned} \tag{13.49}$$

If c_{n+1}, b_{m+1} are chosen so that

$$\begin{aligned} &\text{(a)} \quad b_{m+1} > (\mathbf{A}'\mathbf{U}_0+\mathbf{S}_0-\mathbf{C})'\mathbf{X}_0, \\ &\text{(b)} \quad c_{n+1} > (\mathbf{b}-\mathbf{AX}_0)'\mathbf{U}_0 \end{aligned} \tag{13.50}$$

(here $\mathbf{X} = \mathbf{X}_0$, $\mathbf{U} = \mathbf{U}_0$), then $\mathbf{X}_0, x^0_{n+1}$, and x^0_{n+2} along with $\mathbf{U}_0$, u^0_{m+1}, $\mathbf{S}_0$, s^0_{n+1} and s^0_{n+2} are feasible solutions to (13.48), (13.49) respectively, where

$$\begin{aligned} x^0_{n+1} &= 1 \\ x^0_{n+2} &= b_{m+1} - (\mathbf{A}'\mathbf{U}_0+\mathbf{S}_0-\mathbf{C})'\mathbf{X}_0, \\ u^0_{m+1} &= -1, \\ s^0_{n+1} &= c_{n+1} - (\mathbf{b}-\mathbf{AX}_0)'\mathbf{U}_0, \\ s^0_{n+2} &= 1. \end{aligned}$$

The preceding algorithm can then be applied to the artificial problems (13.48), (13.49) using these initial feasible solutions.

Let $\mathbf{X}_*$ and $(\mathbf{U}_*, \mathbf{S}_*)$ represent optimal solutions to the original primal and dual problems (13.14), (13.17.1) respectively. The following theorem provides us with a sufficient condition on the values chosen for b_{m+1}, c_{n+1} in order for the algorithm applied to (13.48), (13.49) to render approximate solutions to (13.14), (13.17.1). Specifically,

13.8. THEOREM [Kojima, *et al.* (1989b)]. If (13.50.a, b) are satisfied and, in addition,

$$b_{m+1} > (\mathbf{A}'\mathbf{U}_0 + \mathbf{S}_0 - \mathbf{C})'\mathbf{X}_*,$$
$$c_{n+1} > (\mathbf{b} - \mathbf{A}\mathbf{X}_0)'\mathbf{U}_*,$$

then

(i) a feasible solution $(\widehat{\mathbf{X}}, \widehat{x}_{n+1}, \widehat{x}_{n+2})$ to (13.48) is minimal if and only if $\widehat{\mathbf{X}}$ is a minimal solution to (13.14) with $\widehat{x}_{n+1} = 0$;

(ii) a feasible solution $(\widehat{\mathbf{U}}, \widehat{u}_{m+1}, \widehat{\mathbf{S}}, \widehat{s}_{n+1}, \widehat{s}_{n+2})$ to (13.49) is maximal if and only if $(\widehat{\mathbf{U}}, \widehat{\mathbf{S}})$ is a maximal solution to (13.17.1) with $\widehat{u}_{m+1} = 0$.

In practice, McShane, *et al.* (1989) choose

$$c_{n+1} = \widehat{c}n^2 max\,|c_j|,$$
$$b_{m+1} = \widehat{b}n^2 max|b_i|,$$

where $\widehat{c}$, $\widehat{b} = 3$ or 30. Both can be shown to satisfy the conditions for equivalence of the original and artificial primal and dual problems.

We next turn to an additional path-following interior-point algorithm, namely the large step ***predictor-corrector algorithm*** of Mizuno, *et al.* (1993b) (see also Sonnevend, Stoer, and Zhao (1989) and Ye, *et al.* (1993a)). Starting with a primal-dual pair of problems of the form (13.14), (13.17.1) respectively, let $\mathcal{K}_{++} = \{(\mathbf{X}, \mathbf{S}) \mid \mathbf{0} < (\mathbf{X}, \mathbf{S}) \in R^{2n},\ \ \mathbf{A}\mathbf{X} = \mathbf{b},\ \ \mathbf{A}'\mathbf{U} + \mathbf{S} = \mathbf{C}$ for some $\mathbf{U} \in R^m\} \neq \phi$ with the central path specified as $\Gamma(\mu) = \{(\mathbf{X}, \mathbf{S}) \mid (\mathbf{X}, \mathbf{S}) \in \mathcal{K}_{++}, \ \overline{\mathbf{X}}\mathbf{S} = \mu\mathbf{1}\}$ and the duality gap for feasible $(\mathbf{X}, \mathbf{S})$ given by $\mathbf{X}'\mathbf{S} = \mathbf{C}'\mathbf{X} - \mathbf{b}'\mathbf{U}$, where $\overline{\mathbf{X}} = diag(\mathbf{X})$.

The authors pose a path-following algorithm that generates a sequence of iterates in a neighborhood $\delta(\alpha) = \{(\mathbf{X}, \mathbf{S}) \mid \| \overline{\mathbf{X}}\mathbf{S} - \mu\mathbf{1} \| \leq \alpha\mu,\ \mu = \mathbf{X}'\mathbf{S}/n,\ (\mathbf{X}, \mathbf{S}) \in \mathcal{K}_{++}\}$ of the central path. The algorithm takes a single corrector step to the central path after each predictor step in order to decrease μ. In this regard, for a constant $0 < \beta \leq \frac{1}{4}$, the algorithm produces *nearly centered* pairs $(\mathbf{X},\mathbf{S}) \in \delta(2\beta)$ (sets of intermediate iterates) after a predictor step and *very nearly centered* pairs $(\mathbf{X},\mathbf{S}) \in \delta(\beta)$ after a corrector step.

Looking to the motivation behind this predictor-corrector routine, suppose we have a pair $(\mathbf{X},\mathbf{S}) \in \delta(\beta), 0 < \beta \leq \frac{1}{4}$, where $\Gamma(\mu) \subset \delta(\beta) \subset \mathcal{K}_{++}$, $\mu = \mathbf{X}'\mathbf{S}/n$, and $\gamma \in [0,1]$. Then we can generate search directions $\mathbf{d}^X$, $\mathbf{d}^S$, and $\mathbf{d}^U$ via the preceding primal-dual scaling method by solving the linear system

$$\begin{aligned} \bar{\mathbf{S}}\mathbf{d}^X + \bar{\mathbf{X}}\mathbf{d}^S &= \gamma\mu\mathbf{1} - \bar{\mathbf{X}}\mathbf{S} \\ \mathbf{A}\mathbf{d}^X &= \mathbf{0} \\ \mathbf{A}'\mathbf{d}^U + \mathbf{d}^S &= \mathbf{0}, \end{aligned} \tag{13.43.1}$$

with $\gamma = 0$. Given a solution $(\mathbf{d}^X, \mathbf{d}^S)$ to (13.43.1),[8] to undertake the kth ***predictor step*** in the direction $(\mathbf{d}_k^X, \mathbf{d}_k^S)$ from the current iterate $(\mathbf{X},\mathbf{S}) = (\mathbf{X}_k, \mathbf{S}_k)$ to a point near the central path $\Gamma(\mu)$, set

$$\begin{aligned} \mathbf{X}(\theta) &= \mathbf{X}_k + \theta\mathbf{d}_k^X \\ \mathbf{S}(\theta) &= \mathbf{S}_k + \theta\mathbf{d}_k^S \end{aligned}$$

and $\mu(\theta) = \mathbf{X}(\theta)'\mathbf{S}(\theta)/n$ for some steplength $\theta > 0$.

How should θ be chosen? Following Mizuno, *et al.*, we have

13.9. LEMMA [Ye, *et al.* (1993)]. Let $\alpha \in (0,1)$. If there exists a positive θ_k such that

$$\left\| \bar{\mathbf{X}}(\theta)\ \mathbf{S}(\theta) - \mu(\theta)\mathbf{1} \right\| \leq \alpha\mu(\theta) \text{ for all } 0 \leq \theta \leq \theta_k,$$

then $(\mathbf{X}(\theta_k), \mathbf{S}(\theta_k)) \in \delta(\alpha)$.

Since the feasibility of $(\mathbf{X}(\theta_k))$ is guaranteed by this lemma, let us choose the

[8] If $\mathbf{D} = \bar{\mathbf{X}}^{1/2}\bar{\mathbf{S}}^{-1/2}$ and $N(\mathbf{AD})$, $R(\mathbf{DA}')$ depict the null-space of $\mathbf{AD}$ and the range of $\mathbf{DA}'$ respectively, then the directions $\mathbf{d}^X$, $\mathbf{d}^S$ obtained from (13.43.1) with $\gamma=0$ can be expressed as

$$\begin{aligned} \mathbf{d}^X &= -\mathbf{D}\mathbf{P}_{N(\mathbf{AD})}\mathbf{r} \\ \mathbf{d}^S &= -\mathbf{D}^{-1}\mathbf{P}_{R(\mathbf{AD}')}\mathbf{r}, \end{aligned}$$

where $\mathbf{r} = \bar{\mathbf{Z}}^{1/2}\mathbf{1}$ and $\bar{\mathbf{Z}} = diag(\bar{\mathbf{X}}\mathbf{S})$ (Ye, *et al.* (1993a)).

largest steplength $\theta_k \leq 1$ such that $\left\| \bar{\mathbf{X}}(\theta_k)\ \mathbf{S}(\theta_k) - \mu(\theta_k)\mathbf{1} \right\| \leq \alpha\mu(\theta_k)$ is satisfied for $\alpha = 2\beta$, and let $\widehat{\mathbf{X}}_k = \mathbf{X}(\theta_k)$, $\widehat{\mathbf{S}}_k = \mathbf{S}(\theta_k)$.

We now solve (13.43.1) with $(\mathbf{X},\mathbf{S}) = (\widehat{\mathbf{X}}_k, \widehat{\mathbf{S}}_k) \epsilon\, \delta(2\beta)$, $\mu = \widehat{\mathbf{X}}_k' \widehat{\mathbf{S}}_k / n$, and $\gamma = 1$. This enables us to form

$$\begin{aligned} \mathbf{X}_{k+1} &= \widehat{\mathbf{X}}_k + \widehat{\mathbf{d}}^X \\ \mathbf{S}_{k+1} &= \widehat{\mathbf{S}}_k + \widehat{\mathbf{d}}^S \end{aligned}$$

with $(\mathbf{X}_{k+1}, \mathbf{S}_{k+1}) \epsilon\, \delta(\beta)$. This completes the ***corrector step.***

It is evident that the large step predictor-corrector algorithm requires that (13.43.1) be solved twice at each iteration. In addition, Ye, *et al.* (1993a) demonstrate that a lower bound for θ_k can be obtained as follows: if θ_k is the largest θ_k satisfying the conditions of lemma 13.9 with $\alpha = 2\beta$, then

$$\theta_k \geq \frac{2}{\sqrt{1 + 4\|\delta_k\|/\beta + 1}},$$

where $\boldsymbol{\delta}_k = \bar{\mathbf{d}}_k^S \mathbf{d}_k^S / \mu_k$ and $\bar{\mathbf{d}}_k^S = diag(\mathbf{d}_k^S)$. Using this lower bound Ye, *et al.* prove that the large step predictor-corrector algorithm maintains $O(\sqrt{n}\,L)$ iteration complexity and also gives quadratic convergence of the duality gap sequence to zero without assuming nondegeneracy or the convergence of the iteration sequence $\{(\mathbf{X}_k, \mathbf{S}_k)\}$. In this regard, according to Ye, *et al.*: if $\{(\mathbf{X}_k, \mathbf{S}_k)\}$ is the sequence generated by the large step predictor-corrector algorithm with $0 < \beta \leq \frac{1}{4}$ and $\alpha = 2\beta$, then the algorithm has iteration complexity $O(\sqrt{n}\,L)$ and $\mathbf{X}_k' \mathbf{S}_k \rightarrow 0$ *Q*-quadratically, *i.e.*, the duality gap sequence $\{\mathbf{X}_k' \mathbf{S}_k\}$ converges ***Q-quadratically*** to zero if

$$\limsup_{k \rightarrow \infty} \frac{\mathbf{X}_{k+1}' \mathbf{S}_{k+1}}{(\mathbf{X}_k' \mathbf{S}_k)^2} < +\infty.$$

We may now formally state the

13.10. LARGE STEP PREDICTOR-CORRECTOR ALGORITHM

STEP 0. Initialize.

Given $(\mathbf{X}_0, \mathbf{S}_0) \epsilon\, \delta(\frac{1}{4})$ with $\mathbf{X}_0' \mathbf{S}_0 \leq 2^L$ (here we assume integer data), set $k = 0$.

STEP 1. Predictor step.

Set $(\mathbf{X},\mathbf{S})$ to $(\mathbf{X}_k,\mathbf{S}_k)$, $\gamma = 0$, and compute $\mathbf{d}^X$, $\mathbf{d}^S$ from (13.43.1).

Form

$$\begin{aligned}\mathbf{X}(\theta) &= \mathbf{X}_k + \theta \mathbf{d}_k^X \\ \mathbf{S}(\theta) &= \mathbf{S}_k + \theta \mathbf{d}_k^S \\ \mu(\theta) &= \mathbf{X}(\theta)'\mathbf{S}(\theta)/n.\end{aligned}$$

Compute the largest θ_k so that $(\mathbf{X}(\theta),\ \mathbf{S}(\theta)) \in \delta(\frac{1}{2})$ for

$$\theta_k \geq \frac{2}{\sqrt{1+4\,||\delta_k||/\beta+1}},\ 0<\beta\leq\tfrac{1}{4},$$

where $\delta_k = \bar{\mathbf{d}}_k^S \mathbf{d}_k^S/\mu_k$ and $\bar{\mathbf{d}}_k^S = diag(\mathbf{d}_k^S)$.

Set $\widehat{\mathbf{X}}_k = \mathbf{X}(\theta_k)$, $\widehat{\mathbf{S}}_k = \mathbf{S}(\theta_k)$.

STEP 2. Corrector step.

Solve (13.43.1) with $(\mathbf{X},\mathbf{S}) = (\widehat{\mathbf{X}}_k, \widehat{\mathbf{S}}_k) \in \delta(\frac{1}{2})$, $\mu = \widehat{\mathbf{X}}_k'\widehat{\mathbf{S}}_k/n$, and $\gamma = 1$. Form

$$\begin{aligned}\mathbf{X}_{k+1} &= \widehat{\mathbf{X}}_k + \widehat{\mathbf{d}}^X \\ \mathbf{S}_{k+1} &= \widehat{\mathbf{S}}_k + \widehat{\mathbf{d}}^S\end{aligned}$$

with $(\mathbf{X}_{k+1},\mathbf{S}_{k+1}) \in \delta(\frac{1}{4})$.

STEP 3. Check for optimality.

If $\mathbf{X}_k'\,\mathbf{S}_k \leq 2^{-L}$, stop;

otherwise proceed to

STEP 4. Advance k to $k+1$.

Return to step 1.

A ***generic primal-dual interior-point algorithm*** which provides a general framework for many existing primal-dual interior-point methods has been developed by Kojima, Megiddo, and Mizuno (1993b). Their generic routine also generates a sequence $\{(\mathbf{X}_k,\mathbf{U}_k,\mathbf{S}_k)\} \subset \Gamma$, *i.e.*, for each k the point $(\mathbf{X}_k,\mathbf{U}_k,\mathbf{S}_k)$ lies on the central trajectory.

For the kth iterate $(\mathbf{X}_k,\mathbf{U}_k,\mathbf{S}_k)$, let $f^k = \mathbf{X}_k'\mathbf{S}_k/n$. If $(\mathbf{X}_k,\mathbf{U}_k,\mathbf{S}_k)$ lies on the central path Γ, then $x_i^k s_i^k = \mu$, i=1,…,n, for some μ. Hence $f^k = \mu$. Since $nf^k = \mathbf{X}_k'\mathbf{S}_k$ coincides with the duality gap $\mathbf{C}'\mathbf{X}_k - \mathbf{b}'\mathbf{U}_k$, it follows that

$$f^k = \mathbf{X}_k'\mathbf{S}_k/n = (\mathbf{C}'\mathbf{X}_k - \mathbf{b}'\mathbf{U}_k)/n.$$

We next consider the Newton direction $(\mathbf{d}^X, \mathbf{d}^U, \mathbf{d}^S)$ at $(\mathbf{X}_k, \mathbf{U}_k, \mathbf{S}_k) \in \Gamma$ for approximating a point $(\mathbf{X}(\mu), \mathbf{U}(\mu), \mathbf{S}(\mu))$ with $\mu = \beta f^k$ on $\Gamma(\mu)$, where the parameter $\beta \in [0,1]$. In the Newton system (13.43) above we defined $\mathbf{V}(\mu) = \mu\mathbf{1} - \bar{\mathbf{X}}\bar{\mathbf{S}}\mathbf{1}$. If this expression is replaced by $\mathbf{V}_\beta(\mu) = \beta f^k \mathbf{1} - \bar{\mathbf{X}}\bar{\mathbf{S}}\mathbf{1}$ in (13.43), then the direction $(\mathbf{d}^X, \mathbf{d}^U, \mathbf{d}^S)$ emerges as a unique solution of the system

$$\begin{aligned} \bar{\mathbf{S}}_k \mathbf{d}^X + \bar{\mathbf{X}}_k \mathbf{d}^S &= \mathbf{V}_\beta(\mu) \\ \mathbf{A}\mathbf{d}^X &= \mathbf{0} \\ \mathbf{A}'\mathbf{d}^U + \quad \mathbf{d}^S &= \mathbf{0}. \end{aligned} \tag{13.43.2}$$

Finally, step lengths α^P, α^D are chosen to generate a new iterate $(\mathbf{X}_{k+1}, \mathbf{U}_{k+1}, \mathbf{S}_{k+1}) \in \mathcal{W}$ such that

$$\begin{aligned} \mathbf{X}_{k+1} &= \mathbf{X}_k + \alpha^P \mathbf{d}_k^X, \\ \mathbf{U}_{k+1} &= \mathbf{U}_k + \alpha^D \mathbf{d}_k^U, \\ \mathbf{S}_{k+1} &= \mathbf{S}_k + \alpha^D \mathbf{d}_k^S. \end{aligned} \tag{13.46.1}$$

It is evident that this generic primal-dual interior-point algorithm depends upon three parameters: a ***search direction parameter*** β and ***primal and dual steplength parameters*** α^P, α^D which determine steplengths in the primal and dual spaces respectively. Given an initial solution $(\mathbf{X}_0, \mathbf{U}_0, \mathbf{S}_0) \in \mathcal{W}$, if appropriate values are assigned to these parameters at each successive iteration of the generic method, we obtain a particular primal-dual interior-point algorithm.

The search direction parameter β fixes a target point $\left(\mathbf{X}(\beta f^k), \mathbf{U}(\beta f^k), \mathbf{S}(\beta f^k)\right) \in \Gamma$ which is to be approximated by the new iterate $(\mathbf{X}_{k+1}, \mathbf{U}_{k+1}, \mathbf{S}_{k+1}) \in \mathcal{W}$. When $\beta = 1$, the target point $\left(\mathbf{X}(f^k), \mathbf{U}(f^k), \mathbf{S}(f^k)\right) \in \Gamma$ minimizes the distance $\left\| \bar{\mathbf{X}}\bar{\mathbf{S}} - \bar{\mathbf{X}}_k \bar{\mathbf{S}}_k \right\|$ from the current iterate $(\mathbf{X}_k, \mathbf{U}_k, \mathbf{S}_k) \in \mathcal{W}$ to points $(\mathbf{X}, \mathbf{U}, \mathbf{S}) \in \Gamma$. Hence the resulting Newton direction $(\mathbf{d}^X, \mathbf{d}^U, \mathbf{d}^S) = (\mathbf{d}_c^X, \mathbf{d}_c^U, \mathbf{d}_c^S)$ may be regarded as a ***centering direction***. At the other extreme, for $\beta = 0$, we obtain the optimality condition (13.39) with $\mu = \beta f^k = 0$. In this instance the Newton direction $(\mathbf{d}^X, \mathbf{d}^U, \mathbf{d}^S) = (\mathbf{d}_a^X, \mathbf{d}_a^U, \mathbf{d}_a^S)$ from the current point $(\mathbf{X}_k, \mathbf{U}_k, \mathbf{S}_k) \in \mathcal{W}$ aims at optimal solutions $\mathbf{X}$, $(\mathbf{U}, \mathbf{S})$ of the primal and dual problems respectively.

The direction $(\mathbf{d}_a^X, \mathbf{d}_a^U, \mathbf{d}_a^S)$ will be termed an ***affine scaling direction*** since it does not involve any centering direction. In general, each direction $(\mathbf{d}^X, \mathbf{d}^U, \mathbf{d}^S)$ with $\beta \in [0,1]$ can be expressed as a convex combination of the centering direction and the affine scaling direction or

$$\begin{bmatrix} \mathbf{d}^X \\ \mathbf{d}^U \\ \mathbf{d}^S \end{bmatrix} = \beta \begin{bmatrix} \mathbf{d}_c^X \\ \mathbf{d}_c^U \\ \mathbf{d}_c^S \end{bmatrix} + (1-\beta) \begin{bmatrix} \mathbf{d}_a^X \\ \mathbf{d}_a^U \\ \mathbf{d}_a^S \end{bmatrix}, \quad 0 \leq \beta \leq 1.$$

If the same steplength $\alpha^P = \alpha^K = \alpha$ is chosen in the primal and dual spaces, then the new duality gap can be written as

$$\begin{aligned} \mathbf{X}_{k+1}'\mathbf{S}_{k+1} &= \mathbf{C}'\mathbf{X}_{k+1} - \mathbf{b}'\mathbf{U}_{k+1} \\ &= \mathbf{C}'(\mathbf{X}_k + \alpha\mathbf{d}^X) - \mathbf{b}'(\mathbf{U}_k + \alpha\mathbf{d}^U) \\ &= (1 - \alpha(1-\beta))\,(\mathbf{C}'\mathbf{X}_k - \mathbf{b}'\mathbf{U}_k) \end{aligned}$$

(Kojima, Mizuno, and Yoshise (1989b)). Hence it is evident from this expression that:

(i) we must have $\beta \leq 1$ if the duality gap is not to increase; and

(ii) the smaller β and the larger α, the greater the reduction in the duality gap.

In this regard, while any $\beta \in [0,1]$ will satisfy (i) and (ii) above, since it is required that $(\mathbf{X}_{k+1}, \mathbf{U}_{k+1}, \mathbf{S}_{k+1}) \in \mathcal{W}$, we must choose the step lengths α^P, α^D smaller than

$$\begin{aligned} \widehat{\alpha}^P &= max\left\{\alpha \mid \mathbf{X}_k + \alpha\mathbf{d}^X \geqq \mathbf{0}\right\}, \\ \widehat{\alpha}^D &= max\left\{\alpha \mid \mathbf{S}_k + \alpha\mathbf{d}^S \geqq \mathbf{0}\right\}, \end{aligned}$$

respectively. If we select $\alpha^P = \alpha^D = \alpha$, then this common step length is bounded above by

$$\begin{aligned} \widehat{\alpha} &= min\{\widehat{\alpha}^P, \widehat{\alpha}^D\} \\ &= max\left\{\alpha \mid \mathbf{X}_k + \alpha\mathbf{d}^X \geqq \mathbf{0}, \mathbf{S}_k + \alpha\mathbf{d}^S \geqq \mathbf{0}\right\}. \end{aligned}$$

It is known that the bounds $\widehat{\alpha}^P$, $\widehat{\alpha}^D$, and $\widehat{\alpha}$ depend upon the location of the current iterate $(\mathbf{X}_k, \mathbf{U}_k, \mathbf{S}_k,)$ and the value of β. In fact, computational experience indicates that small values of β tend to increase the efficiency of the generic primal-dual method. Moreover, $\widehat{\alpha}$ will tend to be larger when the current point $(\mathbf{X}_k, \mathbf{U}_k, \mathbf{S}_k) \in \mathcal{W}$ is near the central path Γ.

After surveying the assortment of values of the search direction and steplength parameters actually selected by various researchers for the theoretical development and practical implementation of primal-dual interior-point algorithms, Kojima, Megiddo, and Mizuno offer two sets of values for controlling β, α^P, and α^D in the generic primal-dual method. Their first rule (*Rule G*) ensures global convergence while the second (*Rule P*) ensures polynomial-time complexity. Both rules depend only on the steplengths $\widehat{\alpha}^P, \widehat{\alpha}^D$ from the current iterates to the boundaries of the primal and dual feasible regions respectively, and allow taking large steps without performing any line search. (For full details see Kojima, Megiddo, and Mizuno (1993b), pp. 6-8, along with their theorems 3.2 and 4.1, pp. 10-17.)

Let us now examine some important theoretical/computational considerations regarding the preceding primal-dual interior-point methods. One such issue involving the implementation of an interior point algorithm is the preparation of an initial interior point. The usual approach (mentioned above) is to embed the original problem to be solved into an artificial problem having a known initial solution from which a primal-dual interior-point algorithm starts. A drawback of this strategy is that the M-penalty constants must be sufficiently large in order to ensure the equivalence of the original and artificial problems.

To avoid numerical instabilities and computational inefficiencies if the M's are too large, and to circumvent the problem of no solution if they are too small, Kojima, Mizuno, and Yoshise (1993) offer a theorem which makes it possible to design flexible methods for updating M as the iterations of interior-point algorithms progress. Their technique is framed in terms of an artificial self-dual linear program with a single M-penalty that can be applied using practically efficient primal-dual interior-point algorithms which have been developed to handle a pair of primal-dual linear programs in standard form.

A strategy which avoids altogether having to construct an artificial primal-dual set of programs is to admit an infeasibility, *i.e.*, a primal-dual infeasible-interior-point algorithm chooses a new point along the Newton direction

$(\mathbf{d}^X, \mathbf{d}^U, \mathbf{d}^S)$ towards a point on the central trajectory, but does not confine the iterates to lie within the feasible region. While an infeasible-interior-point algorithm exhibits the same basic structure as the primal-dual interior-point algorithm (see (13.46), (13.47)), the generated sequence of iterates $\{(\mathbf{X}_k, \mathbf{U}_k, \mathbf{S}_k)\}$ is not restricted to the interior of the feasible region. Thus an interior $(\mathbf{X}_k, \mathbf{U}_k, \mathbf{S}_k)$ is not required to satisfy the equality constraints of the primal and dual problems ($\mathbf{AX} = \mathbf{b}$, $\mathbf{A}'\mathbf{U} + \mathbf{S} = \mathbf{C}$); only the positivity restrictions $\mathbf{X} > \mathbf{0}$, $\mathbf{S} > \mathbf{0}$ apply. Hence we can approach optimal solutions by moving throughout the interior of the feasible region for the primal-dual pair of problems or along a track outside of the same. In fact, this is a key advantage that an infeasible-interior-point algorithm has over a primal-dual interior-point algorithm when solving the primal-dual pair of problems (13.14), (13.17.1).

To exploit this advantage, Kojima, Megiddo, and Mizuno (1993a), using an infeasible-interior-point algorithm, propose a steplength rule under which the algorithm takes large distinct steplengths in the primal and dual spaces to ensure its global convergence. Their algorithm finds approximate optimal solutions to the primal-dual pair if both problems have interior points, and they detect infeasibility when the sequence of iterates diverges. These authors also claim that computation of the Newton direction at each iteration of the primal-dual interior-point algorithm as applied to the artificial self-dual linear program utilized by Kojima, Mizuno, and Yoshise (1993) is still a bit more expensive than that of the infeasible-interior-point routine.

Mizuno (1994) has proposed a $O(nL)$-iteration modification of the Kojima, Megiddo, and Mizuno (1993a) infeasible-interior-point algorithm based on the $O(\sqrt{n}L)$-iteration interior-point predictor-corrector algorithm of Mizuno, Todd, and Ye (1993b). The primal-dual pair of problems is solved without assuming the existence of feasible or optimal points using a single steplength in primal and dual spaces that is chosen to minimize the duality gap. Mizuno also demonstrates that the Kojima, Megiddo, and Mizuno infeasible-interior-point method terminates in $O(n^2L)$ iterations by imposing certain conditions on the iterates.

In a similar vein, Potra (1994) (see also Mizuno (1992) and Potra (1992)) develop an alternative predictor-corrector method for solving linear programs from infeasible starting points. (If a feasible starting point is used, the algorithm reduces to the Mizuno, *et al.* (1993b) predictor-corrector method considered above.) The new algorithm is well-defined in that at each iteration it generates a positive (α-centered) primal and dual pair of solutions with feasibility and optimality improved at the same rate. It is globally and quadratically convergent when started from any positive primal-dual pair that is approximately centered given that an optimal primal-dual set of solutions exists. For starting points large enough the algorithm has $O(nL)$-iteration complexity and, under certain conditions, displays $O(\sqrt{n}L)$-iteration complexity.

Next, we may readily observe that many interior point methods for solving linear programs solve a simultaneous linear equation system with coefficient matrix $\mathbf{AD^2A'}$, where $\mathbf{D}$ is a diagonal matrix with strictly positive elements and $\mathbf{A}$ is the coefficient matrix for the primal structural constraint system. To avoid the computational difficulties arising when $\mathbf{A}$ is dense, Vanderbei and Carpenter (1993) propose solving instead the related indefinite system

$$\begin{bmatrix} -\mathbf{D}^{-2} & \mathbf{A}' \\ \mathbf{A} & \mathbf{0} \end{bmatrix} \begin{bmatrix} \mathbf{d}^X \\ \mathbf{d}^U \end{bmatrix} = \begin{bmatrix} \tau \\ \rho \end{bmatrix}.$$

In the context of a primal-dual path-following algorithm, Vanderbei and Carpenter set

$$\begin{aligned} \mathbf{D}^2 &= \bar{\mathbf{X}}\bar{\mathbf{S}}^{-1}, \\ \tau &= \mathbf{C} - \mathbf{A}'\mathbf{U} - \mu\bar{\mathbf{X}}^{-1}\mathbf{1}, \\ \rho &= \mathbf{b} - \mathbf{A}\mathbf{X} \end{aligned}$$

and subsequently demonstrate that the strategy based on this system is actually a generalization of the preceding methods involving $\mathbf{AD^2A'}$. At each iteration only $\mathbf{D}^{-2}$, τ, and ρ change. Moreover, $\mathbf{D}^{-2}$ need not be diagonal; it must simply be positive definite and symmetric.

Finally, Choi and Goldfarb (1993) address the issue of *special structure* of a linear program within the context of polynomial-time path-following algorithms.

Specifically, Choi and Goldfarb consider a primal-dual path-following algorithm that applies directly to a linear programming problem of the form $min\{\mathbf{C}'\mathbf{X} \mid \mathbf{A}\mathbf{X} = \mathbf{b},\ \mathbf{H}\mathbf{X} \leqq \mathbf{U},\ \mathbf{X} \geqq \mathbf{0},\ \mathbf{X} \in \mathbf{R}^n\}$. Their algorithm handles, in a unified fashion, upper bounds generalized upper bounds, variable upper bounds, and block diagonal structure.

It was mentioned in section 13.8 that Shaw and Goldfarb (1991) proposed a standard form projective algorithm which achieves $O(\sqrt{n}L)$ step complexity. However, as indicated in Anstreicher (1993), since the primal-dual pair of solutions generated by Shaw and Goldfarb on each iteration satisfy an approximate centering criterion, their algorithm can be strictly interpreted as a path-following technique that uses a projective transformation on each iteration rather than a simple affine rescaling of variables as utilized by the algorithms of this section.

In this same tradition Shaw and Goldfarb (1994) present a new projective interior point routine which actually follows a central trajectory to an optimal solution of a linear program of the form

$$\begin{aligned} &min\ \mathbf{C}'\mathbf{X} \quad \text{s.t.} \\ &\mathbf{A}\mathbf{X} = \mathbf{0},\ x_n = 1,\ \mathbf{X} \geqq \mathbf{0},\ \mathbf{X} \in \mathbf{R}^n \end{aligned}$$

in at most $O(\sqrt{n}L)$ iterations. The algorithm is identical to Anstreicher's (1986) lower bound variant of Karmarkar's (1984) projective algorithm except that it employs a weakened (larger) ball to update the lower bound on the optimal objective function value and starts at a point near the central path. Although the projective method developed by Shaw and Goldfarb is not derived using a path-following framework (*e.g.*, it does not first define the path of centers in the usual path-following format as the path of local minimizers of a logarithmic barrier function or potential function constrained by a set of linear equalities), it operates, for the most part, in a fashion similar to traditional path-following methods. In fact, the Shaw and Goldfarb results underscore the striking similarity that exists between projective and path-following methods.

13.11. Methods Related to Karmarkar's Routine: Potential Reduction Algorithms

Since the introduction of Karmarkar's projective algorithm and its attendant potential function, several ***affine potential reduction algorithms*** have been developed. These algorithms seek reductions in a potential function, like Karmarkar's algorithm, but do so without resorting to the use of the projective transformation. (For the primal case we have Gonzaga (1990); primal-dual variants are provided by, for instance, Anstreicher and Bosch (1992), Freund (1991a), Gonzaga (1991b), Gonzaga and Todd (1992), Kojima, *et al.* (1991, 1992), and Ye (1990a).)

Let us begin with the primal affine potential reduction algorithm introduced by Gonzaga (1988, 1990). For a primal problem of the form

$$min\ \mathbf{C}'\mathbf{X} \quad \text{s.t.} \quad \mathbf{AX} = \mathbf{b}, \quad \mathbf{X} \geqq \mathbf{0}, \tag{(13.14)}$$

let the feasible region $\mathcal{K} = \{\mathbf{X} \mid \mathbf{AX} = \mathbf{b},\ \mathbf{X} \geqq \mathbf{0}\}$ in $\mathbf{R}^n$ be bounded with $\mathcal{K}^0 \neq \phi$. Assume an initial feasible point $\mathbf{X}_0 \in \mathcal{K}^0$ and a lower bound z_0 on the optimal value of the primal objective z_* or $z_0 \leq z_*$.

To solve this primal problem using an interior point method Gonzaga uses, for a given $z \leq z_*$, a potential function $f: \mathbf{R}^n_{++} \to \mathbf{R}$ of the form

$$f_z(\mathbf{X}) = q \log(\mathbf{C}'\mathbf{X} - z) - \Sigma_{i=1}^n \log x_i, \tag{13.51}$$

where $q > n$ is fixed. An $\mathbf{X} \in \mathcal{K}^0$ with $\mathbf{C}'\mathbf{X} - z \leq 2^{-L}$ is taken to be an optimal solution of the primal. Moreover, if for every iteration we find $\mathbf{X}_k \in \mathcal{K}^0$ and $z_k \leq z_*$ such that

$$f_{z_k}(\mathbf{X}_k) \leq f_{z_{k+1}}(\mathbf{X}_{k+1}) - \delta,\ k = 1, 2, \ldots, \tag{13.52}$$

for a fixed $\delta > 0$, then the primal problem is solved in $O\big((q/n)L\big)$ iterations if $f_{z_0}(\mathbf{X}_0) = O(nL)$. In this regard, we must specify a rule for determining $\mathbf{X}_{k+1}$, z_{k+1} such that (13.52) holds for an appropriate δ.

To this end let each iteration employ the affine scaling $\mathbf{Y} = (\bar{\mathbf{X}})^{-1}\mathbf{X}$, where $\bar{\mathbf{X}} = diag(\mathbf{X})$. Then $\mathcal{K}$ is mapped to $\mathcal{K}_{\bar{\mathbf{X}}} = \{\mathbf{Y} \mid \mathbf{A}\bar{\mathbf{X}}\mathbf{Y} = \mathbf{b},\ \mathbf{Y} \geqq \mathbf{0}\}$ in $\mathbf{R}^n$, $\mathbf{X}_k$ is mapped to $\mathbf{Y}_k = \mathbf{1}$, and the scaled potential function $g_z: \mathbf{R}^n_{++} \rightarrow \mathbf{R}$ has the form

$$g_z(\mathbf{Y}) = q \log(\mathbf{C}'\bar{\mathbf{X}}\mathbf{Y} - z) - \Sigma_{i=1}^n \log y_i. \tag{13.53}$$

For $\mathbf{X}_k, \mathbf{X}_{k+1} \in \mathcal{K}^0$ the difference $f_z(\mathbf{X}_{k+1}) - f_z(\mathbf{X}_k)$ equals the difference $g_z(\mathbf{Y}_{k+1}) - g_z(\mathbf{Y}_k)$. Moreover, for a fixed z,

$$\nabla g_z(\mathbf{Y}) = \frac{q}{\mathbf{C}'\bar{\mathbf{X}}\mathbf{Y} - z}\bar{\mathbf{X}}\mathbf{C} - (\bar{\mathbf{Y}})^{-1}\mathbf{1},$$

where $\bar{\mathbf{Y}} = diag(\mathbf{Y})$. For $\mathbf{Y} = \mathbf{1}$,

$$\nabla g_z(\mathbf{1}) = \frac{q}{\mathbf{C}'\bar{\mathbf{X}}\mathbf{1} - z}\bar{\mathbf{X}}\mathbf{C} - \mathbf{1}. \tag{13.54}$$

For $\mathbf{h}$ the search direction and λ the steplength, any interior point $\mathbf{Y} \in \mathcal{K}^0_{\bar{\mathbf{X}}}$ can be expressed as $\mathbf{Y} = \mathbf{1} + \lambda\mathbf{h} > \mathbf{0}$ and must satisfy $\mathbf{A}\bar{\mathbf{X}}\mathbf{h} = \mathbf{0}$ (since $\mathbf{A}\bar{\mathbf{X}}(\mathbf{1}+\lambda\mathbf{h}) = \mathbf{b}$). Also, $\mathbf{h}$ must be chosen so that $\nabla g_z(\mathbf{1})'\mathbf{h} < 0$, *i.e.*, the scaled potential g_z must decrease locally for movements in the $\mathbf{h}$ direction.

Given a fixed search direction $\mathbf{h}$, we may choose the optimal steplength $\lambda = \bar{\lambda}$ according to the following

13.11. THEOREM [Gonzaga (1990)]. Let $\mathbf{h}$ be a search direction satisfying $\nabla g_z(\mathbf{1})'\mathbf{h} < 0$. Then

$$g_z(\mathbf{1}+\bar{\lambda}\mathbf{h}) - g_z(\mathbf{1}) \leq -\tfrac{1}{8} min\left\{\left(\frac{\nabla g_z(\mathbf{1})'\mathbf{h}}{\|\mathbf{h}\|}\right)^2, 4\right\} \tag{13.55}$$

for any $z \leq z_*$, where

$$\bar{\lambda} = \begin{cases} -\nabla g_z(\mathbf{1})'\mathbf{h} \big/ 4\|\mathbf{h}\|^2 & \text{if } -\nabla g_z(\mathbf{1})'\mathbf{h}\|\mathbf{h}\|_\infty \big/ \|\mathbf{h}\|^2 \leq 2; \\ 1/2\|\mathbf{h}\|_\infty & \text{if } -\nabla g_z(\mathbf{1})'\mathbf{h}\|\mathbf{h}\|_\infty \big/ \|\mathbf{h}\|^2 > 2. \end{cases}$$

$\left(\text{Here } \|\mathbf{h}\|_\infty = max\left\{|h_1|, \ldots, |h_n|\right\}.\right)$

What about our choice of $\mathbf{h}$? By virtue of the preceding theorem, we need an $\mathbf{h}$ that satisfies $\nabla g_z(\mathbf{1})'\mathbf{h} < 0$ and which makes $\left(\nabla g_z(\mathbf{1})'\mathbf{h} \big/ \|\mathbf{h}\|\right)^2$ sufficiently large. To accomplish this let us select $\mathbf{h} = -\mathbf{P}_{\mathbf{A}\bar{\mathbf{X}}}\nabla g_z(\mathbf{1})$, where $\mathbf{P}_{\mathbf{A}\bar{\mathbf{X}}}\nabla g_z(\mathbf{1})$ is the projection of $\nabla g(\mathbf{1})$ onto the null space of $\mathbf{A}\bar{\mathbf{X}}$ and $\mathbf{P}_{\mathbf{A}\bar{\mathbf{X}}} = \mathbf{I}_n - \bar{\mathbf{X}}'\mathbf{A}'(\mathbf{A}\bar{\mathbf{X}}^2\mathbf{A}')^{-1}\mathbf{A}\bar{\mathbf{X}}$ is the projection matrix (Gonzaga (1990)). For $\nabla g_z(\mathbf{1})'\mathbf{h} = -\|\mathbf{h}\|^2 < 0$, we obtain, from (13.55),

$$g_z(1+\bar{\lambda}\mathbf{h}) - g_z(1) \leq -\tfrac{1}{8}min\{\|\mathbf{h}\|^2, 4\}$$

for any $z \leq z_*$. Based upon this discussion we see that $-\mathbf{P}_{\mathbf{A}\bar{\mathbf{X}}} \nabla g_z(\mathbf{1})$ is the appropriate search direction when $\|\mathbf{P}_{\mathbf{A}\bar{\mathbf{X}}} \nabla g_z(\mathbf{1})\|$ is sufficiently large; for $\|\mathbf{P}_{\mathbf{A}\bar{\mathbf{X}}} \nabla g_z(\mathbf{1})\|$ too small, we need an improved lower bound z.

For $\bar{z}$ a lower bound satisfying $z < \bar{z} \leq z_*$, it can be shown that

$$g_{\bar{z}}(1) - g_z(1) = q\log\frac{\mathbf{C}'\bar{\mathbf{X}}\mathbf{1} - \bar{z}}{\mathbf{C}'\bar{\mathbf{X}}\mathbf{1} - z} < 0. \tag{13.56}$$

To exploit this relationship we use

13.12. LEMMA [Gonzaga (1990)]. For $z \leq z_*$ and $0 < \in \leq 1$ let $\|\mathbf{P}_{\mathbf{A}\bar{\mathbf{X}}} \nabla g_z(\mathbf{1})\| < \in$. Then

$$\bar{z} = \mathbf{C}'\bar{\mathbf{X}}\mathbf{1} - \frac{n}{q}\left(1 + \frac{\|\mathbf{P}_{\mathbf{A}\bar{\mathbf{X}}} \nabla g_z(\mathbf{1})\|}{\sqrt{n}}\right)(\mathbf{C}'\bar{\mathbf{X}}\mathbf{1} - z) \tag{13.57}$$

is a lower bound with

$$\frac{\mathbf{C}'\bar{\mathbf{X}}\mathbf{1} - \bar{z}}{\mathbf{C}'\bar{\mathbf{X}}\mathbf{1} - z} = \frac{n}{q}\left(1 + \frac{\|\mathbf{P}_{\mathbf{A}\bar{\mathbf{X}}} \nabla g_z(\mathbf{1})\|}{\sqrt{n}}\right).$$

Based upon this result, (13.56) becomes

$$g_{\bar{z}}(1) - g_z(1) = q\log\frac{n + \in\sqrt{n}}{q},$$

where $\in$ has replaced $\|\mathbf{P}_{\mathbf{A}\bar{\mathbf{X}}} \nabla g_z(\mathbf{1})\|$ in (13.57) (Gonzaga (1990)). So for $q > n + \in\sqrt{n}$, $g_{\bar{z}}(1) - g_z(1) < 0$ and thus the scaled potential g decreases when $\bar{z}$ ((13.57)) is used to update z.

We are now in a position to state Gonzaga's (1990, 1992)

13.13. PRIMAL POTENTIAL REDUCTION ALGORITHM.

STEP 0. Initialize.

Choose $\mathbf{X}_0 \in \mathcal{K}^0$, $z_0 \leq z_*$, $0 < \in \leq 1$, $q > n + \in\sqrt{n}$, and set $k = 0$.

STEP 1. Determine the search direction.

a. Find the projection matrix

$$\mathbf{P}_{\mathbf{A}\bar{\mathbf{X}}_k} = \mathbf{I}_n - \bar{\mathbf{X}}_k'\mathbf{A}'(\mathbf{A}\bar{\mathbf{X}}_k^2\mathbf{A}')^{-1}\mathbf{A}\bar{\mathbf{X}}_k,$$

where $\bar{\mathbf{X}}_k = diag(\mathbf{X}_k)$.

b. Find the projection of $\nabla g_{z_k}(1)$ onto the null- space of $\mathbf{A}\bar{\mathbf{X}}_k$ or

$$\begin{aligned}\mathbf{h}_k &= -\mathbf{P}_{\mathbf{A}\bar{\mathbf{X}}_k} \nabla g_{z_k}(\mathbf{1}) \\ &= -\mathbf{P}_{\mathbf{A}\bar{\mathbf{X}}_k}\left(\frac{q}{\mathbf{C}'\bar{\mathbf{X}}_k\mathbf{1}-z_k}\bar{\mathbf{X}}_k\mathbf{C}-\mathbf{1}\right).\end{aligned}$$

STEP 2. Line search.

Find

$$\begin{aligned}\mathbf{X}_{k+1} &= \bar{\mathbf{X}}_k(\mathbf{1}+\lambda_*\mathbf{h}_k) \\ &= \mathbf{X}_k + \lambda_*\bar{\mathbf{X}}_k\mathbf{h}_k\end{aligned}$$

with

$$\lambda_* = \underset{\lambda}{arg\,min}\left\{g_{z_k}(\mathbf{1}+\lambda\mathbf{h}_k) \mid \lambda > 0,\ \mathbf{1}+\lambda\mathbf{h}_k > \mathbf{0}\right\}.$$

STEP 3. Update the lower bound z.

a. If $\| \mathbf{h}_k \| < \in$, compute

$$z_{k+1} = \mathbf{C}'\bar{\mathbf{X}}_k\mathbf{1} - \frac{n}{q}\left(1+\frac{\|\mathbf{h}_k\|}{\sqrt{n}}\right)(\mathbf{C}'\bar{\mathbf{X}}_k\mathbf{1}-z_k);$$

b. Otherwise

$$z_{k+1} = z_k.$$

STEP 4. Check for optimality.

If $\mathbf{C}'\mathbf{X}_k - z_k \leq 2^{-L}$, stop;

otherwise proceed to

STEP 5. Advance k to $k+1$.

Return to step 1.

Under the preceding solution scheme it can be demonstrated that for $\in\ = 1$ and $q = 2n$, the scaled (and the original) potential function decreases by $\delta = \frac{1}{8}$ at each iteration and the algorithm converges to a minimal solution of the above primal problem in $O(nL)$ iterations.

One of the shortcomings of Gonzaga's algorithm is that it requires prior knowledge of the optimal primal objective value z^*. An extension of Gonzaga's algorithm which dispenses with this requirement has been provided by Freund (1991a). Freund's approach is based only on primal affine scaling and the computation of the projected gradient of the potential function

$$F(\mathbf{X},\mathbf{S}) = q\ln(\mathbf{X}'\mathbf{S}) - \Sigma_{j=1}^{n} \ln x_j,\ q = n+\sqrt{n},$$

where $\mathbf{S}$ is a vector of dual slack variables associated with the dual problem:

$max\, \mathbf{b'U}$ s.t. $\mathbf{A'U+S} = \mathbf{C}, \mathbf{S} \geqq \mathbf{0}$. The algorithm computes dual variables and takes either a primal step or recomputes feasible dual variables at each iteration. Moreover, Freund demonstrates that primal iterates decrease the potential function by at least 0.18 whereas the dual iterates decrease the potential function by at least $0.2\sqrt{n}$, thus leading to an overall bound of $O(nL)$ iterations.

Let the primal and dual problems again appear as

$$min\ \mathbf{C'X} \quad \text{s.t.} \quad \mathbf{AX} = \mathbf{b}, \mathbf{X} \geqq \mathbf{0} \tag{(13.14)}$$

and

$$max\, \mathbf{b'U} \quad \text{s.t.} \quad \mathbf{A'U+S} = \mathbf{C}, \mathbf{S} \geqq \mathbf{0} \tag{(13.17.1)}$$

respectively. It is assumed that the primal problem enjoys a strictly positive feasible solution $\mathbf{X} > \mathbf{0}$ and that the set of optimal solutions for the primal is nonempty and bounded (and thus there exists a solution $(\mathbf{U}, \mathbf{S})$ to the dual with $\mathbf{S} > \mathbf{0}$).

For $\mathbf{X} > \mathbf{0}$, let $\bar{\mathbf{X}} = diag(\mathbf{X})$ so that $\mathbf{Y} = \bar{\mathbf{X}}^{-1}\mathbf{X} \geqq \mathbf{0}$. Then the rescaled primal and dual problems appear respectively as

$$min\, \mathbf{C'\bar{X}Y} \quad \text{s.t.} \quad \mathbf{A\bar{X}Y} = \mathbf{b}, \mathbf{Y} \geqq \mathbf{0}, \tag{13.58}$$

$$max\, \mathbf{b'U} \quad \text{s.t.} \quad \mathbf{\bar{X}A'U+t} = \mathbf{\bar{X}C}, \mathbf{t} = \mathbf{\bar{X}S} \geqq \mathbf{0}. \tag{13.59}$$

Then $\mathbf{X}$ is feasible for (13.14) if and only if $\mathbf{Y}$ is feasible for (13.58) while $(\mathbf{U}, \mathbf{S})$ is feasible for (13.17.1) if and only if $(\mathbf{U}, \mathbf{t}) = (\mathbf{U}, \mathbf{\bar{X}S})$ is feasible for (13.59). And for sets of primal and dual variables $\mathbf{X}$, $(\mathbf{U}, \mathbf{S})$ respectively, the duality gap is simply $\mathbf{C'X} = \mathbf{b'U} = \mathbf{X'S}$.

Let us assume that $\mathbf{X}$, $(\mathbf{U}, \mathbf{S})$ are primal and dual feasible solutions respectively with $\mathbf{X} > \mathbf{0}$. Since scaling does not affect the potential function, take $\mathbf{X} = \mathbf{1}$. Then the current duality gap is $\mathbf{1'S}$. In addition, the projection of

$$\nabla_{\mathbf{X}} F(\mathbf{1}, \mathbf{S}) = \left(\frac{q}{\mathbf{1'S}}\right)\mathbf{S} - \mathbf{1}$$

onto the null space of $\mathbf{A}$ is

$$\mathbf{d}(\mathbf{1'S}) = \left[\mathbf{I}_n - \mathbf{A'}(\mathbf{AA'})^{-1}\mathbf{A}\right]\left[\left(\frac{q}{\mathbf{1'S}}\right)\mathbf{S} - \mathbf{1}\right].$$

Given the projected gradient $\mathbf{d}(\mathbf{1}'\mathbf{S}) = \mathbf{d}(\Delta)$, where $\Delta = \mathbf{1}'\mathbf{S}$ is an unknown positive scalar, Freund demonstrates that, for $q = n + \sqrt{n}$:

a. if $\left\| \mathbf{d}(\mathbf{1}'\mathbf{S}) \right\| \geq 0.8$, we can decrease $F(\mathbf{X}, \mathbf{S})$ by 0.18 by taking a step in the primal direction $-\mathbf{d}(\mathbf{1}'\mathbf{S})$;

b. if $\left\| \mathbf{d}(\mathbf{1}'\mathbf{S}) \right\| < 0.8$, we can decrease $F(\mathbf{X}, \mathbf{S})$ by $0.2\sqrt{n}$ by replacing $\mathbf{S}$ by $\widetilde{\mathbf{S}} = \mathbf{S}(\widetilde{\Delta})$, where $\widetilde{\Delta} \in (0, \mathbf{1}'\mathbf{S})$, and $\left\| \mathbf{d}(\widetilde{\Delta}) \right\| = 0.8$.

These remarks thus lead us to Freund's (1991a)

13.14. EXTENDED PRIMAL POTENTIAL REDUCTION ALGORITHM

STEP 0. Initialize.

Choose $\mathbf{X}_0 > \mathbf{0}$, $(\mathbf{U}_0, \mathbf{S}_0)$, $q = n + \sqrt{n} = 0.8$, an optimality tolerance level μ, and set $k = 0$.

STEP 1. Test for Optimality.

If $\mathbf{X}_k'\mathbf{S}_k \leq \mu$, stop.

STEP 2. Rescale.

Set $\bar{\mathbf{A}}_k = \mathbf{A}\bar{\mathbf{X}}_k$, $\bar{\mathbf{C}}_k = \bar{\mathbf{X}}_k\mathbf{C}$, $\bar{\mathbf{t}}_k = \bar{\mathbf{X}}_k\mathbf{S}$, where $\bar{\mathbf{X}}_k = diag(\mathbf{X}_k)$.

STEP 3. Compute the Projected Gradient Direction Function.

$$\mathbf{d}(\Delta) = \left[\mathbf{I}_n \bar{\mathbf{A}}_k'(\bar{\mathbf{A}}_k \mathbf{A}_k')^{-1}\bar{\mathbf{A}}_k\right]\left[\left(\frac{q}{\Delta}\right)\bar{\mathbf{t}}_k - \mathbf{1}\right].$$

If $\left\| \mathbf{d}(\mathbf{1}'\bar{\mathbf{t}}_k) \right\| \geq \gamma$, go to step 5; otherwise go to step 4.

STEP 4. Recompute Dual Variables.

Solve for $\widetilde{\Delta} \in (0, \mathbf{1}'\bar{\mathbf{t}}_k)$ such that $\widetilde{\Delta}$ is the smallest Δ for which $\left\| \mathbf{d}(\widetilde{\Delta}) \right\| = \gamma$. Set

$$\widetilde{\mathbf{t}} = \left(\frac{\widetilde{\Delta}}{q}\right)\left(\mathbf{1} + \mathbf{d}(\widetilde{\Delta})\right)$$

and

$$\widetilde{\mathbf{U}} = (\bar{\mathbf{A}}_k \bar{\mathbf{A}}_k')^{-1}\bar{\mathbf{A}}_k\left[\bar{\mathbf{C}}_k - \left(\frac{\widetilde{\Delta}}{q}\mathbf{1}\right)\right].$$

Set $\mathbf{U}_k = \widetilde{\mathbf{U}}_k$ and $\mathbf{S}_k = \bar{\mathbf{X}}_k^{-1}\widetilde{\mathbf{t}}$, and go to step 1.

STEP 5. Take a step in Primal Variable Space.

Set $\mathbf{d}_k = \mathbf{d}(\mathbf{1}'\bar{\mathbf{t}}_k)$ and $\mathbf{Y}_{k+1} = \mathbf{1} - 0.38\mathbf{d}_k / \left\| \mathbf{d}_k \right\|$.

Set $\mathbf{X}_{k+1} = \bar{\mathbf{X}}_k \mathbf{Y}_{k+1}$, advance k to $k+1$, and return to step 1.

In terms of the implementation of this algorithm, Freund notes that:

1. After the algorithm leaves step 4, the next projected gradient direction function $\mathbf{d}(\Delta)$ will be the same as the current direction function. Thus these is no need to resolve the least squares problem in step 3 at the next iteration.
2. In step 5 we could replace $\mathbf{Y}_{k+1} = \mathbf{1}-0.38\text{k } \mathbf{d}_k/\|\mathbf{d}_k\|$ by $\mathbf{Y}_{k+1} = \mathbf{1}-\alpha\mathbf{d}_k$, where α is determined by a line search that (inexactly) minimizes the potential function $F(\mathbf{1}-\alpha\mathbf{d}_k, \bar{\mathbf{t}}_k)$ over the range $\alpha \in (0.38/\|\mathbf{d}_k\|, \infty)$.
3. The choice of $\tilde{\Delta}$ in step 4 could be accomplished by a line search of Δ with the potential function $F(\mathbf{1}, (\Delta/q)[\mathbf{1}+\mathbf{d}(\Delta)])$, where $\Delta \in (0, \mathbf{1}'\bar{\mathbf{t}}_k)$.
4. The algorithm need not be started with a dual feasible solution. It need only be started with a lower bound z on f. Hence one replaces $\mathbf{d}(\Delta)$ in step 3 by
$$\mathbf{d}(\Delta) = \left[\mathbf{I}_n - \bar{\mathbf{A}}_k'(\bar{\mathbf{A}}_k\bar{\mathbf{A}}_k')^{-1}\bar{\mathbf{A}}_k\right]\left[\left(\tfrac{q}{\Delta}\right)\bar{\mathbf{C}}_k - \mathbf{1}\right].$$
Also, $\mathbf{1}'\bar{\mathbf{t}}_k$ replaced by $\mathbf{C}'\mathbf{X}_k - z$ in steps 3, 4, and 5. The algorithm eventually computes feasible dual variables unless z is the optimal value of the primal problem (13.14).

We next explore the interior-point algorithm offered by Ye (1990a) which achieves $O(\sqrt{n}\,L)$-iteration and $O(n^3L)$-arithmetic-operation convergence like the preceding path-following algorithm 13.7 but avoids actually having to follow the central path by simply relying upon potential reduction alone. Specifically, Ye employs the primal-dual potential function described by Todd and Ye (1990) to measure the solution's progress and develops an interior-point scaled-gradient projection algorithm for a linear program in standard form which directly minimizes the potential function using a line search. Moreover, no step size restriction applies during the iteration process and the potential function is reduced by a constant from any interior point.

Let us depict the primal-dual pair of problems (13.14), (13.17.1) as

a. Primal	b. Dual
$min\, \mathbf{C}'\mathbf{X}$ s.t.	$max\, \mathbf{b}'\mathbf{U}$ s.t.
$\mathbf{AX} = \mathbf{b},\ \mathbf{X} \geqq \mathbf{0}$	$\mathbf{S} = \mathbf{C} - \mathbf{A}'\mathbf{Y} \geqq \mathbf{0}$

where $\mathbf{A}$ is of order $(m \times n)$, $\mathbf{S}$ is an $(n \times 1)$ vector of slack variables, and it is assumed that:

a. the relative interiors of the feasible regions for the above primal and dual problems are nonempty and we have an interior feasible $\mathbf{X}_0$, $\mathbf{U}_0$ for the primal and dual problems respectively such that $\mathbf{AX}_0 = \mathbf{b}$, $\mathbf{X}_0 > \mathbf{0}$ and $\mathbf{S}_0 = \mathbf{C} - \mathbf{A}'\mathbf{U}_0 > \mathbf{0}$.

b. $\rho(\mathbf{A}) = m$;

c. the primal and dual objective functions are not constant on their respective feasible regions.

Consider the following set of ***parametrized potential functions***:

a. PRIMAL: $\bar{\phi}(\mathbf{X}, \underline{z}) = \rho \ln(\mathbf{C}'\mathbf{X} - \underline{z}) - \Sigma_{j=1}^{n} \ln x_j$, where $\underline{z} = \mathbf{b}'\mathbf{U}$ is a ***parametric lower bound***;

b. DUAL: $\underline{\phi}(\mathbf{U}, \bar{z}) = \rho \ln(\bar{z} - \mathbf{b}'\mathbf{U}) - \Sigma_{j=1}^{n} \ln s_j$, (13.60)
where $\bar{z} = \mathbf{C}'\mathbf{X}$ is ***parametric upper bound***; and

c. PRIMAL-DUAL:

$$\phi(\mathbf{X}, \mathbf{S}) = \rho \ln(\mathbf{X}'\mathbf{S}) - \Sigma_{j=1}^{n} \ln(x_j s_j) = (\rho - n) \ln(\mathbf{X}'\mathbf{S}) - \Sigma_{j=1}^{n} \ln(x_j s_j / \mathbf{X}'\mathbf{S}),$$

where $\underline{z} \leq z^* \leq \bar{z}$($z^*$ optimal), $n \leq \rho < +\infty$, and ρ can be thought of as a ***search direction parameter***. Since $\mathbf{X}'\mathbf{S} = \mathbf{X}'(\mathbf{C} - \mathbf{A}'\mathbf{U}) = \mathbf{C}'\mathbf{X} - \mathbf{b}'\mathbf{U}$, (13.60.a, b) are related to (13.60.c) as

$$\text{a. } \bar{\phi}(\mathbf{X}, \underline{z}) = \phi(\mathbf{X}, \mathbf{S}) + \Sigma_{j=1}^{n} \ln s_j; \qquad (13.60.1)$$

$$\text{b. } \underline{\phi}(\mathbf{U}, \bar{z}) = \phi(\mathbf{X}, \mathbf{S}) + \Sigma_{j=1}^{n} \ln x_j$$

with

$$\nabla \bar{\phi} = \nabla_{\mathbf{X}} \phi = \left[\rho / (\mathbf{C}'\mathbf{X} - \underline{z})\right] \mathbf{C} - \bar{\mathbf{X}}^{-1}\mathbf{1};$$

$$\nabla \underline{\phi} = \nabla_{S} \phi = \left[\rho / (\bar{z} - \mathbf{b}'\mathbf{U})\right] \mathbf{b} - \mathbf{A}\bar{\mathbf{S}}^{-1}\mathbf{1},$$

where $\bar{\mathbf{X}} = diag(\mathbf{X})$, $\bar{\mathbf{S}} = diag(\mathbf{S})$.

Based upon the inequality relating the geometric and the arithmetic mean, it can be shown that if ϕ is reduced below $-(\rho-n)\ L$ for a pair of positively feasible $\mathbf{X}$ and $\mathbf{S}$, then $\mathbf{C}'\mathbf{X} - \mathbf{b}'\mathbf{U} \; = \; \mathbf{X}'\mathbf{S} \leq 2^{-L}$.

To determine the primal form of the affine potential reduction algorithm, let $\underline{z}_0 = \mathbf{b}'\mathbf{U}_0$ for some $\mathbf{S}_0 = \mathbf{C} = \mathbf{A}'\mathbf{U}_0 > \mathbf{0}$. Then following Ye (1990a), we seek to minimize the linearized primal potential function subject to the ellipsoid constraint corresponding to the second-order term in a quadratic approximation to the primal problem (13.14), *i.e.*,

$$\begin{aligned} &min\, \nabla\,\bar{\phi}(\mathbf{X}_0, \underline{z}_0)'(\mathbf{X}-\mathbf{X}_0) \qquad \text{s.t.} \\ &\mathbf{A}(\mathbf{X}-\mathbf{X}_0) \;=\; \mathbf{0} \\ &\left\| (\bar{\mathbf{X}}_0)^{-1}(\mathbf{X}-\mathbf{X}_0) \right\| \leq \beta < 1, \end{aligned} \tag{13.61}$$

where $\bar{\mathbf{X}} = diag\,(\mathbf{X}_0)$.

Forming the Lagrangian function

$$\mathcal{L} \;=\; \nabla\,\bar{\phi}'(\mathbf{X}-\mathbf{X}_0) + \lambda_1'(\mathbf{A}(\mathbf{X}-\mathbf{X}_0)) + \lambda_2\Big(\beta - (\mathbf{X}-\mathbf{X}_0)'(\bar{\mathbf{X}}_0)^{-2}(\mathbf{X}-\mathbf{X}_0)\Big)$$

and eliminating λ_1, λ_2 from the Karush-Kuhn-Tucker first-order conditions for a minimum renders

$$\mathbf{X}_1 = \mathbf{X}_0 - \beta\, \frac{\bar{\mathbf{X}}_0\, \mathbf{P}_{\mathbf{A}\bar{\mathbf{X}}_0} \bar{\mathbf{X}}_0\, \nabla\,\bar{\phi}(\mathbf{X}_0, \underline{z}_0)}{\left\| \mathbf{P}_{\mathbf{A}\bar{\mathbf{X}}_0} \bar{\mathbf{X}}_0\, \nabla\,\bar{\phi}(\mathbf{X}_0, \underline{z}_0) \right\|}, \tag{13.62}$$

where the constrained minimum for (13.61) occurs at $\mathbf{X}_1$ and $\mathbf{P}_{\mathbf{A}\bar{\mathbf{X}}_0} = \mathbf{I}_n - \bar{\mathbf{X}}_0\mathbf{A}'(\mathbf{A}\bar{\mathbf{X}}_0^2\mathbf{A}')^{-1}\mathbf{A}\bar{\mathbf{X}}_0$ is the projection matrix. If we set $\mathbf{p}(\underline{z}_0) = \mathbf{P}_{\mathbf{A}\bar{\mathbf{X}}_0}\bar{\mathbf{X}}_0\, \nabla\,\bar{\phi}(\mathbf{X}_0, \underline{z}_0)$, then (13.62) can be rewritten as

$$\mathbf{X}_1 \;=\; \mathbf{X}_0 - \beta\bar{\mathbf{X}}_0\, \frac{\mathbf{p}(\underline{z}_0)}{\left\| \mathbf{p}(\underline{z}_0) \right\|}. \tag{13.62.1}$$

For computational purposes, $\mathbf{p}(\underline{z}_0)$ can be expressed as

$$\mathbf{p}(\underline{z}_0) \;=\; \frac{\rho}{\mathbf{C}'\mathbf{X} - \underline{z}_0}\, \bar{\mathbf{X}}_0\mathbf{S}(\underline{z}_0) - \mathbf{1}, \tag{13.63}$$

where

$$\mathbf{S}(\underline{z}_0) \;=\; \mathbf{C}' - \mathbf{A}'\mathbf{U}(\underline{z}_0) \tag{13.64}$$

and

$$\mathbf{U}(\underline{z}_0) = \mathbf{U}^1 - \frac{\mathbf{C}'\mathbf{X}_0 - \underline{z}_0}{\rho}\,\mathbf{U}^2 \tag{13.65}$$

with

$$\begin{aligned}\mathbf{U}^1 &= (\mathbf{A}\bar{\mathbf{X}}_0^2\mathbf{A}')^{-1}\mathbf{A}\bar{\mathbf{X}}_0^2\mathbf{C},\\ \mathbf{U}^2 &= (\mathbf{A}\bar{\mathbf{X}}_0^2\mathbf{A}')^{-1}\mathbf{b}.\end{aligned}$$

Based upon the preceding developments we have the following

13.15. POTENTIAL REDUCTION THEOREM [Ye (1990a)].

Let $\mathbf{X}_0, \mathbf{U}_0$ be any interior feasible solutions for the primal and dual pair of problems (13.14), (13.17.1) respectively and let $\rho = n + \sqrt{n}$, $\underline{z}_0 = \mathbf{b}'\mathbf{U}_0$, $\mathbf{X}_1$ be given by (13.62.1), and $\mathbf{S}_1 = \mathbf{S}(\underline{z}_0)$ and $\mathbf{U}_1 = \mathbf{U}(\underline{z}_0)$ determined by (13.64), (13.65) respectively. Then either

$$\begin{aligned}&\phi(\mathbf{X}_1, \mathbf{S}_0) \le \phi(\mathbf{X}_0, \mathbf{S}_0) - \delta \quad \text{or}\\ &\phi(\mathbf{X}_0, \mathbf{S}_1) \le \phi(\mathbf{X}_0, \mathbf{S}_0) - \delta,\end{aligned}$$

where $\delta > 0.05$.

This theorem thus establishes the notion that ϕ can be reduced by a constant if (13.61) is solved on the interior of (13.14), (13.17.1) no matter how $\mathbf{X}_0$, $\mathbf{U}_0$ are selected.

We are now in a position to state Ye's (1990a)

13.16. PRIMAL AFFINE POTENTIAL REDUCTION ALGORITHM.

STEP 0. Initialize.

Choose $\mathbf{X}_0 > \mathbf{0}$, $\mathbf{S}_0 = \mathbf{C} - \mathbf{A}'\mathbf{U}_0 > \mathbf{0}$ as a feasible interior solution ($\mathbf{A}\mathbf{X}_0$=$\mathbf{b}$); let $\underline{z}_0 = \mathbf{b}'\mathbf{U}_0$ and set k=0.

STEP 1. Determine the search direction.

a. Find $\mathbf{S}(\underline{z}_k) = \mathbf{C} - \mathbf{A}'\mathbf{U}(\underline{z}_k)$, where

$$\mathbf{U}(\underline{z}_k) = \mathbf{U}^1 - \frac{\mathbf{C}'\mathbf{X}_k - \underline{z}_k}{\rho}\,\mathbf{U}^2$$

and

$$\begin{aligned}\mathbf{U}^1 &= (\mathbf{A}\bar{\mathbf{X}}_k^2\mathbf{A}')^{-1}\mathbf{A}\bar{\mathbf{X}}_k^2\mathbf{C},\\ \mathbf{U}^2 &= (\mathbf{A}\bar{\mathbf{X}}_k^2\mathbf{A}')^{-1}\mathbf{b}.\end{aligned}$$

b. Find

$$\mathbf{p}(\underline{z}_k) = \frac{\rho}{\mathbf{C}'\mathbf{X}_k - \underline{z}_k}\,\bar{\mathbf{X}}_k\mathbf{S}(\underline{z}_k) - \mathbf{1}.$$

STEP 2. Line search.

a. If $\| \mathbf{p}(\underline{z}_k) \| \geq min\Big(\alpha\sqrt{\frac{n}{n+\alpha^2}},\ 1-\alpha\Big)$, then

$$\mathbf{X}_{k+1} = \mathbf{X}_k - \beta^* \bar{\mathbf{X}}_k \mathbf{p}(\underline{z}_k)$$

with

$$\beta^* = arg \min_{\beta \geq 0} \phi(\mathbf{X}_k - \beta \bar{\mathbf{X}}_k \mathbf{p}(\underline{z}_k),\ \mathbf{S}_k);$$

$$\mathbf{S}_{k+1} = \mathbf{S}_k \quad \text{and} \quad \underline{z}_{k+1} = \underline{z}_k$$

b. Otherwise $\mathbf{S}_{k+1} = \mathbf{S}(\underline{z}_*)$ with

$$\underline{z}_* = arg \min_{\underline{z} \geq \underline{z}_k} \phi(\mathbf{X}_k,\ \mathbf{S}(\underline{z}));$$

$$\mathbf{X}_{k+1} = \mathbf{X}_k \quad \text{and} \quad \underline{z}_{k+1} = \mathbf{b}'\mathbf{U}(\underline{z}_*).$$

STEP 3. Check for optimality.

If $\mathbf{C}'\mathbf{X}_k - \mathbf{b}'\mathbf{U}_k < 2^{-L}$, stop;

otherwise proceed to

STEP 4. Advance k to $k+1$.

Return to step 1.

A modification of this algorithm has us update the projection matrix as follows. If we replace $\bar{\mathbf{X}}_0$ in (13.62) by a positive diagonal matrix $\mathbf{D}$ such that $1/1.1 \leq d_j/x_j^0 \leq 1.1$, $j=1,\ldots,n$, then

$$\mathbf{X}_1 = \mathbf{X}_0 - \beta \mathbf{D} \frac{\hat{\mathbf{p}}(\underline{z}_0)}{\| \hat{\mathbf{p}}(\underline{z}_0) \|},$$

where

$$\begin{aligned}\hat{\mathbf{p}}(\underline{z}_0) &= \mathbf{P}_{\mathbf{AD}}\, \mathbf{D} \nabla \bar{\phi}\, (\mathbf{X}_0, \underline{z}_0) \\ &= \frac{\rho}{\mathbf{C}'\mathbf{X} - \underline{z}_0} \mathbf{DS}(\underline{z}_0) - \mathbf{D}(\bar{\mathbf{X}}_0)^{-1}\mathbf{1} \\ &= \mathbf{D}(\bar{\mathbf{X}}_0)^{-1}\mathbf{p}(\underline{z}_0)\end{aligned}$$

and

$$\mathbf{U}(\underline{z}_0) = (\mathbf{AD}^2\mathbf{A}')^{-1}\mathbf{AD}\Big(\mathbf{DC} - \frac{\mathbf{C}'\mathbf{X}_0 - \underline{z}_0}{\rho} \mathbf{D}(\bar{\mathbf{X}}_0)^{-1}\mathbf{1}\Big). \tag{13.65.1}$$

Based upon these developments we now have Ye's (1990a)

13.17. MODIFIED PRIMAL AFFINE POTENTIAL REDUCTION ALGORITHM.

STEP 0. Initialize.

Choose $\mathbf{X}_0 > \mathbf{0}$, $\mathbf{S}_0 = \mathbf{C} - \mathbf{A}'\mathbf{U}_0$ as a feasible interior solution ($\mathbf{AX}_0$=$\mathbf{b}$); let $\underline{z}_0 = \mathbf{b}'\mathbf{U}_0$, $\mathbf{D} = \bar{\mathbf{X}}_0$, and set $\alpha = 0.43$, $\beta = 0.25$, and $k = 0$.

STEP 1. Determine the search direction.

a. If $d_j/x_j^k \notin [1/1.1, 1.1]$, j=1,...,n, then $d_j = x_j^k$. Find

$$\mathbf{S}(\underline{z}_k) = \mathbf{C} - \mathbf{A}'\mathbf{U}(\underline{z}_k), \text{ where now}$$

$$\mathbf{U}\underline{z}_k) = (\mathbf{AD}^2\mathbf{A}')^{-1}\mathbf{AD}(\mathbf{DC} - \frac{\mathbf{C}'\mathbf{X}_k - \underline{z}_k}{\rho}\mathbf{D}(\bar{\mathbf{X}}_k)^{-1}\mathbf{1}.$$

b. Find

$$\widehat{\mathbf{p}}(\underline{z}_k) = \mathbf{D}(\bar{\mathbf{X}}_k)^{-1}\mathbf{p}(\underline{z}_k),$$

$$\mathbf{p}(\underline{z}_k) = \frac{\rho}{\mathbf{C}'\mathbf{X} - \underline{z}_k}\bar{\mathbf{X}}_k\mathbf{S}(\underline{z}_k) - \mathbf{1}.$$

STEP 2. Line search.

a. If $\| \mathbf{p}(\underline{z}_k) \| \geq min\left(\alpha\sqrt{\frac{\alpha}{n+\alpha^2}}, 1-\alpha\right)$, then

$$\mathbf{X}_{k+1} = \mathbf{X}_k - \beta\mathbf{D}\frac{\widehat{\mathbf{p}}(\underline{z}_k)}{\| \widehat{\mathbf{p}}(\underline{z}_k) \|};$$

$$\mathbf{S}_{k+1} = \mathbf{S}_k \text{ and } \underline{z}_{k+1} = \underline{z}_k.$$

b. See step 2.b of algorithm 13.16.

STEP 3. See step 3 of algorithm 13.16.

STEP 4. See step 4 of algorithm 13.16.

The complexity performance of this modified algorithm results from the following

13.18. THEOREM [Ye (1990a)]. Let $\rho = n + \sqrt{n}$ and $\phi(\mathbf{X}_0, \mathbf{S}_0) \leq O(\sqrt{n}L)$. Then the modified primal potential reduction algorithm terminates in $O(\sqrt{n}L)$ iterations and uses a total of $O(\mathrm{n}^3L)$ arithmetic operations.

Let us now consider the dual variant of the affine potential reduction algorithm (Ye (1990a)). For $\mathbf{X}_0 > \mathbf{0}$ ($\mathbf{AX}_0$=$\mathbf{b}$) a feasible interior solution to (13.14) and $\mathbf{C}'\mathbf{X}_0 = \bar{z}_0$, let us minimize the linearized dual potential function subject to the ellipsoid constraint corresponding to the second-order term in a quadratic approximation to the dual problem (13.17.1), *i.e.*,

$$min\ \nabla \underline{\phi}(\mathbf{U}_0, \bar{z}_0)'(\mathbf{U} - \mathbf{U}_0) \qquad \text{s.t.} \tag{13.66}$$

$$\left\| (\bar{\mathbf{S}}_0)^{-1}\mathbf{A}'(\mathbf{U}-\mathbf{U}_0) \right\| \leq \beta.$$

If $\mathbf{U}_1$ denotes the minimal solution to the preceding problem, then

$$\mathbf{U}_1 = \mathbf{U}_0 - \beta \mathbf{d}_0, \tag{13.67}$$

where the direction vector $\mathbf{d}_0$ is defined as

$$\mathbf{d}_0 = \frac{(\mathbf{A}(\bar{\mathbf{S}}_0)^{-2}\mathbf{A}')^{-1} \nabla \underline{\phi}(\mathbf{U}_0, \bar{z}_0)}{\left[\nabla \underline{\phi}(\mathbf{U}_0, \bar{z}_0)'(\mathbf{A}(\bar{\mathbf{S}}_0)^{-2}\mathbf{A}')^{-1} \nabla \underline{\phi}(\mathbf{U}_0, \bar{z}_0) \right]^{1/2}}, \tag{13.68}$$

and the vector of slack variables is given by

$$\mathbf{S}_1 = \mathbf{S}_0 - \mathbf{A}'(\mathbf{U}_1 - \mathbf{U}_0). \tag{13.69}$$

Let

$$\begin{aligned} \mathbf{p}(\bar{z}_0) &= (\bar{\mathbf{S}}_0)^{-1}\mathbf{A}'(\mathbf{A}(\bar{\mathbf{S}}_0)^{-2}\mathbf{A}')^{-1} \nabla \underline{\phi}(\mathbf{U}_0, \bar{z}_0) \\ &= -\frac{\rho}{\bar{z}_0 - \mathbf{b}'\mathbf{U}_0} \bar{\mathbf{S}}_0 \mathbf{X}(\bar{z}_0) + 1, \end{aligned} \tag{13.70}$$

where

$$\mathbf{X}(\bar{z}_0) = \mathbf{X}^1 + \frac{\bar{z}_0 - \mathbf{b}'\mathbf{U}_0}{\rho} \mathbf{X}^2 \tag{13.71}$$

with

$$\begin{aligned} \mathbf{X}^1 &= (\bar{\mathbf{S}}_0)^{-2}\mathbf{A}'(\mathbf{A}(\bar{\mathbf{S}}_0)^{-2}\mathbf{A}')^{-1}\mathbf{b}, \\ \mathbf{X}^2 &= (\bar{\mathbf{S}}_0)^{-1}\mathbf{P}_{\mathbf{A}(\bar{\mathbf{S}}_0)^{-1}}(\bar{\mathbf{S}}_0)^{-1}\mathbf{C}, \end{aligned}$$

and $\mathbf{P}_{\mathbf{A}(\bar{\mathbf{S}}_0)^{-1}} = \mathbf{I}_n - (\bar{\mathbf{S}}_0)^{-1}\mathbf{A}'(\mathbf{A}(\bar{\mathbf{S}}_0)^{-2}\mathbf{A}')^{-1}(\bar{\mathbf{S}}_0)^{-1}$ is the projection matrix.

Based upon this discussion we can now state Ye's (1990a)

13.19. DUAL AFFINE POTENTIAL REDUCTION ALGORITHM.

STEP 0. Initialize.

Choose $\mathbf{X}_0 > \mathbf{0}$, $\mathbf{S}_0 - \mathbf{C} - \mathbf{A}'\mathbf{U}_0 > \mathbf{0}$ as a feasible interior solution ($\mathbf{AX}_0 = \mathbf{b}$); let $\bar{z}_0 = \mathbf{C}'\mathbf{U}_0$ and set $k=0$.

STEP 1. Determine the search direction.

a. Find

$$\mathbf{X}(\bar{z}_k) = \mathbf{X}^1 + \frac{\bar{z}_k - \mathbf{b}'\mathbf{U}_k}{\rho}\mathbf{X}^2,$$

where

$$\mathbf{X}^1 = (\bar{\mathbf{S}}_k)^{-2}\mathbf{A}'(\mathbf{A}(\bar{\mathbf{S}}_k)^{-2}\mathbf{A}')^{-1}\mathbf{b},$$

$$\mathbf{X}^2 = (\bar{\mathbf{S}}_k)^{-1}\mathbf{P}_{\mathbf{A}(\bar{\mathbf{S}}_k)^{-1}}(\bar{\mathbf{S}}_k)^{-1}\mathbf{C}.$$

b. Find

$$\mathbf{p}(\bar{z}_k) = -\frac{\rho}{\bar{z}_k - \mathbf{b}'\mathbf{U}_k}\bar{\mathbf{S}}_k\mathbf{X}(\bar{z}_k) + \mathbf{1}.$$

STEP 2. Line search.

a. If $\| \mathbf{p}(\bar{z}_k) \| \geq min\left(\sqrt{\alpha\frac{n}{n+\alpha^2}}\,,\ 1-\alpha\right)$, then

$$\mathbf{S}_{k+1} = \mathbf{S}_k + \beta^*\bar{\mathbf{S}}_k\mathbf{p}(\bar{z}_k)$$

with

$$\beta^* = arg\ \min_{\beta \geq 0} \phi(\mathbf{X}_k, \mathbf{S}_k + \beta\bar{\mathbf{S}}_k\mathbf{p}(\bar{z}_k));$$

$$\mathbf{X}_{k+1} = \mathbf{X}_k \quad \text{and} \quad \bar{z}_{k+1} = \bar{z}_k.$$

b. Otherwise $\mathbf{X}_{k+1} = \mathbf{X}(\bar{z}_*)$ with

$$\bar{z}_* = arg\ \min_{\bar{z} \leq \bar{z}_k} \phi(\mathbf{X}(\bar{z}), \mathbf{S}_k);$$

$$\mathbf{S}_{k+1} = \mathbf{S}_k \quad \text{and} \quad \bar{z}_{k+1} = \mathbf{C}'\mathbf{X}(\bar{z}_*).$$

STEP 3. Check for optimality.

If $\mathbf{C}'\mathbf{X}_k - \mathbf{b}'\mathbf{U}_k < 2^{-L}$, stop;

otherwise proceed to

STEP 4. Advance k to $k+1$.

Return to step 1.

The performance of this dual algorithm is identical to that of the primal potential reduction algorithm 13.16.

A primal algorithm similar to Ye's (1990a) potential reduction algorithm 13.16 has been proposed by Gonzaga (1992). According to Gonzaga, Ye's method is really a type of path-following algorithm since it only updates the lower bounds or dual variables at *nearly central points.* Anstreicher and Bosch

(1992), however, extend Ye's results by constructing an algorithm whose steps are not restricted to remain approximately centered, *i.e.*, long primal steps can be taken. (These authors use a ***safeguard line search*** or sufficient decrease criterion which controls the partial updating of the diagonal scaling matrix at each iteration,*i.e.*, prevents updating that is excessive relative to the potential reduction obtained.) Moreover, Ye's dual algorithm 13.19 is actually equivalent to his first algorithm 13.16, *i.e.*, the dual update keeps no memory of $\mathbf{S}_k$ and thus the dual algorithm should be viewed as a primal algorithm (Gonzaga (1992)).

A discussion on the near boundary behavior of the primal potential reduction algorithm has been provided by Ye, *et al.* (1993b). These authors find that for any iteration ρ can be chosen within an extremely wide interval $([n+\sqrt{n}, +\infty))$ without compromising the convergence rate of $O(\sqrt{n}L)$ iterations. So as the iterate drifts away from the central path towards the boundary of the positive orthant, ρ need not be set close to $n+\sqrt{n}$. This result is consistent with the recent findings of Zhang, *et al.* (1992) concerning the convergence rate of the algorithm. They conclude that it is desirable to use both a large step size and a large ρ to attain a better local convergence rate.

In addition to providing an extension of Gonzaga's potential reduction algorithm (13.13), Freund (1991a) also introduces an alternative form of Ye's $O(\sqrt{n}\,L)$ primal algorithm (13.16). This involves an extension of algorithm 13.14 by using in place of $F(\mathbf{X}, \mathbf{S})$ the primal-dual potential function

$$G(\mathbf{X}, \mathbf{S}) \;=\; q\ln(\mathbf{X}'\mathbf{S}) - \textstyle\sum_{j=1}^{n} \ln x_j - \sum_{j=1}^{n} \ln s_j$$

with $q \;=\; n+\sqrt{n}$, where s_j is the jth component of the vector of dual slack variables $\mathbf{S}$.

To see exactly how Freund's (1991a) algorithm works, let $\mathbf{X}$ and $(\mathbf{U}, \mathbf{S})$ be primal and dual feasible solution to (13.14), (13.17.1) respectively with $\mathbf{X} > \mathbf{0}$. If (13.14) is rescaled so that $\mathbf{X} = \mathbf{1}$, then the current duality gap is $\mathbf{1}'\mathbf{S}$. As before, let us determine the projected gradient direction function $\mathbf{d}(\Delta)$ as

$$\mathbf{d}(\Delta) \;=\; [\mathbf{I}_n - \mathbf{A}'(\mathbf{A}\mathbf{A}')^{-1}]\left[\left(\tfrac{q}{\Delta}\right)\mathbf{S} - \mathbf{1}\right]$$

and find $\mathbf{d}(\mathbf{1}'\mathbf{S})$, the projected gradient of $G(\mathbf{X},\mathbf{S})$ with respect to $\mathbf{X}$ at $(\mathbf{X},\mathbf{S}) = (\mathbf{1},\mathbf{S})$. Then of $\|\mathbf{d}(\mathbf{1}'\mathbf{S})\| \geq 0.22$, we can achieve a constant decrease in the potential function $G(\mathbf{X},\mathbf{S})$ by taking a step in the direction $-\mathbf{d}(\mathbf{1}'\mathbf{S})$ from the current primal feasible point $\mathbf{1}$. That is, according to Freund, if $q = n+\sqrt{n}$, then:

a. If $\|\mathbf{d}(\mathbf{1}'\mathbf{S})\| \geq 0.22$, we can decrease $G(\mathbf{X},\mathbf{S})$ by 0.02 by taking a step in the primal variables in the direction $-\mathbf{d}(\mathbf{1}'\mathbf{S})$;

b. if $\|\mathbf{d}(\mathbf{1}'\mathbf{S})\| < 0.22$, we can decrease $G(\mathbf{X},\mathbf{S})$ by 0.02 by replacing $\mathbf{S}$ by $\mathbf{S} = \mathbf{S}(\widetilde{\Delta})$, where $\widetilde{\Delta} \epsilon (0, \mathbf{1}'\mathbf{S})$ satisfies $\|\mathbf{d}(\widetilde{\Delta})\| = 0.22$.

These considerations thus lead us to the following alternative form of Ye's (1990a) algorithm for solving (13.14). To this end we state Freund's (1991a)

13.20. ALTERNATIVE FORM OF YE'S ALGORITHM (13.14).

STEP 0. Initialize.

Choose $\mathbf{X}_0 > \mathbf{0}$, $(\mathbf{U}_0, \mathbf{S}_0)$, $q = n+\sqrt{n}$, $\gamma = 0.22$, an optimality tolerance level μ, and set $k = 0$.

STEP 1. Test for Optimality.

If $\mathbf{X}_k'\mathbf{S}_k \leq \mu$, stop.

STEP 2. Rescale.

Set $\bar{\mathbf{A}}_k = \mathbf{A}\bar{\mathbf{X}}_k$, $\bar{\mathbf{C}}_k = \bar{\mathbf{X}}_k\mathbf{C}$, $\bar{\mathbf{t}}_k = \bar{\mathbf{X}}_k\mathbf{S}$, where $\bar{\mathbf{X}}_k = diag(\mathbf{X}_k)$.

STEP 3. Compute the Projected Gradient Direction Function.

$$\mathbf{d}(\Delta) = \left[\mathbf{I}_n - \bar{\mathbf{A}}_k(\bar{\mathbf{A}}_k\bar{\mathbf{A}}_k')^{-1}\bar{\mathbf{A}}_k\right]\left[\left(\frac{q}{\Delta}\right)\bar{\mathbf{t}}_k - \mathbf{1}\right].$$

If $\|\mathbf{d}(\mathbf{1}'\bar{\mathbf{t}}_k)\| \geq \gamma$, go to step 5; otherwise go to step 4.

STEP 4. Recompute Dual Variables.

Solve for $\widetilde{\Delta} \epsilon (0, \mathbf{1}'\bar{\mathbf{t}}_k)$ such that $\widetilde{\Delta}$ is the smallest Δ for which $\|\mathbf{d}(\widetilde{\Delta})\| = \gamma$. Set

$$\widetilde{\mathbf{t}} = \left(\frac{\widetilde{\Delta}}{q}\right)\left(\mathbf{1} + \mathbf{d}(\widetilde{\Delta})\right) \text{ and}$$

$$\widetilde{\mathbf{U}} = (\bar{\mathbf{A}}_k\bar{\mathbf{A}}_k')^{-1}\bar{\mathbf{A}}_k\left[\bar{\mathbf{C}}_k - \left(\frac{\widetilde{\Delta}}{q}\right)\mathbf{1}\right].$$

Set $\mathbf{U}_k = \widetilde{\mathbf{U}}_k$, $\mathbf{S}_k = \bar{\mathbf{X}}_k^{-1}\widetilde{\mathbf{t}}$, go to step 1.

STEP 5. Take a step in Primal Variable Space.

Set $\mathbf{d}_k = \mathbf{d}(\mathbf{1}'\ \bar{\mathbf{t}}_k)$, $\alpha\ =\ 1-\dfrac{1}{\sqrt{1-2\gamma}}$, and $\mathbf{Y}_{k+1}\ =\ \mathbf{1}-\alpha\,\dfrac{\mathbf{d}_k}{\|\mathbf{d}_k\|}$.

Set $\mathbf{X}_{k+1}\ =\ \bar{\mathbf{X}}_k\mathbf{Y}_{k+1}$, advance k to k+1, and return to step 1.

A glance back at Ye's algorithm (13.16) reveals that, except for the specification for when to take primal versus dual steps at each iteration, the two algorithms are identical. Here too Freund's modification at Ye's algorithm does not explicitly require a condition that successive iterates lie near the central path. In addition, both algorithms compute the same primal and/or dual projected gradient directions at each iteration and both employ the same logic for when to take a primal versus dual step at each iteration: either the projected gradient of the potential function is sufficiently large, or the current iterate is sufficiently centered. Thus the choice between either a primal or dual step depends only on $\|\mathbf{d}(\mathbf{1}'\bar{\mathbf{t}}_k)\|$.

Ye's method has precipitated the development of other primal-dual potential reduction routines with low complexity and which are not wedded to following the central path, among which are the algorithms of Kojima, *et al.* (1991) and Gonzaga and Todd (1992). Kojima, *et al.* offer an algorithm which achieves the $O(\sqrt{n}\,L)$ step complexity of Ye's potential reduction method but takes simultaneous primal and dual steps using a primal-dual scaling $\mathbf{D} = (\bar{\mathbf{X}}\bar{\mathbf{S}}^{-1})^{1/2}$, where $\bar{\mathbf{X}}\ = diag(\mathbf{X}_k)$, $\mathbf{S} = diag(\mathbf{S}_k)$, and $\mathbf{S}_k$ is a vector of current dual slack variables. Gonzaga and Todd, however, develop a version of Ye's algorithm which takes either a primal step based on a primal scaling $\mathbf{D} = \bar{\mathbf{X}}$, or a dual step based on a dual scaling $\mathbf{D} = \bar{\mathbf{S}}^{-1}$.

Looking to the Kojima, *et al.* (1991) primal-dual potential reduction algorithm, let the primal and dual problems be given by (13.14), (13.17.1) respectively or

Primal	Dual
$min\ \mathbf{C}'\mathbf{X}$ s.t.	$max\ \mathbf{b}'\mathbf{U}$ s.t.
$\mathbf{AX} = \mathbf{b},\ \ \mathbf{X} \geqq \mathbf{0}$	$\mathbf{A}'\mathbf{U} + \mathbf{S}\ =\ \mathbf{C},\ \mathbf{S} \geqq \mathbf{0}.$

As Kojima, *et al.* demonstrate, the said algorithm is designed to reduce the potential function by at least 0.2 in each iteration requiring $O(n^3)$ arithmetic operations. It starts from an initial feasible solution with the potential function bounded by $O(\sqrt{n}\,L)$ and generates, in at most $O(\sqrt{n}\,L)$ iterations, an approximate solution from which an exact solution can be computed in $O(n^3)$ arithmetic operations.

Let $\mathcal{K}_{++} = \{(\mathbf{X},\mathbf{S}) \mid \mathbf{0} < (\mathbf{X},\mathbf{S}) \in R^{2n},\ \ \mathbf{AX}=\mathbf{b},\ \ \mathbf{A}'\mathbf{U}+\mathbf{S}=\mathbf{C}$ for some $\mathbf{U} \in R^m\} \neq \phi$. To chart the progress of the primal and dual problems, Kojima, *et al.* use the primal-dual potential function

$$f(\mathbf{X},\mathbf{S}) \;=\; (n+\sqrt{n})\ln(\mathbf{X}'\mathbf{S}) - \textstyle\sum_{j=1}^{n} \ln(x_j s_j).$$

Starting from a current point $(\mathbf{X}_k,\mathbf{S}_k) \in \mathcal{K}_{++}$, an iteration of the algorithm determines a direction $(\mathbf{d}_k^X,\mathbf{d}_k^S)$, a steplength $\theta > 0$, and a new iterate $(\mathbf{X}_{k+1},\mathbf{S}_{k+1}) \in \mathcal{K}_{++}$ with $(\mathbf{X}_{k+1},\mathbf{S}_{k+1}) = (\mathbf{X}_k,\mathbf{S}_k) - \theta(\mathbf{d}_k^X,\mathbf{d}_k^S)$. In general, the direction vector $(\mathbf{d}^X,\mathbf{d}^U,\mathbf{d}^S)$ is found by solving the linear system

$$\begin{aligned} \mathbf{A}\mathbf{d}^X \qquad\qquad\qquad &= \mathbf{0} \\ \mathbf{A}'\mathbf{d}^U + \mathbf{d}^S &= \mathbf{0} \\ \mathbf{D}^{-1}\mathbf{d}^X \qquad\quad + \mathbf{D}\mathbf{d}^S &= \frac{\widehat{\mathbf{V}}}{\|\widehat{\mathbf{V}}\|}, \end{aligned} \tag{13.72}$$

where $\bar{\mathbf{X}} = diag(\mathbf{X})$, $\bar{\mathbf{S}} = diag(\mathbf{S})$, $\mathbf{D} = (\bar{\mathbf{X}}\bar{\mathbf{S}}^{-1})^{1/2}$, $\mathbf{V}' = (\sqrt{x_1/s_1},\ldots,\ \sqrt{x_n/s_n})$, $\bar{\mathbf{V}} = diag(\mathbf{V})$, and $\widehat{\mathbf{V}} = \left[(n+\sqrt{n})/\|\mathbf{V}\|^2\right]\mathbf{V} - \bar{\mathbf{V}}^{-1}\mathbf{1}$. The unique solution to the preceding system is given by

$$\begin{aligned} \mathbf{d}^X &= \mathbf{D}(\mathbf{I}_n - \mathbf{P}_{\mathbf{AD}})\frac{\widehat{\mathbf{V}}}{\|\widehat{\mathbf{V}}\|} \\ \mathbf{d}^U &= -(\mathbf{AD}^2\mathbf{A}')^{-1}\mathbf{AD}\frac{\widehat{\mathbf{V}}}{\|\widehat{\mathbf{V}}\|} \\ \mathbf{d}^S &= \mathbf{D}^{-1}\mathbf{P}_{\mathbf{AD}}\frac{\widehat{\mathbf{V}}}{\|\widehat{\mathbf{V}}\|}, \end{aligned} \tag{13.73}$$

where $\mathbf{P_{AD}} = \mathbf{DA'(AD^2A')^{-1}AD}$ is the orthogonal projection matrix onto the range of $(\mathbf{AD})'$ or the subspace $\{(\mathbf{AD})'\mathbf{W} \mid \mathbf{W} \in \mathbf{R}^m\}$ in $\mathbf{R}^n$. (Note the similarity between the directions given in (13.73) and those provided by (13.45).)

To ensure the feasibility of a new point $(\mathbf{X}_{k+1}, \mathbf{S}_{k+1}) \in \mathcal{K}_{++}$, Kojima, *et al.* impose on the direction $(\mathbf{d}^X, \mathbf{d}^S)$ and step parameter θ the requirement that

$$\begin{aligned} &\text{(i)} \quad \mathbf{Ad}^X = \mathbf{0},\ \mathbf{A'd}^U + \mathbf{d}^S = \mathbf{0}; \\ &\text{(ii)} \quad \theta \| \bar{\mathbf{X}}^{-1}\mathbf{d}^X \|_\infty \leq \tau,\ \theta \| \bar{\mathbf{S}}^{-1}\mathbf{d}^S \|_\infty \leq \tau, \end{aligned} \tag{13.74}$$

where $\tau \in (0,1), \theta > 0$, and $\| \mathbf{U} \|_\infty = max\{\, | u_i | , i=1,\ldots,n\}$. They then go on to demonstrate that, for $n \geq 2$, $\tau = 0.4$, and for directions $(\mathbf{d}^X,\ \mathbf{d}^S)$ given by (17.73), there always exists a θ for which their algorithm generates a sequence $\{(\mathbf{X}_k, \mathbf{S}_k) \in \mathcal{K}_{++}\}$ such that $f(\mathbf{X}_{k+1}, \mathbf{S}_{k+1}) \leq f(\mathbf{X}_k, \mathbf{S}_k) - 0.2$, $K = 1, 2, \ldots$, and that one obtains an approximate solution $(\mathbf{X}_p, \mathbf{S}_p)$ of the primal-dual pair of problems for some $p = O(\sqrt{n}\, L)$ such that $\mathbf{X}_p'\mathbf{S}_p \leq 2^{-2L}$, from which an exact solution is reached in $O(n^3)$ arithmetic operations.

Specifically, we have the Kojima, *et al.* (1991)

13.21. PRIMAL-DUAL POTENTIAL REDUCTION ALGORITHM.

STEP 0. Initialize.

Choose $(\mathbf{X}_0, \mathbf{S}_0) > \mathbf{0}$ as a feasible initial solution.

Set $k = 0$ and let $\delta = 0.2$.

STEP 1. Determine the search direction.

Compute the descent direction $(\mathbf{d}^X, \mathbf{d}^S)$ of (13.73).

STEP 2. Line search.

Find

$$(\mathbf{X}_{k+1}, \mathbf{S}_{k+1}) = (\mathbf{X}_k, \mathbf{S}_k) - \bar{\theta}(\mathbf{d}^X, \mathbf{d}^S) \in \mathcal{K}_{++}$$

with

$$\bar{\theta} = arg \min_{\theta} f(\mathbf{X} - \theta\mathbf{d}^X, \mathbf{S} - \theta\mathbf{d}^S).$$

STEP 3. Check for optimality.

At each iteration

$$f\Big((\mathbf{X}_k, \mathbf{S}_k) - \bar{\theta}(\mathbf{d}^X, \mathbf{d}^S)\Big) \leq f(\mathbf{X}_k, \mathbf{S}_k) - \delta.$$

If $\mathbf{X}_k'\mathbf{S}_k \leq 2^{-2L}$ for some $k = O(\sqrt{n}\, L)$, stop and compute an exact solution; otherwise proceed to

STEP 4. Advance k to $k+1$.

Return to step 1.

As mentioned earlier, this algorithm does not invoke any proximity criterion in that it does not have to follow the central path. Moreover, it performs at each iteration a primal-dual affine scaling that does not take the present iterate to **1**.

Kojima, *et al.* offer an interesting interpretation of the descent direction $-(\mathbf{d}^X, \mathbf{d}^S)$ in their algorithm. For a fixed $(\mathbf{X}, \mathbf{S}) \in K_{++}$, the real-valued function $h(\mathbf{d}^X, \mathbf{d}^S) = \| \mathbf{D}^{-1}\mathbf{d}^X + \mathbf{D}\mathbf{d}^S \|$ defines a norm in the linear subspace T spanned by all the feasible directions at $(\mathbf{X}, \mathbf{S})$, *i.e*, $T = \{(\mathbf{d}^X, \mathbf{d}^S) \in \mathrm{R}^{2n} \mid \mathbf{A}\mathbf{d}^X = \mathbf{0},\ \mathbf{A}'\mathbf{d}^U + \mathbf{d}^S = \mathbf{0}\}$. Then the direction $-(\mathbf{d}^X, \mathbf{d}^S)$ determined by (13.72) may be regarded as the steepest descent feasible direction of the potential function f on K_{++} with respect to the metric induced by h, *i.e.*, $-(\mathbf{d}^X, \mathbf{d}^S)$ coincides with the solution $(\tilde{\mathbf{d}}^X, \tilde{\mathbf{d}}^S)$ of the problem

$$\begin{aligned} &min\{ \nabla_{\mathbf{X}} f(\mathbf{X}, \mathbf{S})\mathbf{d}^X + \nabla_{\mathbf{S}} f(\mathbf{X}, \mathbf{S})\mathbf{d}^S \} \quad \text{s.t.} \\ &\mathbf{A}\mathbf{d}^X = \mathbf{0},\ \mathbf{A}'\mathbf{d}^U + \mathbf{d}^S = \mathbf{0},\ \| \mathbf{D}^{-1}\mathbf{d}^X + \mathbf{D}\mathbf{d}^S \|^2 \le 1 \end{aligned}$$

and may be regarded as a affinely scaled steepest descent direction of the potential function f.

Recently Bosch and Anstreicher (1993) introduced a partial updating routine for the Kojima, *et al.* (1991) primal-dual potential reduction algorithm (since convergence of the same is tied to the reduction of a potential function) which employs a safeguard line search condition for controlling the number of updates incurred on primal-dual iterations. Their upgrading is based directly on error in the diagonal scaling matrix **D** and allows for unequal steplengths in the primal and dual variables.

Bosch and Anstreicher demonstrate that their modified algorithm, like the original algorithm of Kojima, *et al.* (1991), requires $O(\sqrt{n}\,L)$ iterations and offer a condition which is sufficient to exert control over the total number of updates, thus leading to an overall complexity of $O(n^3L)$ iterations.

Gonzaga and Todd (1992) present a primal-dual interior point algorithm for the standard form primal linear programming problem

$$\begin{aligned} &min\ \mathbf{C}'\mathbf{X} \quad \text{s.t.} \\ &\mathbf{AX} = \mathbf{b},\ \mathbf{X} \geqq \mathbf{0}. \end{aligned}$$

The algorithm reduces a primal-dual potential function at each iteration by applying separate scalings to the primal and dual problems and takes a scaled projected steepest-descent step in either primal or dual space. The authors demonstrate that a constant reduction in the potential function can always be achieved in either primal or dual variables, thus leading to a bound of $O(\sqrt{n}\,L)$ iterations given that the matrices **A**, **C**, and **b** contain all integer elements.

While the conventional dual of the preceding primal problem is

$$\begin{aligned} &max\ \mathbf{b}'\mathbf{U} \quad \text{s.t.} \\ &\mathbf{A}'\mathbf{U} + \mathbf{S} = \mathbf{C},\ \mathbf{S} \geqq \mathbf{0},\ \mathbf{U}\ \text{unrestricted}, \end{aligned}$$

Gonzaga and Todd work with the dual problem also in standard form involving only the dual slack variables $\mathbf{S} \geqq \mathbf{0}$. To this end, let $\mathbf{F}$ be a matrix whose rows span the null-space of $\mathbf{A}$ ($N(\mathbf{A}) = \{\mathbf{X} \mid \mathbf{AX} = \mathbf{0},\ \mathbf{X} \epsilon R^n\}$) or $\mathbf{FA}' = \mathbf{0}$ and let $\mathbf{FC} = \mathbf{g}$. Then $\mathbf{A}'\mathbf{U} + \mathbf{S} = \mathbf{C}$ for some $\mathbf{U}$ if and only if $\mathbf{FS} = \mathbf{g}$. For $\mathbf{d}$ any vector satisfying $\mathbf{Ad} = \mathbf{b}$, $\mathbf{b}'\mathbf{U} = \mathbf{d}'\mathbf{A}'\mathbf{U} = \mathbf{d}'\mathbf{C} - \mathbf{d}'\mathbf{S}$. Hence the conventional dual can be written in terms of $\mathbf{S}$ alone as

$$\begin{aligned} &min\ \mathbf{d}'\mathbf{S} \quad \text{s.t.} \\ &\mathbf{FS} = \mathbf{g},\quad \mathbf{S} \geqq \mathbf{0}. \end{aligned} \tag{13.75}$$

Here weak duality assumes the form $\mathbf{C}'\mathbf{X} \geq \mathbf{d}'\mathbf{C} - \mathbf{d}'\mathbf{S}$ for all feasible $\mathbf{X}, \mathbf{S}$ and strong duality states that equality holds if and only if $\mathbf{X}, \mathbf{S}$ are optimal. Moreover, the duality gap $\mathbf{C}'\mathbf{X} - \mathbf{b}'\mathbf{U}$ equals $\mathbf{C}'\mathbf{X} + \mathbf{d}'\mathbf{S} - \mathbf{d}'\mathbf{C}$ (and also $\mathbf{X}'\mathbf{S}$).

Let $\mathcal{H}_P^+$, $\mathcal{H}_D^+$ denote the set of strictly positive feasible solutions to the primal and dual problems respectively. For $(\mathbf{X}, \mathbf{S}) \epsilon\, \mathcal{U}^+ = \mathcal{H}_P^+ \times \mathcal{H}_D^+$ let the primal-dual potential function be

$$\phi_\rho(\mathbf{X}, \mathbf{S}) = \rho \ln \mathbf{X}'\mathbf{S} - \textstyle\sum_{j=1}^n \ln x_j s_j,\ \rho > n. \tag{13.76}$$

Also, let the primal and dual penalized functions appear as

$$\begin{aligned} &\text{(a)}\ f_\alpha^P(\mathbf{X}) = \alpha \mathbf{C}'\mathbf{X} - \textstyle\sum_{j=1}^n \ln x_j,\ \alpha > 0; \\ &\text{(b)}\ f_\alpha^D(\mathbf{S}) = \alpha \mathbf{d}'\mathbf{S} - \textstyle\sum_{j=1}^n \ln s_j,\ \alpha > 0, \end{aligned} \tag{13.77}$$

respectively.

To ascertain the relationship between the potential function ϕ_ρ and the penalized functions $f_\alpha^P(\mathbf{X})$ and $f_\alpha^D(\mathbf{S})$, let $\mathbf{P_A}$ denote the projection onto the null space of $\mathbf{A}$. Then for any $\mathbf{S}\in\mathcal{K}_D^+$, $\mathbf{P_A S} = \mathbf{P_A C}$ (the rows of $\mathbf{P_A}$ span $N(\mathbf{A})$). Also, it can easily be shown that

$$\begin{aligned}
&\text{(a) } \mathbf{P_A}\nabla_{\mathbf{X}}\,\phi_\rho(\mathbf{X},\mathbf{S}) \;=\; \mathbf{P_A}\nabla f_\alpha^P(\mathbf{X}),\;\; \alpha = \frac{\rho}{\mathbf{X'S}};\\
&\text{(b) } \mathbf{P_F}\nabla_{\mathbf{S}}\,\phi_\rho(\mathbf{X},\mathbf{S}) \;=\; \mathbf{P_F}\nabla f_\alpha^D(\mathbf{S}),\;\; \alpha = \frac{\rho}{\mathbf{X'S}}.
\end{aligned} \tag{13.78}$$

Let the minimizer of f_α^P over $\mathcal{K}_P^+$ be called the ***α-center of*** $\mathcal{K}_P^+$ and denoted $\mathbf{X}(\alpha)$. At any such point $\mathbf{P_A}\nabla f_\alpha^P(\mathbf{X}) = \mathbf{0}$. Likewise, the minimizer of f_α^D over $\mathcal{K}_D^+$ is termed the ***α-center of*** $\mathcal{K}_D^+$ and denoted $\mathbf{S}(\alpha)$. Here too $\mathbf{P_A}\nabla f_\alpha^D(\mathbf{S}) = \mathbf{0}$ at $\mathbf{S}(\alpha)$. Thus the equations relating the α-centers $\mathcal{K}_P^+, K_D^+$ are

$$\bar{\mathbf{X}}(\alpha)\bar{\mathbf{S}}(\alpha)\mathbf{1} \;=\; \alpha^{-1}\mathbf{1},\; \mathbf{X}(\alpha)'\mathbf{S}(\alpha) \;=\; \tfrac{n}{\alpha},$$

where $\bar{\mathbf{X}}$ and $\bar{\mathbf{S}}$ are respectively diagonal matrices containing the components of $\mathbf{X}$ and $\mathbf{S}$. The path $\{\mathbf{X}(\alpha),\ \mathbf{S}(\alpha)\}$ is thus the central trajectory (see the discussion underlying (13.40)).

In order for their algorithm to effectively minimize the primal-dual potential function ϕ_ρ, Gonzaga and Todd note that the projected gradient of ϕ_ρ must be large so that ϕ_ρ decreases substantially at each iteration. And this will occur if the algorithm takes us far from the central path where the projected gradients of the primal and dual penalized functions tend to be small. If fact, the authors verify for potential function (13.76) that the primal and dual projected gradients can never be small. That is, they prove that, given $(\mathbf{X},\mathbf{S})\in\mathcal{U}^+$, we cannot simultaneously have $\mathbf{X}$ close to $\mathbf{X}(\alpha)$ and $\mathbf{S}$ close to $\mathbf{S}(\alpha)$ for $\alpha = \frac{\rho}{\mathbf{X'S}}$. Hence constant reduction in ϕ_ρ can be achieved at each iteration by taking a step in either primal or dual space using separate steplengths in each.

Two important elements of the Gonzaga and Todd algorithm are *scaling* and *measuring centrality*. First, for $(\mathbf{X},\mathbf{S})\in\mathcal{U}^+$, let $\bar{\mathbf{X}} = diag(\mathbf{X})$ so that $\mathbf{X}$ is transformed into $\mathbf{1}$. Hence $\bar{\mathbf{X}}$ is used to scale $\mathbf{X}$ when considering changes in $\mathbf{X}$ and $\bar{\mathbf{S}}^{-1}$ scales $\mathbf{S}$ when contemplating changes in $\mathbf{S}$. (Hence scaling allows us to assume that the current $\mathbf{X}$ (or $\mathbf{S}$) iterate is $\mathbf{1}$.)

Next, we can effectively measure the distance from $\mathbf{X} \in \mathcal{K}_P^+$ to the α-center of $\mathcal{K}_P^+$ by considering the norm of the (scaled) projected gradient of f_α^P. Specifically, the measure of centrality of $\mathbf{X} \in \mathcal{K}_P^+$ to the α-center $\mathbf{X}(\alpha)$ is

$$\delta^P(\mathbf{X}, \alpha) = \left\| \mathbf{P}_{\mathbf{A}\bar{\mathbf{X}}} \bar{\mathbf{X}} \nabla f_\alpha^P(\mathbf{X}) \right\|;$$

and the measure of centrality of $\mathbf{S} \in \mathcal{K}_D^+$ to the α-center $\mathbf{S}(\alpha)$ is

$$\delta^D(\mathbf{S}, \alpha) = \left\| \mathbf{P}_{\mathbf{F}\bar{\mathbf{S}}} \bar{\mathbf{S}} \nabla f_\alpha^D(\mathbf{S}) \right\|.$$

It was mentioned above that, for a suitable choice of α, we cannot simultaneously have $\mathbf{X}$ close to $\mathbf{X}(\alpha)$ and $\mathbf{S}$ close to $\mathbf{S}(\alpha)$. In fact, according to Gonzaga and Todd (1992) (their corollary 2.1, p. 354): let $(\mathbf{X}, \mathbf{S}) \in \mathcal{U}^+$ and let $\alpha = \rho / \mathbf{X}'\mathbf{S}$, where $\rho = n + v\sqrt{n}$, $v > 0$, and let $\Delta > 0$ be such that $\Delta \leq \frac{1}{4}$ and $\Delta < \frac{v}{6}$. Then we cannot have both $\delta^P(\mathbf{X}, \alpha) \leq \Delta$ and $\delta^D(\mathbf{S}, \alpha) \leq \Delta$. This result implies that, when $\rho = n + v\sqrt{n}$ with $v \geq 2$, we cannot have both $\mathbf{P}_{\mathbf{A}\bar{\mathbf{X}}} \bar{\mathbf{X}} \nabla_{\mathbf{X}} \phi_\rho(\mathbf{X}, \mathbf{S})$ and $\mathbf{P}_{\mathbf{F}\bar{\mathbf{S}}} \bar{\mathbf{S}} \nabla_{\mathbf{S}} \phi_\rho(\mathbf{X}, \mathbf{S})$ with norms at most $\frac{1}{4}$. Thus the potential function ϕ_ρ can be decreased substantially by taking a step in either primal space or dual space. This observation consequently leads to the Gonzaga and Todd (1992)

13.22. ALGORITHM.

STEP 0. Initialize.

Choose $(\mathbf{X}_0, \mathbf{S}_0) \in \mathcal{U}^+$ and termination parameter $\eta > 0$ (to obtain the promised polynomial bound, $(\mathbf{X}_0, \mathbf{S}_0)$ must be chosen with $\phi_\rho(\mathbf{X}_0, \mathbf{S}_0) = O(\sqrt{n}\, L)$ and η should be 2^{-2L}). Let $\rho = n + v\sqrt{n}$ for $v \geq 2$ and set $k = 0$.

STEP 1. Set $(\mathbf{X}, \mathbf{S})$ to $(\mathbf{X}_k, \mathbf{S}_k)$ and select either a primal or dual step. A primal step can be chosen only if

$$\left\| \mathbf{P}_{\mathbf{A}\bar{\mathbf{X}}} \bar{\mathbf{X}} \nabla_{\mathbf{X}} \phi_\rho(\mathbf{X}, \mathbf{S}) \right\| \geq \tfrac{1}{4};$$

a dual step only if

$$\left\| \mathbf{P}_{\mathbf{F}\bar{\mathbf{S}}} \bar{\mathbf{S}} \nabla_{\mathbf{S}} \phi_\rho(\mathbf{X}, \mathbf{S}) \right\| \geq \tfrac{1}{4}.$$

Choose either:

(a) Primal Step

$\widehat{\mathbf{X}} = \mathbf{X} - \beta\bar{\mathbf{X}}\mathbf{P}_{\mathbf{A}\bar{\mathbf{X}}}\bar{\mathbf{X}}\,\nabla_{\mathbf{X}}\phi_\rho(\mathbf{X},\mathbf{S}),$

$\widehat{\mathbf{S}} = \mathbf{S}$; or

(b) Dual Step

$\widehat{\mathbf{X}} = \mathbf{X}$

$\widehat{\mathbf{S}} = \mathbf{S} - \beta\bar{\mathbf{S}}\mathbf{P}_{\mathbf{F}\bar{\mathbf{S}}}\bar{\mathbf{S}}\,\nabla_{\mathbf{S}}\phi_\rho(\mathbf{X},\mathbf{S}),$

where $\beta > 0$ is chosen as

$$\left(6\left\| \mathbf{P}_{\mathbf{A}\bar{\mathbf{X}}}\bar{\mathbf{X}}\,\nabla_{\mathbf{X}}\phi_\rho(\mathbf{X},\mathbf{S})\right\|\right)^{-1} \text{ or } \left(6\left\| \mathbf{P}_{\mathbf{F}\bar{\mathbf{S}}}\bar{\mathbf{S}}\,\nabla_{\mathbf{S}}\phi_\rho(\mathbf{X},\mathbf{S})\right\|\right)^{-1}$$

according to whether a primal or dual step is taken.

STEP 2. Check for optimality.

If $\mathbf{X}_k'\mathbf{S}_k \leq \eta$, stop; otherwise proceed to

STEP 3. Advance K. to $K+1$

Set $(\widehat{\mathbf{X}},\widehat{\mathbf{S}})$ to $(\mathbf{X}_{k+1},\mathbf{S}_{k+1})$.

Return to step 1.

As a practical matter, since the conditions for taking a primal or a dual step can always be met (by virtue of the aforementioned corollary 2.1 of Gonzaga and Todd (1992)), one can choose a primal or dual step corresponding to the larger projected gradient in step 1. In addition, if β is selected according to the criterion stated in step 1, then $\phi_\rho(\widehat{\mathbf{X}},\widehat{\mathbf{S}}) \leq \phi_\rho(\mathbf{X},\mathbf{S}) - \frac{1}{40}$.

Finally, Gonzaga and Todd note that their algorithm is more symmetric than that of Ye (1990a) (his $O(n^3L)$ potential reduction algorithm) which either takes a primal step or updates the dual, but it is less so relative to the primal dual potential reduction algorithm of Kojima, *et al.* (1991).

An affine potential reduction algorithm for solving linear programs that simultaneously seeks feasibility and optimality is provided by Todd (1993). Based upon the affine potential reduction methods of Anstreicher (1989), Gonzaga (1990), Ye (1992), and Freund (1991a), Todd's routine utilizes two potential functions — one driving an artificial variable to zero in order to attain feasibility and the other moving the objective value toward a lower bound — which are decreased simultaneously. Specifically, Todd uses a two-variable linear program at each iteration to obtain a better current lower bound on the objective value. Moreover, his method does not require that an initial lower bound be given since the algorithm actually generates one. In fact, progress is made in the phase-II potential function even before such a lower bound is attained. By using a subproblem to find a search direction that treats the phase-I and phase-II objectives more symmetrically than similar algorithms, the phase-I objective decreases faster than its phase-II counterpart, so that feasibility can be attained (phase-I is terminated) exactly before optimality (Todd(1993)).

A rather novel algorithm that solves a standard form primal linear program ((13.14)) directly from an infeasible *warm start*, *i.e.*, directly from an infeasible solution $\widehat{\mathbf{X}}$ satisfying $\mathbf{A}\widehat{\mathbf{X}} = \mathbf{b}$ but not $\widehat{\mathbf{X}} \geqq \mathbf{0}$, has been developed by Freund (1991b). By starting from an initial infeasible solution, Freund's algorithm dispenses with the need for a phase-I procedure involving the introduction of artificial variables and generates a sequence of primal solutions that become increasingly less infeasible and increasingly more optimal, and that coverage to an optimal feasible solution. Derived as an altered version of Ye's algorithm (13.14) (Ye (1990a), Freund (1991a)), Freund offers a potential reduction routine that utilizes affine scaling and employs a potential function of the form

$$F(\mathbf{X}, B) = (n + \sqrt{n}) \ln(\mathbf{C}'\mathbf{X} - B) - \Sigma_{j=1}^{n} \ln\Big(x_j + h_j(\mathbf{C}'\mathbf{X} - B)\Big),$$

where B is a lower bound on the optimal value of the primal objective function and h_j is the jth component of a strictly positive ***shift vector*** $\mathbf{h} \epsilon \mathbf{R}^n$ which, with the aid of the duality gap $\mathbf{C}'\mathbf{X} - B$, is used to shift the non-negativity conditions from $\mathbf{X} \geqq \mathbf{0}$ to $\mathbf{X} + \in \mathbf{h} \geqq \mathbf{0}$, $0 < \in \epsilon R$. Under suitable conditions (*e.g.*, $\mathbf{A}$ has full row rank; sets of optimal primal and dual solutions are nonempty and bounded; we have an initial $\widehat{\mathbf{X}}$ for which $\mathbf{A}\widehat{\mathbf{X}} = \mathbf{0}$, $\widehat{\mathbf{X}} \not\geqq \mathbf{0}$; we know a lower bound $\widehat{B}$ on z^*, the optimal primal objective value; and the dual feasible region is bounded and that a lower bound on the size of the dual feasible region is known in advance), Freund's algorithm achieves a constant decrease in the duality gap (and in infeasibility) in $O(n)$ iterations and is designed to solve the potential function minimization problem

$$\begin{aligned} \min_{\mathbf{X},B} F(\mathbf{X},B) &= (n+\sqrt{n}) \ln(\mathbf{C}'\mathbf{X}-B) - \Sigma_{j=1}^{n} \ln\Big(x_j + h_j(\mathbf{C}'\mathbf{X}-B)\Big) \quad \text{s.t.} \\ \mathbf{AX} &= \mathbf{b} \\ & \mathbf{X} + \mathbf{h}(\mathbf{C}'\mathbf{X}-B) > \mathbf{0} \\ & B \leq z^*. \end{aligned} \qquad (13.79)$$

(The last inequality in (13.79) is equivalent to the condition that $B \leq \mathbf{b}'\mathbf{U}$ for some dual feasible solution $(\mathbf{U},\mathbf{S})$.)

The algorithm computes a projected gradient $\bar{\mathbf{d}}$ that is used to take a step in the negative of the primal direction, where the steplength α can be determined analytically or by a line search of the potential function. As in Ye's $O(n^3L)$ potential reduction algorithm (Ye (1990a)), if $\bar{\mathbf{d}}$ is large $\Big(e.g., \|\bar{\mathbf{d}}\| \geq \gamma\ (=0.5)\Big)$, the algorithm takes a primal step; otherwise, if $\|\bar{\mathbf{d}}\| < \gamma$, the algorithm updates the lower bound $\bar{B}$ by taking a dual step and thus computing new dual variables. Whenever the algorithm takes its first dual step it generates a dual feasible solution. The algorithm is designed to reduce the potential function by at least $\frac{1}{12}$ at each iteration.

A modification of (13.79) that includes a barrier term for the dual slack variables is

$$\begin{aligned} \min_{\mathbf{X},\mathbf{U},\mathbf{S},B} H(\mathbf{X},\mathbf{S},B) &= (n+\sqrt{n}) \ln(\mathbf{C}'\mathbf{X}-B) \\ & - \Sigma_{j=1}^{n} \ln\Big(x_j + h_j(\mathbf{C}'\mathbf{X}-B)\Big) - \Sigma_{j=1}^{n} \ln s_j \quad \text{s.t.} \\ \mathbf{AX} &= \mathbf{b} \\ & \mathbf{X} + \mathbf{h}(\mathbf{C}'\mathbf{X}-B) > \mathbf{0} \\ \mathbf{A}'\mathbf{U} + \mathbf{S} &= \mathbf{C} \\ & \mathbf{S} > \mathbf{0} \\ \mathbf{b}'\mathbf{U} &= B. \end{aligned} \qquad (13.80)$$

To solve this revised potential function minimization problem Freund introduces a modified algorithm requiring an initial dual feasible solution $(\widehat{\mathbf{U}},\widehat{\mathbf{S}})$ and an initial primal lower bound $\widehat{B} = \mathbf{b}'\widehat{\mathbf{U}}$. Under a set of assumptions more restrictive than those given above to solve (13.79) (*i.e.*, we have a known initial interior dual feasible solution $(\widehat{\mathbf{U}}, \widehat{\mathbf{S}})$ for which $\widehat{\mathbf{U}}, \widehat{\mathbf{S}}$ are feasible in (13.80) so that $\widehat{\mathbf{S}} > \mathbf{0}$, $\widehat{\mathbf{X}} + \mathbf{h}(\widehat{\mathbf{X}}'\widehat{\mathbf{S}}) > \mathbf{0}$; and a bound on the set of all dual feasible slack vectors $\mathbf{S}$ is known), the iterates of Freund's refined algorithm, corresponding to improved values of both primal and dual variables, decrease H by at least 0.4 at each iteration, thus leading to a complexity of $O(\sqrt{n})$ iterations to achieve a constant decrease in the duality gap $\mathbf{C}'\mathbf{X} - B$.

We close this section by mentioning that Mizuno and Nagasawa (1993) offer a potential-reduction algorithm which employs the primal-dual affine scaling direction as a search direction. Using the potential function of Todd and Ye (1990), the authors choose a steplength at each iteration of the algorithm such that the potential function does not increase, thus admitting a step size longer than that which minimizes the potential function. Computational complexity is based on the reduction of the duality gap at each iteration and their algorithm is proved to be polynomial-time bounded. Mizuno and Nagasawa also propose a low-complexity routine in which a centering direction is used whenever an iteration is far from the central path.

13.12. Methods Related to Karmarkar's Routine: A Homogeneous and Self-Dual Interior Point Method

Given a pair of primal and dual problems of the (standard) form

$$\text{PRIMAL: } \min\ \mathbf{C}'\mathbf{X} \ \text{ s.t. } \ \mathbf{A}\mathbf{X} = \mathbf{b},\ \mathbf{X} \geqq \mathbf{0}; \tag{(13.14)}$$

$$\text{DUAL: } \max\ \mathbf{b}'\mathbf{U} \ \text{ s.t. } \ \mathbf{A}'\mathbf{U} \leqq \mathbf{C},\ \mathbf{U} \text{ unrestricted} \tag{(13.17.1)}$$

respectively, virtually all of the interior point algorithms examined above solve, say, the primal problem under the regularity assumption $\mathcal{K}_{++} = \{(\mathbf{X}, \mathbf{S}) \mid \mathbf{0} < (\mathbf{X}, \mathbf{S}) \in R^{2n},\ \ \mathbf{A}\mathbf{X} = \mathbf{b},\ \ \mathbf{A}'\mathbf{U} + \mathbf{S} = \mathbf{C} \text{ for some } \mathbf{U} \in R^m\} \neq \phi$ or the optimal solution set for both the primal and dual problems is bounded. However, since no *a priori* information on the status of $\mathcal{K}_{++}$ is available, the usual procedure is to explicitly bound the feasible region of the linear program. (With integer data, a worst-case bound of 2^L is chosen in practice.) This issue, along with the parallel problem of finding ways to accomplish the difficult task of obtaining a strictly feasible initial point for starting an interior-point algorithm, has been addressed by Ye, Todd, and Mizuno (1994).

Beginning with the idealized notion that a complete linear programming algorithm should be versatile enough to: (1) detect whether a linear program is infeasible or unbounded; and (2) generate a finite optimal solution if one exists, these authors present a homogeneous and self-dual linear programming algorithm having $O(\sqrt{n}\,L)$ iteration complexity. Specifically, the algorithm: (a) solves a linear program without regularity assumptions concerning the existence of optimal, feasible, or interior feasible solutions; (b) can start at any positive primal-dual pair, feasible or infeasible, near the central ray of the positive orthant, and does not use any M-penalty parameter or lower bound; (c) generates a sequence of iterates that approaches feasibility and optimality simultaneously (provided a solution exists) and if the problem is infeasible or unbounded, correctly detects infeasibility for at least one of the primal and dual problems; and (d) at each iteration solves a linear equation system whose dimension is virtually the same as those solved by the standard primal-dual iterior-point routines.

In particular, for a self-dual linear program, its dual is equivalent to its primal. That is, if $\mathbf{A}$ is a $(p \times p)$ skew-symmetric matrix (*i.e.*, $\mathbf{A} = -\mathbf{A}'$) and $\mathbf{b} = -\mathbf{C} \in \mathbf{R}^p$, then it is easily shown that the program

$$\begin{aligned} &min\,\mathbf{C}'\mathbf{U} \quad \text{s.t.} \\ &\mathbf{AU} \geqq \mathbf{b},\ \mathbf{U} \geqq \mathbf{0} \end{aligned}$$

is equivalent to its dual. In this regard, if $\mathbf{U}$ is feasible in the preceding problem, then it is also feasible in the dual problem, with the two objective values summing to zero. Hence this program has an optimal solution value of zero. Clearly a distinct advantage of self-duality is that we can apply a primal-dual interior point algorithm to solve a self-dual linear program without doubling the dimension of the linear equation system solved at each iteration.

Ye, *et al.* (1994) consider a homogeneous and self-dual artificial linear program relating the above primal and dual problems. Given any $\mathbf{X}_0 > \mathbf{0}$, $\mathbf{S}_0 > \mathbf{0}$, and $\mathbf{U}_0$, they formulate the problem

$$
\begin{array}{llr}
 & min(\mathbf{X}_0'\mathbf{S}_0 + 1)\,\theta \quad \text{s.t.} & \\
\text{(a)} & \mathbf{A}\mathbf{X} - \mathbf{b}\tau + \bar{\mathbf{b}}\theta = \mathbf{0} & \\
\text{(b)} & -\mathbf{A}'\mathbf{U} + \mathbf{C}\tau - \bar{\mathbf{C}}\theta \geqq \mathbf{0} & (13.81)\\
\text{(c)} & \mathbf{b}'\mathbf{U} - \mathbf{C}'\mathbf{X} + \bar{z}\theta \geq 0 & \\
\text{(d)} & -\bar{\mathbf{b}}'\mathbf{U} + \bar{\mathbf{C}}'\mathbf{X} - \bar{z}\tau = -\mathbf{X}_0'\mathbf{S}_0 - 1 & \\
 & \mathbf{U} \text{ unrestricted, } \mathbf{X} \geqq \mathbf{0},\ \tau \geq 0,\ \theta \text{ unrestricted,} &
\end{array}
$$

where

$$\bar{\mathbf{b}} = \mathbf{b} - \mathbf{A}\mathbf{X}_0,\ \bar{\mathbf{C}} = \mathbf{C} - \mathbf{A}'\mathbf{U}_0 - \mathbf{S}_0,\ \bar{z} = \mathbf{C}'\mathbf{X}_0 + 1 - \mathbf{b}'\mathbf{U}_0.$$

Here $\bar{\mathbf{b}}$, $\bar{\mathbf{C}}$, and $\bar{z}$ represent, respectively, the *infeasibility* of the initial primal point, dual point, and duality gap. It is easily shown that the constraints of (13.81) constitute a skew-symmetric system.

Expressions (13.81.a-c) with $\tau = 1$, $\theta = 0$, and $\mathbf{X} \geqq \mathbf{0}$ depict primal and dual feasibility and reversed weak duality and, taken together, define primal and dual optimality. Making τ a variable, thereby homogenizing (13.81.a,b), adds the requisite variable complementary to (13.81.c). To achieve feasibility for $\mathbf{X} = \mathbf{X}_0$ and $(\mathbf{U},\mathbf{S}) = (\mathbf{U}_0,\mathbf{S}_0)$, Ye, *et al.* introduce the artificial variable θ with *infeasibility coefficients* $\bar{\mathbf{b}}$, $-\bar{\mathbf{C}}$, and $\bar{z}$. Finally, (13.81.d) is a normalizing constraint added to achieve self-duality. So while (13.81) is characterized as homogeneous, it nevertheless admits (13.81.d) as a single inhomogeneous constraint.

Let $\mathbf{S}$ depict the $(n \times 1)$ vector of slack variables for (13.81.b) and κ represent the slack scalar for (13.81.c). In addition, let $\mathcal{K}_h$ denote the set of feasible points $(\mathbf{U},\mathbf{X},\tau,\theta,\mathbf{S},\kappa)$ for (13.81) with $\mathcal{K}_h^0 = \{(\mathbf{X},\tau,\mathbf{S},\kappa) \mid \mathbf{0} < (\mathbf{X},\tau,\mathbf{S},\kappa) \in \mathcal{K}_h\}$ the set of strictly feasible points of $\mathcal{K}_h$.

Looking to the dual of (13.81), let $\widehat{\mathbf{U}}$ and $\widehat{\mathbf{X}}$ be the vectors of dual variables complementary to constraints (13.81.a,b) respectively and let $\hat{\tau}$ and $\hat{\theta}$ be the dual variables for constraints (13.81.c,d) respectively. By definition, the dual of (13.81) must simply be (13.81) itself expressed in terms of $\widehat{\mathbf{U}}$, $\widehat{\mathbf{X}}$, $\widehat{\tau}$, and $\widehat{\theta}$. Trivially, any feasible point of (13.81) must be a feasible point of its dual and vice versa.

To establish the connection between (13.81) and (13.14) (respectively (13.17.1)), Ye, *et al.* (1994) point out that if $\mathbf{X}_0$, $(\mathbf{U}_0, \mathbf{S}_0)$ is feasible in (13.14) ((13.17.1)), then $\bar{\mathbf{b}}$ $(\bar{\mathbf{C}})$ is zero, and thus every feasible solution to (13.81) with $\tau > 0$ has $\mathbf{X}/\tau$ feasible in (13.14) ($(\mathbf{U}, \mathbf{S})/\tau$ is feasible in (13.17.1)). Conversely, if $\bar{z} < 0$, then every feasible solution to (13.81) with $\theta > 0$ and $\tau > 0$ has $\mathbf{C}'\mathbf{X} - \mathbf{b}'\mathbf{U} \leq \bar{z}\theta < 0$ so that either $\mathbf{X}/\tau$ or $(\mathbf{U}, \mathbf{S})/\tau$ must be infeasible. Moreover, Ye, *et al.* also verify (essentially their theorem 2, pp. 56-57) that:

a. (13.81) has a strictly feasible point $\mathbf{U} = \mathbf{U}_0$, $\mathbf{X} = \mathbf{X}_0 > \mathbf{0}$, $\tau = 1$, $\theta = 1$, $\mathbf{S} = \mathbf{S}_0 > \mathbf{0}$, and $\kappa = 1$.

b. (13.81) has an optimal solution and its optimal solution set is bounded.

c. The optimal value of (13.81) is zero, and for any feasible point $(\mathbf{U}, \mathbf{X}, \tau, \theta, \mathbf{S}, \kappa) \in \mathcal{K}_h$, $(\mathbf{X}_0'\mathbf{S}_0 + 1)\theta = \mathbf{X}'\mathbf{S} + \tau\kappa$.

d. There is an optimal solution $(\mathbf{U}_*, \mathbf{X}_*, \tau_*, \theta_* = 0, \mathbf{S}_*, \kappa_*) \in \mathcal{K}_h$ that is ***strictly self-complementary***, *i.e.*,

$$\begin{pmatrix} \mathbf{X}_* + \mathbf{S}_* \\ \tau_* + \kappa_* \end{pmatrix} > \mathbf{0}$$

(any such solution for (13.81) must also be one for its dual and vice versa).

Based upon these considerations, Ye, *et al.* choose $\mathbf{U}_0 = \mathbf{0}$, $\mathbf{X}_0 = \mathbf{1}$, and $\mathbf{S}_0 = \mathbf{1}$, to wit (13.81) becomes

$$\begin{array}{llr} & min\ (n+1)\theta \quad \text{s.t.} & \\ \text{(a)} & \mathbf{AX} - \mathbf{b}\tau + \bar{\mathbf{b}}\theta = \mathbf{0} & \\ \text{(b)} & -\mathbf{A}'\mathbf{U} + \mathbf{C}\tau - \bar{\mathbf{C}}\theta - \mathbf{S} = \mathbf{0} & \\ \text{(c)} & \mathbf{b}'\mathbf{U} - \mathbf{C}'\mathbf{X} + \bar{z}\theta - \kappa = 0 & \\ \text{(d)} & -\bar{\mathbf{b}}'\mathbf{U} + \bar{\mathbf{C}}'\mathbf{X} - \bar{z}\tau = -(n+1) & \end{array} \tag{13.82}$$

$\mathbf{U}$ unrestricted, $\mathbf{X} \geqq \mathbf{0}$, $\tau \geq 0$, θ unrestricted, $\mathbf{S} \geqq \mathbf{0}$, $\kappa \geq 0$,

where

$$\bar{\mathbf{b}} = \mathbf{b} - \mathbf{A1},\ \bar{\mathbf{C}} = \mathbf{C} - \mathbf{1},\ \text{and } \bar{z} = \mathbf{C}'\mathbf{1} + 1.$$

We may combine the constraints of (13.82) and conveniently rewrite (13.82.d) as

$$\mathbf{1}'\mathbf{X} + \mathbf{1}'\mathbf{S} + \tau + \kappa - (n+1)\theta = n + 1.$$

We can now relate optimal solutions for (13.82) to those of (13.14), (13.17.1). Specifically, we have

13.23. THEOREM [Ye, *et al.* (1994)]. Let $(\mathbf{U}_*, \mathbf{X}_*, \tau_*, \theta_* = 0,\ \mathbf{S}_*, \kappa_*)$ be a strictly self-complementary solution for (13.81). Then:

(i) (13.14) has a feasible and bounded solution if and only if $\tau_* > 0$ so that $\mathbf{X}_*/\tau_*$ is an optimal solution for (13.14) and $(\mathbf{U}_*/\tau_*, \mathbf{S}_*/\tau_*)$ is an optimal solution for (13.17.1).

(ii) If $\tau_* = 0$, then $\kappa_* > 0$ and thus $\mathbf{C}'\mathbf{X}_* - \mathbf{b}'\mathbf{U}_* < 0$, *i.e.*, at least one of $\mathbf{C}'\mathbf{X}$ and $-\mathbf{b}'\mathbf{U}_*$ is strictly less than zero. If $\mathbf{C}'\mathbf{X}_* < 0$, then (13.17.1) is infeasible; if $-\mathbf{b}'\mathbf{U}_* < 0$, then (13.14) is infeasible; and if both $\mathbf{C}'\mathbf{X}_*, -\mathbf{b}'\mathbf{U}_* < 0$, then both (13.14) and (13.17.1) are infeasible.

Moreover,

13.24. COROLLARY [Ye, *et al.* (1994)]. Let $(\bar{\mathbf{U}}, \bar{\mathbf{X}}, \bar{\tau}, \bar{\theta} = 0,\ \bar{\mathbf{S}}, \bar{\kappa})$ be any optimal solution for (13.82). Then if $\bar{\kappa} > 0$, either (13.14) or (13.17.1) is infeasible.

If the primal and dual pair of problems is written in symmetric form as

$$\begin{array}{lll} \text{PRIMAL: } & \min \mathbf{C}'\mathbf{X} & \text{s.t. } \mathbf{AX} - \mathbf{t} = \mathbf{b},\ \mathbf{X} \geqq \mathbf{0},\ \mathbf{t} \geqq \mathbf{0}; \\ \text{DUAL: } & \max \mathbf{b}'\mathbf{U} & \text{s.t. } \mathbf{A}'\mathbf{U} + \mathbf{S} = \mathbf{C},\ \mathbf{U} \geqq \mathbf{0},\ \mathbf{S} \geqq \mathbf{0}, \end{array}$$

where $\mathbf{t}, \mathbf{S}$ denote $(m \times 1)$ and $(n \times 1)$ vectors of primal surplus and dual slack variables respectively, then, given any $\mathbf{X}_0 > \mathbf{0}$, $\mathbf{t}_0 > \mathbf{0}$, $\mathbf{S}_0 > \mathbf{0}$, and $\mathbf{U}_0 > \mathbf{0}$, we may write the homogeneous and self-dual program as

$$\min\ (\mathbf{X}_0'\mathbf{S}_0 + \mathbf{t}_0'\mathbf{U}_0 + 1)\,\theta \qquad \text{s.t.}$$

$$\begin{array}{lllll} \text{(a)} & & \mathbf{AX} - \mathbf{b}\tau + \bar{\mathbf{b}}\theta & \geqq & \mathbf{0} \\ \text{(b)} & -\mathbf{A}'\mathbf{U} & + \mathbf{C}\tau - \bar{\mathbf{C}}\theta & \geqq & \mathbf{0} \\ \text{(c)} & \mathbf{b}'\mathbf{U} - \mathbf{C}'\mathbf{X} & + \bar{z}\theta & \geq & 0 \\ \text{(d)} & -\bar{\mathbf{b}}'\mathbf{U} + \bar{\mathbf{C}}'\mathbf{X} - \bar{z}\tau & & = & -\mathbf{X}_0'\mathbf{S}_0 - \mathbf{t}_0'\mathbf{U}_0 - 1 \\ & \multicolumn{4}{l}{\mathbf{U} \geqq \mathbf{0},\ \mathbf{X} \geqq \mathbf{0},\ \tau \geq 0,\ \theta \text{ unrestricted,}} \end{array} \qquad (13.83)$$

where $\bar{\mathbf{b}} = \mathbf{b} - \mathbf{A}\mathbf{X}_0 + \mathbf{t}_0$, $\bar{\mathbf{C}} = \mathbf{C} - \mathbf{A}'\mathbf{U}_0 - \mathbf{S}_0$, and $\bar{z}_0 = \mathbf{C}'\mathbf{X}_0 + 1 - \mathbf{b}'\mathbf{U}_0$.

The preceding collection of homogeneous and self-dual programs do not require any particular or stylized algorithm for their solution. However, Ye, *et al.* (1994) indicate that interior-point algorithms that generate a strictly self-complementary solution (see Güler and Ye (1993)) can be utilized. For instance, one can conveniently adapt the preceding central path routine (see section 13.10) to, say, (13.82) by employing

13.25. THEOREM [Ye, *et al.* (1994)]. (i) For any $\mu > 0$, there is a unique point $(\mathbf{U}, \mathbf{X}, \tau, \theta, \mathbf{S}, \kappa) \in \mathcal{K}_h^0$ such that

$$\begin{pmatrix} \bar{\mathbf{X}}\mathbf{S} \\ \tau\kappa \end{pmatrix} > \mu\mathbf{1},$$

where $\bar{\mathbf{X}} = diag(\mathbf{X})$. (ii) Let $\mathbf{M}$ be the coefficient matrix of (13.82.a-d) and let $N(\mathbf{M})$ denote its null-space. If the direction vector $\mathbf{d} = (\mathbf{d}^U, \mathbf{d}^X, d^\tau, d^\theta, \mathbf{d}^S, \mathbf{d}^\kappa) \in N(\mathbf{M})$, *i.e.*, $\mathbf{Md} = \mathbf{0}$, then

$$(\mathbf{d}^X)'\mathbf{d}^S + d^\tau d^\kappa = 0.$$

The importance of this theorem is that it defines the ***self-central path*** for (13.82) or

$$\Gamma_h = \left\{ (\mathbf{U}, \mathbf{X}, \tau, \theta, \mathbf{S}, \kappa) \in \mathcal{K}_h^0 \,\middle|\, \begin{pmatrix} \bar{\mathbf{X}}\mathbf{S} \\ \tau\kappa \end{pmatrix} = \frac{\mathbf{X}'\mathbf{S} + \tau\kappa}{n+1}\mathbf{1} \right\}.$$

If $\bar{\mathbf{X}}_0\mathbf{S}_0 = \mathbf{1}$, $\bar{\mathbf{X}}_0 = diag(\mathbf{X}_0)$, then the initial interior feasible point $\mathbf{U} = \mathbf{U}_0$, $\mathbf{X} = \mathbf{X}_0 > \mathbf{0}$, $\mathbf{S} = \mathbf{S}_0 > \mathbf{0}$, $\tau = \theta = \kappa = 1$ is on Γ_h with $\mu = 1$. Our earlier selection of $\mathbf{U}_0 = \mathbf{0}$, $\mathbf{X}_0 = \mathbf{S}_0 = \mathbf{1}$ meets this requirement. A neighborhood of the self-central path can be defined, for some $\beta \in (0, 1)$, as

$$\delta(\beta) = \left\{ (\mathbf{U}, \mathbf{X}, \tau, \theta, \mathbf{S}, \kappa) \in \mathcal{K}_h^0 \,\middle|\, \left\| \begin{pmatrix} \bar{\mathbf{X}}\mathbf{S} \\ \tau\kappa \end{pmatrix} - \mu\mathbf{1} \right\| \le \beta\mu,\ \mu = \frac{\mathbf{X}'\mathbf{S} + \tau\kappa}{n+1} \right\}.$$

Armed with the results of the preceding theorem, Ye, *et al.* (1994) apply the predictor-corrector algorithm (Mizuno, *et al.* (1993)) to (13.82) along with the termination technique introduced by Ye (1992) (see also Mehrotra and Ye (1991)) and a least squares projection that produces a strictly self-complementary optimal solution. Specifically, given an interior feasible point $(\mathbf{U}_k, \mathbf{X}_k, \tau_k, \theta_k, \mathbf{S}_k, \kappa_k) \in \mathcal{K}_h^0$ and the linear system:

$$
\begin{aligned}
&\text{(a)}\ (\mathbf{d}^U,\ \mathbf{d}^X,\ d^\tau,\ d^\theta,\ \mathbf{d}^S,\ d^\kappa) \in N(\mathbf{M}), \\
&\text{(b)}\ \begin{pmatrix} \bar{\mathbf{X}}_k \mathbf{d}^S + \bar{\mathbf{S}}_k \mathbf{d}^X \\ \tau_k d^\kappa + \kappa_k d^\tau \end{pmatrix} = \gamma \mu_k \mathbf{1} - \begin{pmatrix} \bar{\mathbf{X}}_k \mathbf{S}_k \\ \tau_k \kappa_k \end{pmatrix},
\end{aligned} \tag{13.84}
$$

where $\bar{\mathbf{X}}_k = diag(\mathbf{X}_k)$ and $\bar{\mathbf{S}}_k = diag(\mathbf{S}_k)$, let us consider the

PREDICTOR STEP. For any even number k let $(\mathbf{U}_k, \mathbf{X}_k, \tau_k, \theta_k, \mathbf{S}_k, \kappa_k) \in \delta(\beta)$ with $\beta = \frac{1}{4}$. Solve (13.84) with $\gamma = 0$ and let

$$
\begin{aligned}
\mathbf{U}(\alpha) &= \mathbf{U}_k + \alpha \mathbf{d}^U & \theta(\alpha) &= \theta_k + \alpha d^\theta \\
\mathbf{X}(\alpha) &= \mathbf{X}_k + \alpha \mathbf{d}^X & \mathbf{S}(\alpha) &= \mathbf{S}_k + \alpha \mathbf{d}^S \\
\tau(\alpha) &= \tau_k + \alpha d^\tau & \kappa(\alpha) &= \kappa_k + \alpha d^\kappa.
\end{aligned}
$$

Determine the steplength adaptively from

$$
\bar{\alpha} = max\{\alpha \mid (\mathbf{U}(\alpha), \mathbf{X}(\alpha), \tau(\alpha), \theta(\alpha), \mathbf{S}(\alpha), \kappa(\alpha)) \in \delta(2\beta)\} \tag{13.85}
$$

and compute the next points from

$$
\begin{aligned}
\widehat{\mathbf{U}}_k &= \mathbf{U}(\bar{\alpha}) & \widehat{\theta}_k &= \theta(\bar{\alpha}) \\
\widehat{\mathbf{X}}_k &= \mathbf{X}(\bar{\alpha}) & \widehat{\mathbf{S}}_k &= \mathbf{S}(\bar{\alpha}) \\
\widehat{\tau}_k &= \tau(\bar{\alpha}) & \widehat{\kappa}_k &= \kappa(\bar{\alpha}).
\end{aligned}
$$

Next comes the

CORRECTOR STEP. For any odd number k solve (13.84) with $(\widehat{\mathbf{U}}_k, \widehat{\mathbf{X}}_k, \widehat{\tau}_k, \widehat{\theta}_k, \widehat{\mathbf{S}}_k, \widehat{\kappa}_k) \in \delta(2\beta)$ and $\gamma = 1$. Then form

$$
\begin{aligned}
\mathbf{U}_{k+1} &= \widehat{\mathbf{U}}_k + \widehat{\mathbf{d}}^U & \theta_{k+1} &= \widehat{\theta}_k + \widehat{d}^\theta \\
\mathbf{X}_{k+1} &= \widehat{\mathbf{X}}_k + \widehat{\mathbf{d}}^X & \mathbf{S}_{k+1} &= \widehat{\mathbf{S}}_k + \widehat{\mathbf{d}}^S \\
\tau_{k+1} &= \widehat{\tau}_k + \widehat{d}^\tau & \kappa_{k+1} &= \widehat{\kappa}_k + \widehat{d}^\kappa
\end{aligned}
$$

with $(\mathbf{U}_{k+1}, \mathbf{X}_{k+1}, \tau_{k+1}, \theta_{k+1}, \mathbf{S}_{k+1}, \kappa_{k+1}) \in \delta(\beta)$.

TERMINATION. Let index set $\mathcal{I}_k = \{j \mid x_j^k \geq s_j^k, j{=}1,\ldots,n\}$ and let $\mathbf{B}$ contain the columns $\mathbf{a}_j$ of $\mathbf{A}$ with $j \in \mathcal{I}_k$ and let $\mathbf{R}$ contain the remaining columns of $\mathbf{A}$. Then to obtain a solution that is strictly self-complementary and optimal for (13.82), consider:

Case 1. If $\tau_k \geq \kappa_k$, solve for $\mathbf{U}, \mathbf{X}_B$, and τ from

$$min \| \mathbf{U}_k - \mathbf{U} \|^2 + \| \mathbf{X}_{Bk} - \mathbf{X}_B \|^2 + (\tau_k - \tau)^2 \quad \text{s.t.}$$

$$\mathbf{B}\mathbf{X}_B - \mathbf{b}\tau = \mathbf{0}$$

$$-\mathbf{B}'\mathbf{U} + \mathbf{C}_B\tau = \mathbf{0}$$

$$\mathbf{b}'\mathbf{U} - \mathbf{C}_B'\mathbf{X}_B = 0;$$

Case 2. If $\tau_k < \kappa_k$, solve for $\mathbf{U}, \mathbf{X}_B$, and κ from

$$min \| \mathbf{U}_k - \mathbf{U} \|^2 + \| \mathbf{X}_{Bk} - \mathbf{X}_B \|^2 + (\kappa_k - \kappa)^2 \quad \text{s.t.}$$

$$\mathbf{B}\mathbf{X}_B = \mathbf{0}$$

$$-\mathbf{B}'\mathbf{U} = \mathbf{0}$$

$$\mathbf{b}'\mathbf{U} - \mathbf{C}_B'\mathbf{X}_B - \kappa = 0.$$

This projection guarantees that solutions $\mathbf{X}_B^*$ and $\mathbf{S}_R^*$ (where $\mathbf{S}_R^* = \mathbf{C}_R'\tau^* - \mathbf{R}'\mathbf{U}^*$ for case 1 and $\mathbf{S}_R^* = -\mathbf{R}'\mathbf{U}^*$ in case 2) are positive, and τ^* is positive in case 1 and κ^* is positive for case 2, as long as $\mathbf{X}_k'\mathbf{S}_k + \tau_k\kappa_k$ is sufficiently small.

If (13.14) and (13.17.1) have integer data with bit length L, then the process just described renders a strictly self-complementary solution for (13.82) in $O(\sqrt{n}\,L)$ iterations. But this implies that, within $O(\sqrt{n}\,L)$ iterations, this solution scheme also generates either optimal solutions to (13.14) and (13.17.1) or indicates that one or the other of these problems is infeasible. Problem (13.14) has an optimal solution if τ_k remains bounded away from zero for all k; it has no optimal solution if both τ_k and θ_k converge to zero at the same rate. In fact, if both (13.14) and (13.17.1) are feasible with nonempty interior, then $\tau_* > 0$ at *any* self-complementary solution $(\mathbf{U}_*, \mathbf{X}_*, \tau_*, \theta_* = 0,\ \mathbf{S}_*, \kappa_*)$ for (13.82) (Ye, *et al.* (1994)).

14 INTERIOR POINT ALGORITHMS FOR SOLVING LINEAR COMPLEMENTARITY PROBLEMS

14.1. Introduction

In chapter ten we employed a variant of the standard simplex routine, called the complementary pivot method, to generate a solution to the linear complementarity problem $LCP(\mathbf{q}, \mathbf{M})$, which we shall now express as: find an $(\mathbf{X}, \mathbf{Y}) \in R^{2n}$ satisfying

$$\begin{aligned} &(a)\ \ \mathbf{Y} = \mathbf{M}\mathbf{X} + \mathbf{q} \\ &(b)\ \ (\mathbf{X}, \mathbf{Y}) \geqq (\mathbf{0}, \mathbf{0}) \\ &(c)\ \ x_i y_i = 0, i=1,\ldots,n, \end{aligned} \tag{14.1}$$

where $\mathbf{M}$ is of order $(n \times n)$ and $\mathbf{q} \in R^n$. Here the feasible region $\mathcal{K}$ associated with (14.1), its relative interior $\mathcal{K}^0$, and the set of all solutions to (14.1) will be denoted as:

$$\mathcal{K} = \{(\mathbf{X}, \mathbf{Y}) \mid (\mathbf{X}, \mathbf{Y}) \geqq (\mathbf{0}, \mathbf{0}),\ \mathbf{Y} = \mathbf{M}\mathbf{X} + \mathbf{q}\},$$

$$\mathcal{K}^0 = \mathcal{K} \cap R^{2n}_{++} = \{(\mathbf{X}, \mathbf{Y}) \in R^{2n}_{++} \mid \mathbf{Y} = \mathbf{M}\mathbf{X} + \mathbf{q}\},$$

and

$$\mathcal{K}_{LCP} = \{(\mathbf{X}, \mathbf{Y}) \in \mathcal{K} \mid x_i y_i = 0,\ i=1,\ldots,n\}$$

respectively.

Our objective in this chapter is to explore several recent interior point algorithms for solving (14.1) or its equivalent. These algorithms are based upon the techniques developed in the preceding chapter and are designed to generate an approximate solution from which an exact solution can readily be determined. An overview of the principal approaches to solving $LCP(\mathbf{q}, \mathbf{M})$ will be given, followed by the particulars of the same. As we shall now see, virtually all of the theoretical and algorithmic machinery developed in the preceding chapter nicely carries over to the material presented herein.

One of the earliest adaptations of an interior point method to solving linear complementarity problems was introduced by Kojima, Mizuno, and Yoshise (1989a). Starting from a known initial point in a neighborhood of the central trajectory, the algorithm generates a sequence $\{(\mathbf{X}_k, \mathbf{Y}_k)\}$ near the same until we attain a sufficiently small value for a parameter μ. For some positive constant δ, each iteration decreases the error $\mathbf{X}_k'\mathbf{Y}_k$ at least linearly by $(1-\delta/\sqrt{n})$, so that the algorithm terminates within $O(\sqrt{n}L)$ iterations when an approximate solution satisfying $(\widehat{\mathbf{X}}, \widehat{\mathbf{Y}}) \in \mathcal{K}$, and $\widehat{\mathbf{X}}'\widehat{\mathbf{Y}} < 2^{-2L}$ is attained, where the size L of $LCP(\mathbf{q}, \mathbf{M})$, $\mathbf{M}$ positive semidefinite, is

$$L = \left\lfloor \sum_{i=1}^{n} \sum_{j=1}^{n} \log(\mid a_{ij} \mid +1) + \log(n^2) \right\rfloor + 1,$$

where $\lfloor x \rfloor$ is the ***largest integer*** not greater than x and a_{ij} is the i, j^{th} element of the $(n \times n+1)$ matrix $\mathbf{A} = [\mathbf{M}, \mathbf{q}]$. Each iteration computes a new point near the central trajectory by using a Newton direction obtained as the solution of a linear equation system. This output requires $O(n^3)$ arithmetic operations. Hence the total number of arithmetic operations taken over $O(\sqrt{n}L)$ iterations is $O(n^{3.5}L)$.

An algorithm closely related to the preceding central path-following algorithm is the $O(\sqrt{n}L)$ potential reduction algorithm given by Kojima, Mizuno, and Yoshise (1991b). Again considering a positive semidefinite linear complementarity problem, this algorithm reduces a potential function by a constant amount at each iteration requiring $O(n^3)$ arithmetic operations. Starting from an interior feasible solution with the potential function value bounded by $O(\sqrt{n}L)$, the algorithm generates, in at most $O(\sqrt{n}L)$ iterations, an approximate solution with potential function value equal to $-O(\sqrt{n}L)$. From this an exact solution can be found in $O(n^3)$ additional arithmetic operations. An important feature of this algorithm relative to the preceding path-following one is that the former can be initiated from an arbitrary interior feasible point, and then readily implemented using a line search along a descent direction of the potential function.

In a similar vein, Kojima, Megiddo, and Ye (1992) also offer an interior point potential reduction algorithm for solving $LCP(\mathbf{q},\mathbf{M})$ with $\mathbf{M}$ positive semidefinite. Starting with an equivalent quadratic program, these authors employ a linear scaling transformation to map a point $(\mathbf{X},\mathbf{Y})$ from original space to $(\mathbf{1},\mathbf{1})$, take a step in transformed space in the direction of steepest descent of the potential function, and then transform back to original space. This process yields a polynomial time algorithm which converges in $O(n^2L)$ iterations.

A third class of algorithms for solving linear complementarity problems, termed predictor-corrector routines, have been offered by Ye and Anstreicher (1993) and Mehrotra and Stubbs (1994). Ye and Anstreicher observe that the predictor-corrector algorithm for solving linear programs exhibits asymptotic quadratic convergence of the duality gap $\mathbf{X}_k'\mathbf{Y}_k$ to zero without any assumptions concerning nondegeneracy (a requirement for a unique solution) or the convergence of the iteration sequence $\{(\mathbf{X}_k,\mathbf{Y}_k)\}$. They then proceed to prove a similar result for the linear complementarity problem assuming only that: $\mathcal{K}^0 \neq \phi$; $\mathbf{M}$ is positive semidefinite; and that a strictly complementary solution exists (this last condition appears to be necessary for achieving superlinear convergence of the algorithm). By extending the predictor-corrector method of Mizuno, *et al.* (1993b) to the linear complementarity problem (as suggested by Ji, *et al.* (1991)), Ye and Anstreicher demonstrate that their algorithm has iteration complexity $O(\sqrt{n}L)$ and $\mathbf{X}_k'\mathbf{Y}_k \to 0$ Q-quadratically.[1]

---

[1]For a sequence of points $\{(\mathbf{X}_k,\mathbf{Y}_k)\}$ such that $(\mathbf{X}_k,\mathbf{Y}_k)\in\mathcal{K}^0$ and $\mathbf{X}_k'\mathbf{Y}_k\to 0$, the duality gap sequence $\{\mathbf{X}_k'\mathbf{Y}_k\}$ converges Q-superlinearly to zero if

$$\lim_{k\to\infty} \frac{\mathbf{X}_{k+1}'\mathbf{Y}_{k+1}}{\mathbf{X}_k'\mathbf{Y}_k} = 0;$$

and Q-quadratically to zero if

$$\limsup_{k\to\infty} \frac{\mathbf{X}_{k+1}'\mathbf{Y}_{k+1}}{(\mathbf{X}_k'\mathbf{Y}_k)^2} < +\infty.$$

Neither of these notions of convergence of $\{(\mathbf{X}_k'\mathbf{Y}_k)\}$ requires the convergence of the iteration sequence $\{(\mathbf{X}_k,\mathbf{Y}_k)\}$.

Several new predictor-corrector methods for solving $LCP(\mathbf{q}, \mathbf{M})$ have recently been developed by Mehrotra and Stubbs (1994). These methods replace the Newton predictor step with directions computed from certain specialized ball optimization and trust region problems. Although these new methods retain the basic predictor-corrector strategy, they compute alternative predictor directions at each iteration by solving subproblems defined over ellipsoidal approximations of the feasible region and solve the linear complementary problem in $O(\sqrt{n}L)$ iterations. As the authors note, since the effectiveness of a predictor-corrector method is essentially determined by the progress made in the predictor step, using alternative predictor directions to secure better approximations can ostensibly lead to improved interior point algorithms for solving linear complementarity problems.

We noted in the preceding chapter that Kojima, Megiddo, and Mizuno (1993b) demonstrate theoretical convergence of primal-dual interior point algorithms for linear programs by considering improved steplength rules (their rules G and P) which admit large steplengths independent of any line search and which rely only on the lengths of steps taken from the current iterates in primal and dual spaces to the boundaries of the primal and dual feasible regions respectively. Kojima, Kurita, and Mizuno (1993) modify the aforementioned rules and extend the analysis of Kojima, *et al.* (1993b) and Megiddo (1989b) to the construction of interior point algorithms for solving $LCP(\mathbf{q}, \mathbf{M})$, $\mathbf{M}$ positive semidefinite, on the central trajectory by moving in the Newton direction towards said path at each iteration. Their new steplength rules, Rules G' and P', ensure global convergence and $O(nL)$ iteration polynomial-time complexity respectively.

14.2. An Interior-Point, Path-Following Algorithm for $LCP(\mathbf{q}, \mathbf{M})$

To set the stage for the development of the Kojima, Mizuno, and Yoshise (1989a) path-following algorithm, let us represent the path of centers $\mathcal{K}_{cen}$ as

$$\mathcal{K}_{cen} = \{(\mathbf{X},\mathbf{Y}) \in \mathcal{K} \mid x_i y_i = \mu,\ i=1,\dots,n, \text{ for some } \mu > 0\},$$

$0 \leq \mu \in \mathbf{R}$ a parameter. In addition, it is assumed that:

(i) $n \geq 2$;

(ii) all elements of $\mathbf{M}, \mathbf{q}$ are integers;

(iii) $\mathbf{M}$ is positive semidefinite (*i.e.*, $\mathbf{X}'\mathbf{M}\mathbf{X} \geq 0$ for every $\mathbf{X} \in \mathbf{R}^n$); and

(iv) each row of $\mathbf{M}$ has at least one nonzero element.

Consider the mapping $H: \mathbf{R}_+^{1+2n} \to \mathbf{R}^{2n}$ defined by

$$H(\mu,\mathbf{X},\mathbf{Y}) = (\bar{\mathbf{X}}\bar{\mathbf{Y}}\mathbf{1}-\mu\mathbf{1},\ \mathbf{Y}-\mathbf{M}\mathbf{X}-\mathbf{q})$$

for every $(\mu,\mathbf{X},\mathbf{Y}) \in \mathbf{R}_+^{1+2n}$, where $\bar{\mathbf{X}} = diag(\mathbf{X}), \bar{\mathbf{Y}} = diag(\mathbf{Y})$. Then

$$H(\mu,\mathbf{X},\mathbf{Y}) = \mathbf{0},\ (\mathbf{X},\mathbf{Y}) \in \mathbf{R}_+^{2n},\ \mu \geq 0, \tag{14.2}$$

represents a family of systems of equations with parameter μ and thus (14.1) is equivalent to the system of equations

$$H(0,\mathbf{X},\mathbf{Y}) = \mathbf{0},\ (\mathbf{X},\mathbf{Y}) \in \mathbf{R}_+^{2n}, \tag{14.2.1}$$

i.e., $(\mathbf{X},\mathbf{Y}) \in \mathbf{R}_+^{2n}$ is a solution of $LCP(\mathbf{q},\mathbf{M})$ if and only if it is a solution of (14.2.1). Moreover, a point $(\mathbf{X},\mathbf{Y})$ satisfying (14.2) for some $\mu \geq 0$ will be termed a ***center of*** $\mathcal{K}$, and the path of centers $\mathcal{K}_{cen}$ can now be reexpressed as

$$\begin{aligned}\mathcal{K}_{cen} &= \{(\mathbf{X},\mathbf{Y}) \in \mathbf{R}_+^{2n} \mid H(\mu,\mathbf{X},\mathbf{Y}) = \mathbf{0} \text{ for some } \mu > 0\} \\ &= \{(\mathbf{X},\mathbf{Y}) \in \mathcal{K}^0 \mid \bar{\mathbf{X}}\bar{\mathbf{Y}}\mathbf{1} = \mu\mathbf{1} \text{ for some } \mu > 0\}.\end{aligned}$$

When $\mathcal{K}^0 \neq \phi$ and $\mathbf{M}$ is positive semidefinite, (14.1) has a unique solution for each $\mu > 0$ and $\mathcal{K}_{cen}$ is a smooth curve running through $\mathcal{K}^0$ to a point on the boundary of $\mathcal{K}$.

Let us next define an ***α-center neighborhood*** of the path of centers as

$$\mathcal{K}_{cen}(\alpha) = \left\{(\mathbf{X},\mathbf{Y}) \in \mathcal{K}^0 \middle| \left\| \bar{\mathbf{X}}\bar{\mathbf{Y}}\mathbf{1} - (\mathbf{X}'\mathbf{Y}/n)\mathbf{1} \right\| \leq \alpha(\mathbf{X}'\mathbf{Y}/n)\right\},$$

where $\| \bar{\mathbf{X}}\bar{\mathbf{Y}}\mathbf{1} - (\mathbf{X}'\mathbf{Y}/n)\mathbf{1} \|$ serves as a measure of the deviation of each point $(\mathbf{X},\mathbf{Y}) \in \mathcal{K}^0$ from the central trajectory. Clearly $(\mathbf{X},\mathbf{Y}) \in \mathcal{K}^0$ lies on $\mathcal{K}_{cen}$ if and only if $\| \bar{\mathbf{X}}\bar{\mathbf{Y}}\mathbf{1} - (\mathbf{X}'\mathbf{Y}/n)\mathbf{1} \| = 0$. Hence we seek to refine our approximation to $\mathcal{K}_{cen}$ in a fashion such that $\| \bar{\mathbf{X}}\bar{\mathbf{Y}}\mathbf{1} - (\mathbf{X}'\mathbf{Y}/n)\mathbf{1} \| \to 0$ as $\mathbf{X}'\mathbf{Y} \to 0$.

Given an initial (known in advance) point $(\mathbf{X},\mathbf{Y})\in\mathcal{H}_{cen}(\alpha)$ with $\mathbf{X}'\mathbf{Y}\leq 2^{O(L)}$ and a parameter $\mu>0$, let us form a local linearization of (14.2) as

$$\partial H/\partial(\mathbf{X},\mathbf{Y})\begin{bmatrix}\mathbf{d}^X\\ \mathbf{d}^Y\end{bmatrix} = H(\mu,\mathbf{X},\mathbf{Y})$$

with Jacobian matrix $\partial H/\partial(\mathbf{X},\mathbf{Y})$. Then the Newton direction at $(\mathbf{X},\mathbf{Y})$ is the solution $(\mathbf{d}^X,\mathbf{d}^Y)$ of the system

$$\begin{aligned}\bar{\mathbf{Y}}\mathbf{d}^X+\bar{\mathbf{X}}\mathbf{d}^Y &= \bar{\mathbf{X}}\bar{\mathbf{Y}}\mathbf{1}-\mu\mathbf{1}\\ -\mathbf{M}\mathbf{d}^X+\mathbf{d}^Y &= \mathbf{0}\end{aligned}\tag{14.3}$$

or

$$\begin{aligned}\mathbf{d}^X &= (\mathbf{M}+\bar{\mathbf{X}}^{-1}\bar{\mathbf{Y}})^{-1}(\bar{\mathbf{Y}}\mathbf{1}-\mu\bar{\mathbf{X}}^{-1}\mathbf{1})\\ \mathbf{d}^Y &= \mathbf{M}\mathbf{d}^X.\end{aligned}\tag{14.4}$$

(Since $\mathbf{M}$ is positive semidefinite, the directions $\mathbf{d}^X$, $\mathbf{Y}^Y$ form an acute angle, *i.e.*, $(\mathbf{d}^X)'\mathbf{M}\mathbf{d}^X\geq 0$.) The new point is then given by

$$(\widehat{\mathbf{X}},\widehat{\mathbf{Y}}) = (\mathbf{X},\mathbf{Y})-(\mathbf{d}^X,\mathbf{d}^Y)\tag{14.5}$$

with $\widehat{\mathbf{Y}} = \mathbf{M}\widehat{\mathbf{X}}+\mathbf{q}$ for any $(\mathbf{X},\mathbf{Y})\in\mathcal{H}$ and any $\mu>0$. Moreover, if $(\mathbf{X},\mathbf{Y})\in\mathcal{H}_{cen}(\alpha)$ and we choose a suitable $\mu>0$, then $(\widehat{\mathbf{X}},\widehat{\mathbf{Y}})$ remains in $\mathcal{H}_{cen}(\alpha)$. Formally, we have

14.1. THEOREM [Kojima, *et al.* (1989a)]. Let $\alpha\leq 0.2$ and $\delta=\alpha/(1-\alpha)$. Suppose that $(\mathbf{X},\mathbf{Y})\in\mathcal{H}_{cen}(\alpha)$ and $\mu = (1-\delta/\sqrt{n})\mathbf{X}'\mathbf{Y}/n$. Then the point $(\widehat{\mathbf{X}},\widehat{\mathbf{Y}})$ given by (14.4), (14.5) satisfies

$$(\widehat{\mathbf{X}},\widehat{\mathbf{Y}})\in\mathcal{H}_{cen}(\alpha),$$
$$\widehat{\mathbf{X}}'\widehat{\mathbf{Y}}\leq(1-\delta/2\sqrt{n})\mathbf{X}'\mathbf{Y}.$$

Armed with these considerations, we may now state the Kojima, *et al.* (1989a) interior point

14.2. PATH-FOLLOWING ALGORITHM FOR $LCP(\mathbf{q},\mathbf{M})$.

STEP 0. Initialize.

Choose an initial point $(\mathbf{X}_0,\mathbf{Y}_0)\in\mathcal{H}_{cen}(\alpha)$ with $\mathbf{X}_0'\mathbf{Y}_0\leq 2^{O(L)}$. Choose $0<\alpha\leq 0.2$ and $\delta=\alpha/(1-\alpha)$ and set $k=0$.

STEP 1. Let $\mu = (1-\delta/\sqrt{n})\mathbf{X}_k'\mathbf{Y}_k/n$ and set $(\mathbf{X},\mathbf{Y}) = (\mathbf{X}_k,\mathbf{Y}_k)$.

STEP 2. Determine the Search Direction.

Compute the Newton direction $(\mathbf{d}^X, \mathbf{d}^Y)$ as the solution to (14.4).

STEP 3. Take a step in the Newton Direction.

Determine the new point

$$(\mathbf{X}_{k+1}\mathbf{Y}_{k+1}) = (\mathbf{X}_k, \mathbf{Y}_k) - (\mathbf{d}^X, \mathbf{d}^Y).$$

STEP 4. Test for Optimality.

If $\mathbf{X}_{k+1}'\mathbf{Y}_{k+1} < 2^{-2L}$, stop. Otherwise set $k = k+1$ and return to step 1.

As required, the sequence $\{(\mathbf{X}_k, \mathbf{Y}_k)\}$ generated by the algorithm lies in the α-center neighborhood and the error $\mathbf{X}_k'\mathbf{Y}_k$ decreases at least linearly along the central trajectory. Once an approximate solution $(\widehat{\mathbf{X}}, \widehat{\mathbf{Y}}) \in \mathcal{K}$ satisfying $\widehat{\mathbf{X}}'\widehat{\mathbf{Y}} < 2^{-2L}$ has been found, the authors provide a mechanism for determining an exact solution of $LCP(\mathbf{q}, \mathbf{M})$ in $O(n^3)$ additional arithmetic operations (their appendix B).

While algorithm 14.2 may be viewed as an $O(n^{3.5}L)$ algorithm, Kojima, *et al.* (1989a) seek to improve its computational complexity to $O(n^3L)$ by modifying the Newton direction using approximations to the diagonal matrices $\bar{\mathbf{X}}$, $\bar{\mathbf{Y}}$. Their modification requires an average of $O(n^{2.5})$ arithmetic operations per iteration and preserves the linear convergence rate of $1-\delta/\sqrt{n}$ for some $\delta > 0$. (On all this see Kojima, *et al.* (1989a, pp. 7-10).)

Finally, Kojima, *et al.* (1989a, pp. 17-20) offer a technique for preparing the initial point $(\mathbf{X}_0, \mathbf{Y}_0)$ from which their algorithm(s) start. Then demonstrate how to construct an artificial linear complementarity problem which satisfies assumptions (i)-(iii) above and has a trivial initial point.

14.3. An Interior-Point, Potential-Reduction Algorithm for $LCP(\mathbf{q}, \mathbf{M})$

Given (14.1), let $\mathbf{M}$ be positive semidefinite with $\mathcal{K}^0 \neq \phi$ and let the potential function be of the form

$$f(\mathbf{X},\mathbf{Y}) = \sqrt{n}\log \mathbf{X}'\mathbf{Y} - \textstyle\sum_{i=1}^{n} \log(nx_iy_i/\mathbf{X}'\mathbf{Y})$$

for every $(\mathbf{X},\mathbf{Y}) \epsilon \mathcal{H}^0$. (Clearly $f(\mathbf{X},\mathbf{Y}) \geq \sqrt{n}\log \mathbf{X}'\mathbf{Y}$ for every $(\mathbf{X},\mathbf{Y}) \epsilon \mathcal{H}^0$ and thus $\mathbf{X}'\mathbf{Y} \leq 2^{-O(L)}$ if $f(\mathbf{X},\mathbf{Y}) = -O(\sqrt{n}\ L)$.)

Let us assume that a point $(\mathbf{X},\mathbf{Y}) \epsilon \mathcal{H}^0$ satisfying $f(\mathbf{X},\mathbf{Y}) \leq O(\sqrt{n}L)$ is known in advance. (If such a point is not readily available, then we can again invoke the procedure outlined in Kojima, *et al.* (1989a).) Given this point, let us determine a feasible descent direction $(\mathbf{d}^X,\mathbf{d}^Y)$ and a step parameter θ such that

$$f((\mathbf{X},\mathbf{Y}) - \theta(\mathbf{d}^X,\mathbf{d}^Y)) \leq f(\mathbf{X},\mathbf{Y}) - \delta.$$

(To accomplish this, Kojima, Mizuno, and Yoshise (1991b) suggest undertaking a line search along the direction $-(\mathbf{d}^X,\mathbf{d}^Y)$ to find either a new solution $(\mathbf{X},\mathbf{Y}) - \theta^*(\mathbf{d}^X,\mathbf{d}^Y)$, where $\theta^* = sup\{\theta \mid (\mathbf{X},\mathbf{Y}) - \theta(\mathbf{d}^X,\mathbf{d}^Y)\ \epsilon\, \mathcal{H}\}$, or an approximation to the optimal step parameter $\hat{\theta}\,\epsilon\,(0,\theta^*)$ such that

$$f(\mathbf{X},\mathbf{Y}) - \hat{\theta}(\mathbf{d}^X,\mathbf{d}^Y) \leq f(\mathbf{X},\mathbf{Y}) - \theta(\mathbf{d}^X,\mathbf{d}^Y)$$

for every $\theta\,\epsilon\,(0,\theta^*)$.) Let the new point rendered be $(\hat{\mathbf{X}},\hat{\mathbf{Y}}) = (\mathbf{X},\mathbf{Y}) - \theta(\mathbf{d}^X,\mathbf{d}^Y)\,\epsilon\,\mathcal{H}^0$. To ensure the feasibility of $(\hat{\mathbf{X}},\hat{\mathbf{Y}})$, Kojima, *et al.* (1991b) impose on the direction $(\mathbf{d}^X,\mathbf{d}^Y)$ and steplength θ the requirement that

$$\begin{gathered} \mathbf{d}^Y = \mathbf{M}\mathbf{d}^X \quad \text{and} \\ \theta\| \bar{\mathbf{X}}^{-1}\mathbf{d}^X \|_\infty \leq \tau,\ \theta\| \bar{\mathbf{Y}}^{-1}\mathbf{d}^Y \|_\infty \leq \tau \end{gathered} \tag{14.6}$$

for some $\tau\,\epsilon\,(0,1)$ and $\bar{\mathbf{X}} = diag(\mathbf{X})$, $\bar{\mathbf{Y}} = diag(\mathbf{Y})$. (Remember that for $\mathbf{V}\,\epsilon\,\mathbf{R}^n$, $\|\mathbf{V}\|_\infty = max\{\,|v_i|\,, i{=}1,\ldots,n\}$.)

How should the descent direction $(\mathbf{d}^X,\mathbf{d}^Y)$ be chosen? Kojima, *et al.* (1991b) start by demonstrating that the difference between the potential function values at the new and old points is expressible as

$$f(\hat{\mathbf{X}},\hat{\mathbf{Y}}) - f(\mathbf{X},\mathbf{Y}) \leq \theta g_1(\mathbf{d}^X,\mathbf{d}^Y) + \theta^2 g_2(\mathbf{d}^X,\mathbf{d}^Y),$$

where

$$g_1(\mathbf{d}^X,\mathbf{d}^Y) = -\frac{n+\sqrt{n}}{\mathbf{X}'\mathbf{Y}}(\mathbf{Y}'\mathbf{d}^X + \mathbf{X}'\mathbf{d}^Y) + \mathbf{1}'(\bar{\mathbf{X}}^{-1}\mathbf{d}^X + \bar{\mathbf{Y}}^{-1}\mathbf{d}^Y),$$

$$g_2(\mathbf{d}^X,\mathbf{d}^Y) = \frac{n+\sqrt{n}}{\mathbf{X}'\mathbf{Y}}(\mathbf{d}^X)'\mathbf{d}^Y + \frac{\|\bar{\mathbf{X}}^{-1}\mathbf{d}^X\|^2 + \|\bar{\mathbf{Y}}^{-1}\mathbf{d}^Y\|^2}{2(1-\tau)}.$$

Let us restrict our attention to the term $\theta g_1(\mathbf{d}^X, \mathbf{d}^Y)$ which is linear in the step parameter θ. Then $g_1(\mathbf{d}^X, \mathbf{d}^Y)$ can be rewritten as

$$g_1(\mathbf{d}^X, \mathbf{d}^Y) \;=\; (\bar{\mathbf{V}}^{-1}\mathbf{1} - \frac{n+\sqrt{n}}{\|\mathbf{V}\|^2}\,\bar{\mathbf{V}}\mathbf{1})'\mathbf{d}^U,$$

where

$$\mathbf{V}' \;=\; ((x_1y_1)^{1/2}, \ldots, (x_ny_n)^{1/2}),$$

$$\bar{\mathbf{V}} \;=\; diag(\mathbf{V}) \;=\; (\bar{\mathbf{X}}\bar{\mathbf{Y}})^{1/2}, \text{ and}$$

$$\mathbf{d}^U \;=\; \bar{\mathbf{V}}^{-1}(\bar{\mathbf{Y}}\mathbf{d}^X + \bar{\mathbf{X}}\mathbf{d}^Y).$$

If $\mathbf{d}^U$ is taken as the steepest descent direction for $g_1(\mathbf{d}^X, \mathbf{d}^Y)$, *i.e.*,

$$\mathbf{d}^U \;=\; -\frac{\bar{\mathbf{V}}^{-1}\mathbf{1} - ((n+\sqrt{n})/\|\mathbf{V}\|^2)\mathbf{V}}{\left\|\bar{\mathbf{V}}^{-1}\mathbf{1} - ((n+\sqrt{n})/\|\mathbf{V}\|^2)\;\mathbf{V}\right\|}$$

(here $\mathbf{d}^U$ is the negative of the $\bar{\mathbf{V}}^{-1}\mathbf{1} - ((n+\sqrt{n})/\|\mathbf{V}\|^2)\mathbf{V})$, then we can choose the direction $(\mathbf{d}^X, \mathbf{d}^Y)$ as the unique solution of the linear system

$$\mathbf{d}^Y \;=\; \mathbf{M}\mathbf{d}^X,\; \bar{\mathbf{V}}^{-1}(\bar{\mathbf{Y}}\mathbf{d}^X + \bar{\mathbf{X}}\mathbf{d}^Y) \;=\; \mathbf{d}^U. \tag{14.7}$$

On the basis of these developments, Kojima, *et al.* (1991b) proceed to establish

14.3. THEOREM [Kojima, *et al.* (1991b)]. Let $v_{min} = min\{v_1, \ldots, v_n\}$ and let $\theta = v_{min}\tau$ for some $\tau \in (0,1)$. If $(\mathbf{d}^X, \mathbf{d}^Y)$ satisfies (14.7), then (14.6) holds and

$$\theta g_1 \le -\tfrac{1}{2}\sqrt{3}\,\tau$$

$$\theta^2 g_2 \le max\left\{\frac{n+\sqrt{n}}{2n}, \frac{1}{2(1-\tau)}\right\}\tau^2.$$

In the light of these results it is easily shown that if $n \ge 2$ and $\tau = 0.4$, $f(\hat{\mathbf{X}}, \hat{\mathbf{Y}}) - f(\mathbf{X}, \mathbf{Y}) \le -0.2$.

We can now summarize the Kojima, *et al.* (1991b)

14.4. POTENTIAL-REDUCTION ALGORITHM FOR $LCP(\mathbf{q}, \mathbf{M})$.

STEP 0. Initialize.

Choose an initial point $(\mathbf{X}_0, \mathbf{Y}_0) \in \mathcal{K}^0$ with $f(\mathbf{X}_0, \mathbf{Y}_0) \le O(\sqrt{n}\,L)$. Choose $\delta = 0.2$ and set $k = 0$.

STEP 1. Determine the Search Direction.

Compute $(\mathbf{d}^X, \mathbf{d}^Y)$ as the solution to (14.7).

STEP 2. Determine the Search Parameter.

Compute a step parameter $\overline{\theta}$ via linesearch along $-(\mathbf{d}^X, \mathbf{d}^Y)$ or use an approximation of the optimal steplength (see above).

STEP 3. Take a step in the $(\mathbf{d}^X, \mathbf{d}^Y)$ Direction.

Determine the new point$(\mathbf{X}_{k+1}, \mathbf{Y}_{k+1}) - \overline{\theta}(\mathbf{d}^X, \mathbf{d}^Y) \epsilon \mathcal{K}^0$ such that

$$f((\mathbf{X}_{k+1}, \mathbf{Y}_{k+1}) - \overline{\theta}(\mathbf{d}^X, \mathbf{d}^Y)) \leq f(\mathbf{X}_k, \mathbf{Y}_k) - \delta.$$

STEP 4. Test for Optimality.

If $\mathbf{X}_k'\mathbf{Y}_k \leq 2^{-2L}$, stop. Otherwise set $k = k+1$ and return to step 1.

The algorithm generates a sequence $\{(\mathbf{X}_k, \mathbf{Y}_k)\} \epsilon \mathcal{K}^0$ such that

$$f(\mathbf{X}_{k+1}, \mathbf{Y}_{k+1}) \leq f(\mathbf{X}_k, \mathbf{Y}_k) - \delta,\ k = 0, 1, \ldots,$$

and ultimately an approximate solution $(\mathbf{X}_p, \mathbf{Y}_p)$ of $LCP(\mathbf{q}, \mathbf{M})$ for some $p = O(\sqrt{n}\,L)$ such that $\mathbf{X}_p'\mathbf{Y}_p \leq 2^{-2L}$. An exact solution is then obtainable in $O(n^3)$ arithmetic operations.

To establish the connection between algorithms 14.2 and 14.4, let $(\mathbf{X}, \mathbf{Y}) \epsilon \mathcal{K}^0$. If we take $\overline{\mu} = (1 - \sqrt{n}/(n+\sqrt{n}))(\mathbf{X}'\mathbf{Y}/n)$, then the solution of system (14.7) gives the same direction $(\mathbf{d}^X, \mathbf{d}^Y)$ as the solution of the Newton system (14.3) at $(\mathbf{X}, \mathbf{Y})$.

An alternative interior-point potential-reduction algorithm for solving a linear complementarity problem of the form (14.1) has been offered by Kojima, Megiddo, and Ye (1992). Working with its equivalent quadratic programming formulation which seeks to minimize the total complementarity over $\mathcal{K}$,

$$\begin{aligned} &min\, \mathbf{X}'\mathbf{Y} \quad \text{s.t.} \\ &\mathbf{Y} \ = \ \mathbf{M}\mathbf{X} + \mathbf{q},\ (\mathbf{X}, \mathbf{Y}) \geqq \ (\mathbf{0}, \mathbf{0}) \end{aligned} \tag{14.8}$$

(note that if $min\, \mathbf{X}'\mathbf{Y} = 0$, (14.1) has a solution), the authors define an ***ϵ*-*complementary solution*** as a pair $(\mathbf{X}, \mathbf{Y}) \geqq (\mathbf{0}, \mathbf{0})$ such that $\mathbf{Y} = \mathbf{M}\mathbf{X} + \mathbf{q}$ and

$\mathbf{X}'\mathbf{Y} < \epsilon$. In this regard, Kojima, *et al.* (1992) are interested in finding a feasible point $(\mathbf{X},\mathbf{Y})$ with $\mathbf{X}'\mathbf{Y} < \epsilon$ for a given $\epsilon > 0$. Their goal is to develop a polynomial time algorithm for finding an ϵ-complementary solution to (14.8). Once such a solution is determined, an exact solution can readily by obtained from it by following the $O(n^3)$ arithmetic operation procedure of Kojima, *et al.* (1989a).

For $(\mathbf{X},\mathbf{Y}) \epsilon \mathcal{K}^0$, let the potential function assume the form

$$f(\mathbf{X},\mathbf{Y}) = \rho \ln \mathbf{X}'\mathbf{Y} - \sum_{i=1}^{n} \ln x_i y_i, \; \rho = \rho(n) > n.$$

In addition, let $p(L)$ depict any polynomial in input length L. If the potential function value at some initial point is $p(L)$ and if the potential value decreases during each iteration by at least n^{-k} for some k, then the algorithm takes a polynomial number of steps to reach an ϵ-complementary solution. In fact, if $\mathbf{M}$ is positive semidefinite and $\rho = 2n + \sqrt{2n}$, a polynomial time algorithm obtains, *i.e.*, if $\mathbf{M}$ is positive semidefinite, then starting at any point $(\mathbf{X},\mathbf{Y}) \epsilon \mathcal{K}^0$ where the potential value is $O(nL)$, the algorithm converges in $O(n^2L)$ iterations.

Kojima, *et al.* (1992) respecify problem (14.8) under a scaling of the variables. For $(\mathbf{X}_0,\mathbf{Y}_0) \epsilon \mathcal{K}^0$, $\bar{\mathbf{X}} = diag(\mathbf{X})$, and $\bar{\mathbf{Y}} = diag(\mathbf{Y})$, a linear transformation of the original space is defined as

$$\hat{\mathbf{X}} = \bar{\mathbf{X}}^{-1}\mathbf{X}, \; \hat{\mathbf{Y}} = \bar{\mathbf{Y}}^{-1}\mathbf{Y}.$$

If we let

$$\hat{\mathbf{M}} = \bar{\mathbf{Y}}^{-1}\mathbf{M}\bar{\mathbf{X}}, \; \hat{\mathbf{q}} = \bar{\mathbf{Y}}^{-1}\mathbf{q},$$
$$w_i = x_i^0 y_i^0, i=1\ldots,n,$$

and, for $\mathbf{W} \epsilon \mathbf{R}^n$, $\bar{\mathbf{W}} = \bar{\mathbf{X}}\bar{\mathbf{Y}}$, then the transformed problem is

$$\begin{aligned} &min\, \hat{\mathbf{X}}'\bar{\mathbf{W}}\hat{\mathbf{Y}} \quad \text{s.t.} \\ &\hat{\mathbf{Y}} = \hat{\mathbf{M}}\hat{\mathbf{X}} + \hat{\mathbf{q}}, \; (\hat{\mathbf{X}},\hat{\mathbf{Y}}) \geqq (\mathbf{0},\mathbf{0}). \end{aligned} \tag{14.8.1}$$

(Here $\hat{\mathbf{M}}$ is positive semidefinite if $\mathbf{M}$ is.) Under this transformation, feasible solutions of (14.8) are mapped into feasible solutions of (14.8.1), *i.e.*, $\hat{\mathbf{Y}} = \bar{\mathbf{Y}}^{-1}\mathbf{Y} = \bar{\mathbf{Y}}^{-1}(\mathbf{M}\mathbf{X}+\mathbf{q}) = \hat{\mathbf{M}}\hat{\mathbf{X}}+\hat{\mathbf{q}}$, and the point $(\mathbf{X}_0,\mathbf{Y}_0)$ is mapped into $(\mathbf{1},\mathbf{1})$. In addition, the transformed potential function appears as

$$\widehat{f}(\widehat{\mathbf{X}},\widehat{\mathbf{Y}}) = \rho \ln \widehat{\mathbf{X}}'\overline{\mathbf{W}}\widehat{\mathbf{Y}} - \sum_{i=1}^{n} \ln \widehat{x}_i w_i \widehat{y}_i. \tag{14.9}$$

Since $\widehat{f} = f$, any reduction in one of these functions precipitates the same reduction in the other.

To guarantee that their algorithm takes a polynomial number of steps, Kojima, *et al.* (1992) suggest starting the algorithm at an interior feasible point where the value of the potential function is bounded by a polynomial $p(L)$ in size L. To obtain an initial feasible interior point, let us consider the modified linear complementarity problem

$$\begin{aligned} \mathbf{Y} &= \mathbf{MX} + (\mathbf{1}-\mathbf{M1}-\mathbf{q})x_0 + \mathbf{q} \\ y_0 &= x_0 \\ &(\mathbf{X},\mathbf{Y}) \geqq (\mathbf{0},\mathbf{0}),\ (x_0, y_0) \geq (0,0) \\ &x_i y_i = 0,\ i=0,1,\ldots,n. \end{aligned} \tag{14.10}$$

Setting $\mathbf{X} = \mathbf{Y} = \mathbf{1}$ and $x_0 = y_0 = 1$ thus renders an interior feasible solution with potential function value $\rho \ln(n+1)$. Then an ϵ-complementary solution of (14.10) is also an approximate solution to (14.8) so that following the procedure outlined by Kojima, *et al.* (1989a) will yield an exact initial solution.

To execute an iteration in original space, we must: (1) apply the preceding linear scaling transformation in order to map the current point $(\mathbf{X}_k, \mathbf{Y}_k)$ to $(\mathbf{1},\mathbf{1})$ and f to $\widehat{f}$; (2) take a step in transformed space in the direction of the gradient of $\widehat{f}$, where the steplength $\theta > 0$ is either prespecified by the algorithm or determined by a line search which minimizes the value of $\widehat{f}$; (3) determine the new point in transformed space according to $(\widehat{\mathbf{X}}_k, \widehat{\mathbf{Y}}_k) = (\mathbf{1},\mathbf{1}) - \theta \mathbf{d}^X, \mathbf{d}^Y)$; and (4) map the new point in transformed space back to original space to obtain $(\mathbf{X}_{k+1}, \mathbf{Y}_{k+1})$.

How may we specify the direction vector $(\mathbf{d}^X, \mathbf{d}^Y)$? Let us assume that the current point is $(\mathbf{1},\mathbf{1})$ and the potential function is of the form given by (14.9). Given that it is to be understood that we are working in the transformed space, the circumflex "$\widehat{\ }$" above **X**, **Y**, **M**, and f will be omitted for convenience. Then from (14.9),

$$\nabla_{\mathbf{X}} f(\mathbf{X},\mathbf{Y}) = \frac{\rho}{\mathbf{X}'\bar{\mathbf{W}}\mathbf{X}}\,\bar{\mathbf{W}}\mathbf{Y} - \bar{\mathbf{X}}^{-1}\mathbf{1},$$

$$\nabla_{\mathbf{Y}} f(\mathbf{X},\mathbf{Y}) = \frac{\rho}{\mathbf{X}'\bar{\mathbf{W}}\mathbf{X}}\,\bar{\mathbf{W}}\mathbf{Y} - \bar{\mathbf{Y}}^{-1}\mathbf{1}.$$

Let

$$\mathbf{g} = \frac{\rho}{\mathbf{1}'\mathbf{W}}\,\mathbf{W} - \mathbf{1}$$

and, for $(\mathbf{X},\mathbf{Y}) = (\mathbf{1},\mathbf{1})$, it can readily be shown that

$$\nabla_{\mathbf{X}} f(\mathbf{1},\mathbf{1}) = \nabla_{\mathbf{Y}} f(\mathbf{1},\mathbf{1}) = \frac{\rho}{\mathbf{1}'\bar{\mathbf{W}}\mathbf{1}}\,\bar{\mathbf{W}}\mathbf{1} - \mathbf{1} = \mathbf{g}.$$

Hence we may view $(\mathbf{d}^X, \mathbf{d}^Y)$ as the orthogonal projection of $\nabla f(\mathbf{1},\mathbf{1})' = (\mathbf{g}',\mathbf{g}')$ on the linear space $\mathbf{d}^Y = \mathbf{M}\mathbf{d}^X$, *i.e.*, $(\mathbf{d}^X, \mathbf{d}^Y)$ is determined as the solution of the constrained optimization problem

$$\min_{\mathbf{d}^X,\mathbf{d}^Y} \left\{ \left\| \mathbf{d}^X - \mathbf{g} \right\|^2 + \left\| \mathbf{d}^Y - \mathbf{g} \right\|^2 \right\} \quad \text{s.t.} \quad \mathbf{d}^Y = \mathbf{M}\mathbf{d}^X.$$

From the Lagrangian

$$\mathcal{L} = (\mathbf{d}^X - \mathbf{g})'(\mathbf{d}^X - \mathbf{g}) + (\mathbf{d}^Y - \mathbf{g})'(\mathbf{d}^Y - \mathbf{g}) + \boldsymbol{\lambda}'(\mathbf{d}^Y - \mathbf{M}\mathbf{d}^X)$$

we can obtain the first-order optimality conditions

$$\begin{aligned} \mathbf{d}^X - \mathbf{g} - \mathbf{M}'\boldsymbol{\lambda} &= \mathbf{0}, \\ \mathbf{d}^Y - \mathbf{g} + \boldsymbol{\lambda} &= \mathbf{0}, \\ \mathbf{d}^Y - \mathbf{M}\mathbf{d}^X &= \mathbf{0}. \end{aligned}$$

Upon eliminating $\boldsymbol{\lambda}$ we have

$$\begin{aligned} \mathbf{d}^X &= (\mathbf{I}_n - \mathbf{M}'\mathbf{M})^{-1}(\mathbf{I}_n + \mathbf{M}')\mathbf{g}, \\ \mathbf{d}^Y &= \mathbf{M}\mathbf{d}^X \end{aligned}$$

(where the **M**'s in this system are actually $\widehat{\mathbf{M}}$'s since we are operating in transformed space).

14.4. A Predictor-Corrector Algorithm for Solving *LCP*(q, M)

The predictor-corrector algorithm of Ye and Anstreicher (1993a) generates a solution to the linear complementarity problem (14.8) under the assumptions that:

(i) the feasible region for (14.8) has a nonempty relative interior or $\mathcal{K}^0 \neq \phi$;

(ii) $\mathbf{M}$ is positive semidefinite; and

(iii) (14.8) has a strictly complementary solution,

i.e., while $(\mathbf{X}_*, \mathbf{Y}_*) \epsilon \mathcal{K}$ is ***optimal*** (or ***complementary***) if and only if $x_i^* y_i^* = 0$, i=1,...,n, a ***strictly complementary solution*** is an optimal solution satisfying the condition $x_i^* + y_i^* > 0$, i=1,...,n.

Define a neighborhood $\delta(\alpha) = \{(\mathbf{X},\mathbf{Y}) \epsilon \mathcal{K} \mid \| \bar{\mathbf{X}}\mathbf{Y}/\mu - \mathbf{1} \| \leq \alpha, \mu = \mathbf{X}'\mathbf{Y}/n, 0 \leq \alpha \leq 1\}$ and take $0 < \beta \leq \frac{1}{4}$ (typically $\beta = \frac{1}{4}$). In addition, the search direction $(\mathbf{d}^X, \mathbf{d}^Y)$ is determined as the solution of the linear system

$$\begin{aligned} \bar{\mathbf{X}}\mathbf{d}^Y + \bar{\mathbf{Y}}\mathbf{d}^X &= \gamma\mu\mathbf{1} - \bar{\mathbf{X}}\mathbf{Y} \\ \mathbf{d}^Y - \mathbf{M}\mathbf{d}^X &= \mathbf{0}, \end{aligned} \tag{14.11}$$

$0 < \gamma < 1$ (Kojima, *et al.* (1989a)). As we shall now see, (14.11) must be solved twice at each iteration of the predictor-corrector routine — once to find the Newton direction for the predictor step and again to determine a centering direction for the corrector step.

To undertake the kth ***predictor step***, given $(\mathbf{X}_k, \mathbf{Y}_k) \epsilon \delta(\beta)$, solve (14.11) with $(\mathbf{X},\mathbf{Y}) = (\mathbf{X}_k, \mathbf{Y}_k)$ and $\gamma = 0$. Denote the resulting direction by $(\mathbf{d}^X, \mathbf{d}^Y)$ and, for some steplength $\theta \geq 0$, let

$$\begin{aligned} \mathbf{X}(\theta) &= \mathbf{X}_k + \theta\mathbf{d}_k^X \\ \mathbf{Y}(\theta) &= \mathbf{Y}_k + \theta\mathbf{d}_k^Y \end{aligned}$$

and $\mu(\theta) = \mathbf{X}(\theta)'\mathbf{Y}(\theta)/n$. A guide to the choice of θ is provided by

14.5. LEMMA [Ye and Anstreicher (1993a)]. If for some $0 < \theta_k < 1$ we have

$$\| \bar{\mathbf{X}}(\theta)\mathbf{Y}(\theta)/\mu(\theta) - \mathbf{1} \| \leq \alpha < 1 \text{ for all } 0 \leq \theta \leq \theta_k, \tag{14.12}$$

then $(\mathbf{X}(\theta_k), \mathbf{Y}(\theta_k)) \epsilon \delta(\alpha)\mathcal{K}$.

Since the feasibility of $(\mathbf{X}(\theta_k), \mathbf{Y}(\theta_k))$ is guaranteed if (14.12) is satisfied, let us choose the *largest* steplength $\theta_k \leq 1$ such that $\| \bar{\mathbf{X}}(\theta_k)\mathbf{Y}(\theta_k)/\mu(\theta_k) - \mathbf{1} \| \leq \alpha \leq 1$ holds for $\alpha = \beta + \tau$, $0 < \tau \leq \beta$ and let $\hat{\mathbf{X}}_k = \mathbf{X}(\theta_k)$, $\hat{\mathbf{Y}}_k = \mathbf{Y}(\theta_k)$.

To execute the kth ***corrector step***, solve (14.11) with $(\mathbf{X},\mathbf{Y}) = (\widehat{\mathbf{X}}_k, \widehat{\mathbf{Y}}_k) \in \delta(\beta+\tau)$, $\mu = \widehat{\mathbf{X}}_k'\widehat{\mathbf{Y}}_k/n$ and $\gamma = 1$. If the resulting solution is denoted as $(\widehat{\mathbf{d}}^X, \widehat{\mathbf{d}}^Y)$, then we may form

$$\begin{aligned} \mathbf{X}_{k+1} &= \widehat{\mathbf{X}}_k + \widehat{\mathbf{d}}^X \\ \mathbf{Y}_{k+1} &= \widehat{\mathbf{Y}}_k + \widehat{\mathbf{d}}^Y \end{aligned}$$

with $(\mathbf{X}_{k+1}, \mathbf{Y}_{k+1}) \in \delta(\beta)$ as long as $0 < \beta \leq \frac{1}{4}$ and $0 < \tau \leq \beta$. This process, involving a single corrector or centering step to the central trajectory after each predictor step in order to decrease μ, generates a feasible sequence of points $\{(\mathbf{Y}_k, \mathbf{Y}_k)\}$ satisfying $\| \bar{\mathbf{X}}_k\mathbf{Y}_k/\mu_k - \mathbf{1} \| \leq \beta$ and, for $0 < \beta \leq \frac{1}{4}$ and $\alpha = 2\beta$, has $O(\sqrt{n}\,L)$ iteration complexity with $\mathbf{X}_k'\mathbf{Y}_k \rightarrow 0$ Q-quadratically.

Ye and Anstreicher point out that there is no need to choose the largest steplength θ_k in the predictor step — a lower bound for θ_k will suffice, *i.e.*, if θ_k is that largest $\theta_k \leq 1$ satisfying the preceding lemma with $\alpha = \beta + \tau$ and $0 < \tau \leq \beta$, then $\theta_k \geq 2\tau/((\tau^2 + 4\tau\| \boldsymbol{\delta}_k \|)^{1/2} + \tau)$, where $\boldsymbol{\delta}_k = \bar{\mathbf{d}}_k^X \mathbf{d}_k^Y/\mu_k$ and $\bar{\mathbf{d}}_k^X = diag(\mathbf{d}_k^X)$.

Based upon the preceding considerations, Ye and Anstreicher formulate the

14.6. PREDICTOR-CORRECTOR ALGORITHM FOR $LCP(\mathbf{q},\mathbf{M})$.

STEP 0. Initialize.

Given $(\mathbf{X}_0, \mathbf{Y}_0) \in \delta(\frac{1}{4})$, set $k = 0$.

STEP 1. Predictor step.

Set $(\mathbf{X},\mathbf{Y})$ to $(\mathbf{X}_k, \mathbf{Y}_k)$, $\gamma = 0$, and compute $\mathbf{d}_k^X, \mathbf{d}_k^Y$ from (14.11). Form

$$\begin{aligned} \mathbf{X}(\theta) &= \mathbf{X}_k + \theta \mathbf{d}_k^X \\ \mathbf{Y}(\theta) &= \mathbf{Y}_k + \theta \mathbf{d}_k^Y \\ \mu(\theta) &= \mathbf{X}(\theta)'\mathbf{Y}(\theta)/n. \end{aligned}$$

Compute the largest θ_k so that $(\mathbf{X}(\theta), \mathbf{Y}(\theta)) \in \delta(\frac{1}{2})$ for

$$\theta_k \geq \frac{2\tau}{\left((\tau^2 + 4\tau\| \boldsymbol{\delta}_k \|)^{1/2} + \tau\right)}, \; 0 < \tau < \tfrac{1}{4},$$

where $\boldsymbol{\delta}_k = \bar{\mathbf{d}}_k^X \mathbf{d}_k^Y/\mu_k$ and $\bar{\mathbf{d}}_k^X = diag(\mathbf{d}_k^X)$.

Set $\widehat{\mathbf{X}}_k = \mathbf{X}(\theta_k)$, $\widehat{\mathbf{Y}}_k = \mathbf{Y}(\theta_k)$.

STEP 2. Corrector Step.

Solve (14.11) with $(\mathbf{X}, \mathbf{Y}) = (\widehat{\mathbf{X}}_k, \widehat{\mathbf{Y}}_k) \epsilon\, \delta(\frac{1}{2}), \mu = \widehat{\mathbf{X}}_k' \widehat{\mathbf{Y}}_k / n$, and $\gamma = 1$. Form

$$\begin{aligned} \mathbf{X}_{k+1} &= \widehat{\mathbf{X}}_k + \widehat{\mathbf{d}}^X \\ \mathbf{Y}_{k+1} &= \widehat{\mathbf{Y}}_k + \widehat{\mathbf{d}}^Y \end{aligned}$$

with $(\mathbf{X}_{k+1}, \mathbf{Y}_{k+1}) \epsilon\, (\frac{1}{4})$.

STEP 3. Check for Optimality.

If $\mathbf{X}_k' \mathbf{Y}_k \leq 2^{-2L}$, stop. Otherwise proceed to

STEP 4. Advance k to $k+1$.

Return to step 1.

The preceding predictor-corrector methodology has been modified somewhat by Mehrotra and Stubbs (1994). As mentioned earlier, these authors modify the predictor step so as to determine alternative predictor directions as solutions to subproblems defined over ellipsoidal approximations of $\mathcal{K}$. Two broad categories of subproblems are considered: (a) methods using primal-dual scaling ellipsoids; and (b) methods employing affine scaling ellipsoids. Within these, both ball optimization and trust region approximations are considered.

Relative to primal-dual scaling, the ball optimization sub-problem may be derived by first letting $\mathbf{D} = \bar{\mathbf{X}}^{1/2} \bar{\mathbf{Y}}^{-1/2}$. Then the predictor direction in step 1 of algorithm 14.6 emerges as the solution to

<u>**Ball Optimization Subproblem I**</u>

$$\begin{aligned} & max\{\mathbf{Y}'\mathbf{d}^X + \mathbf{X}'\mathbf{d}^Y\} \quad \text{s.t.} \\ & \mathbf{M}\mathbf{d}^X - \mathbf{d}^Y = \mathbf{0} \\ & \left\| \begin{matrix} \mathbf{D}^{-1}\mathbf{d}^X \\ \mathbf{D}\mathbf{d}^Y \end{matrix} \right\|^2 \leq \mathbf{X}'\mathbf{Y}. \end{aligned} \tag{14.13}$$

Here we seek to maximize a linear function over a lower-dimensional ball. If we let $\mathbf{V} = \bar{\mathbf{X}}^{1/2} \bar{\mathbf{Y}}^{1/2} \mathbf{1}, \mathbf{g} = \mathbf{D}^{-1}\mathbf{d}^X$, and $\mathbf{g}^Y = \mathbf{D}\mathbf{d}^Y$, then (14.13) is equivalent to

$$\begin{aligned} & max\{\mathbf{V}'\mathbf{g}^X + \mathbf{V}'\mathbf{g}^Y\} \quad \text{s.t.} \\ & \mathbf{M}\mathbf{D}\mathbf{g}^X - \mathbf{D}^{-1}\mathbf{g}^Y = \mathbf{0} \\ & \left\| \begin{matrix} \mathbf{g}^X \\ \mathbf{g}^Y \end{matrix} \right\|^2 \leq \mathbf{X}'\mathbf{Y}.^2 \end{aligned} \tag{14.13.1}$$

Based upon these considerations, the Mehrotra-Stubbs modification of predictor-corrector algorithm 14.6 is to replace the Newton direction $(\mathbf{d}^X, \mathbf{d}^Y)$ determined in step 1 by the direction obtained by solving (14.13). Under this substitution, Mehrotra and Stubbs then verify that if we start at an initial point $(\mathbf{X}_0, \mathbf{Y}_0) \in \delta(\beta)$, $0 \leq \beta \leq \frac{1}{6}$, the iterates $(\mathbf{X}_k, \mathbf{Y}_k)$ generated by their modified algorithm also satisfy $(\mathbf{X}_k, \mathbf{Y}_k) \in \delta(\beta)$ so that an $O(\sqrt{n}\,L)$ iteration algorithm emerges. Move formally, they prove

14.7. THEOREM [Mehrotra and Stubbs (1994)]. Let $(\mathbf{X}_k, \mathbf{Y}_k) \in \delta(\beta)$, $0 \leq \beta \leq \frac{1}{6}$, and $(\mathbf{d}^X, \mathbf{d}^Y)$ be the optimal solution of (14.13) for predictor step 1 of algorithm 14.6. Let $\theta = 0.073/\sqrt{n}$, and $\widehat{\mathbf{X}}_k = \mathbf{X}_k - \theta\mathbf{d}^X$, $\widehat{\mathbf{Y}}_k = \mathbf{Y}_k - \theta\mathbf{d}^Y$. Then under corrector step 2 of algorithm 14.6, with $\mathbf{X}_{k+1} = \widehat{\mathbf{X}}_k - \widehat{\mathbf{d}}^X$, $\mathbf{Y}_{k+1} = \widehat{\mathbf{Y}}_k - \widehat{\mathbf{d}}^Y$, we have $\mathbf{X}_{k+1}'\mathbf{Y}_{k+1}/\mathbf{X}_k'\mathbf{Y}_k \leq 1 - (0.013)/\sqrt{n})$ and $(\mathbf{X}_{k+1}, \mathbf{Y}_{k+1}) \in \delta(\frac{1}{6})$.

Alternatively, the predictor step required at each iteration of algorithm 14.6 can be undertaken by solving a trust region subproblem that minimizes a convex quadratic objective over an ellipsoidal region. In this regard, we now determine the predictor direction by solving

Trust Region Subproblem I

$$\begin{aligned} min(\mathbf{X} - \mathbf{d}^X)'\,(\mathbf{Y} - \mathbf{d}^Y) \quad &\text{s.t.} \\ \mathbf{M}\mathbf{d}^X - \mathbf{d}^Y &= \mathbf{0} \\ \left\| \begin{matrix} \mathbf{D}^{-1}\mathbf{d}^X \\ \mathbf{D}\mathbf{d}^Y \end{matrix} \right\|^2 &\leq \alpha^2 \mathbf{X}'\mathbf{Y} \end{aligned} \tag{14.14}$$

or, since $\mathbf{d}^Y = \mathbf{M}\mathbf{d}^X$,

---

[2]For $\mathbf{W}' = (\mathbf{V}', \mathbf{V}')$, the solution direction $(\mathbf{g}^X, \mathbf{g}^Y)$ for (14.13.1) can be obtained by projecting $\mathbf{W}$ orthogonally onto the null-space of $\mathbf{B} = [\mathbf{MD}, -\mathbf{D}^{-1}]$ or

$$\begin{pmatrix} \mathbf{g}^X \\ \mathbf{g}^Y \end{pmatrix} = \rho \mathbf{P}_B \mathbf{W},$$

where $\mathbf{P}_B = \mathbf{I}_n - \mathbf{B}'(\mathbf{BB}')^{-1}\mathbf{B}$ and $\rho = (\mathbf{X}'\mathbf{Y})^{1/2}/\|\mathbf{P}_B\mathbf{W}\|$, $\mathbf{P}_B\mathbf{W} \neq \mathbf{0}$.

$$
\begin{aligned}
&min\{\mathbf{X}'\mathbf{Y} + (\mathbf{d}^X)'\mathbf{M}\mathbf{d}^X - (\mathbf{Y}+\mathbf{M}'\mathbf{X})'\mathbf{d}^X\} \quad \text{s.t.} \\
&(\mathbf{d}^X)'[\mathbf{M}'\mathbf{D}^2\mathbf{M}+\mathbf{D}^{-2}]\mathbf{d}^X \leq \alpha^2\mathbf{X}'\mathbf{Y}.
\end{aligned}
\tag{14.14.1}
$$

Here too Mehrotra and Stubbs verify that if (14.14) is solved in the predictor step of algorithm 14.6, then $(\mathbf{X}_k, \mathbf{Y}_k) \epsilon\, \delta(\beta)$, $0 \leq \beta \leq \frac{1}{6}$, and convergence of the predictor-corrector algorithm in $O\sqrt{n}$ iterations results. That is,

14.8. THEOREM [Mehrotra and Stubbs (1994)]. Let $(\mathbf{X}_k, \mathbf{Y}_k) \epsilon\ \delta(\beta)$, $0 \leq \beta \leq \frac{1}{6}$, with $(\mathbf{d}^X, \mathbf{d}^Y)$ the optimal solution of (14.14) in the predictor step 1 of algorithm 14.6. Let $\widehat{\mathbf{X}}_k = \mathbf{X}_k - \mathbf{d}^X$, $\widehat{\mathbf{Y}}_k = \mathbf{Y}_k - \mathbf{d}^Y$ and, at corrector step 2, let $\mathbf{X}_{k+1} = \widehat{\mathbf{X}}_k - \widehat{\mathbf{d}}^X$, $\mathbf{Y}_{k+1} = \widehat{\mathbf{Y}}_k - \widehat{\mathbf{d}}^Y$. Then for $\alpha = 0.073/\sqrt{n}$, $\mathbf{X}'_{k+1}\mathbf{Y}_{k+1}/\mathbf{X}'_k\mathbf{Y}_k \leq 1 - (0.013\sqrt{n})$ and $(\mathbf{X}_{k+1}, \mathbf{Y}_{k+1}) \epsilon\, \delta(\frac{1}{6})$.

Next, subproblems that employ ellipsoids obtained by an affine scaling of the feasible region have also been utilized by Mehrotra and Stubbs to construct on $O(\sqrt{n}\,L)$ iteration predictor-corrector algorithm for solving the linear complementarity problem. For instance, we may determine the predictor direction by solving either

Ball Optimization Subproblem II

$$
\begin{aligned}
&max\{\mathbf{Y}'\mathbf{d}^X + \mathbf{X}'\mathbf{d}^Y\} \quad \text{s.t.} \\
&\mathbf{M}\mathbf{d}^X - \mathbf{d}^Y = \mathbf{0} \\
&\left\| \begin{matrix} \bar{\mathbf{X}}^{-1}\mathbf{d}^X \\ \bar{\mathbf{Y}}^{-1}\mathbf{d}^Y \end{matrix} \right\|^2 \leq \frac{n}{1-\beta}
\end{aligned}
\tag{14.15}
$$

or

Trust Region Subproblem II

$$
\begin{aligned}
&min(\mathbf{X} - \mathbf{d}^X)'\,(\mathbf{Y} - \mathbf{d}^Y) \quad \text{s.t.} \\
&\mathbf{M}\mathbf{d}^X - \mathbf{d}^Y = \mathbf{0} \\
&\left\| \begin{matrix} \bar{\mathbf{X}}^{-1}\mathbf{d}^X \\ \bar{\mathbf{Y}}^{-1}\mathbf{d}^Y \end{matrix} \right\|^2 \leq \frac{\alpha^2 n}{1-\beta}.
\end{aligned}
\tag{14.16}
$$

Then theorems analogous to theorems 14.7, 14.8 given in support of ball optimization subproblem I and trust region subproblem I respectively can now be offered for this second wave of subproblems. Specifically, relative to ball optimization subproblem II,

14.9. THEOREM [Mehrotra and Stubbs (1994)]. Let $(\mathbf{X}_k, \mathbf{Y}_k) \in \delta(\beta)$, $0 \leq \beta \leq \frac{1}{6}$, and $(\mathbf{d}^X, \mathbf{d}^Y)$ be the optimal solution of (14.15) for predictor step 1 of algorithm 14.6. Let $\widehat{\mathbf{X}}_k = \mathbf{X}_k - \theta \mathbf{d}^X$, $\widehat{\mathbf{Y}}_k = \mathbf{Y}_k - \theta \mathbf{d}^Y$. Then for corrector step 2 of algorithm 14.6, let $\mathbf{X}_{k+1} = \widehat{\mathbf{X}}_k - \widehat{\mathbf{d}}^X$, $\mathbf{Y}_{k+1} = \widehat{\mathbf{Y}}_k - \widehat{\mathbf{d}}^Y$. Then for $\theta = 0.06185/\sqrt{n}$, $\mathbf{X}'_{k+1}\mathbf{Y}_{k+1}/\ \mathbf{X}'_k\mathbf{Y}_k \leq 1 - (0.0015/\sqrt{n})$ and $(\mathbf{X}_{k+1}, \mathbf{Y}_{k+1}) \in \delta(\beta)$.

And for trust region subproblem II,

14.10. THEOREM [Mehrotra and Stubbs (1994)]. Let $(\mathbf{X}_k, \mathbf{Y}_k) \in \delta(\beta)$, $0 \leq \beta \leq \frac{1}{6}$, and $(\mathbf{d}^X, \mathbf{d}^Y)$ be the optimal solution of (14.16) in the predictor step 1 of algorithm 14.6. Let $\widehat{\mathbf{X}}_k = \mathbf{X}_k - \mathbf{d}^X, \widehat{\mathbf{Y}}_k = \mathbf{Y}_k - \mathbf{d}^Y$ and, in corrector step 2 of algorithm 14.6, let $\mathbf{X}_{k+1} = \widehat{\mathbf{X}}_k - \widehat{\mathbf{d}}^X, \mathbf{Y}_{k+1} = \widehat{\mathbf{Y}}_k - \widehat{\mathbf{d}}^Y$. Then for $\theta = 0.06185/\sqrt{n}$, $\mathbf{X}'_{k+1}\mathbf{Y}_{k+1}/\mathbf{X}'_k\mathbf{Y}_k \leq 1 - (0.0015/\sqrt{n})$ and $(\mathbf{X}_{k+1}, \mathbf{Y}_{k+1}) \in \delta(\beta)$.

14.5. Large-Step Interior-Point Algorithms for Solving $LCP(\mathbf{q}, \mathbf{M})$

Kojima, Kurita, and Mizuno (1993) offer a set of improved steplength rules in a path-following setting for finding a solution to an equivalent representation of $LCP(\mathbf{q}, \mathbf{M})$ (14.1) or to the problem: find a point $(\mathbf{X}, \mathbf{Y}) \in \mathbf{R}^{2n}$ such that

$$\begin{aligned} \mathbf{Y} &= \mathbf{XM} + \mathbf{q} \\ (\mathbf{X}, \mathbf{Y}) &\geqq (\mathbf{0}, \mathbf{0}) \\ \bar{\mathbf{X}}\mathbf{Y} &= \mathbf{0}, \end{aligned} \tag{14.1.1}$$

where $\bar{\mathbf{X}} = diag(\mathbf{X})$. It is assumed that: (a) an initial point $(\mathbf{X}_0, \mathbf{Y}_0) \in \mathcal{K}^0$ is known in advance and; (b) $\mathbf{M}$ is positive semidefinite. In addition, for parameter $\mu > 0$, let the ***central path*** for (14.1.1) be defined as

$$\Gamma_{LCP}(\mu) = \{(\mathbf{X}(\mu), \mathbf{Y}(\mu)) \mid \mathbf{Y} = \mathbf{MX} + \mathbf{q},\ \bar{\mathbf{X}}\mathbf{Y} = \mu\mathbf{1},\ (\mathbf{X}, \mathbf{Y}) \geqq (\mathbf{0}, \mathbf{0})\}.$$

Then as $\mu \to 0$, $\Gamma_{LCP}(\mu)$ leads to a solution of (14.1.1) under assumptions (a), (b) above.

Given the current iterate $(\mathbf{X}_k, \mathbf{Y}_k)$, let $f_k = \mathbf{X}_k'\mathbf{Y}_k/n$ with $\beta \in [0,1]$ the search parameter. To find a point $(\mathbf{X}(\mu), \mathbf{Y}(\mu)) \in \Gamma_{LCP}(\mu)$, $\mu = \beta f_k$, let us determine the Newton search direction $(\mathbf{d}^X, \mathbf{d}^Y)$ at $(\mathbf{X}_k, \mathbf{Y}_k)$ by solving the linear system

$$\begin{aligned} \bar{\mathbf{Y}}_k\mathbf{d}^X + \bar{\mathbf{X}}_k\mathbf{d}^Y &= \mu\mathbf{1} - \bar{\mathbf{X}}_k\mathbf{Y}_k \\ -\mathbf{Md}^X + \mathbf{d}^Y &= \mathbf{0}. \end{aligned} \tag{14.17}$$

Then we may take a single steplength α_k over $\mathcal{K}$ and obtain a new iterate

$$(\mathbf{X}_{k+1}, \mathbf{Y}_{k+1}) = (\mathbf{X}_k, \mathbf{Y}_k) + \alpha_k(\mathbf{d}^X, \mathbf{d}^Y),\ 0 \leq \alpha_k < \bar{\alpha}_k, \tag{14.18}$$

where $\bar{\alpha}_k = max\{\alpha \mid (\mathbf{X}_k + \alpha\mathbf{d}^X, \mathbf{Y}_k + \alpha\mathbf{d}^Y) \geqq (\mathbf{0},\mathbf{0})\}$. (Note that α_k is always finite under assumptions (a), (b).)

To obtain their modified Rules G' and P', Kojima, *et al.* (1993) let $0 \leq \beta = \beta_k < 1$ and set $\alpha = \alpha_k$ in (14.18) so as to express total complementarity at the new iterate $\mathbf{X}_{k+1}'\mathbf{Y}_{k+1}$ as a quadratic function of α or

$$\begin{aligned} \mathbf{X}_{k+1}'\mathbf{Y}_{k+1} &= \mathbf{X}_k'\mathbf{Y}_k + \alpha(\mathbf{X}_k'\mathbf{d}^Y + \mathbf{Y}_k'\mathbf{d}^X) + \alpha^2(\mathbf{d}^X)'\mathbf{d}^Y \\ &= \mathbf{X}_k'\mathbf{Y}_k - \alpha(1-\beta_k)\mathbf{X}_k'\mathbf{Y}_k + \alpha^2(\mathbf{d}^X)'\mathbf{d}^Y. \end{aligned}$$

Since $-\alpha(1-\beta_k)\mathbf{X}_k'\mathbf{Y}_k < 0$ and $\alpha^2(\mathbf{d}^X)'\mathbf{d}^Y > 0$ (by virtue of the fact that $\mathbf{M}$ is positive semidefinite), it is evident that the total complementarity for the linear complementarity problem is not a monotonic decreasing function of α if $(\mathbf{d}^X)'\mathbf{d}^Y > 0$. Minimizing $\mathbf{X}_{k+1}'\mathbf{Y}_{k+1}$ with respect to α consequently renders

$$\alpha_k^{min} = \begin{cases} \dfrac{(1-\beta_k)\mathbf{X}_k'\mathbf{Y}_k}{2(\mathbf{d}^X)'\mathbf{d}^Y} & \text{if } (\mathbf{d}^X)'\mathbf{d}^Y > 0; \\ +\infty & \text{otherwise.} \end{cases}$$

Hence we must choose a steplength $\alpha = \alpha_k$ satisfying $0 < \alpha < \bar{\alpha}_k$ and $\alpha \leq {}_k^{min}$.

Based upon these results Kojima, *et al.* (1993) state their Rules G', P' for controlling the parameters α, β in an interior-point, path-following algorithm for solving $LCP(\mathbf{q}, \mathbf{M})$. That is,

Rule G': Let $0 \leq \bar{\beta} < 1$, $0 < \bar{\theta} < 1$, and $0 < \alpha^*$ be fixed. At each iteration choose a search direction parameter $\beta = \beta_k \in [0, \bar{\beta}]$ and a steplength $\alpha = \alpha_k = min\{\alpha', \alpha_k^{min}\}$, where

$$\alpha' \in [\bar{\theta}\alpha^*, \bar{\alpha}_k] \text{ if } \bar{\alpha}_k \geq \alpha^*;$$
$$\alpha' \in [\bar{\theta}(\bar{\alpha}_k)^2/\alpha^*,\ (\bar{\alpha}_k)^2/\alpha^*] \text{ if } \bar{\alpha}_k < \alpha^*.$$

Rule P': Let $0 < \beta^* \leq \bar{\beta} = \frac{1}{2}$, $\bar{\theta} = \frac{3}{4} \leq \theta^* < 1$ and $0 < \alpha^* \leq 1$ be fixed. At each iteration choose a search direction parameter $\beta = \beta_k \in [\beta^*, \bar{\beta}]$ and a steplength parameter $\alpha = \alpha_k = min\{\alpha', \alpha_k^{min}\}$, where

$$\alpha' \in [\bar{\theta}\alpha^*, \theta^*\bar{\alpha}_k] \text{ if } \bar{\alpha}_k \geq \alpha^*;$$
$$\alpha' \in [\bar{\theta}(\bar{\alpha}_k)^2/\alpha^*,\ \theta^*(\bar{\alpha}_k)^2/\alpha^*] \text{ if } \bar{\alpha}_k < \alpha^*.$$

(Here Rule P' is a special case of Rule G'.) Using either of these rules in the following interior-point, path-following algorithm generates a sequence $\{(\mathbf{X}_k, \mathbf{Y}_k)\} \in \mathcal{K}^0$ such that $\mathbf{X}_{k+1}'\mathbf{Y}_{k+1} < \mathbf{X}_k'\mathbf{Y}_k$. As determined by Kojima, *et al.* (1993), the algorithm using Rule G' ensures global convergence, *i.e.*, it generates a bounded sequence $\{(\mathbf{X}_k, \mathbf{Y}_k)\}$ such that $\mathbf{X}_k'\mathbf{Y}_k \to 0$ as $k \to \infty$ while Rule P' ensures $O(nL)$ iteration polynomial-time convergence.

We may now state the Kojima, *et al.* (1993) modified interior-point

14.11. PATH-FOLLOWING ALGORITHM FOR $LPC(\mathbf{q}, \mathbf{M})$

STEP 0. Initialize.

Let $(\mathbf{X}_0, \mathbf{Y}_0) \in \mathcal{K}^0$ be a known initial feasible point with $\mathbf{X}_0'\mathbf{Y}_0 \leq 2^{O(L)}$.

STEP 1. Determine the Search Parameter, Steplength, and Search Direction.

Choose either Rule G' or Rule P' with $(\mathbf{d}^X, \mathbf{d}^Y)$ determined from (14.17).

STEP 2. Take a Step in the Newton Direction.

Compute a new iterate $(\mathbf{X}_{k+1}, \mathbf{Y}_{k+1})$ according to (14.18) with steplength $\alpha = \alpha_k$ satisfying $0 < \; min\{\bar{\theta}\alpha^*, \bar{\theta}(\bar{\alpha}_k)^2/\alpha^*, \alpha_k^{min}\} \leq \alpha < \bar{\alpha}_k$ and $\alpha \leq \alpha_k^{min}$ regardless of whether Rule G' or Rule P' is chosen.

STEP 3. Test for Optimality.

If $\mathbf{X}_{k+1}'\mathbf{Y}_{k+1} < 2^{-2L}$, stop. Otherwise set $k = k+1$ and return to step 1.

As the iterations progress, the quantity

$$0 < \delta_k = min\Big\{x_i^k y_i^k / f_k,\ i{=}1,\ldots,n\Big\} \leq 1$$

measures the extent of the deviation of the current iterate $(\mathbf{X}_k, \mathbf{Y}_k) \in \mathcal{K}^0$ from the central trajectory (Kojima, *et al.* (1991b); Kojima, *et al.* (1993)). Here too we can start the algorithm from a known interior feasible solution of an equivalent artificial linear complementarity problem. This then produces, in $O(nL)$ iterations, an approximate solution to the artificial problem satisfying the optimality criterion in step 3, from which an exact solution can be refined (provided one exists) in $O(n^3)$ arithmetic operations.

As a final point, Kojima, *et al.* (1993) indicate that the predictor-corrector algorithm for solving $LCP(\mathbf{q},\mathbf{M})$ offered by Ye and Anstreicher (1993) is a special case of their algorithm 14.11. In particular, they point out that under certain restrictions on α_k and β_k and if the linear complementarity problem has a unique strictly complementary solution, then under their computational scheme the complementarity sequence $\{\mathbf{X}_k'\mathbf{Y}_k\}$ also converges to zero superlinearly as $k \rightarrow \infty$.

14.6. Related Methods for Solving LCP(q, M)

Other approaches to solving $LCP(\mathbf{q},\mathbf{M})$ have been advanced in the literature. For instance, McShane (1994) presents an interior-point adaptive path-following algorithm for solving (14.8) (with $\mathbf{M}$ positive semidefinite) which requires at most $O(\sqrt{n}\,L)$ iterations to determine feasible vectors $\mathbf{X}, \mathbf{Y}$ such that $\mathbf{X}'\mathbf{Y} \leq 2^{-2L}$ and from which an optimal solution can be derived in $O(n^3)$ arithmetic operations. The algorithm is based on an adaptive path-following scheme in that the relative weights on the search directions are not fixed from iteration to iteration. For (14.8) nondegenerate, if there exists a unique strictly complementary solution $(\mathbf{X}_*, \mathbf{Y}_*)$, then the iterates $(\mathbf{X}_k, \mathbf{Y}_k)$ converge to $(\mathbf{X}_*, \mathbf{Y}_*)$, and the duality gap and each component of the sequence $\{\bar{\mathbf{X}}_k \bar{\mathbf{Y}}_k \mathbf{1}\}$ converges to zero Q-quadratically. Under an *approximately centered condition*, the iterates are forced to stay close to the central path, thus allowing for a sufficient decrease in the objective function between iterations and guaranteeing strict complementarity.

An approach for finding a solution to a linear complementarity problem which departs significantly from the preceding collection of algorithms has been proposed by Zhang (1994). Observing that finding an initial interior feasible solution to $LCP(\mathbf{q}, \mathbf{M})$ can, at times, involve as much work as solving the original problem itself, Zhang suggests an approach which involves neither problem reformulation (*e.g.*, one need not introduce artificial variables and embed $\mathcal{K}$ into a new feasible region of higher dimension) nor a feasible starting point. The method can start from any positive but otherwise infeasible point (*i.e.*, $(\mathbf{X}_0, \mathbf{Y}_0) > (\mathbf{0},\ \mathbf{0})$ but $\mathbf{Y}_0 \neq \mathbf{M}\mathbf{X}_0 + \mathbf{q}$) and seeks to achieve feasibility and optimality simultaneously.

The infeasible interior-point algorithm developed by Zhang is essentially a centered and damped Newton method with positive starting point and is applied to a general model called the ***horizontal linear complementarity problem*** that includes linear and quadratic programs as well as problems (14.1) and (14.8). The structure of the horizontal linear complementarity problem is depicted by the following nonlinear system:

$$\begin{pmatrix} \mathbf{MX} + \mathbf{NY} - \mathbf{h} \\ \bar{\mathbf{X}}\bar{\mathbf{Y}}\mathbf{1} \end{pmatrix} = \mathbf{0},\ (\mathbf{X}, \mathbf{Y}) \geqq (\mathbf{0}, \mathbf{0}),$$

where $\mathbf{X}, \mathbf{Y}, \mathbf{h} \in R^n$; $\mathbf{M}, \mathbf{N}$ are both of order $(n \times n)$; and $\bar{\mathbf{X}} = diag(\mathbf{X})$, $\bar{\mathbf{Y}} = diag(\mathbf{Y})$. (Here the term *horizontal* is used to convey the notion that the matrix $[\mathbf{M}, \mathbf{N}]$ has fewer rows than columns.) At each iteration a steplength is chosen to ensure that the next iterate remains positive and a centering step is added to the Newton step in order to prevent the iterates from approaching the boundary of the positive orthant too early. As the iterates approach a solution, the centering restriction becomes progressively weaker. Given an appropriate starting point, Zhang establishes an $O(n^2 \ln(1/\epsilon))$ – iteration convergence bound for the algorithm to attain ϵ-feasibility and ϵ-complementarity. Moreover, under a set of fairly unrestrictive assumptions and properly controlled steps, the algorithm converges at a global Q-linear rate.

Recently, Wright (1994) has proposed a modification of Zhang's algorithm to obtain an $O(n^2L)$ infeasible-interior-point algorithm for solving (14.1) that exhibits an asymptotic Q-subquadratic convergence rate under the assumptions that (14.1) has a strictly feasible point and exhibits a strictly complementary solution. The iterates follow the Newton direction and, as in Zhang's procedure, are subject to a centering condition that keeps them from approaching the boundary of the positive orthant too soon. However, unlike Zhang's algorithm, Wright also imposes a condition that gives a higher priority to feasibility rather than complementarity. The desirability of each point is described by a ***merit function*** $\phi(\mathbf{X}, \mathbf{Y}) = \mathbf{X}'\mathbf{Y} + \|\mathbf{Y} - \mathbf{M}\mathbf{X} - \mathbf{q}\|$ whose first term measures uncomplementarity and whose second term depicts infeasibility. Hence a point $(\mathbf{X}_*, \mathbf{Y}_*)$ is a solution of (14.1) if and only if $(\mathbf{X}_*, \mathbf{Y}_*) \geqq (\mathbf{0}, \mathbf{0})$ and $\phi(\mathbf{X}_*, \mathbf{Y}_*) = 0$. As the iterations progress and the merit function exceeds a certain threshold level, Wright's algorithm is identical to Zhang's algorithm. At each iteration a linear system of equations is solved to yield a so-called *safe step*. If the value of the merit function falls below the threshold, a *fast step* that may yield rapid convergence is determined by the algorithm. If the fast step does not render a significant reduction in the merit function, it is discarded in favor of a safe step.

Additional algorithms designed to solve linear complementarity problems that are slightly more general than (14.1) have been proposed by Mizuno (1992); Kojima, Mizuno, and Yoshise (1991b); and Kojima, Megiddo, Noma, and Yoshise (1991a). Indeed, unified models involving various path-following and potential reduction models as special cases have been developed.

For instance, Mizuno (1992) presents a new polynomial-time algorithm for solving a linear complementarity problem with $\mathbf{M}$ positive semidefinite and which *follows a sequence of points* to a solution of a linear complementarity problem more general than, say, (14.1.1). That is, if the sequence falls on a path, it constitutes a path-following method; and if the sequence is based on a potential function, it amounts to a potential-reduction method. Given an initial feasible point whose components are bounded between $2^{-O(L)}$, the algorithm solves (14.1.1) in $O(\sqrt{n}\,L)$ iterations.

In this regard, for each $\mathbf{V} \in \mathbf{R}^n_{++}$ and $\alpha \geq 0$, let us define a neighborhood $\delta(\mathbf{V}, \alpha) = \{\mathbf{U} \,\|\, \mathbf{V} - \mathbf{U} \,\| \leq \alpha v_{min}\}$ in $\mathbf{R}^n$, where $v_{min} = min\{v_i, i=1,\ldots,n\}$. Then for each $\epsilon \geq 0$, $\mathbf{V}^* \in \mathbf{R}^n_{++}$, and $(\mathbf{X}_0, \mathbf{Y}_0) \in \mathbf{R}^{2n}_{++}$, Mizuno considers the problem:

$$
\begin{aligned}
&\text{Find an } (\mathbf{X},\mathbf{Y}) \in S_{++} \text{ such that} \\
&\bar{\mathbf{X}}\mathbf{Y} \in \delta(\mathbf{V}^*, \epsilon), \text{ where} \\
&S = \{(\mathbf{X},\mathbf{Y}) \in \mathbf{R}^{2n} \mid \mathbf{Y} - \mathbf{Y}_0 = \mathbf{M}(\mathbf{X} - \mathbf{X}_0)\} \\
&S_{++} = S \cap \mathbf{R}^{2n}_{++}.
\end{aligned}
\tag{14.19}
$$

Mizuno's technique solves (14.19) for each $\epsilon > 0$ and, if $(\mathbf{X}_0, \mathbf{Y}_0)$ is a feasible solution of (14.1.1), then any solution of (14.19) for $\epsilon = 1$ and $\mathbf{V}^* = 2^{-2(L+1)}\mathbf{1}$ is a solution of the problem:

$$
\begin{aligned}
&\text{Find an } (\mathbf{X},\mathbf{Y}) \in \mathbf{R}_{++} \text{ such that} \\
&\mathbf{Y} = \mathbf{M}\mathbf{X} + \mathbf{q},\ \bar{\mathbf{X}}\mathbf{Y} \leq 2^{-2L}\mathbf{1}.
\end{aligned}
\tag{14.20}
$$

Moreover, Kojima, *et al.* (1989a) demonstrate that the solution of (14.1.1) can be obtained from the solution of (14.20) in $O(n^3)$ arithmetic operations. (For details on implementing Mizuno's method for finding a solution to (14.19) see corollaries 4, 5 in Mizuno (1992, pp. 36-37).) Mizuno then goes on to relate his technique to more traditional methods for solving linear complementarity problems by constructing $O(\sqrt{n}\,L)$ path-following and potential-reduction methods for solving (14.20) that are based on his more general method (see Mizuno (1992, pp. 39-42)).

Next, we note briefly that Kojima, Mizuno, and Yoshise (1991b) offer a unified model for both the path-following and potential reduction algorithms (algorithms 14.2 and 14.4 respectively) presented earlier. To obtain their unified model, the authors generalize the potential function f given above by replacing $\sqrt{n}$ by a parameter $v \geq 0$, *i.e.*, now

$$f(\mathbf{X},\mathbf{Y}) \; = \; (n+v)\log\mathbf{X}'\mathbf{Y} - \sum_{i=1}^{n} \log x_i y_i - n \log n.$$

Then for each $(\mathbf{X},\mathbf{Y}) \in \mathcal{K}_{++}$, consider the Newton equations

$$\bar{\mathbf{Y}}\mathbf{d}^X + \bar{\mathbf{X}}\mathbf{d}^Y \; = \; \bar{\mathbf{X}}\bar{\mathbf{Y}}\mathbf{1} - \Big(1 - \tfrac{v}{n+v}\Big)(\mathbf{X}'\mathbf{Y}/n)\mathbf{1} \tag{14.21}$$

obtained from the system $H(\bar{\mu},\mathbf{X},\mathbf{Y}) \; = \; (\bar{\mathbf{X}}\mathbf{Y} - \mu\mathbf{1}, \mathbf{Y} - \mathbf{M}\mathbf{X} - \mathbf{q}) \; = \; \mathbf{0}, \bar{\mu} \; = \; \{1 - v/(n+v)\}(\mathbf{X}'\mathbf{Y}/n)$. Given that the Newton direction $-(\mathbf{d}^X, \mathbf{d}^Y)$ obtained from (14.21) is a continuous function in $(\mathbf{X},\mathbf{Y}) \in \mathcal{K}_{++}$, each resulting trajectory in $\mathcal{K}_{++}$ converges to a solution of $LCP(\mathbf{q},\mathbf{M})$ when $v > 0$, or to a point on the central path $\mathcal{K}_{cen}$ when $v = 0$, with the value of f decreasing monotonically along each trajectory. Kojima, Mizuno, and Yoshise (1991b) note that the path-following algorithm given earlier by Kojima, Mizuno, and Yoshise (1989a) may be regarded as a special case of the continuous model given above, *i.e.*, $v = O(\sqrt{n})$ with $v/(n+v) \; = \; \delta/\sqrt{n}$, $\delta > 0$, was used in the path-following algorithm of Kojima, *et al.* (1989a).

Finally, Kojima, Megiddo, Noma, and Yoshise (1991a) extend the unified approach of Kojima, Mizuno,and Yoshise (1991b) to a large class of interior point potential-reduction algorithms that: (a) move in a Newton direction toward the central trajectory; and (b) choose a new point along the said direction in a certain neighborhood of the central trajectory such that a significant reduction in the potential function occurs. Their chapter two provides an excellent summary of the unified interior point method along with global and polynomial-time convergence results for some special cases of the same such as potential-reduction and path-following algorithms and a damped Newton method for solving linear complementarity problems starting from an interior feasible solution. On all this see Kojima, Megiddo, Noma, and Yoshise (1991a, pp. 7-23).

APPENDIX A:

UPDATING THE BASIS INVERSE

It was noted in section 5.5 above that to obtain a new basic feasible solution to a linear programming problem via the simplex method, we need change only a single vector in the basic matrix at each iteration. Hence we are faced with the problem of finding the inverse of a matrix $\widehat{\mathbf{B}}$ which has only one column different from that of a matrix $\mathbf{B}$ whose inverse is known. In this regard, given $\mathbf{B} = [\mathbf{b}_1, \ldots, \mathbf{b}_m]$ with $\mathbf{B}^{-1}$ known, we wish to replace the rth column $\mathbf{b}_r$ of $\mathbf{B}$ by $\mathbf{r}_j$ so as obtain the inverse of $\widehat{\mathbf{B}} = [\mathbf{b}_1, \ldots, \mathbf{b}_{r-1}, \mathbf{r}_j, \mathbf{b}_{r+1}, \ldots, \mathbf{b}_m]$. Since the columns of $\mathbf{B}$ form a basis for $\mathbf{R}^m$, $\mathbf{r}_j = \mathbf{B}\mathbf{Y}_j$. If the columns of $\widehat{\mathbf{B}}$ are also to yield a basis for $\mathbf{R}^m$, then, by theorem 2.33, it must be true that $y_{rj} \neq 0$. So if

$$\begin{aligned}\mathbf{r}_j &= \sum_{i=1}^{m} y_{ij}\mathbf{b}_i \\ &= y_{1j}\mathbf{b}_1 + \cdots + y_{r-1,j}\mathbf{b}_{r-1} + y_{rj}\mathbf{b}_r + y_{r+1,j}\mathbf{b}_{r+1} + \cdots + y_{mj}\mathbf{b}_m\end{aligned}$$

with $y_{rj} \neq 0$, then

$$\begin{aligned}\mathbf{b}_r &= -\frac{y_{1j}}{y_{rj}}\mathbf{b}_1 - \cdots - \frac{y_{r-1,j}}{y_{rj}}\mathbf{b}_{r-1} + \frac{1}{y_{rj}}\mathbf{r}_j - \frac{y_{r+1,j}}{y_{rk}}\mathbf{b}_{r+1} - \cdots - \frac{y_{mj}}{y_{rj}}\mathbf{b}_m \\ &= \widehat{\mathbf{B}}\widehat{\mathbf{Y}}_j,\end{aligned}$$

where $\widehat{\mathbf{Y}}'_j = \left(-\frac{y_{1j}}{y_{rj}}, \ldots, -\frac{y_{r-1,j}}{y_{rj}}, \frac{1}{y_{rj}}, -\frac{y_{r+1,j}}{y_{rj}}, \ldots, -\frac{y_{mj}}{y_{rj}}\right)$. Hence $\mathbf{b}_r$

has been removed from the basis with $\mathbf{r}_j$ entered in its stead. If we next replace the rth column of the mth order identity matrix $\mathbf{I}_m$ by $\widehat{\mathbf{Y}}_j$, we obtain

$$\widehat{\mathbf{I}}_m = [\mathbf{e}_1,\ldots,\mathbf{e}_{r-1},\widehat{\mathbf{Y}}_j,\mathbf{e}_{r+1},\ldots,\mathbf{e}_m], \tag{A.1}$$

where, as previously defined, $\mathbf{e}_j$ is the jth unit column vector. Now it is easily demonstrated that $\mathbf{B} = \widehat{\mathbf{B}}\widehat{\mathbf{I}}_m$ so that

$$\widehat{\mathbf{B}}^{-1} = \widehat{\mathbf{I}}_m\mathbf{B}^{-1}. \tag{A.2}$$

APPENDIX B:

STEEPEST EDGE SIMPLEX METHODS

We noted in section 5.1 that, in the absence of degeneracy, the standard (primal) simplex method starts from an initial basic feasible solution and proceeds along a path to the optimum for which the value of the objective function at any extreme point is greater than the objective value at an adjacent extreme point. If the primal problem is a minimization program of the form

$$\begin{aligned} \min f &= \mathbf{C}'\mathbf{X} \quad \text{s.t.} \\ \mathbf{AX} &= \mathbf{b},\ \mathbf{X} \geqq \mathbf{0} \end{aligned}$$

(an assumption consistent with the development of this body of theory), where $\mathbf{A}$ is of order $(m \times n)$ with $\rho(\mathbf{A}) = m < n$, then the simplex method traverses a path to the minimum from an initial extreme point to an adjacent extreme point along edges of the polytope that render the objective function strictly decreasing. Hence under minimization, the direction taken is simply *downhill*. However, in the ***steepest edge simplex routine*** (see Kuhn and Quandt (1963), Harris (1973), Goldfarb (1976), Goldfarb and Reid (1977), and Forrest and Goldfarb (1992)), the direction selected at each iteration is the one that is *most downhill*. That is, the direction edge chosen is the one which makes the most obtuse angle with $\mathbf{C}$, *i.e.*, the edge direction η for which the directional derivative of f in the η direction or $\mathbf{C}'(\eta/\|\eta\|)$ is smallest.

Following Forrest and Goldfarb, let us assume that at the start of a single step of the simplex method the first m components of $\mathbf{X}$ (and thus the first m columns of $\mathbf{A}$) are basic. Then $\mathbf{X}, \mathbf{A}$ may be partitioned as $\mathbf{X}' = (\mathbf{X}'_B, \mathbf{X}'_R)$, $\mathbf{A} = [\mathbf{B}, \mathbf{R}]$ respectively so that a simplex step has the form

$$\bar{\mathbf{X}} = \mathbf{X} + \theta \boldsymbol{\eta}_q = \begin{bmatrix} \mathbf{B}^{-1}\mathbf{b} \\ \mathbf{0} \end{bmatrix} + \theta \boldsymbol{\eta}_q, \tag{B.1}$$

where $\mathbf{X}$ is the basic solution at the start of the step, θ is the steplength, and $\boldsymbol{\eta}_q$ is one of the edge directions

$$\boldsymbol{\eta}_j = \begin{bmatrix} -\mathbf{B}^{-1}\mathbf{R} \\ \mathbf{I}_p \end{bmatrix} \mathbf{e}_{j-m}, \quad j = m+1, \ldots, n = p+m, \tag{B.2}$$

emanating from vertex $\mathbf{X}$. If $-\mathbf{B}^{-1}\mathbf{R} = [-\mathbf{Y}_1, \ldots, -\mathbf{Y}_p]$, then, from (B.2), the collection of edge directions appears as

$$\boldsymbol{\eta}_{m+1} = \begin{bmatrix} -\mathbf{Y}_1 \\ \mathbf{e}_1 \end{bmatrix}, \boldsymbol{\eta}_{m+2} = \begin{bmatrix} -\mathbf{Y}_2 \\ \mathbf{e}_2 \end{bmatrix}, \ldots, \boldsymbol{\eta}_{m+p} = \begin{bmatrix} -\mathbf{Y}_p \\ \mathbf{e}_p \end{bmatrix}. \tag{B.2.1}$$

Hence the current basic solution $\mathbf{X}$ is a vertex of the polytope $\mathcal{K} = \{\mathbf{X} \mid \mathbf{AX} = \mathbf{b}, \mathbf{X} \geqq \mathbf{0}, \mathbf{X} \in \mathbf{R}^n\}$ and the p vectors $\boldsymbol{\eta}_{m+1}, \ldots, \boldsymbol{\eta}_{m+p}$ point down the p edges of $\mathcal{K}$ that emanate from $\mathbf{X}$. In the light of (B.2.1), (B.1) becomes

$$\bar{\mathbf{X}} = \begin{bmatrix} \mathbf{B}^{-1}\mathbf{b} \\ \mathbf{0} \end{bmatrix} + \theta \begin{bmatrix} -\mathbf{Y}_q \\ \mathbf{e}_q \end{bmatrix}. \tag{B.1.1}$$

In the standard primal simplex routine, if basic x_{Bt} is to be replaced by nonbasic x_{Rq}, then θ is a scalar (determined by a ratio test) which makes $x_{Bt} = 0$. Moreover, since the angle between $\boldsymbol{\eta}_q$ and $\mathbf{C}$ must be obtuse, the *reduced cost*

$$\mathbf{C}'\boldsymbol{\eta}_q = (\mathbf{C}'_B, \mathbf{C}'_R)\begin{bmatrix} -\mathbf{Y}_q \\ \mathbf{e}_q \end{bmatrix} = -\mathbf{C}'_B \mathbf{Y}_q + \mathbf{C}_{Rq} = \bar{c}_q$$

must be negative, *i.e.*, $\boldsymbol{\eta}_q$ points *downhill*. By virtue of the usual primal entry criterion,

$$\bar{c}_q = \mathbf{C}'\boldsymbol{\eta}_q = \min_{j > m} \{\bar{c}_j, \bar{c}_j < 0\}.$$

In the steepest-edge algorithm, the direction edge $\boldsymbol{\eta}_q$ chosen (or the nonbasic variable x_{Rq} selected to become basic) is the edge that is *most downhill* relative to $\mathbf{C}$ or

$$\frac{\bar{c}_q}{\|\eta_q\|} = \frac{C'\eta_q}{\|\eta_q\|} = \min_{j>m}\left\{\frac{C'\eta_j}{\|\eta_j\|},\ C'\eta_j < 0\right\}. \tag{B.3}$$

So in the standard simplex method, we minimize the reduced cost $\bar{c}_j$ (the change in the objective function per unit change in x_{Rj}) while the steepest-edge method has us minimize *normalized reduced cost* $C'\eta_j/\|\eta_j\|$. The latter entry criterion corresponds to choosing the edge of steepest descent of f.

To expedite pivoting, Goldfarb and Reid (1977) (see also Forrest and Goldfarb (1992)) derive recursive formulas for updating after each pivot the norms $\|\eta_j\|$ or the quantities $\gamma_j = \|\eta_j\|^2 = \eta_j'\eta_j$ for all nonbasic subscripts $j > m$. In this regard, if the nonbasic variable x_{Rq} replaces x_{Bt} in the set of basic variables, then the edge direction recurrence formulas appear as

$$\begin{aligned} &(a)\ \bar{\eta}_t = -\eta_q/\alpha_q, \\ &(b)\ \bar{\eta}_j = \eta_j - \eta_q\bar{\alpha}_j, j > m, j \neq q, \end{aligned} \tag{B.4}$$

where $\alpha_j = e_t'B^{-1}a_j$ (the components of row t of $B^{-1}a_j$), $\bar{\alpha}_j = \alpha_j/\alpha_q$ (the components of row t of $\bar{B}^{-1}a_j$, where $\bar{B}$ is the new basis matrix after pivoting); and, from (B.4), we can obtain the recurrence formulas

$$\begin{aligned} &(a)\ \bar{\gamma}_t = \gamma_q/\alpha_q^2, \\ &(b)\ \bar{\gamma}_j = \gamma_j - 2\bar{\alpha}_j a_j'V + \bar{\alpha}_j^2\gamma_q, j > m, j \neq q, \end{aligned}$$

where $BW = a_q$ and $B'V = W$ (note that $\alpha_q = w_t$).

We can also update the reduced costs by obtaining, from (B.4), a recurrence of the form

$$\begin{aligned} &(a)\quad \bar{\bar{c}}_q = C'\bar{\eta}_t = -C'\eta_q/\alpha_q = -\bar{c}_q/\alpha_q, \\ &(b)\quad \bar{\bar{c}}_j = C'\bar{\eta}_j = C'\bar{\eta}_j - C'\eta_q\bar{\alpha}_j = \bar{c}_j - \bar{c}_q\bar{\alpha}_j, j > m, j \neq q. \end{aligned} \tag{B.5}$$

Based upon the preceding discussion let us now formally offer a single step of the Goldfarb and Reid (1977)

B.1. PRIMAL STEEPEST EDGE ALGORITHM.

STEP 1. Check optimality.

If $\bar{c}_j > 0$, $j > m$, the current solution is optimal and the algorithm terminates. Otherwise proceed to

STEP 2. Determine the pivot column.

Find $\bar{c}_q / \| \eta_q \|$ from (B.3) or equivalently find

$$\max_{j > m}\left\{\bar{c}_j^2/\gamma_j \mid \bar{c}_j < 0\right\} = \bar{c}_q^2/\gamma_q \tag{B.3.1}$$

and compute $\mathbf{W} = \mathbf{B}^{-1}\mathbf{a}_q$.

STEP 3. Check for an unbounded solution.

If $w_i \leq 0$, $i=1,\ldots,m$, then the problem is unbounded and the algorithm terminates. Otherwise proceed to

STEP 4. Determine the pivot row.

Find
$$\min_{1 \leq i \leq m}\{x_{Bi}/w_i \mid w_i > 0\} = x_{Bt}/w_t$$

and store w_t separately.

STEP 5. Update the basic solution.

$$\begin{aligned}\bar{x}_{Bt} &= x_{Bt}/w_t,\\ \bar{x}_{Bi} &= x_{Bi} - w_i\bar{x}_{Bt},\ i=1,\ldots,m;\ i \neq t.\end{aligned}$$

STEP 6. Update $\bar{c}, \gamma$, and $\mathbf{B}^{-1}$: compute $\bar{c}_q, \gamma_q$ afresh as

$$\begin{aligned}\bar{c}_q &= \mathbf{C}'\mathbf{e}_q - \mathbf{C}_B'\,\mathbf{W},\\ \gamma_q &= 1 + \mathbf{W}'\mathbf{W}.\end{aligned}$$

Let $\mathbf{V} = (\mathbf{B}')^{-1}\mathbf{W}$.

Update $\mathbf{B}^{-1}$ to $\bar{\mathbf{B}}^{-1}$ and compute

$$\begin{aligned}\bar{\alpha}_j &= \mathbf{e}_t'\bar{\mathbf{B}}^{-1}\mathbf{a}_j \ (\text{or } \bar{\alpha}_j = \mathbf{e}_t'\mathbf{B}^{-1}\mathbf{a}_j/w_t),\\ \bar{\bar{c}}_j &= \bar{c}_j - \bar{\alpha}_j\bar{c}_q,\\ \bar{\gamma}_j &= \max\left(\gamma_j - 2\bar{\alpha}_j\mathbf{a}_j'\mathbf{V} + \bar{\alpha}_j^2\gamma_q, 1 + \bar{\alpha}_j^2\right),\\ \bar{\bar{c}}_q &= -\bar{c}_j/w_t,\\ \bar{\gamma}_q &= \gamma_q/w_t^2,\end{aligned}$$

$j > m, j \neq q$.

Let us next consider a steepest-edge modification of the dual simplex method (Forrest and Goldfarb (1992)). To apply the dual simplex routine to the preceding primal problem ($\min f = \mathbf{C}'\mathbf{X}$ s.t. $\mathbf{AX} = \mathbf{b}$, $\mathbf{X} \geqq \mathbf{0}$), let us consider the associated dual program

$$\begin{gathered} max\, g = \mathbf{b}'\mathbf{U} \quad \text{s.t.} \\ \mathbf{A}'\mathbf{U} \leqq \mathbf{C},\ \mathbf{U} \text{ unrestricted.} \end{gathered}$$

Then the dual solution $\mathbf{U}$ corresponding to the primal basic feasible solution $\mathbf{X}_B = \mathbf{B}^{-1}\mathbf{b}$, $\mathbf{X}_R = \mathbf{0}$ must satisfy, under weak complementary slackness,

$$\begin{aligned} &\text{(a)} \quad \mathbf{B}'\mathbf{U} = \mathbf{C}_B, \\ &\text{(b)} \quad \mathbf{R}'\mathbf{U} \leqq \mathbf{C}_R. \end{aligned} \tag{B.6}$$

In this regard, on a dual simplex step, selecting x_{Bt} to leave the current set of basic variables amounts to relaxing the tth equation in (B.6.a), *i.e.*, the tth constraint $\mathbf{a}_t'\mathbf{U} \leq c_t$ can now hold as a strict inequality, where $\mathbf{a}_t$ is the tth column of $\mathbf{A}$.

For a dual solution of the form

$$\bar{\mathbf{U}} = \mathbf{U} + \theta\rho_t = (\mathbf{B}')^{-1}\mathbf{C}_B + \theta\rho_t,$$

where $\mathbf{B}'\rho_t' = -\mathbf{e}_t$, steplength $\theta \geq 0$, $\mathbf{a}_i'\bar{\mathbf{U}} = c_i, 1 \leq i \leq m,\ i \neq t$, and $\mathbf{a}_t'\bar{\mathbf{U}} \leq c_t$, the rows $\rho_i, i=1,\ldots,m$, of $-\mathbf{B}^{-1}$ are taken as the edge directions emanating from the vertex $\mathbf{U}$ of the dual polyhedron $\{\mathbf{U} \mid \mathbf{A}'\mathbf{U} \leqq \mathbf{C},\ \mathbf{U} \in R^m\}$. So in the dual simplex method the edge direction ρ_t selected (or the basic variable x_{Bt} chosen to become nonbasic) must make an acute angle with the gradient $\mathbf{b}$ of the dual objective function or $\rho_t\mathbf{b} = -\mathbf{e}_t'\mathbf{B}^{-1}\mathbf{b} = -x_{Bt}$ must be positive (or $x_{Bt} < 0$). Moreover, if ρ_t is a steepest edge, then it must be the one that points *most downhill.* Hence the direction edge chosen is the one which makes the smallest acute angle with $\mathbf{b}$, *i.e.*, the ρ direction or $(\rho/\|\rho\|)\mathbf{b}$ is smallest. Thus the direction edge ρ_t selected is the edge that is most downhill relative to $\mathbf{b}$ or

$$\frac{\rho_t\mathbf{b}}{\|\rho_t\|} = \min_{i \leq m}\left\{\frac{\rho_i\mathbf{b}}{\|\rho_i\|},\ \rho_i\mathbf{b} > 0\right\}.$$

As was the case with the primal steepest edge algorithm given above, a key element underlying steepest-edge pivoting in the dual simplex method is updating after each iteration the norms $\|\rho_i\|$ or the quantities $\beta_i = \rho_i\rho_i' = \|\rho_i\|^2$, $i=1,\ldots,m$. In this regard, if variable x_{Rq} replaces variable x_{Bt} in the set of basic variables, then the updated basis matrix inverse is

$$\bar{\mathbf{B}}^{-1} = \mathbf{B}^{-1} - \frac{\mathbf{W} - \mathbf{e}_t}{w_t}\mathbf{e}_t'\mathbf{B}^{-1},$$

where $\mathbf{W} = \mathbf{B}^{-1}\mathbf{a}_q$. Then

$$\begin{aligned} &\text{(a)}\ \bar{\rho}_t = (1/w_t)\rho_t, \\ &\text{(b)}\ \bar{\rho}_i = \rho_i - (w_i/w_t)\rho_t, i=1,\ldots,m,\ i \neq t, \end{aligned} \tag{B.7}$$

and

$$\begin{aligned}
&\text{(a) } \bar{\beta}_t = (1/w_t)^2\beta_t, \\
&\text{(b) } \bar{\beta}_i = \beta_i - 2(w_i/w_t)\tau_i + (w_i/w_t)^2\beta_t,\ i=1,\ldots,m,\ i \neq t,
\end{aligned} \tag{B.8}$$

where $\mathbf{B}'\boldsymbol{\sigma} = \mathbf{e}_t, \mathbf{B}\boldsymbol{\tau} = \boldsymbol{\sigma}$, and $\beta_t = \boldsymbol{\sigma}'\boldsymbol{\sigma}$.

Forrest and Goldfarb extend their steepest edge simplex algorithm to the case where upper bounds on the primal variable are present, *i.e.*, to solving a problem of the form

$$\begin{aligned}
\min f &= \mathbf{C}'\mathbf{X} \quad \text{s.t.} \\
\mathbf{AX} &\leqq \mathbf{b},\ \mathbf{0} \leqq \mathbf{X} \leqq \mathbf{U}.
\end{aligned}$$

By working with the dual variant of this program or with

$$\begin{aligned}
\max g &= \mathbf{b}'\mathbf{Y} - \mathbf{U}'\mathbf{V} \quad \text{s.t.} \\
\mathbf{A}'\mathbf{Y} - \mathbf{V} + \mathbf{S} &= \mathbf{C} \\
\mathbf{Y} \text{ unrestricted},\ &\mathbf{V} \geqq \mathbf{0},\ \mathbf{S} \geqq \mathbf{0},
\end{aligned}$$

where $\mathbf{S}$ is a vector of dual slack variables, Forrest and Goldfarb apply the (dual) steepest edge simplex routine and demonstrate that the preceding recursive formulas (B.8) with $\boldsymbol{\tau}$ appropriately defined (computed) still apply. (For details see Forrest and Goldfarb (1992, pp. 348-350).)

In addition to the basic primal and dual steepest edge simplex algorithms, Forrest and Goldfarb offer additional algorithmic embellishments involving a discussion of ***approximate steepest-edge algorithms*** (variants of steepest edge algorithms in which the edge direction norms used in pivoting are replaced by approximate weights) as well as ***projected steepest edge algorithms*** (modifications of steepest edge algorithms that are based on, say, the exact norms of subvectors $\tilde{\boldsymbol{\eta}}_j$ comprising only those components of the $\boldsymbol{\eta}_j$'s in a certain *reference framework* which is initially and periodically identified with the current set of nonbasic variables.)

APPENDIX C:

DERIVATION OF THE PROJECTION MATRIX[1]

We mentioned in section 13.3 above that in order to solve problem (13.5.2) we had to move from $\mathbf{Y}_0 = (\frac{1}{n})\mathbf{1}$ in the direction of steepest descent $-\mathbf{C}_p$ to the boundary of the sphere $B(\mathbf{Y}_0, \alpha r)$. Here $\mathbf{C}_p$ depicts the orthogonal projection of the transformed cost vector $\bar{\mathbf{C}}$ onto the null-space of $\mathbf{P}$ (or onto the constraint surface of $\mathbf{PY}=\mathbf{P}_0$), where $\bar{\mathbf{C}}, \mathbf{P}$ are specified in (13.5.1). Thus at a feasible point $\mathbf{Y}_0$ we seek a feasible (descent) direction $\mathbf{d}$ satisfying $\mathbf{C}'\mathbf{d} < 0$ given that $\mathbf{d}$ must also satisfy $\mathbf{Pd} = \mathbf{0}$ with $\rho(\mathbf{P}) = m+1$, *i.e.*, $\mathbf{d}$ is an element of the tangent subspace $\mathcal{M} = \{\mathbf{d} \mid \mathbf{Pd} = \mathbf{0}, \mathbf{d} \in \mathbf{R}^n\}$ of dimension $n-m-1$ in $\mathbf{R}^n$ spanned by the gradients (the rows of $\mathbf{P}$) of the transformed constraint equations. Consistent with this requirement, the direction vector selected must be the orthogonal projection of the gradient $\bar{\mathbf{C}}$ onto the subspace $\mathcal{M}$.

To obtain this projection let us represent the $(m-1)$-dimension subspace orthogonal to $\mathcal{M}$ as $\mathcal{M}^{\perp} = \{\mathbf{X} \mid \mathbf{X} = \mathbf{P}'\boldsymbol{\lambda}, \boldsymbol{\lambda} \in \mathbf{R}^{m+1}\}$ in $\mathbf{R}^n$. Since any vector in $\mathbf{R}^n$ can be expressed as the sum of vectors from each of these two complementary subspaces, we may write $\bar{\mathbf{C}} = \mathbf{d} + \mathbf{P}'\boldsymbol{\lambda}$, where $\mathbf{d} \in \mathcal{M}$, $\mathbf{P}'\boldsymbol{\lambda} \in \mathcal{M}^{\perp}$. From this representation of $\bar{\mathbf{C}}$ we may solve for $\boldsymbol{\lambda}$ as follows. Since $\mathbf{d} = \bar{\mathbf{C}} - \mathbf{P}'\boldsymbol{\lambda}$,

$$\mathbf{Pd} = \mathbf{P}\bar{\mathbf{C}} - \mathbf{PP}'\boldsymbol{\lambda} = \mathbf{0}$$

[1] The reader is encouraged to become familiar with the concepts developed in sections 2.2-2.4 before attempting to read this material.

and thus

$$\boldsymbol{\lambda} = (\mathbf{PP}')^{-1}\mathbf{P}\bar{\mathbf{C}}$$

($(\mathbf{PP}')^{-1}$ exists since the rows of $\mathbf{P}$ are linearly independent). Then

$$\begin{aligned}\mathbf{d} = \mathbf{C}_p &= \bar{\mathbf{C}} - \mathbf{P}'\boldsymbol{\lambda} \\ &= \bar{\mathbf{C}} - \mathbf{P}'(\mathbf{PP}')^{-1}\mathbf{P}\bar{\mathbf{C}} \\ &= [\mathbf{I}_n - \mathbf{P}'(\mathbf{PP}')^{-1}\mathbf{P}]\bar{\mathbf{C}} = \mathbf{P}_r\bar{\mathbf{C}}. \qquad \text{(C.1)}\end{aligned}$$

Here

$$\mathbf{P}_r = [\mathbf{I}_n - \mathbf{P}'(\mathbf{PP}')^{-1}\mathbf{P}] \qquad \text{(C.2)}$$

is termed the ***projection operator*** or ***projection matrix*** associated with subspace $\mathcal{M}$, *i.e.*, the operation of $\mathbf{P}_r$ on any vector in $\mathbf{R}^n$ yields the projection of that vector onto $\mathcal{M}$. Note that $\mathbf{P}_r = \mathbf{P}_r'$, $\mathbf{P}_r = \mathbf{P}_r^2$, and $\mathbf{PP}_r = \mathbf{0}$. If $\mathbf{C}_p = \mathbf{P}_r\bar{\mathbf{C}} \neq \mathbf{0}$, then $\mathbf{PC}_p = \mathbf{0}$ and $-\mathbf{C}_p$ serves as a descent direction on the constraint surface $\{\mathbf{Y} \mid \mathbf{PY} = \mathbf{P}_0\}$. That is, movement along $-\mathbf{C}_p$ improves the objective value since $\bar{\mathbf{C}}'(-\mathbf{d}) = \bar{\mathbf{C}}'(-\mathbf{C}_p) = (\mathbf{C}_p+\mathbf{V})'(-\mathbf{C}_p) = -\mathbf{C}_p'\mathbf{C}_p < 0$.

Let us examine the geometry involved in the derivation of equation (C.1). Again $\mathcal{M} = \{\mathbf{Z} \mid \mathbf{PZ} = \mathbf{0}, \mathbf{Z} \in \mathbf{R}^n\}$ depicts the null-space of $\mathbf{P}$ while its complementary subspace $\mathcal{M}^{\perp} = \{\mathbf{X} \mid \mathbf{X} = \mathbf{P}'\boldsymbol{\lambda}, \boldsymbol{\lambda} \in \mathbf{R}^{m+1}\}$. Let

$$\bar{\mathbf{C}} = \mathbf{C}_p + \mathbf{V} = \mathbf{C}_p + (\bar{\mathbf{C}} - \mathbf{C}_p) \notin \mathcal{M},$$

where $\mathbf{C}_p$ is the orthogonal projection of $\bar{\mathbf{C}}$ on $\mathcal{M}$ and $\bar{\mathbf{C}} - \mathbf{C}_p$ is orthogonal to all vectors in $\mathcal{M}$ or $(\bar{\mathbf{C}} - \mathbf{C}_p) \in \mathcal{M}^{\perp}$ (Figure (C.1). Hence

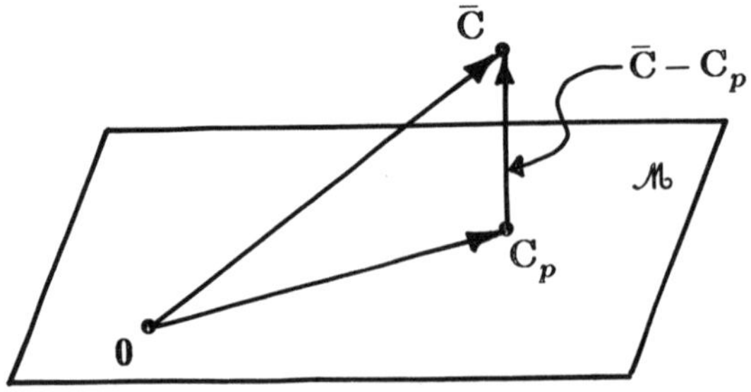

Figure C.1

$\bar{\mathbf{C}}-\mathbf{C}_p$ must be a linear combination of the columns of $\mathbf{P}'$ or $\bar{\mathbf{C}}-\mathbf{C}_p = \mathbf{P}'\boldsymbol{\lambda}$. Since $\mathbf{C}_p$ is orthogonal to the columns of $\mathbf{P}'$, it follows that

$$\mathbf{P}(\bar{\mathbf{C}}-\mathbf{P}'\boldsymbol{\lambda}) = \mathbf{0}.$$

Given that $\mathbf{PP}'$ has rank $m+1$ ($(\mathbf{PP}')^{-1}$ exists), the preceding expression yields

$$\boldsymbol{\lambda} = (\mathbf{PP}')^{-1}\mathbf{P}\bar{\mathbf{C}}.$$

Since $\mathbf{P}'\boldsymbol{\lambda} = \mathbf{P}'(\mathbf{PP}')^{-1}\mathbf{P}\bar{\mathbf{C}}$ is a vector in $\mathcal{M}^{\perp}$, we define

$$\mathbf{M} = \mathbf{P}'(\mathbf{PP}')^{-1}\mathbf{P}$$

to be the orthogonal projection operator or projection matrix onto $\mathcal{M}^{\perp}$, *i.e.*, onto the linear subspace spanned by the columns of $\mathbf{P}'$. Then $\mathbf{I}_n - \mathbf{M}$ is the projection operator orthogonal to $\mathcal{M}^{\perp}$ or onto $\mathcal{M}$. Hence $\mathbf{C}_p = (\mathbf{I}_n-\mathbf{M})\bar{\mathbf{C}} = \mathbf{P}_r\bar{\mathbf{C}}$ or (C.1) above.

Let us consider two alternative approaches to the derivation of (C.1) which involve the direct solution of a constrained optimization problem. First, consider

$$min\,\bar{\mathbf{C}}'\mathbf{h} \quad \text{s.t.} \quad \mathbf{Ph} = \mathbf{0},\ \mathbf{h}'\mathbf{h} = 1, \tag{C.3}$$

where $\mathbf{P}$ is of order $(m+1 \times n)$ and of full rank $m+1 \leq n$ with $\mathbf{h}$ a (unit) direction vector. From the (primal) Lagrangian function

$$\mathcal{L} = \bar{\mathbf{C}}'\mathbf{h} + \boldsymbol{\lambda}_1'(-\mathbf{P}\boldsymbol{\lambda}) + \lambda_2(1-\mathbf{h}'\mathbf{h})$$

we may write the first-order conditions for a minimum as

$$\begin{aligned} &\text{(a)}\ \mathcal{L}_{\mathbf{h}} = \bar{\mathbf{C}} - \mathbf{P}'\boldsymbol{\lambda}_1 - 2\lambda_2\mathbf{h} = \mathbf{0}, \\ &\text{(b)}\ \mathcal{L}_{\boldsymbol{\lambda}_1} = -\mathbf{Ph} = \mathbf{0} \\ &\text{(c)}\ \mathcal{L}_{\lambda_2} = 1 - \mathbf{d}'\mathbf{d} = 0. \end{aligned} \tag{C.4}$$

From (C.4.a) we have

$$\mathbf{h} = \frac{1}{2\lambda_2}(\bar{\mathbf{C}}-\mathbf{P}'\boldsymbol{\lambda}_1). \tag{C.5}$$

Substituting this expression into (C.4.b) yields

$$\boldsymbol{\lambda}_1 = (\mathbf{PP}')^{-1}\mathbf{P}\bar{\mathbf{C}}. \tag{C.6}$$

Both (C.5), (C.6) together allow us to write, using (C.4.C), $2\lambda_2 = \bar{\mathbf{C}}'\mathbf{h} = -1$ or $\lambda_2 = -\frac{1}{2}$ (since $\bar{\mathbf{C}}+\mathbf{h},\mathbf{h}$ are orthogonal). Combining this result with (C.5), (C.6) yields

$$\mathbf{h} = -[\mathbf{I}_n - \mathbf{P}'(\mathbf{PP}')^{-1}\mathbf{P}]\,\bar{\mathbf{C}} = -\hat{\mathbf{C}} = -\mathbf{C}_p/\|\mathbf{C}_p\|, \tag{C.7}$$

i.e. $\mathbf{h}$ is a unit vector in the direction of $-\mathbf{C}_p$. In sum, the optimal solution to problem (C.3) is $\mathbf{h} = -\mathbf{C}_p/\|\mathbf{C}_p\|$, where $-\mathbf{C}_p$ is the projection of $-\bar{\mathbf{C}}$ onto the null-space of $\mathbf{P}$.

Next, an alternative track to the derivation of (C.2) has us solve the least squares problem

$$min\|\bar{\mathbf{C}} - \mathbf{C}_p\|^2 \quad \text{s.t.} \quad \mathbf{PC}_p = \mathbf{0}. \tag{C.8}$$

Here we seek to make the distance between $\bar{\mathbf{C}}$ and $\mathbf{C}_p$ as small as possible given that $\mathbf{C}_p$ is an element of the null-space of $\mathbf{P}$. From the Lagrangian

$$\mathcal{L} = (\bar{\mathbf{C}} - \mathbf{C}_p)'(\bar{\mathbf{C}} - \mathbf{C}_p) + \lambda(-\mathbf{PC}_p)$$

we find

$$\begin{aligned} &\text{(a)}\ \mathcal{L}_{\mathbf{C}_p} = 2(\bar{\mathbf{C}} - \mathbf{C}_p) - \mathbf{P}'\lambda' = \mathbf{0}, \\ &\text{(b)}\ \mathcal{L}_{\lambda} = -\mathbf{PC}_p = \mathbf{0}. \end{aligned} \tag{C.9}$$

From (C.9.a),

$$\mathbf{C}_p = \bar{\mathbf{C}} - \tfrac{1}{2}\mathbf{P}'\lambda'. \tag{C.10}$$

Upon substituting this expression into (C.9.b) we have

$$\lambda' = 2(\mathbf{PP}')^{-1}\mathbf{P}\bar{\mathbf{C}}. \tag{C.11}$$

Then inserting (C.11) into (C.10) yields

$$\mathbf{C}_p = [\mathbf{I}_n - \mathbf{P}'(\mathbf{PP}')^{-1}\mathbf{P}]\bar{\mathbf{C}}.$$

An alternative least squares approach to finding the (negative) projected gradient is offered by Todd (1988a). Specifically, for

$$\mathbf{P} = \begin{bmatrix} \mathbf{AD}_k \\ \mathbf{1}' \end{bmatrix} \text{ and } \bar{\mathbf{C}} = \mathbf{D}_k\mathbf{C},$$

$\mathbf{AD}_k\mathbf{1} = \mathbf{0}$ (the current solution is feasible) implies that

$$\mathbf{PP'} = \begin{bmatrix} \mathbf{AD}_k^2\mathbf{A'} & \mathbf{0} \\ \mathbf{0'} & n \end{bmatrix}.$$

Hence it is sufficient to find a vector $\bar{\mathbf{U}} \epsilon \mathbf{R}^m$ satisfying

$$(\mathbf{AD}_k^2\mathbf{A'})\mathbf{U} = \mathbf{AD}_k^2\mathbf{C},$$

which are the normal equations for the weighted least squares problem

$$\min_{\mathbf{U}} \| \mathbf{D}_k(\mathbf{A'U} - \mathbf{C}) \|^2.$$

Having obtained $\mathbf{U}$, let $z_k = \mathbf{C'D}_k\mathbf{1}/n$. Then the negative projected gradient is provided by

$$-\mathbf{C}_p = -\mathbf{D}_k(\mathbf{C} - \mathbf{A'U}) - z_k\mathbf{1}.$$

REFERENCES

Adler, I., N. Karmarkar, M.G.C. Resende, G. Veiga. 1989. "An Implementation of Karmarkar's Algorithm for Linear Programming." *Mathematical Programming*, 44, 297-335.

Anstreicher, K.M. 1986. "A Monotonic Projective Algorithm for Fractional Linear Programming." *Algorithmica*, 1, 483-494.

Anstreicher, K.M. 1989. "A Combined Phase I - Phase II Projective Algorithm for Linear Programming." *Mathematical Programming*, 43, 209-223.

Anstreicher, K.M. 1991. "On the performance of Karmarkar's Algorithm Over a Sequence of Iterations." *SIAM Journal on Optimization*, 1, 22-29.

Anstreicher, K.M. 1993. "Strict Monotonicity and Improved Complexity in the Standard Form Projective Algorithm for Linear Programming." *Mathematical Programming*, 62, 517-535.

Anstreicher, K., R. Bosch. 1992. "Long Steps in an $O(n^3L)$ Algorithm for Linear Programming." *Mathematical Programming*, 54, 251-265.

Arbel, A. 1993. *Exploring Interior-Point Linear Programming.* Cambridge, MA: The MIT Press.

Ašić, M.D., V.V. Kovacevic-Vujcic, M.D. Radosavljevic-Nikolic. 1990. "Asymptotic Behavior of Karmarkar's Method for Linear Programming." *Mathematical Programming*, 46, 173-190.

Azpeitia, A.G., D.J. Dickinson. 1964. "A Decision Rule in the Simplex Method that Avoids Cycling." *Numerische Mathematik*, 6, 324-331.

Balinski, M.L., W.J. Baumol. 1968. "The Dual in Nonlinear Programming and its Economic Interpretation." *Review of Economic Studies*, 35, 237-256.

Balinski, M.L., A.W. Tucker. 1969. "Duality Theory of Linear Programs: A Constructive Approach with Applications." *SIAM Review*, 11, 347-377.

Barnes, E.R. 1986. "A Variation of Karmarkar's Algorithm for Solving Linear Programming Problems." *Mathematical Programming*, 36, 174-182.

Barnes, E., S. Chopra, D. Jensen. 1988. "A Polynomial Time Version of the Affine-Scaling Algorithm." Manuscript, IBM T.J. Watson Research Center, Yorktown Heights, N.Y.

Bayer, D., J.C. Lagarias. 1991. "Karmarkar's Linear Programming Algorithm and Newton's Method." *Mathematical Programming*, 50, 291-330.

Bazaraa, M., J.Jarvis, H. Sherali. 1990. *Linear Programming and Network Flows*, 2nd ed. New York: John Wiley & Sons, Inc.

Beale, E.M.L. 1955. "Cycling in the Dual Simplex Algorithm." *Naval Research Logistics Quarterly*, 2, 269-275.

Ben-Israel, A., A. Charnes. 1968. "An Explicit Solution of a Special Class of Linear Programming Problems." *Operations Research*, 16, 1166-1175.

Blair, C.E. 1986. "The Iterative Step in the Linear Programming Algorithm of N. Karmarkar." *Algorithmica*, 1, 537-539.

Bland, R.G. 1977. "New Finite Pivot Rules for the Simplex Method." *Mathematics of Operations Research*, 2, 103-107.

Bosch, R.A., K.M. Anstreicher. 1993. "On Partial Updating in a Potential Reduction Linear Programming Algorithm of Kojima, Mizuno, and Yoshise." *Algorithmica*, 9, 184-197.

Charnes, A. 1952. "Optimality and Degeneracy in Linear Programming." *Econometrica*, 20, 160-170.

Charnes, A., W.W. Cooper, A. Henderson. 1953. *An Introduction to Linear Programming.* New York: John Wiley & Sons.

Choi, I.C., D. Goldfarb, 1993. "Exploiting Special Structure in a Primal-Dual Path-Following Algorithm." *Mathematical Programming*, 58, 33-52.

Choi, I.C., C.L. Monma, D.F. Shanno. 1990. "Further Development of a Primal-Dual Interior Point Method." *ORSA Journal on Computing*, 2, 304-311.

Dantzig, G.B. 1951a. "Maximization of a Linear Function of Variables Subject to Linear Inequalities." In Koopmans, T.C. (ed). *Activity Analysis of Production and Allocation*, Cowles Commission Monograph 13. New York: John Wiley & Sons, Inc.

Dantzig, G.B. 1951b. "Application of the Simplex Method to a Transportation Problem." In Koopmans, T.C. (ed). *Activity Analysis of Production and Allocation*, Cowles Commission Monograph 13. New York: John Wiley & Sons, Inc.

Dantzig, G.B. 1963. *Linear Programming and Extensions.* Princeton: Princeton University Press.

Dantzig, G.B., A. Orden, and P. Wolfe. 1954. *Generalized Simplex Method for Minimizing a Linear Form Under Linear Inequality Restraints*, RAND Report RM1264. Santa Monica, CA: The RAND Corp.

Dantzig, G.B., A. Orden, and P. Wolfe. 1955. "The Generalized Simplex Method for Minimizing a Linear Form Under Linear Inequality Restraints." *Pacific Journal of Mathematics*, 5, 183-195.

Dikin, I.I. 1967. "Iterative Solution of Problems of Linear and Quadratic Programming." *Soviet Mathematics Doklady*, 8, 674-675.

Ding, J., T.Y. Li. 1991. "A Polynomial-Time Predictor-Corrector Algorithm for a Class of Linear Complementarity Problems." *SIAM Journal of Optimization*, 1, 83-92.

Forrest, J., D.Goldfarb, 1992. "Steepest Edge Simplex Algorithms for Linear Programming." *Mathematical Programming*, 59, 341-374.

Fraley, C. 1990. "Linear Updates for a Single-Phase Projective Method." *Operations Research Letters*, 9, 169-174.

Freund, R. 1988. "Projective Transformations for Interior Point Methods, Part I: Basic Theory an Linear Programming." *MIT Operations Research Center Working Paper*, OR, 179-188.

Freund, R.M. 1991a. "Polynomial-Time Algorithm for Linear Programming Based Only on Primal Affine Scaling and Projected Gradients of a Potential Function." *Mathematical Programming*, 51, 203-222.

Freund, R.M. 1991b. "A Potential-Function Reduction Algorithm for Solving a Linear Program Directly From an Infeasible 'Warm Start'." *Mathematical Programming*, 52, 441-466.

Freund, R.M. 1993. "Projective Transformations for Interior-Point Algorithms, and a Superlinearly Convergent Algorithm for the W-Center Problem." *Mathematical Programming*, 58, 385-414.

Gale, D. 1960. *The Theory of Linear Economic Models.* New York: McGrawHill Book Co., Inc.

Gay, D. 1987. "A Variant of Karmarkar's Linear Programming Algorithm for Problems in Standard Form." *Mathematical Programming*, 37, 81-90.

de Ghellinck, G., J.P. Vial. 1986. "A Polynomial Newton Method for Linear Programming." *Algorithmica*, 1, 425-453.

Gass, S.I. 1985. *Linear Programming*, 5th ed. New York: McGrawHill Book Co.

Gill, P.E., W. Murray, M.A. Saunders, J.A. Tomlin, M.H. Wright. 1986. "On Projected Barrier Methods for Linear Programming and an Equivalence to Karmarkar's Projective Method." *Mathematical Programming*, 36, 183-209.

Goldfarb, D. 1976. "Using the Steepest Edge Simplex Algorithm to Solve Sparse Linear Programs." In Bunch, J., D. Rose (eds.). *Sparse Matrix Computations.* New York: Academic Press.

Goldfarb, D., J.K. Reid. 1977. "A Practicable Steepest Edge Simplex Algorithm." *Mathematical Programming*, 12, 361-371.

Goldfarb, D., S. Mehrotra. 1988a. "Relaxed Variants of Karmarkar's Algorithm for Linear Programming with Unknown Optimal Objective Value." *Mathematical Programming*, 40, 183-196.

Goldfarb, D., S. Mehrotra. 1988b. "A Relaxed Version of Karmarkar's Method." *Mathematical Programming*, 40, 289-315.

Goldfarb, D., M.J. Todd. 1989. "Linear Programming." In Nemhauser, G.L., A.H.G. Rinnooy Kan (eds). Optimization, *Handbook in Operations Research and Management Science*, Vol. 1. Amsterdam: Elsevier/North-Holland Publishing Co.

Goldman, A.J., A.W. Tucker. 1956. "Theory of Linear Programming." In Kuhn, H.W., A.W. Tucker (eds). *Linear Inequalities and Relaxed Systems*, Annals of Mathematical Studies 38. New Jersey: Princeton University Press.

Gonzaga, C. 1988. "Polynomial Affine Algorithms for Linear Programming." Dept. of Systems Engineering and Computer Science, COPPE-Federal University of Rio de Janeiro, ES-139/88. Rio de Janeiro, RJ, Brazil.

Gonzaga, C. 1989a. "A Conical Projection Algorithm for Linear Programming." *Mathematical Programming*, 43, 151-173.

Gonzaga, C. 1989b. "An Algorithm for Solving Linear Programming Problems in $O(n^3L)$ Operations." In Megiddo, N. (ed). *Progress in Mathematical Programming: Interior Point and Related Methods.* New York: Springer-Verlag.

Gonzaga, C. 1990. "Polynomial Affine Algorithms for Linear Programming." *Mathematical Programming*, 49, 7-21.

Gonzaga, C. 1991a. "Large-Step Path-Following Methods for Linear Programming, Part 1: Barrier Function Method." *SIAM Journal on Opti-mization*, 1, 268-279.

Gonzaga, C. 1991b. "Large-Step Path-Following Methods for Linear Programming, Part 2: Potential Reduction Method." *SIAM Journal on Optimization*, 1, 280-292.

Gonzaga, C. 1991c. "Search Directions for Interior Linear Programming Methods." *Algorithmica*, 6, 153-181.

Gonzaga, C. 1992. "Path-Following Methods for Linear Programming." *SIAM Review*, 34, 167-224.

Gonzaga, C., M.J. Todd. 1992. "An $O(nL)$ Iteration Large-Step Primal-Dual Affine Algorithm for Linear Programming." *SIAM Journal on Optimization*, 2, 349-359.

Harris, P.M.J. 1973. "Pivot Selection Methods of the Devex LP Code." *Mathematical Programming*, 5, 1-28.

den Hertog, D., C. Roos, T. Terlaky. 1992. "A Complexity Reduction for the Long-Step Path-Following Algorithm for Linear Programming." *SIAM Journal on Optimization*, 2, 71-87.

Hoffman, A. 1953. *Cycling in the Simplex Algorithm.* Washington, D.C.: National Bureau of Standards.

Ishihara, T., M. Kojima. 1993. "On the Big M in the Affine Scaling Algorithm." *Mathematical Programming*, 62, 85-93.

Jan, G.M., S.C. Fang. 1991. "A Variant of Primal Affine Scaling Algorithm for Linear Programs." *Optimization*, 22, 681-715.

Ji, J., F.Potra, S. Huang. 1991. "A Predictor-Corrector Method for Linear Complementarity Problems with Polynomial Complexity and Superlinear Convergence." Report No. 18, University of Iowa, Iowa City, Iowa.

Karmarkar, N. 1984. "A New Polynomial Time Algorithm for Linear Programming." *Combinatorica*, 4, 373-395.

Karush, W. 1939. "Minima of Functions of Several Variables with Inequalities as Side Conditions." M.S. Thesis, Department of Mathematics, Univeristy of Chicago.

Klee, V., G.L. Minty. 1992. "How Good is the Simplex Algorithm?" In Shisha, O. (ed). *Inequalities III.* New York: Academic Press.

Kojima, M. 1986. "Determining Basic Variables of Optional Solutions in Karmarkar's New Algorithm." *Algorithmica*, 1, 499-515.

Kojima, M., N. Megiddo, T. Noma, A. Yoshise. 1991a. *A Unified Approach to Interior Point Algorithms for Linear Complementarity Problems.* New York: Springer-Verlag.

Kojima, M., N. Megiddo, Y.Ye. 1992. "An Interior Point Potential Reduction Algorithm for the Linear Complementarity Problem." *Mathematical Programming*, 54, 267-279.

Kojima, M., S. Mizuno, A. Yoshise. 1989a. "A Polynomial-Time Algorithm for a Class of Linear Complementarity Problems." *Mathematical Programming*, 44, 1-26.

Kojima, M., S. Mizuno, A. Yoshise. 1989b. "A Primal-Dual Interior Point Method for Linear Programming." In Megiddo, N. (ed). *Progress in Mathematical Programming: Interior Point and Related Methods.* New York: Springer-Verlag.

Kojima, M., Y. Kurita, S. Mizuno. 1993. "Large-Step Interior Point Algorithms for Linear Complementarity Problems." *SIAM Journal on Optimization*, 3, 398-412.

Kojima, M., N. Megiddo, S. Mizuno. 1993a. "A Primal-Dual Infeasible Interior Point Algorithm for Linear Programming." *Mathematical Programming*, 61, 263-280.

Kojima, M., N. Megiddo, S. Mizuno. 1993b. "Theoretical Convergence of Large-Step Primal-Dual Interior Point Algorithms for Linear Programming." *Mathematical Programming*, 59, 1-22.

Kojima, M., S. Mizuno, A. Yoshise. 1991b. "An $O(nL)$ Iteration Potential Reduction Algorithm for Linear Complementarity Problems." *Mathematical Programming*, 50, 331-342.

Kojima, M., S. Mizuno, A. Yoshise. 1993b. "A Little Theorem of the Big M in Interior Point Algorithms." *Mathematical Programming*, 59, 361-375.

Kortanek, K.O., J. Zhu. 1988. "New Purification Algorithms for Linear Programming." *Naval Research Logistics*, 35, 571-583.

Kuhn, H., R. Quandt. 1963. "An Experimental Study of the Simplex Method." *Proceedings of the Symposia in Applied Mathematics*, Vol. XV, American Mathematical Society.

Lemke, C.E. 1954. "The Dual Method for Solving Linear Programming Problem." *Naval Research Logistics Quarterly*, 1, 48-54.

Lemke, C.E. 1965. "Bimatrix Equilibrium Points and Mathematical Programming." *Management Science*, 11, 681-689.

Luenberger, D.G. 1984. *Linear and Nonlinear Programming*, 2nd ed. Reading MA: Addison-Wesley Publishing Co., Inc.

Lustig, I.J., R.E. Marsten, D.F. Shanno. 1991. "Computational Experience with a Primal-Dual Interior Point Method for Linear Programming." *Linear Algebra and Its Applications*, 152, 191-222.

Marshall, K.T., J.W. Suurballe. 1969. "A Note on Cycling in the Simplex Method." *Naval Research Logistics Quarterly*, 16, 121-137.

Marsten, R., M. Saltzman, M. Shanno, D. Pierce, J. Ballintijn. 1989. "Implementation of a Dual Affine Interior Point Algorithm for Linear Programming." *ORSA Journal on Computing*, 1, 287-297.

McShane, K. 1994. "Superlinearly Convergent $O(nL)$ Iteration Interior-Point Algorithms for Linear Programming and the Monotone Linear Complementarity Problem." *SIAM Journal on Optimization*, 4, 247-261.

McShane, K.A., C.L. Monma, D. Shanno. 1989. "An Implementation of a Primal-Dual Interior Point Method for Linear Programming." *ORSA Journal on Computing*, 1, 70-83.

Megiddo, N. 1986. "Introduction: New Approaches to Linear Programming." *Algorithmica*, 1, 387-394.

Megiddo, N. 1989a. *Progress in Mathematical Programming: Interior-Point and Related Methods.* New York: Springer-Verlag.

Megiddo, N. 1989b. "Pathways to the Optimal Set in Linear Programming." In Megiddo, N. (ed). *Progress in Mathematical Programming: Interior-Point and Related Methods.* New York: Springer-Verlag.

Megiddo, N., M. Shub. 1989. "Boundary Behavior of Interior Point Algorithms in Linear Programming." *Mathematics of Operations Research*, 14, 97-146.

Mehrotra, S. 1991. "On Finding a Vertex Solution Using Interior Point Methods." *Linear Algebra and its Applications*, 152, 106-111.

Mehrotra, S., R. Stubbs. 1994. "Predictor-Corrector Methods for a Class of Linear Complementarity Problems." *SIAM Journal on Optimization*, 4, 441-453.

Mizuno, S. 1992. "A New Polynomial Time Method for a Linear Complementarity Problem." *Mathematical Programming*, 56, 31-43.

Mizuno, S., A. Nagasawa. 1993a. "A Primal-Dual Affine-Scaling Potential-Reduction Algorithm for Linear Programming." *Mathematical Programming*, 62, 119-131.

Mizuno, S., M.J. Todd, Y. Ye. 1993b. "On Adaptive-Step Primal-Dual Interior Point Algorithms for Linear Programming." *Mathematics of Operations Research*, 18, 964-981.

Mizuno, S. 1994. "Polynomiality of Infeasible Interior-Point Algorithms for Linear Programming." *Mathematical Programming*, 67, 109-119.

Monma, C.L., A.J. Morton. 1987. "Computational Experience with a Dual Affine Variant of Karmarkar's Method for Linear Programming." *Operations Research Letters*, 6, 261-267.

Monteiro, R.C. 1991. "Convergence and Boundary Behavior of the Projective Scaling Trajectories for Linear Programming." *Mathematics of Operations Research*, 16, 842-858.

Monteiro, R.C., I. Adler. 1989. "Interior Path Following Primal-Dual Algorithms. Part 1: Linear Programming." *Mathematical Programming*, 44, 27-42.

Monteiro, R.D.C., I. Adler, M.G.C. Resende. 1990. "A Polynomial-Time Primal-Dual Affine Scaling Algorithm for Linear and Convex Quadratic Programming and Its Power Series Extension." *Mathematics of Operations Research*, 15, 191-214.

Padberg, M.W. 1986. "A Different Convergence Proof of the Projective Method for Linear Programming." *Operations Research Letters*, 4, 253-257.

Panik, M.J. 1976. *Classical Optimization: Foundations and Extensions.* The Netherlands: North-Holland Publishing Co.

Panik, M.J. 1993. *Fundamentals of Convex Analysis.* Boston MA: Kluwer Academic Publishers.

Perold, A.F. 1980. "A Degeneracy Exploiting LU Factorization for the Simplex Method." *Mathematical Programming*, 19, 239-254.

Potra, F. 1992. "On a Predictor-Corrector Method for Solving Linear Programs From Infeasible Starting Points." Reports on Computational Mathematics No. 34, Dept. of Mathematics, University of Iowa, Iowa City, Iowa.

Potra, F. 1994. "A Quadratically Convergent Predictor-Corrector Method for Solving Linear Programs from Infeasible Starting Points." *Mathematical Programming*, 67, 383-406.

Renegar, J. 1988. "A Polynomial-Time Algorithm Based on Newton's Method for Linear Programming." *Mathematical Programming*, 40, 59-93.

Rockafellar, R.T. 1970. *Convex Analysis.* Princeton, NJ: Princeton University Press.

Roos, C. 1989. "A New Trajectory Following Polynomial-Time Algorithm for Linear Programming Problem." *Journal of Optimization Theory and Applications*, 63, 433-458.

Shanno, D.F. 1988. "Computing Karmarkar Projection Quickly." *Mathematical Programming*, 41, 61-71.

Shanno, D.F., R.L. Weil. 1971. "Linear Programming with Absolute Value Functionals." *Management Science*, 16, 408.

Shapiro, J. 1979. *Mathematical Programming: Structures and Algorithms.* New York: John Wiley & Sons, Inc.

Shaw, D., D. Goldfarb. 1991. "A Primal Projective Interior Point Method for Linear Programming." *Mathematical Programming*, 51, 17-41.

Shaw, D., D. Goldfarb. 1994. "A Path-Following Projective Interior Point Method for Linear Programming." *SIAM Journal on Optimization*, 4, 65-85.

Shub, M. 1987. "On the Asymptotic Behavior of the Projective Rescaling Algorithm for Linear Programming." *Journal of Complexity*, 3, 258-269.

Sonnevend, G., J. Stoer, G. Zhao. 1989. "On the Complexity of Following the Central Path of Linear Programs by Linear Extrapolation." *Methods of Operations Research*, 63, 19-31.

Tapia, R.A., Y. Zhang. 1990. "Cubically Convergent Method for Locating a Nearby Vertex in Linear Programming." *Journal of Optimization Theory and Applications*, 67, 217-225.

Tardos, E. 1986. "A Strongly Polynomial Algorithm to Solve Combinatorial Linear Programs." *Operations Research*, 34, 250-256.

Todd, M.J. 1986a. "Large Scale Linear Programming-Geometry, Working Bases, and Factorization." *Mathematical Programming*, 26, 1-20.

Todd, M.J. 1986b. "Polynomial Expected Behavior of a Pivoting Algorithm for Linear Complementarity and Linear Programming Problems." *Mathematical Programming*, 35, 173-192.

Todd, M.J. 1998a. "Improved Bounds and Containing Ellipsoids in Karmarkar's Linear Programming Algorithm." *Mathematics of Operations Research*, 13, 650-659.

Todd, M.J. 1988b. "Exploring Special Structure in Karmarkar's Linear Programming Algorithm." *Mathematical Programming*, 41, 97-113.

Todd, M.J. 1992. "On Anstreicher's Combined Phase I - Phase II Projec-tive Algorithm for Linear Programming." *Mathematical Programming*, 55, 1-16.

Todd, M.J. 1993. "Combining Phase I and Phase II in a Potential Reduction Algorithm for Linear Programming." *Mathematical Programming*, 59, 133-150.

Todd, M.J., B.P. Burrell. 1986. "An Extension to Karmarkar's Algorithm for Linear Programming Using Dual Variables." *Algorithmica*, 1, 409-424.

Todd, M.J., Y. Wang. 1993. "On Combined Phase 1 - Phase 2 Projective Methods for Linear Programming." *Algorithmica*, 9, 64-83.

Todd, M.J., Y. Ye. 1990. "A Centered Projective Algorithm for Linear Programming." *Mathematics of Operations Research*, 15, 508-529.

Tomlin, M.J. 1987. "An Experimental Approach to Karmarkar's Projective Method for Linear Programming." *Mathematical Programming Studies*, 31, 175-191.

Tseng, P., Z-Q. Luo. 1992. "On the Convergence of the Affine-Scaling Algorithm." *Mathematical Programming*, 56, 301-319.

Tsuchiya, T. 1991. "Global Convergence of the Affine Scaling Methods for Degenerate Linear Programming Problems." *Mathematical Programming*, 52, 377-404.

Vanderbei, R.J. 1989. "Karmarkar's Algorithm and Problems With Free Variables." *Mathematical Programming*, 43, 31-44.

Vanderbei, R.J., T. Carpenter. 1993. "Symmetric Indefinite Systems for Interior Point Methods." *Mathematical Programming*, 58, 1-32.

Vanderbei, R.J., J.C. Lagarias. 1990. "I.I. Dikin's Convergence Result for the Affine-Saling Algorithm." *Contemporary Mathematics*, 114, 109-119.

Vanderbei, R.J., M.S. Meketon, B.A. Freedman. 1986. "A Modification of Karmarkar's Linear Programming Algorithm." *Algorithmica*, 1, 395-407.

Vial, JP. 1989. "A Unified Approach to Projective Algorithms for Linear Programming." *Optimization Fifth French-German Conference, Castel Novel 1988, Lecture Notes in Mathematics 1405.* New York: Springer-Verlag, 191-220.

Williams, A.C. 1963. "Marginal Values in Linear Programming." *SIAM Journal of Applied Mathematics*, 11, 82-94.

Wolfe, P. 1963. "A Technique for Resolving Degeneracy in Linear Programming." *SIAM Journal*, 11, 205-211.

Wolfe, P. 1965. "The Composite Simplex Algorithm." *SIAM Review*, 7, 42-54.

Wright, S. 1994. "An Infeasible-Interior-Point Algorithm for Linear Complementarity Problems." *Mathematical Programming*, 67, 29-51.

Ye, Y. 1987. "Karmarkar's Algorithm and the Ellipsoid Method." *Operations Research Letters*, 4, 177-182.

Ye, Y. 1988. "A Class of Potential Functions for Linear Programming." Manuscript, Department of Management Sciences, The University of Iowa, Iowa City, Iowa.

Ye, Y. 1990a. "An $O(n^3L)$ Potential Reduction Algorithm for Linear Programming." *Contemporary Mathematics*, 114, 91-107.

Ye, Y. 1990b. "Recovering Optimal Basic Variables in Karmarkar's Polynomial Algorithm for Linear Programming." *Mathematics of Operations Research*, 15, 564-572.

Ye, Y. 1990c. "A Class of Projective Transformations for Linear Programming." *SIAM Journal on Computing*, 19, 457-466.

Ye, Y. 1992. "A Potential Reduction Algorithm Allowing Column Generations." *SIAM Journal on Optimization*, 2, 7-20.

Ye, Y., K. Anstreicher. 1993a. "On Quadratic and $O(n^3L)$ Convergence of a Predictor-Corrector Algorithm for LCP." *Mathematical Programming*, 62, 537-551.

Ye, Y., O. Güler, R.A. Tapia, Y. Zhang. 1993b. "A Quadratically Convergent $O(nL)$ Iteration Algorithm for Linear Programming." *Mathematical Programming*, 59, 151-162.

Ye, Y., M. Kojima. 1987. "Recovering Optimal Dual Solutions in Karmarkar's Polynomial Algorithm for Linear Programming." *Mathematical Programming*, 39, 305-317.

Ye, Y., P. Pardalos. 1989. "A Class of Linear Complementarity Problems Solvable in Polynomial Time." Manuscript, Department of Management Sciences, The University of Iowa, Iowa City, Iowa.

Ye, Y., R.A. Tapia, Y. Zhang. 1991. "A Superlinearly Convergent $O(nL)$ Iteration Algorithm for Linear Programming." Technical Report TR9122, Department of Mathematical Sciences, Rice University, Houston, TX.

Ye, Y., M.J. Todd. 1990. "Containing and Shrinking Ellipsoids in the Path-Following Algorithm." *Mathematical Programming*, 42, 1-9.

Ye, Y., M.J. Todd, S. Mizuno. 1994. "An $O(nL)$ Iteration Homogeneous and Self-Dual Linear Programming Algorithm." *Mathematics of Operations Research*, 19, 53-68.

Yudin, D.B., E.G. Gol'shtein. 1965. *Linear Programming*. Jerusalem: Israel Program for Scientific Translations Ltd.

Zhang, Y. 1994. "On the Convergence of a Class of Infeasible Interior Point Methods for the Horizontal Linear Complementarity Problem." *SIAM Journal on Optimization*, 4, 208-227.

Zhang, Y., R.A. Tapia, J.E. Dennis. 1992. "On the Superlinear and Quadratic Convergence of Primal-Dual Interior-Point Linear Programming Algorithms." *SIAM Journal on Optimization*, 2, 304-324.

NOTATION INDEX

$\mathbf{R}$	set of real numbers
ϕ	null or empty set
$\mathbf{X} = (x_i)$	the vector whose i^{th} component is x_i
$\mathbf{X} \geqq \mathbf{0}$	$x_i \geq 0$ for all i
$\mathbf{X} > \mathbf{0}$	$x_i > 0$ for all i
$\mathbf{1}$	the sum vector containing all 1's
$\mathbf{0}$	the null vector/matrix containing all 0's
$\mathbf{I}_n$	identity matrix of order n
$\mathbf{e}_i$	i^{th} unit column vector
$\mathbf{X}'$	transpose of a vector $\mathbf{X}$
$\mathbf{X} \cdot \mathbf{Y}$	scalar product of vectors $\mathbf{X}$, $\mathbf{Y}$
$\Vert \cdot \Vert$	Euclidean norm
$\mathbf{X} / \Vert \mathbf{X} \Vert$	direction of vector $\mathbf{X}$
$\Vert \mathbf{X} - \mathbf{Y} \Vert$	distance between points $\mathbf{X}$, $\mathbf{Y}$
$\mathbf{A}$	an $(m \times n)$ matrix $\mathbf{A}$
$\rho(\mathbf{A})$	rank of a matrix $\mathbf{A}$
$\vert \mathbf{A} \vert$	determinant of an nth order matrix $\mathbf{A}$
$diag(\mathbf{X})$	a diagonal matrix whose elements are the components of $\mathbf{X}$
$Col(\mathbf{A})$	column space of $\mathbf{A}$
$Row(\mathbf{A})$	row space of $\mathbf{A}$
$N(\mathbf{A})$	null-space of $\mathbf{A}$
$n(\mathbf{A})$	nullity of $\mathbf{A}$
$\mathbf{A}^{\dagger}$	pseudo-inverse of $\mathbf{A}$
$\mathbf{A}^{-1}$	inverse of $\mathbf{A}$
$\mathbf{X} \in \mathcal{S}$	$\mathbf{X}$ is an element of a set $\mathcal{S}$
$\mathbf{X} \notin \mathcal{S}$	$\mathbf{X}$ is not an element of a set $\mathcal{S}$
$\mathcal{A} \subseteq \mathcal{B}$	set $\mathcal{A}$ is a subset of set $\mathcal{B}$ (possibly $\mathcal{A} = \mathcal{B}$)
$\mathcal{A} \subset \mathcal{B}$	set $\mathcal{A}$ is a proper subset of set $\mathcal{B}$

$\mathbf{R}^n$	n-space or the set of all real n-component vectors
$\mathbf{R}^n_{++}$	set of all real n-component vectors with strictly positive components
$\mathcal{M}$	subspace of $\mathbf{R}^n$
$dim(\mathcal{M})$	dimension of subspace $\mathcal{M}$
$\ell(\mathcal{M})$	linear hull of subspace $\mathcal{M}$
$\mathcal{M}_1 \oplus \mathcal{M}_2$	direct sum of subspaces $\mathcal{M}_1$, $\mathcal{M}_2$
$\mathcal{M}^{\perp}$	orthogonal complement or dual subspace of $\mathcal{M}$
$\lceil x \rceil$	smallest integer $\geq x$
$\lfloor x \rfloor$	largest integer not greater than x
$O(n^r)$	r^{th} order polynomial of problem size n
$arg\ min$	minimization with respect to a particular argument
$d(\mathcal{S})$	diameter of a set $\mathcal{S}$
$d(\mathbf{X}, \mathcal{S})$	distance between vector $\mathbf{X}$ and set $\mathcal{S}$
$d(\mathcal{S}, \mathcal{T})$	distance between sets $\mathcal{S}, \mathcal{T}$
$B(\mathbf{X}_0, \delta)$	δ-neighborhood of a point $\mathbf{X}_0 \in \mathbf{R}^n$ (or an open ball or sphere of radius δ about $\mathbf{X}_0$)
$\mathcal{C}$	cone
$\mathcal{C}^+$	polar cone
$\mathcal{C}^*$	dual cone
$(\mathbf{a})$	ray or halfline
$(\mathbf{a})^{\perp}$	orthogonal cone to $(\mathbf{a})$
$(\mathbf{a})^+$	polar of $(\mathbf{a})$
$(\mathbf{a})^*$	dual of $(\mathbf{a})$
$\mathcal{S}^0$	interior of set $\mathcal{S}$
$\mathcal{S}^e$	exterior of set $\mathcal{S}$
$\mathcal{S}^c$	complementary set of $\mathcal{S}$
$\overline{\mathcal{S}}$	closure of set $\mathcal{S}$
$\partial(\mathcal{S})$	boundary of set $\mathcal{S}$
$coni(\mathcal{S})$	conical hull of set $\mathcal{S}$

$aff(\mathcal{S})$	affine hull of set $\mathcal{S}$
$co(\mathcal{S})$	convex hull of set $\mathcal{S}$
$ri(\mathcal{S})$	relative interior of convex set $\mathcal{S}$
$r\partial(\mathcal{S})$	relative boundary of set $\mathcal{S}$
sup	supremum or least upper bound
inf	infimum or greatest lower bound
$lim\,sup$	limit superior (or upper limit)
$\mathcal{A} \times \mathcal{B}$	Cartesian product of sets $\mathcal{A}$, $\mathcal{B}$
s.t.	subject to
∇	gradient (*del* operator)
$\{\mathbf{X}_k\}$	seqeunce of points
$\mathcal{H}$	hyperplane
$[\mathcal{H}^+], [\mathcal{H}^-]$	closed half-planes
$(\mathcal{H}^+), (\mathcal{H}^-)$	open half-planes
σ^k	k-dimensional simplex
$f: \mathbf{R}^n \rightarrow \mathbf{R}$	function or mapping from $\mathbf{R}^n$ to $\mathbf{R}$
f^{-1}	single-valued inverse mapping
$T: \mathbf{R}^n \rightarrow \mathbf{R}^m$	transformation from $\mathbf{R}^n$ to $\mathbf{R}^m$
T^{-1}	inverse transformation
$\mathbf{Y} = T(\mathbf{X})$	image of $\mathbf{X}$ under transformation T
$D(T)$	domain of T
$R(T)$	range of T
$\mathcal{D}(\mathbf{X}_0)$	set of feasible directions at $\mathbf{X}$
$\mathcal{D}(\mathbf{X}_0)^+$	polar support cone
$\mathcal{D}(\mathbf{X}_0)^*$	dual support cone
$\mathcal{K}_P$ (or $\mathcal{K}$)	primal feasible region
$\mathcal{K}_D$	dual feasible region
$LCP(\mathbf{q}, \mathbf{M})$	linear complementarity problem
$\overline{lim}$	upper limit
$\underline{lim}$	lower limit

INDEX